Polymorphism

Edited by
Rolf Hilfiker

Related Titles

R. J. D. Tilley

Understanding Solids
The Science of Materials

2005
ISBN 0-470-85276-3

H. U. Blaser, E. Schmidt (Eds.)

Asymmetric Catalysis on Industrial Scale
Challenges, Approaches and Solutions

2004
ISBN 3-527-30631-5

H. J. Scheel, T. Fukuda

Crystal Growth Technology

2004
ISBN 0-471-49524-7

P. Bamfield

Research and Development Management in the Chemical and Pharmaceutical Industry

2003
ISBN 3-527-30667-6

F. Schüth, K.S.W. Sing, J. Weitkamp (Eds.)

Handbook of Porous Solids
5 Volumes

2002
ISBN 3-527-30246-8

Polymorphism

in the Pharmaceutical Industry

Edited by
Rolf Hilfiker

WILEY-VCH Verlag GmbH & Co. KGaA

Editor

Dr. Rolf Hilfiker
Solvias AG
Klybeckstrasse 191
4002 Basel
Switzerland

1st Edition 2006
1st Reprint 2006

Library of Congress Card No.: applied for

British Library Cataloguing-in-Publication Data:
A catalogue record for this book is available from the British Library

Bibliographic information published by the Deutsche Nationalbibliothek
The Deutsche Nationalbibliothek lists this publication in the Deutsche Nationalbibliografie; detailed bibliographic data are available in the Internet at http://dnb.d-nb.de.

Cover illustration SCHULZ Grafik-Design, Fußgönheim
Typesetting K+V Fotosatz GmbH, Beerfelden
Printing betz-druck GmbH, Darmstadt
Binding Litges & Dopf Buchbinderei GmbH, Heppenheim

Printed on acid-free paper
Printed in the Federal Republic of Germany

ISBN-13: 978-3-527-31146-0
ISBN-10: 3-527-31146-7

Contents

Polymorphism: in the Pharmaceutical Industry. Edited by Rolf Hilfiker

ISBN: 3-527-31146-7

Preface

Polymorphism, a term derived from the Greek words for "much/many" (poly, πολύ) and "form" (morphē, μορφή), is used in disciplines as diverse as linguistics, computer science, biology, genetics, and crystallography. In the life sciences industry, two completely different types of polymorphism play a major role: polymorphisms in DNA sequence and polymorphs of crystalline substances. In the former, great strides are being made using polymorphisms in their DNA sequence to predict an individual's susceptibility to disease and response to drugs, making it possible to design and select appropriate drugs. In the latter, the polymorphic form of a drug substance or excipient can have a profound impact on a spectrum of aspects, such as biological action, production, formulation and intellectual property protection. This book deals exclusively with the polymorphs of solids, covering not only polymorphs in the narrow sense, i.e., different crystalline forms of the same molecular entity, but also other solid-state forms relevant to industry, such as solvates, salts, and the amorphous form.

Interest in the solid-state properties of drugs has grown tremendously in recent decades as can be seen, for example, by the numerous conferences and workshops organized by various scientific and commercial institutions. This interest is well deserved. Anyone who has worked in the field for some time can point to examples where insufficient understanding of solid-state properties has led to serious setbacks. Problems encountered range from the sudden unexpected inability to produce reliably a form that has been used for pivotal clinical studies and is the basis for registration documents to variations in the drug product properties due to seemingly random changes of the solid form during processing or storage. Conversely, a thorough understanding of solid-state properties can create opportunities, which are increasingly being exploited for the benefit of both the company and the patient. Not only can patent protection be broadened or prolonged, and production made more efficient and cheaper, but the properties of the drug can also be improved to the advantage of the patient.

Increasing recognition of the importance of polymorphism to the life sciences industry has generated a great deal of interest and the field has been evolving rapidly. Given the pace of recent developments, an update is both useful and timely. This book discusses the whole breadth of the subject, covering all relevant aspects of solid-state issues for the pharmaceutical industry. It should act as a manual and a guideline for scientists dealing with solid-state issues, and

Polymorphism: in the Pharmaceutical Industry. Edited by Rolf Hilfiker

ISBN: 3-527-31146-7

serve both as an introduction to people new to the field and as a source for experts to round off their knowledge. It also provides valuable information for scientists working in other areas where solid-state issues are important, such as animal health, agrochemical, and specialty chemical industries.

Chapters are organized according to the following aspects of polymorphism: relevance, tools, properties, practical approaches, and legal issues. Chapter 1 discusses the relevance of solid-state forms in the pharmaceutical industry and makes recommendations on how best to approach solid-state issues. Chapter 2, on the thermodynamics of polymorphs, provides the theoretical tools needed to understand solid-state behavior. Chapters 3 to 7 give detailed descriptions, instructions, and hints on how to characterize solids, since solid-state behavior can only be understood after thorough characterization. Such an understanding is crucial to making the right decisions at key stages of drug development and production. Chapters 8 to 10 highlight the properties and importance of solid-state forms that are not included in the narrow definition of polymorphism, namely, solvates, hydrates and the amorphous form. Essential practical aspects for development scientists are described in Chapters 11 to 13, which deal with identifying relevant polymorphs, finding optimal salts and controlling solid-state behavior during processing. The last two chapters discuss legislative aspects of solid-state properties. Often, solid-state forms can be protected by patents, which may create significant financial benefits. Chapter 14 outlines the principles of intellectual property protection and provides relevant examples. Finally, since the solid form can have an impact on the safety and efficacy of drugs, Chapter 15 explains regulatory issues in connection with solid-state behavior. Rules, based on scientific considerations, are elucidated.

The broad range of topics discussed in this text, from thermodynamics to legal issues, emphasizes the complexity of the subject. It also demonstrates that the challenges and opportunities connected with solid-state properties can only be addressed successfully through an integral approach that considers all these aspects.

The strength of this volume lies in the quality of its contributions. My sincere thanks go to every author for the excellent standard of their submissions and their engaged cooperation. The balance of contributions from industry, academia and government highlights the far-reaching importance of the subject. From a personal perspective, I very much appreciated the fact that after developing the concept for this book and inviting authors to submit chapters on specific themes, colleagues willingly agreed to do so despite their very busy schedules. Finally, I thank Wiley-VCH for recognizing the timeliness of such a volume and Dr. Elke Maase and Dr. Bettina Bems for an enjoyable collaboration in the preparation of this book.

Basel, January 2006 *Rolf Hilfiker*

List of Contributors

Joel Bernstein
Department of Chemistry
Ben-Gurion University of the Negev
Beer Sheva 84105
Israel

Fritz Blatter
Solvias AG
Klybeckstrasse 191
4002 Basel
Switzerland

John M. Chalmers
VS Consulting
14 Croft Hills
Tame Bridge, Stokesley TS9 5NW
United Kingdom

Gérard Coquerel
Sciences et Méthodes Séparatives,
EA 3233
Université de Rouen
IRCOF, Rue Tesnière
76821 Mont Saint Aignan Cedex
France

Duncan Q. M. Craig
School of Chemical Sciences and
Pharmacy
University of East Anglia
Norwich, Norfolk NR4 7TJ
United Kingdom

Susan M. De Paul
Solvias AG
Klybeckstrasse 191
4002 Basel
Switzerland

Geoffrey Dent
GD Analytical Consulting
53 Nudger Green
Dobcross, Oldham OL3 5AW
United Kingdom

Ramprakash Govindarajan
Department of Pharmaceutics
University of Minnesota
308 Harvard Street SE
Minneapolis, MN 55455
USA

David J. W. Grant
Department of Pharmaceutics
College of Pharmacy
University of Minnesota
308 Harvard Street SE
Minneapolis, MN 55455
USA

Ulrich J. Griesser
Institute of Pharmacy
University of Innsbruck
Innrain 52
6020 Innsbruck
Austria

Polymorphism: in the Pharmaceutical Industry. Edited by Rolf Hilfiker

ISBN: 3-527-31146-7

Rolf Hilfiker
Solvias AG
Klybeckstrasse 191
4002 Basel
Switzerland

Sachin Lohani
Department of Pharmaceutics
College of Pharmacy
University of Minnesota
308 Harvard Street SE
Minneapolis, MN 55455-0343
USA

Joseph W. Lubach
Department of Pharmaceutical
Chemistry
University of Kansas
2095 Constant Ave.
Lawrence, KS 66047
USA

Stephen P. F. Miller
Office of New Drug Quality
Assessment
Food and Drug Administration –
Center for Drug Evaluation
and Research
Bld. 22, Mail Stop 2411
10903 New Hampshire Ave.
Silver Spring, MD 20993
USA

Eric J. Munson
Department of Pharmaceutical
Chemistry
University of Kansas
2095 Constant Ave.
Lawrence, KS 66047
USA

Ann W. Newman
SSCI, Inc.
3065 Kent Avenue
West Lafayette, IN 47906
USA

Gary Nichols
Pharmaceutical Sciences
Pfizer Global R&D
Ramsgate Road
Sandwich, Kent CT1 3NN
United Kingdom

Philippe Ochsenbein
Sanofi-Aventis
371, rue du Professeur J. Blayac
34184 Montpellier Cedex 04
France

Samuel Petit
Sciences et Méthodes Séparatives,
EA 3233
Université de Rouen
IRCOF, Rue Tesnière
76821 Mont Saint Aignan Cedex
France

Markus von Raumer
Solvias AG
Klybeckstrasse 191
4002 Basel
Switzerland

Andre S. Raw
Office of Generics Drugs
Food and Drug Administration –
Center for Drug Evaluation
and Research
Metro Park North II, Room E204
7500 Standish Place
Rockville, MD 20855
USA

Susan M. Reutzel-Edens
Eli Lilly & Company
Pharmaceutical Research &
Development
Lilly Corporate Center
Indianapolis, IN 46285
USA

Kurt J. Schenk
École Polytechnique Fédérale
de Lausanne
LCr1–IPMC–FSB
BSP-521 Dorigny
1015 Lausanne
Switzerland

Peter Heinrich Stahl
Private Consultant
Lerchenstrasse 28
79104 Freiburg
Germany
Former business address:
CIBA-GEIGY/ Novartis Pharma
Basel, Switzerland

Raj Suryanarayanan
Department of Pharmaceutics
University of Minnesota
308 Harvard Street SE
Minneapolis, MN 55455
USA

Bertrand Sutter
Novartis Pharma AG
PHAD Analytical R&D
WSJ-360.508
Lichtstrasse 35
4056 Basel
Switzerland

Martin Szelagiewicz
Solvias AG
Klybeckstrasse 191
4002 Basel
Switzerland

Lawrence X. Yu
Office of Generics Drugs
Food and Drug Administration –
Center for Drug Evaluation
and Research
Metro Park North II, Room 285
7500 Standish Place
Rockville, MD 20855
USA

1
Relevance of Solid-state Properties for Pharmaceutical Products

Rolf Hilfiker, Fritz Blatter, and Markus von Raumer

1.1
Introduction

Many organic and inorganic compounds can exist in different solid forms [1–6]. They can be in the amorphous (Chapter 10), i.e., disordered, or in the crystalline, i.e., ordered, state. According to McCrone's definition [2], "The polymorphism of any element or compound is its ability to crystallize as more than one distinct crystal species", we will call different crystal arrangements of the same chemical composition polymorphs. Other authors use the term "polymorph" more broadly, including both the amorphous state and solvates (Chapter 15). Since different inter- and intramolecular interactions such as van der Waals interactions and hydrogen bonds will be present in different crystal structures, different polymorphs will have different free energies and therefore different physical properties such as solubility, chemical stability, melting point, density, etc. (Chapter 2). Also of practical importance are solvates (Chapter 8), sometimes called pseudopolymorphs, where solvent molecules are incorporated in the crystal lattice in a stoichiometric or non-stoichiometric [6, 7] way. Hydrates (Chapter 9), where the solvent is water, are of particular interest. If non-volatile molecules play the same role, the solids are called co-crystals. Solvates and co-crystals can also exist as different polymorphs, of course.

In addition to the crystalline, amorphous and liquid states, condensed matter can exist in various mesophases. These mesophases are characterized by exhibiting partial order between that of a crystalline and an amorphous state [8, 9]. Several drug substances form liquid crystalline phases, which can be either thermotropic, where liquid crystal formation is induced by temperature, or lyotropic, where the transition is solvent induced [10–12].

Polymorphism is very common in connection with drug substances, which are mostly (about 90%) small organic molecules with molecular weights below 600 g mol^{-1} [13, 14]. Literature values concerning the prevalence of true polymorphs range from 32% [15] to 51% [16, 17] of small organic molecules. According to the same references, 56 and 87%, respectively, have more than one

Polymorphism: in the Pharmaceutical Industry. Edited by Rolf Hilfiker

ISBN: 3-527-31146-7

solid form if solvates are included. When a compound is acidic or basic, it is often possible to create a salt (Chapter 12) with a suitable base or acid, and such a salt can in turn often be crystallized. Such crystalline salts may also exist as various polymorphs or solvates. Obviously, solvates, co-crystals and salts will have different properties from the polymorphs of the active molecule. Since salts generally have higher water solubility and bioavailability than the corresponding uncharged molecule, they are popular choices for drug substances. About half of all active molecules are marketed as salts [14, 18]. Polymorphs, solvates, salts, and co-crystals are schematically depicted in Fig. 1.1. We will use the term "drug substance" for the therapeutic moiety, which may be a solvate, salt or a co-crystal, while the single, uncharged molecule will be called the "active molecule".

Most drug products (formulated drug substances) are administered as oral dosage forms, and by far the most popular oral dosage forms are tablets and other solid forms such as capsules. Drugs for parenteral application are also often stored as solids (mainly as lyophilized products) and dissolved just prior to use since in general the chemical stability of a molecule in the solid form is much higher than in solution. Drugs administered by inhalation have become increasingly popular, and dry powder inhalers are now commonly in use. Evidently, therefore, both the solid form of the drug substance and the selected excipients have a strong impact on the properties of the formulated drug. Even if the envisaged market form of the drug is a solution, information about the solid-state properties of the drug substance may still be necessary [19]. If different forms have significantly different solubilities, it may be possible to unintentionally create a supersaturated solution with respect to the least soluble form by creating a concentrated solution of a metastable form. Also, the drug substance will in most cases be handled as a solid in some stages of the manufacturing process, and its handling and stability properties may depend critically on the solid form.

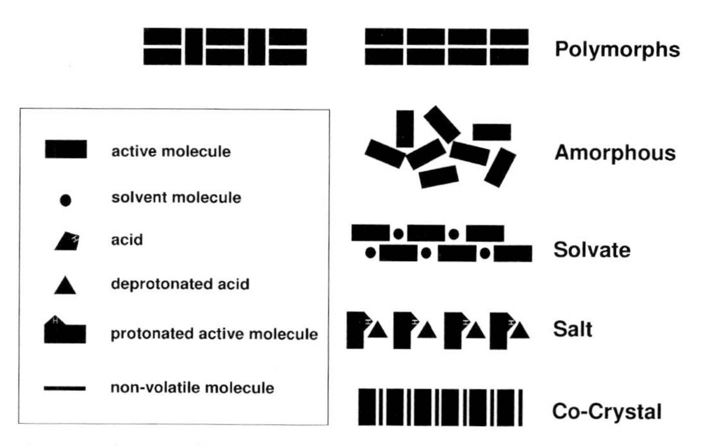

Fig. 1.1 Schematic depiction of various types of solid forms.

In fact, the whole existence of a drug is affected by the properties of the solid form, and the final goal of solid form development is to find and select the solid with the optimal characteristics for the intended use.

Initially, when the drug substance is first produced, one has to be certain that the desired solid form is obtained in a consistent, pure and reproducible manner. Subsequently, when it is formulated to obtain the drug product, one has to make sure that no undesired transitions occur (Chapter 13). For this phase, a profound knowledge of potential solvate formation is especially useful. It is highly advisable to avoid using solvents that can form solvates with the drug substance in the formulation process. Otherwise, such solvates might be generated during formulation and subsequently desolvated in a final drying step. In such a situation the final polymorph would probably differ from the initial one – an undesirable effect in most cases. Similarly, the energy–temperature diagram (Chapter 2) of the polymorphs and the kinetics of the change from one polymorph into another should be known so that one can be sure that temperature variations during the formulation process will not lead to an unacceptable degree of change in the solid form.

In the next step, when the drug substance or drug product is stored during its shelf-life, it is imperative that the solid form does not transform over time. Otherwise, important properties of the drug might change drastically. Stability properties have to be evaluated with respect to ambient conditions, storage, and packaging. Thermodynamic stability depends on the environment. A solvate, for example, represents a metastable form under ambient conditions but is likely to be the most stable form in its solvent. Thermodynamically, any metastable form will eventually transform into a more stable form. The kinetics under which this transformation occurs, however, are polymorph specific. Therefore, the existence of a more stable polymorph does not necessarily imply that a metastable polymorph cannot be developed.

In the final step, when the patient takes the drug, the solubility and dissolution rate of the drug substance will be influenced by its solid form. This will affect the bioavailability if solubility is a rate-limiting step, i.e., if the drug belongs to class 2 or 4 of the biopharmaceutics classification system (BCS) [20]. Because a change of solid form may render a drug ineffective or toxic, regulatory authorities demand elucidation and control of solid-state behavior (Chapter 15).

Finally, thorough, experimentally obtained knowledge of the solid-state behavior also has the advantages that a good patent situation for a drug substance can be obtained and that valuable intellectual property can be generated (Chapter 14). Although in hindsight everything may appear to be easy and straightforward, crystalline molecular solid-state forms are non-obvious, novel and require inventiveness. For instance, typically, many attempts to crystallize an amorphous drug substance fail until, suddenly, a stable crystalline form is obtained. Once seed crystals are available, the crystallization becomes the simple last step of a production process.

1.2 Drug Discovery and Development

Typically, it takes eight to twelve years, or sometimes even longer, for a molecule with biological activity to progress from its first synthesis to market introduction as an efficacious, formulated drug [21]. This process is normally divided into two main phases: (a) research or discovery and (b) development [22]. In the research phase, the appropriate target for a particular disease model is identified and validated, and candidate molecules are synthesized or chosen from libraries. They are primarily tested with respect to binding affinity to the target or, if possible, directly for their potential to alter a target's activity. Sometimes other parameters, such as selectivity, are also considered. Promising candidates are usually termed "hits". As a rule at this stage, limited attention is paid to the possibility to formulate a drug for a certain administration route. Often, from a drug delivery aspect, simple vehicles like DMSO solutions are used. As a result, the activity of especially poorly water-soluble drugs may not be identified at all because they precipitate under the used *in vitro* conditions [23]. In a medicinal chemistry program the "hits" are then modified to improve physicochemical parameters such as solubility and partition coefficient. This is the first time that solid-state properties come into play. When solubility is evaluated, it is critical to know whether the solubility of an amorphous or crystalline substance was measured. Permeation measurements are performed using, e.g., Caco-2 [24], PAMPA [25] or MDCK [26] assays, and dose–response studies are conducted in *in vitro* models. Selectivity is assessed in counter screens. At the same time, preliminary safety studies are carried out, and IP opportunities are assessed. Structure–activity relationship (SAR) considerations play a large role at this stage.

Molecules that show promise in all important aspects are called "leads". Often several series of leads are identified and are then further optimized and scrutinized in more sophisticated models, including early metabolic and *in vivo* studies. Both pharmacokinetics (PK, the quantitative relationship between the administered dose and the observed concentration of the drug and its metabolites in the body, i.e., plasma and/or tissue) and pharmacodynamics (PD, the quantitative relationship between the drug concentration in plasma and/or tissue and the magnitude of the observed pharmacological effect) are studied in animal models to predict bioavailability and dose in humans. Simultaneously with characterization of the drug substance, a proper dosage form needs to be designed, enabling the drug substance to exert its maximum effect. For freely water-soluble drugs this is less critical than for poorly water-soluble drugs, which without the aid of an adequate dosage form cannot be properly investigated in the research stage. In the discovery phase, high-throughput methods play an increasingly important role in many aspects, such as target identification, synthesis of potential candidate molecules, and screening of candidate molecules. Considering that only about 1 out of 10 000 synthesized molecules will reach the market [21], high-throughput approaches are a necessity. The optimal molecule arising from these assessments is then promoted to the next stage, i.e., development.

	non-clinical		clinical			
		IND				NDA Approval
	Early Devel-op-ment	Phase 0	Phase I	Phase II	Phase III	Submis-sion and Approval
description	pre-form-ulation	short term toxi-cology	first in humans, safety, PK long term toxicology	efficacy, dose finding synthesis redesign, process development	efficacy and safety comparison against standard, data for registration	-
# patients	-	-	10-100 healthy volunteers	100-500 patients	300-3000+ patients	-
duration	0.5 to 1 year	0.5 to 1 year	1 to 2 years	1 to 2 years	2 to 4 years	1 year
# compounds at beginning of phase (per approved compound)	9 to 20	7 to 15	5 to 12	3 to 7	1.5 to 3	1.1

Fig. 1.2 Drug development process with a description of respective phases, approximate number of test persons, timelines and attrition rates. These numbers are a rough guideline only and can differ significantly according to the specific indication, the characteristics of the drug substance, etc.

The development process of a pharmaceutical product is depicted in Fig. 1.2. It consists of a non-clinical and a clinical phase. While drug companies' approaches to the non-clinical phase can differ somewhat, the clinical phase is treated very similarly due to regulatory requirements. In the non-clinical phase enough data is gathered to compile an Investigational New Drug Application (IND) in the US or a Clinical Trial Application (CTA) in the European Union, which is the prerequisite for the first use of the substance in humans. For obvious reasons, particular emphasis is placed on toxicology studies during this phase, including assessment of toxicity by single-dose and repeated-dose administration and evaluation of carcinogenicity, mutagenicity and reproductive toxicity. An absolute necessity at this stage is that the drug is maximally bioavailable, resulting in sufficient exposure of the animals to the drug to obtain an adequate assessment of its toxicity profile. Whenever possible, the need for animal studies is reduced by using, e.g., human cell *in vitro* tests. The non-clinical development phase lasts between one and two years, and the attrition rate is ca. 50% (Fig. 1.2). At the end of the non-clinical phase, the decision has to be made whether the neutral molecule, a salt, or a co-crystal will be developed. If a salt form or co-crystal is chosen, it has to be clear which salt (Section 1.4.1) or co-crystal is optimal. In the clinical phases the product is first tested on healthy volunteers and then on small and large patient populations. For certain disease indications, like oncology, Phase I studies are performed directly on patients. Approximate population sizes are given in Fig. 1.2. One has to bear in mind, however, that these numbers depend significantly on the indication the drug is intended to treat. Attrition rates during the clinical phases are between 80 and 90%. During the clinical phases, analytical, process and dosage-form development continues in parallel with long-term toxicology studies. Of course, solid-state properties continue to play a crucial role during both chemical development of the drug substance and pharmaceutical development of the dosage form.

1.3 Bioavailability of Solids

An issue that has to be addressed for every drug product, and which is closely related to its solid-state properties, is whether its solubility and dissolution rate are sufficiently high. This leads to the question of what the minimal acceptable solubility and dissolution rates are.

Bioavailability essentially depends on three factors: solubility, permeability and dose [27], and the question of minimal acceptable solubility can only be answered if the other two factors are known. According to the BCS a drug substance is considered highly soluble when the highest strength dosage is soluble in 250 mL of aqueous media over the pH range 1.0–7.5 [28].

A valuable concept for estimating what the minimum solubility of a drug substance for development purposes should be uses the maximum absorbable dose (MAD) [29, 30]. MAD corresponds to the maximum dose that could be absorbed if there were a saturated solution of the drug in the small intestine during the small intestinal transit time (SITT $\approx$ 270 min). The bioavailable dose is smaller than MAD due to metabolism of components in the portal blood in the liver (first pass effect) and in the intestinal mucosal tissue [20]. MAD can be calculated from the solubility, S, at pH 6.5 (corresponding to typical conditions in the small intestine), the transintestinal absorption rate (Ka), the small intestinal water volume (SIWV $\approx$ 250 mL) and the SITT.

$$\text{MAD (mg)} = S\ (\text{mg mL}^{-1}) \times Ka\ (\text{min}^{-1}) \times \text{SIWV (mL)} \times \text{SITT (min)} \tag{1}$$

Human Ka can be estimated from measured rat intestinal perfusion experiments [30, 31]. It is related to the permeability (P) through SIWV and the effective surface of absorption (S_{abs}) [20].

$$Ka\ (\text{min}^{-1}) = P\ (\text{cm min}^{-1}) \times S_{abs}\ (\text{cm}^2)/\text{SIWV (mL)} \tag{2}$$

In the absence of active diffusion, permeability is related to the diffusion coefficient (D), the partition coefficient K ($=c_{\text{in membrane}}/c_{\text{in solution}}$) and the membrane thickness (δ).

$$P\ (\text{cm min}^{-1}) = D\ (\text{cm}^2\ \text{min}^{-1}) \times K/\delta\ (\text{cm}) \tag{3}$$

In reality, proportionality between the partition coefficient and the permeability is only found for a rather small range of partition coefficients [24, 32]. This is because the model of a single homogeneous membrane is an oversimplification. The intestinal wall is better represented by a bilayer membrane consisting of an aqueous and an adjoining lipid region. Therefore, for highly lipophilic substances, the water layer becomes the limiting factor and leads to a decrease in permeability as K is increased [33].

Implicit in Eq. (1) is that the solution stays saturated during the SITT and therefore that there is a large excess of solid drug in the small intestine. In deriving this equation as a limiting case, the authors [29] took into account the dissolution kinetics of a polydisperse powder and showed how the percentage of the dose that is absorbed is influenced by solubility, particle size and permeability. They showed that for highly soluble drugs, as defined above, the percentage of dose absorbed is only limited by permeability. For smaller solubilities, the dissolution rate and hence the particle size become important factors as well. The influence of particle size is greatest for low-solubility and low-dose drugs.

MAD readily translates into minimal acceptable solubility [30].

$$\text{Minimal acceptable solubility} = S \times \{\text{target dose (mg)}/\text{MAD}\}$$
$$= \text{target dose}/\{Ka \times \text{SIWV} \times \text{SITT}\} \qquad (4)$$

Realistic values for Ka lie between 0.001 and 0.05 min^{-1} and vary over a much narrower range than typical solubilities (0.1 $\mu g\ mL^{-1}$ to 100 $mg\ mL^{-1}$) [30]. Considering these facts and assuming a typical dose of 70 mg, i.e., 1 $mg\ kg^{-1}$, minimal acceptable solubilities between 20 $\mu g\ mL^{-1}$ and 1 $mg\ mL^{-1}$ are obtained. When making these estimates, one has to keep in mind that the assumptions of the model break down if there is possible absorption in other parts of the gastrointestinal tract or if the diffusivity of the drug is changed due to the meal effect, etc. [34]. Furthermore, it is important to realize that S represents a "kinetic" solubility. A weakly basic drug might be freely soluble in the stomach while its equilibrium solubility in the small intestine might be very low. Nevertheless, it may remain in the supersaturated state in the small intestine, in which case that "kinetic" solubility would be the relevant one for calculating the MAD.

1.4
Phases of Development and Solid-state Research

Normally, solid-state research and development involves the following stages, which may also overlap:

- deciding whether the uncharged molecule or a salt should be developed;
- identifying the optimal salt;
- identifying and characterizing all relevant solid forms of the chosen drug substance;
- patenting new forms;
- choosing a form for chemical and pharmaceutical development;
- developing a scalable crystallization process to obtain the desired form of the drug substance;

- developing a method to determine the polymorphic purity of the drug substance;
- formulating the drug substance to obtain the drug product;
- developing a method to determine the polymorphic purity of the drug substance in the drug product.

Not all of these stages may be necessary for every drug substance, and the order of the stages may be varied according to the specific properties and behavior of the drug. Particularly for drugs that are poorly water soluble, polymorphism in formulations can play a crucial role since it could significantly influence the dissolution rate and degree of dissolution required to achieve adequate bioavailability.

1.4.1 Salt Selection

Clearly, the first decision is whether it is more desirable to develop the uncharged molecule or, if possible, a salt thereof (Chapter 12). In general, salt formation will be possible if the molecule contains acidic or basic groups, which is the case for most active molecules. Since making a salt will normally involve an additional step in the synthesis and since the molecular weight of a salt will always be higher than that of the neutral molecule, salts will only be chosen if they promise to have clear advantages compared with the free acid/base. As a rule, a salt is chosen if the free acid/base has at least one of the following undesirable properties:

- very low solubility in water;
- apparently not crystallizable;
- low melting point (typical cutoff 80 °C [35]);
- high hygroscopicity;
- low chemical stability, etc.;
- IP issues.

Low water solubility is relative and always has to be assessed in the context of dose and permeability (Section 1.3). A very low water solubility may mean a high lipophilicity, enabling efficient passage through membranes, or a very large binding constant with the receptor, allowing a low dose. Also, the amorphous state of a neutral molecule may be the best option to get high oral bioavailability, provided the amorphous form can be kinetically stabilized over a reasonable time scale. Therefore, the decision to develop a salt should be based on a head-to-head broad comparison, taking into consideration both *in vivo* performance and physicochemical properties. If the decision has been made to develop a salt, it is obviously important to carry out a broad salt screening and salt selection process to identify the optimal salt. Potential counterions are chosen based on pK_a differences, counterion toxicity (preferably GRAS status [18, 36]), etc. (Chapter 12). Desirable properties of the salts include crystallinity, high water solubility, low hygroscopicity, good chemical stability, and high melting

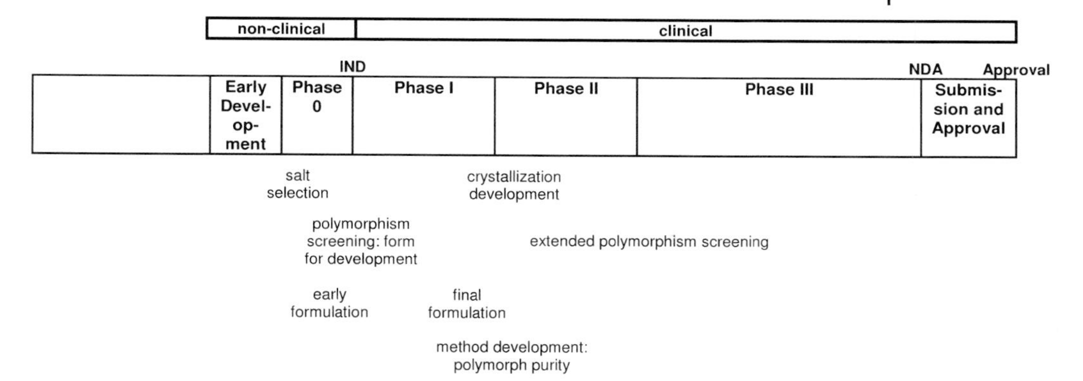

Fig. 1.3 Rough guideline of when the various issues related to solid-state properties generally should be taken care of in the development process. Depending on company policies, obtained results and other circumstances, large shifts are possible. In particular, certain steps may have to be repeated due to unanticipated experimental results.

point. The relative importance of these properties may vary from project to project. At this stage it also has to be decided whether co-crystals are to be considered. Co-crystals can offer valuable alternatives, especially for very weak bases or acids. Very often, salt screening and salt selection are performed in stages: first a large number of salts is produced on a microscopic scale and characterized with a limited number of methods (e.g., birefringence, Raman, XRD) to identify a few promising candidates, which are then produced on a scale of a few 100 mg and characterized in more detail.

Different companies perform salt screening in different phases of development. Some even move the salt selection process to the research phase [35], but clearly the decision on the salt form should ideally be made no later than the beginning of the long-term toxicology studies, i.e., at the start of Phase I (Fig. 1.3).

1.4.2
Polymorph Screening

The objective of the next important step with respect to solid-state development is identifying all relevant polymorphs and solvates (Chapter 11), characterizing them (Chapters 3 to 7), and choosing the optimal form for further chemical and pharmaceutical development. In the absence of solvents and humidity, the thermodynamically stable polymorph is the only one that is guaranteed not to convert into another polymorphic form. This is why this form is most often chosen for the drug product [31]. The disadvantage of the thermodynamically stable form is, of course, that it is always the least soluble polymorph (Chapter 2) and therefore has the lowest bioavailability. But in most cases this is a small price to pay for the very large advantage of absolute kinetic stability. Differences in the solubility of various polymorphs are typically lower than a factor of 2 (see Ref. [37] for a review of literature data), but sometimes as much as a five-fold difference can be observed [38]. In cases where several enantiotropically related forms exist and where the transition temperature is around room temperature, the choice may be difficult, but it is based on the same criteria as for all solid forms. The kinetics of interconversion from one form into the other and the reproducibility of producing consistently the same ratio of polymorphs are important.

Apart from the thermodynamically stable polymorph of a drug substance, hydrates are also very popular components of the final dosage form. Owing to the ubiquity of water vapor, hydrates are often the thermodynamically stable form at ambient conditions. If a certain hydrate is stable within a rather large range of humidities, it may therefore be much easier to formulate the hydrate in a controlled way and to subsequently store and package it.

In a few cases, a metastable form might be preferable [31], normally for one of the following reasons:

1. too low a solubility (and bioavailability) of the stable form;
2. high dissolution rate needed for quick-relief formulations;
3. manufacturing difficulties;
4. IP issues;
5. chemical instability of the thermodynamically stable form due to topochemical factors.

(1) If the solubility of the stable polymorph is critically low (Section 1.3) and no salt is feasible, several options exist [39]. Liquid-like formulations (emulsions, microemulsions, liposomal formulations) or soft gelatin capsules filled with solutions of the drug in a non-aqueous solvent may be used. Alternatively, a metastable solid form, a solvate or a co-crystal might be selected for development. If a solid form with a higher solubility than the thermodynamically stable form is desired, it is often better to use the amorphous form rather than a metastable polymorph, provided that the glass transition temperature (T_g) of the amorphous form is sufficiently high (Chapter 10) [40]. Firstly, the amorphous form often has a ten-fold or higher increased solubility relative to the stable form [41], while metastable polymorphs typically have a less than a two-fold higher solubility, as mentioned above. Secondly, it is normally impossible to stabilize a metastable form reliably by excipients, since they can only interact with the surface of the crystals of the metastable drug substance. This will change the surface free energy, but for crystal sizes larger than some tens of nanometers, the contribution of the surface free energy to the total free energy is negligible. The best way to stabilize a metastable form kinetically is to ensure the absence of any seeds of the stable form because such seeds have a very large effect on the kinetics of transformation [42]. The amorphous form, however, can be stabilized, for example, by creating a solid dispersion with a polymer [43, 44]. Such a dispersion will be highly kinetically stable if two conditions are fulfilled: if it remains in the glassy state under the storage conditions, thus blocking all translational diffusion, and if the

drug substance molecules are molecularly dispersed within the matrix. In any case, irrespective of whether a crystalline or disordered metastable form is to be developed, very careful kinetic stability studies will be necessary. For amorphous solids, particular attention has to be paid to the lowering of the glass transition temperature due to humidity.

(2) In some instances, quick onset of action of a drug is of particular importance. In such cases, metastable forms with a higher dissolution rate may accelerate the uptake of the drug and may therefore act faster.

(3) Different polymorphs will also have different mechanical properties, such as hardness, powder flow properties, compressibility and bonding strength. A well-known example is acetaminophen (also known as paracetamol), where the thermodynamically stable form (monoclinic form I) cannot be compressed into stable tablets while the metastable form II (orthorhombic) can as it shows more favorable properties with respect to plastic deformation [45]. In very rare cases, this might lead to a decision to develop a metastable form.

(4) If the thermodynamically stable polymorph is protected by patents, while other forms are free, the respective drug substance can be marketed as a metastable form without obtaining a license from the patent owner (Chapter 14) [5].

(5) Generally, the thermodynamically most stable polymorph is also the most stable chemically (Chapter 2) [31]. This has been attributed to the fact that its density is typically higher, but it could also be explained by its lower free energy. Only in extremely rare cases, where the arrangement of atoms in the stable polymorph favors an intermolecular chemical reaction, could its chemical stability be lower. In such cases, development of a metastable form might be advisable.

A very important question is, of course, when a polymorphism screening should be carried out and when the choice of form to develop should be made. Since different solid forms have different properties and may have different bioavailabilities, it is definitely advisable to select the final form together with the accompanying formulation before carrying out pivotal clinical studies [19, 46]. It is, therefore, critical to have at least identified the thermodynamically stable form along with important hydrates by the end of Phase I at the latest (Fig. 1.3). Accordingly, by that time a polymorphism screening that is primarily designed to identify these forms with a large probability should have been completed. Owing to economic reasons and the expected attrition rate of up to 90% of potential drug candidates after this stage, a full polymorphic screening, which identifies all relevant metastable forms as well, may need to be deferred. However, this should only be the exception because knowledge of metastable phases, thermodynamic stability as a function of temperature and conditions for solvate formation is crucial for the design of crystallization and formulation processes.

While the kinetic stability of dry metastable forms is not much influenced by additives, as mentioned above, additives and impurities can influence their kinetic stability in solutions and suspensions [47] by affecting both nucleation and growth rates. Therefore, a polymorphism screening that is performed with an early batch of drug substance still containing many impurities may provide different results from a screening performed with a later, purer batch. In particularly unfortunate cases, important forms may not be discovered in the initial screening. Therefore, it is highly advisable to repeat at least a limited polymorphism screening with a batch of drug substance produced with the final GMP procedure, which has the impurity profile of the product to be marketed.

Clearly, the unexpected appearance of a new form at a late stage can be disastrous. A very well publicized example is that of ritonavir (Norvir) [38, 48]. When it was launched on the market, only form I was known. One marketed formulation consisted of soft gelatin capsules filled with a nearly saturated solution of form I. About two years after market introduction, some capsules failed the dissolution test due to precipitation of a new, thermodynamically more stable form of ritonavir (form II). The solubility difference between forms I and II is about a factor of 5 [38], which is unusually high. In the end, the original formulation had to be taken off the market, and a new formulation had to be developed with considerable effort and expense [38]. While this is certainly an extreme case, there are many instances of new polymorphs appearing in Phase II and Phase III studies, leading to considerable difficulties [49].

1.4.3
Crystallization Process Development

After selecting the appropriate solid form for the drug substance, a reliable large-scale process to produce that form has to be developed. Parameters such as yield, chemical purity, polymorphic purity, solvent class (preferably Class 3 solvents according to ICH Q3C [50]), residual solvent content and cost need to be optimized. As a rule, it is also necessary to obtain solids with a consistent particle size and morphology (external shape, habit). The crystal habit can have a profound impact on important processing parameters such as filterability, flowability [46] and bulk density. It can sometimes be controlled by choosing the appropriate solvent and method for crystallization [51].

Crystallization, even of a drug substance precursor, is generally by far the most efficient and economic way of obtaining chemically pure compounds. Solvates can also be useful for obtaining crystalline material with increased purity if a drug substance is difficult to crystallize in a solvent-free form. The formation of a solvate with subsequent drying to produce the desired form by desolvation might be feasible as an intended process. However, this usually corresponds to a rearrangement of the lattice, which is generally susceptible to loss of crystallinity.

Precise knowledge of the thermodynamic stability relationships among the various forms as a function of temperature (ET diagrams, Chapter 2) is a prerequisite for designing reliable crystallization processes [52, 53], where parameters such as solvent composition, concentration, cooling rate, etc. are optimized [54]. In addition, the metastable zone width of all relevant forms has to be known [55–58]. Often a seeding process provides the only reliable way to obtain the desired form [1, 59]. Even if a drug substance does not show polymorphism, seeding is often applied to control the crystallization process. Seeding can also be very useful for controlling the crystal size. Other ways to control crystal size include the use of ultrasound [60] in the crystallization process and, of course, milling as a processing step. In milling processes, care must be taken that no phase changes are induced due to increases in pressure or temperature although an exact understanding of phase changes induced by milling is still incomplete [61]. A common phenomenon is amorphization upon micronization. Samples with several percent of amorphous content are frequently produced. Depending on the intended use, e.g., for inhalation purposes, such amorphous parts have to be quantified, as requested by regulatory authorities. Particular attention also has to be paid to drying processes. It must be assured that, at the drying temperature used, no conversion into an undesired form takes place. Regulatory authorities like to know the rationale for the choice of a particular condition [19]. Again, ET diagrams are very helpful for establishing such criteria.

Crystallization development normally is carried out as a part of synthesis process development (see Figs. 1.2 and 1.3) [62, 63] .

1.4.4 Formulation

Different formulations are used at the various stages of drug development. The first formulations of the drug substance are made for pharmacokinetics (PK) and toxicology studies. At this stage, it is important that the formulation is quick and easy to produce, and other aspects such as shelf-life or ease of application play a minor role. Often, a tiered approach is used to test the drugs orally. Drug suspensions are the first choice [35], followed by pH-adjusted aqueous solutions, solutions in non-aqueous solvents and self-emulsifying lipid-based systems [41]. When using suspensions, it is very important to control particle size as this might have an effect on bioavailability. In many cases, parenteral administration of test compounds is a better method because the resulting absolute bioavailability information allows a better assessment of the efficacy of the lead compounds. The oral route may be hampered by first-pass effects and/or low oral absorption. Evidently, therefore, during the early research phase, the drug can be properly profiled only by using adequate formulations with both the oral and the parenteral route. In general, the preclinical screening of poorly water-soluble compounds is more challenging than for freely water-soluble compounds. The formulation that is used for these preclinical studies has implications for the possible final formulations [41]. For Phase II or, at the very latest, for Phase III, the final formulation must be developed. Final oral formulation types include tablets as the most popular form, capsules, syrups and solutions.

Other possible formulations include solids for inhalation, creams, gels, patches, nasal sprays, suppositories, solids for reconstitution prior to injection, etc. [64]. Choosing the final formulation can be difficult, and the solid-state properties of both excipients and the drug substance again play a key role. As mentioned in the Introduction, phase changes in the formulation process induced by solvate/hydrate formation or temperature must be avoided. This can be particularly difficult if wet granulation is used with a substance that can form hydrates [65]. In such a case, one may have to assure by post-process controls that the desired form is present in the formulation. Solvation/desolvation processes during formulation may also change particle size [65, 66]. The influence of the process parameters temperature and pressure on solid-state properties has to be monitored carefully. Of particular concern is unintended formation of amorphous parts due to their generally much lower chemical stability and higher solubility [19]. Of course, chemical compatibility between excipients and the drug substances must be checked as well. Particularly challenging are formulations intended for inhalation [60]. There, particle size is important not only for dissolution kinetics but also for absorption. Only particles within a narrow size range of about 1 to 5 μm can be deposited in the lungs.

1.4.5 Method Development

In cases where differences in polymorphic form affect drug performance, bioavailability or stability, the appropriate solid state form must be specified (Chapter 15, ICH Q6A) [67]. It may even be necessary to specify acceptable levels of undesired forms mixed with the desired form. In such cases the crucial question is what the acceptable level is. It depends both on solubility differences and chemical stability differences between the possible forms. From the production process it is generally known which forms can be present as “phase impurities” in the selected form. Other forms can often be regarded as uncritical or very unlikely to be formed by the chosen crystallization process, and method development can be focused on critical forms. For instance, the amorphous form is normally the solid form that shows the most pronounced differences to the most stable crystalline form. Therefore, requests to assess the content of amorphous form have often been made by regulatory authorities.

Suitable methods to determine solid-state compositions include differential scanning calorimetry, microcalorimetry, solution calorimetry, thermogravimetry, moisture sorption, IR, Raman, powder X-ray diffraction, solid-state NMR, solubility and dissolution rate measurements, and light and electron microscopy (Chapters 3 to 7) [42, 68]. Which method is optimal depends on both the drug substance and the excipients. If the polymorph composition is used as a release parameter, the appropriate method has to be validated [42] with respect to linearity, accuracy, precision, intermediate precision, limit of quantitation and limit of detection (ICH Q2A, Q2B) [69, 70].

1.5
Solid-state and Life Cycle Management

Exploiting superior properties of new solid forms of a certain active molecule may also be used for life cycle management. An example is the sodium salt of diclofenac. It was marketed as Voltaren® by Ciba-Geigy. Before the patent expired, other salts with properties enabling substantially better penetration of the skin were discovered and patented [71]. These salts, in appropriate formulations, are particularly suitable for topical applications. So discovering and patenting new salts and formulations enabled retention of an exclusive position in this market segment.

1.6
Conclusions

The solid-state form can drastically alter the properties of a pharmaceutical product. It may change its effectiveness, stability and suitability for a particular formulation. Therefore, developing the "right" solid form is critical for the success of a product. Finding this form and assuring that it is successfully delivered is part of an integrated approach to solid-state issues, all the way from salt screening to quality control. The ultimate goal of solid form screening (free molecule, salt or co-crystal) is to identify and to select the optimal solids for the intended use. This is independent of whether amorphous or crystalline solids are to be used. Necessary controls for different solid forms need to be established on a case-by-case basis.

The solid-state behavior of a drug plays an important role during the whole life span of a drug, from discovery through to the life cycle management stage. While understanding the solid-state behavior is particularly an issue in development, there are increasing efforts to carry out preliminary solid-state investigations already in the research phase [35]. Timing, available amount of substance and attrition rate suggest that the effort of solid-state development should be staged. It makes sense to adapt the solid-state development effort to the preclinical and clinical development phases.

Furthermore, once the product is on the market, concerns about solid-state issues do not end. Discovering new forms, possibly in combination with new formulations, may provide opportunities for the life cycle management of the product. Also, if changes in the manufacturing process are made, consistent quality in terms of solid-state properties and adequate quality control method development must be ensured [19].

Acknowledgments

We acknowledge the valuable comments and suggestions of Peter van Hoogevest.

Abbreviations

BCS	Biopharmaceutics classification system
c	Concentration
Caco-2	Human colon adenocarcinoma cell line
CTA	Clinical trial application
δ	Membrane thickness
D	Diffusion coefficient
DMSO	Dimethyl sulfoxide
ET diagram	Energy–temperature diagram
GMP	Good manufacturing practice
GRAS	Generally regarded as safe
ICH	International Conference on Harmonisation
IND	Investigational new drug application
IP	Intellectual property
IR	Infrared
K	Partition coefficient
Ka	Transintestinal absorption rate
MAD	Maximum absorbable dose
MDCK	Madin Darby Canine Kidney
NDA	New drug application
NMR	Nuclear magnetic resonance
P	Permeability
PAMPA	Parallel artificial membrane permeability assay
PD	Pharmacodynamics
PK	Pharmacokinetics
S	Solubility
S_{abs}	Effective surface of absorption
SAR	Structure–activity relationship
SITT	Small intestinal transit time
SIWV	Small intestinal water volume
XRD	X-ray diffraction

References

1 Ostwald, W., *Z. Phys. Chem.*, 22 (**1897**) 289–330.
2 McCrone, W.C., *Phys. Chem. Org. Solid State*, 2 (**1965**) 725–767.
3 Brittain, H.G., *Polymorphism in Pharmaceutical Solids*, Marcel Dekker, Inc., New York (**1999**).
4 Byrn, S.R., Pfeiffer, R., Stowell, J.G., *Solid State Chemistry of Drugs*, 2nd edn., SSCI Inc., West Lafayette (**1999**).
5 Bernstein, J. *Polymorphism in Molecular Crystals*, Oxford Science Publications, Oxford (**2002**).
6 Datta, S., Grant, D.J.W., *Nat. Rev.*, 3 (**2004**) 42–57.
7 Morris, K.R. in *Polymorphism in Pharmaceutical Solids* (ed. Brittain, H.G.), Marcel Dekker, New York (**1999**) pp. 125–181.
8 Wunderlich, B., Grebowicz, J., *Adv. Polym. Sci.*, 60/61 (**1984**) 1–59.
9 Wunderlich, B., *Thermochim. Acta*, 340 (**1999**) 37–52.
10 Vadas, E.B., Toma, P., Zografi, G., *Pharm. Res.*, 8 (**1991**) 148–155.
11 Morris, K.R., Newman, A.W., Bugay, D.E., Ranadive, S.A., Singh, A.K., Szyper, M., Varia, S.A., Brittain, H.G., Serajuddin, A.T.M., *Int. J. Pharm.*, 108 (**1994**) 195–206.
12 Stevenson, C.L., Bennett, D.B., Lechuga-Ballesteros, D., *J. Pharm. Sci.*, 94 (**2005**) 1861–1880.
13 Lipinski, C.A., Lombarda, F., Dominy, B.W., Feeney, P.J., *Adv. Drug. Deliv. Rev.*, 46 (**2001**) 3–26.
14 Griesser, Ulrich J., Stowell, J.G. in *Pharmaceutical Analysis*, (eds. Lee, David C., Webb, Michael L.), Blackwell Publishing Ltd., Oxford (**2003**).
15 Henck, J.-O., Griesser, U.J., Burger, A., *Pharm. Ind.*, 59 (**1997**) 165–169.
16 Stahly, G.P., at the American Chemical Society ProSpectives *Polymorphism in Crystals: Fundamentals, Predictions and Industrial Practice*, Tampa, FL, Feb 23–26 (**2003**).
17 Storey, R., Docherty, R., Higginson, P., Dallman, C., Gilmore, C., Barr, G., Dong, W., *Crystallogr. Rev.*, 10 (**2004**) 45–56.
18 Stahl, P.H., Wermuth, C.G. (eds.), *Handbook of Pharmaceutical Salts: Properties, Selection, and Use*, Wiley-VCH, Weinheim (**2002**).
19 DeCamp, W.H., *Am. Pharm. Rev.*, 4 (**2001**) 70–77.
20 Amidon, G.L., Lennernas, H., Shah, V.P., Crison, J.R., *Pharm. Res.*, 12 (**1995**) 413–420.
21 Harman, R.J., *Pharm. J.*, 262 (**1999**) 334–337.
22 Stenberg, P., Bergström, C.A.S., Luthman, K., Artursson, P., *Clin. Pharmacokinet.*, 41 (**2002**) 877–899.
23 McGovern, S.L., Caselli, E., Grigorieff, N., Shoichet, B.K., *J. Med. Chem.*, 45 (**2002**) 1712–1722.
24 Ren, S., Lien, E.J., in *Progress in Drug Research*, Vol. 54, (ed. Jucker, E.), Birkhäuser Verlag, Basel (**2000**) pp. 1–23.
25 Kansy, M., Senner, F., Gubernator, K., *J. Med. Chem.*, 41 (**1998**) 1007–1010.
26 Irvine, J.D., Takahashi, L., Lockhart, K., Cheong, J., Tolan, J.W., Selick, H.E., Grove, J.R., *J. Pharm. Sci.*, 88 (**1999**) 28–33.
27 Lipinski, C.A., *Adv. Drug. Deliv. Rev.*, 23 (**1997**) 3–25.
28 Yu, L.X., Amidon, G.L., Polli, J.E., Zhao, H., Mehta, M., Conner, D.P., Shah, V.P., Lesko, L.J., Chen, M.-L., Lee, V.H.L., Hussain, A.S., *Pharm. Res.*, 19 (**2002**) 921–925.
29 Johnson, K., Swindell, A., *Pharm. Res.*, 13 (**1996**) 1795–1798.
30 Curatolo, W., *Pharm. Sci. Technol. Today*, 1 (**1998**) 387–393.
31 Singhal, D., Curatolo, W., *Adv. Drug Deliv. Rev.*, 56 (**2004**) 335–347.
32 Artursson, P., *J. Pharm. Sci.*, 79 (**1990**) 476–482.
33 Crison, J.R., in *Water-Insoluble Drug Formulation* (ed. Liu, R.), CRC Press, Boca Raton, FL (**2000**), 97–110.
34 Martinez, M.N., Amidon, G.L., *J. Clin. Pharmacol.*, 42 (**2002**) 620–643.
35 Balbach, S., Korn, C., *Int. J. Pharmaceutics*, 275 (**2004**) 1–12.

36 http://www.cfsan.fda.gov/~dms/opa-noti.html.
37 Pudipeddi, M., Serajuddin, A.T.M., *J. Pharm. Sci.*, 94 (**2005**) 929–939.
38 Chemburkar, S.R, Bauer, J., Deming, K., Spiwek, H., Patel, K., Morris, J., Henry, R., Spanton, S., Dziki, W., Porter, W., Quick, J., Bauer, P., Donaubauer, J., Narayanan, B.A., Soldani, M., Riley, D., McFarland, K., *Org. Process Res. Dev.*, 4 (**2000**) 413–417.
39 Löbenberg, R., Amidon, G.L., *Eur. J. Pharm. Biopharm.*, 50 (**2000**) 3–12.
40 Clas, S.-D., Cotton, M., Moran, E., Spagnoli, S., Zografi, G., Vadas, E.B., *Thermochim. Acta*, 288 (**1996**) 83–96.
41 Huang, L.-F., Tong, W.-Q., *Adv. Drug Deliv. Rev.*, 56 (**2004**) 321–334.
42 Giron, D., Mutz, M., Garnier, S., *J. Therm. Anal. Cal.* 77 (**2004**) 709–747.
43 Imaizumi, H., Nambu, N., Nagai, T., *Chem. Pharm. Bull.*, 31 (**1983**) 2510–2512.
44 Miyazaki, T., Yoshioka, S., Aso, Y., Kojima, S., *J. Pharm. Sci.*, 93 (**2004**) 2710–2717.
45 Nichols, G., Frampton, C.S., *J. Pharm. Sci.*, 87 (**1998**) 684–693.
46 Snider, D.A., Addicks, W., Owens, W., *Adv. Drug Deliv. Rev.*, 56 (**2004**) 391–395.
47 Gu, C.-H., Chatterjee, K., Young Jr., V., Grant, D.J.W., *J. Crystal Growth*, 235 (**2002**) 471–481.
48 Bauer, J., Spanton, R., Henry, R., Quick, J., Dziki, W., Porter, W., Morris, J., *Pharm. Res.*, 18 (**2001**) 859–866.
49 Laird, T., *Org. Process Res. Dev.*, 8 (**2004**) 301–302.
50 International Conference on Harmonisation of Technical Requirements for Registration of Pharmaceuticals for Human Use, ICH Harmonised Tripartite Guideline, Impurities: Guideline for Residual Solvents, Q3C (www.ich.org).
51 Stoica, C., Verwe, P., Meekes, H., van Hoof, P.J.C.M., Kaspersen, F.M., Vlieg, E., *Crystal Growth Design*, 4 (**2004**) 765–768.
52 Marti, E., *J. Therm. Anal. Cal.*, 33 (**1988**) 37–45.
53 Giron, D., *Eng. Life Sci.*, 3 (**2003**) 103–112.
54 Hulliger, J., *Angew. Chem., Int. Ed. Engl.*, 33 (**1994**) 143–162.
55 Beckmann, W., Nickisch, K., Budde, U., *Org. Process Res. Dev.*, 5 (**1998**) 298–304.
56 Beckmann, W., *Org. Process Res. Dev.*, 4 (**2000**) 372–383.
57 Beckmann, W., Otto, W., Budde, U., *Org. Process Res. Dev.*, 5 (**2001**) 387–392.
58 Myerson, A.S. (ed.), *Handbook of Industrial Crystallization*, Butterworth-Heinemann, Woburn, MA (**2002**).
59 Heffels, S.K., Kind, M., *14th International Symposium on Industrial Crystallization*, Institution of Chemical Engineers, Rugby (UK), (**1999**) 2234–2246.
60 Dennehy, R.D., *Org. Process Res. Dev.*, 7 (**2003**) 1002–1006.
61 Trask, A.V., Shan, N., Motherwell, W.D.S., Jones, W., Feng, S., Tan, R.B.H., Carpenter, K.J., *Chem. Commun.*, (**2005**) 880–882.
62 Yu, L.X., Lionberger, R.A., Raw, A.S., D'Costa, R., Wu, H., Hussain, A.S., *Adv. Drug Delivery Rev.*, 56 (**2004**) 349–369.
63 Birch, M., Fussell, S.J., Higginson, P.D., McDowall, N., Marziano, I., *Org. Process Res. Dev.*, 9 (**2005**) 360–364.
64 Giron, D., *J. Therm. Anal. Cal.*, 68 (**2002**) 335–357.
65 Byrn, S., Pfeiffer, R., Ganey, M., Hoiberg, C., Poochikan, G., *Pharm. Res.*, 12 (**1995**) 945–954.
66 Himuro, I., Tsuda, Y., Sekiguchi, K., Horikoshi, I., Kanke, M., *Chem. Pharm. Bull.*, 19 (**1971**) 1034–1040.
67 International Conference on Harmonisation of Technical Requirements for Registration of Pharmaceuticals for Human Use, ICH Harmonised Tripartite Guideline, Specifications: Test Procedures and Acceptance Criteria for New Drug Substances and New Drug Products: Chemical Substances: Q6A (www.ich.org).
68 Yu, L., Reutzel, S.M., Stephenson, G.A., *Pharmaceutical Sci. Technol. Today*, 1 (**1998**) 118–127.
69 International Conference on Harmonisation of Technical Requirements for Registration of Pharmaceuticals for Human Use, ICH Harmonised Tripartite Guideline, Text on Validation of Analytical Procedures: Q2A (www.ich.org).

70 International Conference on Harmonisation of Technical Requirements for Registration of Pharmaceuticals for Human Use, ICH Harmonised Tripartite Guideline, Validation on Analytical Procedures: Methology: Q2B (www.ich.org).

71 Foraita, H.-G. in *Handbook of Pharmaceutical Salts: Properties, Selection, and Use* (eds. Stahl, P. H., Wermuth, C. G.), Wiley-VCH, Weinheim (**2002**) pp. 221–235.

2
Thermodynamics of Polymorphs

Sachin Lohani and David J. W. Grant

2.1
Introduction

Polymorphism may be defined as the ability of a compound to crystallize in two or more crystalline phases with different arrangements and/or conformations of the molecules in the crystal lattice [1]. Hence, polymorphs are different crystalline forms of the same pure chemical compound. The phenomenon of polymorphism in molecular crystals is analogous to allotropism among elements. Approximately one-third of organic compounds and about 80% of marketed pharmaceuticals exhibit polymorphism under experimentally accessible conditions [2–4]. On account of the different arrangement and/or conformation of molecules, polymorphs exhibit different physical and chemical properties [5]. These differences disappear in the liquid and vapor phases.

The relative stability of polymorphs depends on their free energies, such that a more stable polymorph has a lower free energy. Under a defined set of experimental conditions (with the exception of transition points) only one polymorph has the lowest free energy. This polymorph is the thermodynamically stable form and the other polymorph(s) is termed a metastable form(s). A metastable form is one that is unstable thermodynamically but has a finite existence as a result of relatively slow rate of transformation. In the pharmaceutical industry the metastable form is sometimes desirable on account of its special properties, such as higher bioavailability, better behavior during grinding and compression, or lower hygroscopicity. However, a metastable form has a thermodynamic tendency to reduce its free energy by transforming into the stable form. Such a polymorphic transformation is often detrimental to the efficacy of the formulation. Examples include chloramphenicol palmitate [6] and ritonavir [7]. Furthermore, manufacturing processes and pharmaceutical processing, such as compaction, milling, wet granulation, and freeze drying, can also result in polymorphic transitions [8–11]. The extent of polymorphic transition depends on the processing conditions and the relative stability of the polymorphs in question. Thermodynamics provides us with a framework within which the relative stability of the polymorphs can be determined.

Polymorphism: in the Pharmaceutical Industry. Edited by Rolf Hilfiker

ISBN: 3-527-31146-7

2.2 Structural Origin of Polymorphism

Structural differences between the crystalline lattices of polymorphs originate through two mechanisms, namely, *packing polymorphism* and *conformational polymorphism*. Packing polymorphism is a mechanism by which molecules that are conformationally relatively rigid can be packed into different three-dimensional structures. Conformational polymorphism is a mechanism by which conformationally flexible molecules can fold into different shapes that can pack into different three-dimensional structures [12]. This mechanism does not adequately describe cases for which isomers of organic molecules crystallize in different lattices. The term *configurational polymorphism* is sometimes used to describe cases in which the constituent molecules in the polymorphs exist as different isomers, such as geometric isomers or tautomers [13]. However, because different types of molecules are involved, the term polymorphism is not strictly applicable in these cases. The distinction between packing and conformational polymorphism is not clear-cut. In a few known examples of conformational polymorphism the molecules exist in different conformations while retaining the same crystalline lattice (or packing arrangement). Examples include 2,2′,4,4′-hexanitroazobenzene [14] and the anti-viral agent virazole [15]. Similarly, in a few known examples of packing polymorphism the molecules show different packing arrangements while retaining the same conformation. Examples include *p*-nitrophenol [16], *p*-chlorophenol [17], chlordiazepoxide [18], and sulfathiazole [19]. In general, the differences in packing arrangements invariably affect the molecular geometry and, conversely, the differences in molecular geometry cause the molecules to pack differently. As a result, most examples of polymorphism in organic crystals have a mixed origin and exhibit differences in both the conformation and packing arrangement of the constituent molecules [4, 20].

2.3 Thermodynamic Theory of Polymorphism

The constituent molecules in the crystal lattice of a polymorph differ in the nature of the non-covalent interactions, such as hydrogen bonds, van der Waals forces, π–π stacking, and electrostatic interactions. The difference in the nature of the non-covalent interactions often results in structural differences in the crystalline lattices of the polymorphs. Furthermore, the non-covalent interactions also influence the mechanisms by which heat is dissipated by the molecules in the crystal lattice. As a result, each polymorph has its own characteristic molar heat capacity [3], C_m. Molar heat capacity is a fundamental thermodynamic property of crystalline solids and may be regarded as the energy required to overcome molecular friction [21]. Heat capacity can be measured calorimetrically under constant pressure, P, or under constant volume, V, which are represented by $C_{P,m}$, and $C_{V,m}$, respectively. These quantities are defined as follows:

$$C_{P,\mathrm{m}} = \left(\frac{\partial H}{\partial T}\right)_P \quad \text{and} \quad C_{V,\mathrm{m}} = \left(\frac{\partial U}{\partial T}\right)_V \tag{1}$$

where H is the enthalpy, U is the internal energy and T is the absolute temperature. Conversely, integration of the heat capacity at constant pressure ($C_{P,\mathrm{m}}$) with respect to temperature can be used to calculate the enthalpy (H^{T_1}) and entropy (S^{T_1}) at temperature T_1, (Eqs. 2 and 3, respectively).

$$H^{T_1} = \int_0^{T_1} C_{P,\mathrm{m}} \mathrm{d}T + H^0 = U^{T_1} + PV = E^{T_1}_{\text{Lattice-energy}} + E^0_{\text{Zero-point energy}} + PV \tag{2}$$

$$S^{T_1} = \int_0^{T_1} \frac{C_{P,\mathrm{m}}}{T} \mathrm{d}T + S^0 \tag{3}$$

The internal energy of a crystal has contributions from lattice energy ($E^{T_1}_{\text{Lattice-energy}}$) and zero-point energy ($E^0_{\text{Zero-point energy}}$) [22]. The third law of thermodynamics states that for a perfectly ordered crystal the entropy term will vanish at absolute zero (i.e., $S^0=0$). Burger and Ramberger [23] applied Eqs. (2) and (3) to a system with two polymorphs A (more stable at absolute zero) and B. They derived the following equations for the differences in enthalpy ($\Delta H^{T_1}_{B\to A} = H^{T_1}_A - H^{T_1}_B$) and entropy ($\Delta S^{T_1}_{B\to A} = S^{T_1}_A - S^{T_1}_B$) between the two polymorphs at T_1:

$$\Delta H^{T_1}_{B\to A} = \int_0^{T_1} \Delta C_{P,\mathrm{m(B\to A)}} \mathrm{d}T + \Delta H^0_{\mathrm{B\to A}} \tag{4}$$

$$\Delta S^{T_1}_{B\to A} = \int_0^{T_1} \frac{\Delta C_{P,\mathrm{m(B\to A)}}}{T} \mathrm{d}T + \Delta S^0_{\mathrm{B\to A}} \tag{5}$$

where $\Delta S^0_{B\to A}$ and $\Delta H^0_{B\to A}$ are, respectively, the entropy and enthalpy differences between polymorphs at $T=0$ K, while, $\Delta C_{P,\mathrm{m(B\to A)}}$ is the molar heat capacity difference between the polymorphs. Again, we note that the entropy difference between the perfect crystals of forms A and B will vanish at absolute zero ($\Delta S^0_{\mathrm{B\to A}} = 0$). If the molecular conformations in the lattice of the two polymorphs are similar, the major contributor to $\Delta H^0_{B\to A}$ will be the difference between their $E^0_{\text{Lattice-energy}}$. Due to the similar molecular conformation, the contribution from $E^0_{\text{Zero-point energy}}$ will be small. Using Eqs. (4) and (5) the Gibbs free energy difference ($\Delta G^{T_1}_{\mathrm{B\to A}}$) between the polymorphs at T_1 can be expressed as follows:

$$\Delta G^{T_1}_{\mathrm{B\to A}} = \Delta H^{T_1}_{\mathrm{B\to A}} - T_1 \Delta S^{T_1}_{B\to A} \tag{6a}$$

$$\Delta G^{T_1}_{\mathrm{B\to A}} = \left(\Delta H^0_{\mathrm{B\to A}} + \int_0^{T_1} \Delta C_{P,\mathrm{m(B\to A)}} \mathrm{d}T\right) - T_1 \left(\int_0^{T_1} \frac{\Delta C_{P,\mathrm{m(B\to A)}}}{T} \mathrm{d}T\right) \tag{6b}$$

Eq. (6b) assumes that no phase transition has occurred in the system between 0 and T_1 K. For cases involving phase transitions, Eq. (6b) will contain additional terms, corresponding to enthalpy and entropy changes associated with each phase transition [24]. From Eqs. (4) to (6), the value of the heat capacity difference between two polymorphs ($\Delta C_{P,\mathrm{m(B \to A)}}$) as a function of temperature enables us to calculate the difference in enthalpy and in entropy between the polymorphs. Einstein developed a theory to estimate the relationship between $C_{V,\mathrm{m}}$ and T for monoatomic crystals [25]. He assumed that each atom in the lattice oscillates about its equilibrium position with a frequency (ν). The following expression of $C_{V,\mathrm{m}}$ can be derived after modifying Einstein's theory by taking into account vibrations over a range of frequencies [23]:

$$C_{V,\mathrm{m}} = \frac{k}{n} \sum_{\nu} \left\{ \frac{(h\nu/kT)^2 \exp(h\nu/kT)}{[\exp(h\nu/kT) - 1]^2} \right] \quad (7)$$

where k is Boltzmann's constant, n is the number of moles in the crystal, and h is Planck's constant. Using thermodynamic arguments, it can be shown for any substance that [26]

$$C_{P,\mathrm{m}} - C_{V,\mathrm{m}} = \alpha TV(\partial V/\partial T)_P \quad (8)$$

where $\alpha = [(\partial V/\partial T)_P]/V$, which is the thermal expansibility at constant pressure. Most crystalline solids have a small value of α [26], therefore $C_{V,\mathrm{m}} \approx C_{P,\mathrm{m}}$. Thus, for a crystalline solid, the expression in Eq. (7) also corresponds to $C_{P,\mathrm{m}}$ to a good approximation.

2.4 Thermodynamic Relationship Between Polymorphs: Enantiotropy and Monotropy

Lehmann, in 1888, coined the terms enantiotropy and monotropy to distinguish two different types of polymorphic behavior [27]. Two types of graphs are commonly used to describe the thermodynamic behavior of polymorphs, namely, energy–temperature diagrams and pressure–temperature diagrams. We shall describe polymorphic behavior using each type of diagram.

2.4.1 Energy–Temperature Diagrams

Energy–temperature diagrams were first introduced in crystallography by Buerger et al. in 1937 [22, 28]. These authors used schematic plots of internal energy (U) and Helmholtz free energy (A) versus temperature to represent the phase transformations in crystalline solids. Buerger et al. argued that the enthalpy of crystalline solids under normal pressure conditions has a negligible contribution from pressure–volume energy (PV). Therefore, for crystalline solids at ambient pressure:

$$H = U + PV \approx U \tag{9}$$

$$G = H - TS \approx U - TS = A \tag{10}$$

Nowadays, widespread use of thermal analytical techniques, such as differential scanning calorimetry (DSC), has made it easier to generate plots of H vs. T and G vs. T than plots of U vs. T and A vs. T for organic crystals [3, 29]. Hence, H vs. T and G vs. T plots are becoming increasingly common in publications on the thermodynamics of pharmaceutical compounds. Figure 2.1 shows a typical energy–temperature diagram of a crystalline solid.

The molar heat capacity of a crystalline substance increases with increasing temperature [3, 21]. Therefore, the mathematical expression for $C_{V,\text{m}} \approx C_{P,\text{m}}$ in Eq. (7) also increases monotonically with temperature. Thus, in Fig. 2.1, the enthalpy (H) isobar is depicted as increasing with increasing temperature, because the slope of the curve is given by $C_{P,\text{m}}$. The entropy term (TS) is shown as increasing with temperature, because entropy is always positive as result of the Third law of thermodynamics. Furthermore, the free energy isobar (G) decreases with increasing temperature, because the slope of the curve is equal to the negative value of the entropy. Let us reconsider a system with two polymorphs A (more stable at absolute zero) and B. For this system Fig. 2.2 can be drawn (similar to Fig. 2.1).

Figure 2.2 (a) represents an enantiotropic system. A pair of polymorphs is said to have an enantiotropic relationship if there exists a transition point at which the two polymorphs can undergo reversible solid–solid transformation. The defining feature for such a system is the existence of a *transition point* (T_t) below the melting point of both polymorphs. In Fig. 2.2, the melting point of a polymorph can be defined as the temperature at which the free energy isobar of the polymorph intersects the free energy isobar of the liquid. However, the transition temperature is defined as the temperature at which the free energy isobar of polymorph A in-

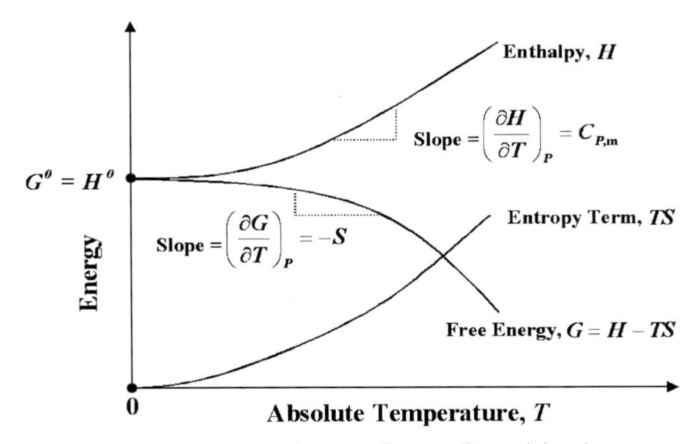

Fig. 2.1 Energy–temperature diagram of a crystalline solid under constant pressure. (Adapted from [1]).

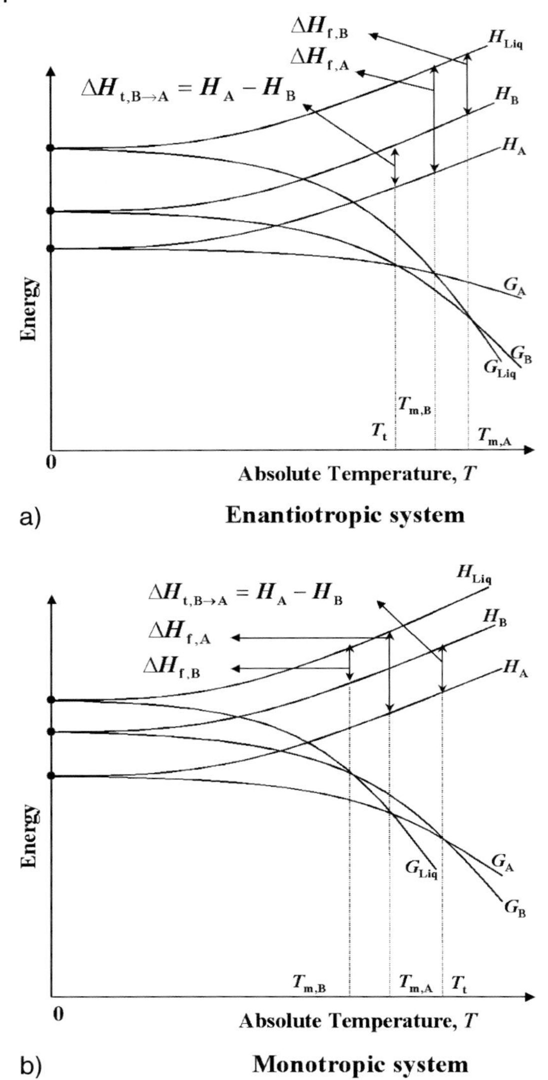

Fig. 2.2 Energy–temperature plots for an enantiotropic and a monotropic system. G is the free energy, H is the enthalpy, T is the temperature, subscripts A and B and Liq refer to the polymorph A, polymorph B, and liquid phase, respectively, while subscripts f, t and m, refer to fusion, transition point, and melting point, respectively. (Adapted from [23]).

tersects the free energy isobar of polymorph B. Thus, at T_t both polymorphs have equal free energy, i.e., $G_A = G_B$ and consequently are in equilibrium with each other. Below T_t, polymorph A is the stable solid phase because the free energy of A is lower than that of B, i.e., $G_A < G_B$. Consequently, below T_t, polymorph B can undergo spontaneous exothermic transformation into polymorph A. Above T_t, polymorph B is the stable solid phase because its free energy is lower than that of polymorph A, i.e., $G_B < G_A$. Therefore, above T_t polymorph A can undergo spontaneous endothermic transformation into polymorph B.

Figure 2.2 (b) represents a monotropic system. A pair of polymorphs has a monotropic relationship if one of the polymorphs is always stable below the melting point of both polymorphs. As illustrated in the diagram, the free energy of polymorph A is always less than that of polymorph B, i.e., $G_A < G_B$, at all temperatures below $T_{m,A}$. Consequently, in this case, polymorph B can undergo a spontaneous exothermic transformation into polymorph A. For a monotropic system, this transformation is thermodynamically feasible at any temperature, because $G_A < G_B$ at all temperatures. However, solid–solid transformations are kinetically hindered because of the activation energy associated with them. In general, solid–solid transformation occurs at a temperature that provides the system with sufficient thermal energy to cross the activation energy barrier. The transition point (T_t) in a monotropic system is a *virtual transition point*, because it lies above the melting points of both polymorphs. This notion assumes that the free energy curves of the two polymorphs converge beyond their melting point. Another possible situation corresponding to monotropic behavior is the divergence of free energy curves of the polymorphs. In this case, the virtual transition point lies below absolute zero [30, 31]. Based on aforementioned ideas, Burger and Ramberger have formulated the heat of transition rule (Section 2.5.1) and the heat of fusion rule (Section 2.5.2) to determine whether the relationship between a pair of polymorphs is enantiotropic or monotropic [23].

The Gibbs free energy difference between two polymorphs reflects the ratio of their fugacities (escaping tendency). The fugacity (f) can be approximated by the vapor pressure (p). Therefore,

$$\Delta G_{B \to A} = RT \ln\left(\frac{f_A}{f_B}\right) \approx RT \ln\left(\frac{p_A}{p_B}\right) \tag{11}$$

The fugacity is proportional to the thermodynamic activity (a), where the constant of proportionality depends on the choice of standard state [1]. Furthermore, the thermodynamic activity can be approximated by the solubility (s). Thus,

$$\Delta G_{B \to A} = RT \ln\left(\frac{a_A}{a_B}\right) \approx RT \ln\left(\frac{s_A}{s_B}\right) \tag{12}$$

For a transport-controlled dissolution process, under sink conditions and under constant hydrodynamic flow, the dissolution rate per unit area (J) is proportional to the solubility according to the Noyes-Whitney equation. Thus,

$$\Delta G_{B \to A} = RT \ln\left(\frac{J_A}{J_B}\right) \tag{13}$$

Furthermore, according to the law of mass action, the rate of chemical reaction (r) is proportional to the thermodynamic activity of the reacting polymorph. Therefore,

$$\Delta G_{\mathrm{B}\rightarrow\mathrm{A}} = RT \ln\left(\frac{r_{\mathrm{A}}}{r_{\mathrm{B}}}\right) \tag{14}$$

Eqs. (11) to (14) indicate that the high energy polymorph will have a higher fugacity, vapor pressure, thermodynamic activity, solubility, dissolution rate per unit surface area, and rate of reaction [1].

2.4.2
Pressure–Temperature Diagrams

Figure 2.3 shows a typical pressure–temperature diagram of a one-component system for which only one solid phase exists, corresponding to the absence of polymorphism. The curve S-V represents the vapor pressure vs. temperature curve of the solid and shows that the vapor pressure of a solid increases with the increasing temperature. The curve L-V represents the vapor pressure vs. temperature curve of the liquid and shows that the vapor pressure of a liquid increases with increasing temperature. The curve S-L represents the melting point curve of the solid. This curve has been drawn assuming that the molar volume of the liquid exceeds that of the solid.

The effect of increasing temperature and/or pressure on the equilibrium can be qualitatively predicted by application of the Le Chatelier principle, which states that if a constraint is imposed on a system at equilibrium, the equilibrium will shift in such a direction as to relieve the effect of the constraint. The Le Chatelier principle can be expressed quantitatively in the form of the Clapeyron equation shown below, which can be derived directly from the Maxwell equations in thermodynamics:

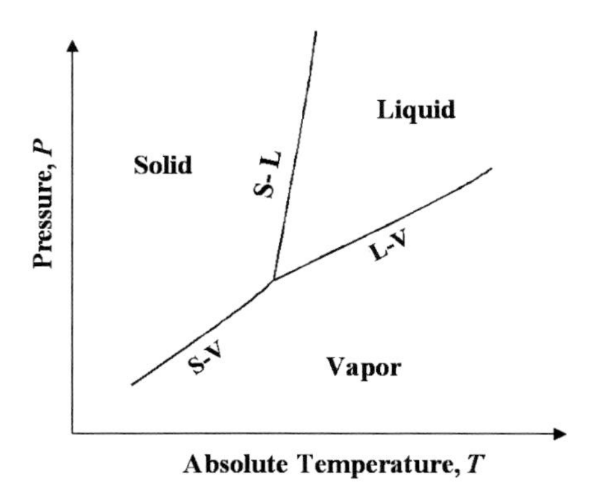

Fig. 2.3 Pressure–temperature diagram of a crystalline solid. S, L and V refer to the solid, liquid and vapor phase, respectively. The intersection of the three lines is the solid↔liquid↔vapor triple point. (Adapted from [32]).

$$\frac{dP}{dT} = \frac{\Delta H_{B \to A}}{T(V_{A,m} - V_{B,m})} \tag{15}$$

where subscripts A and B refer to the two phases in equilibrium, V_m is the molar volume and $\Delta H_{B \to A}$ is the heat of transition between phases A and B.

Figure 2.4 is a pressure-temperature diagram for a single component system that exhibits polymorphism. Here, S_A-V is the vapor pressure vs. temperature curve for polymorph A, while S_B-V is the curve of polymorph B. S_A-L is the melting point curve of polymorph A while S_B-L is that of polymorph B. The melting points of the polymorphs under atmospheric pressure ($T_{m,A}$ and $T_{m,B}$) are also shown in the Fig. 2.4, with brackets denoting the melting points of the metastable forms. The curve L-V represents the vapor pressure vs. temperature curve of the liquid. Only one liquid phase and only one vapor phase are possible for both polymorphs, because the difference between the polymorphs disappears in vapor and liquid phases. Figure 2.4 a and b are pressure–temperature diagrams of polymorphic systems showing enantiotropic and monotropic behavior, respectively (Fig. 2.4).

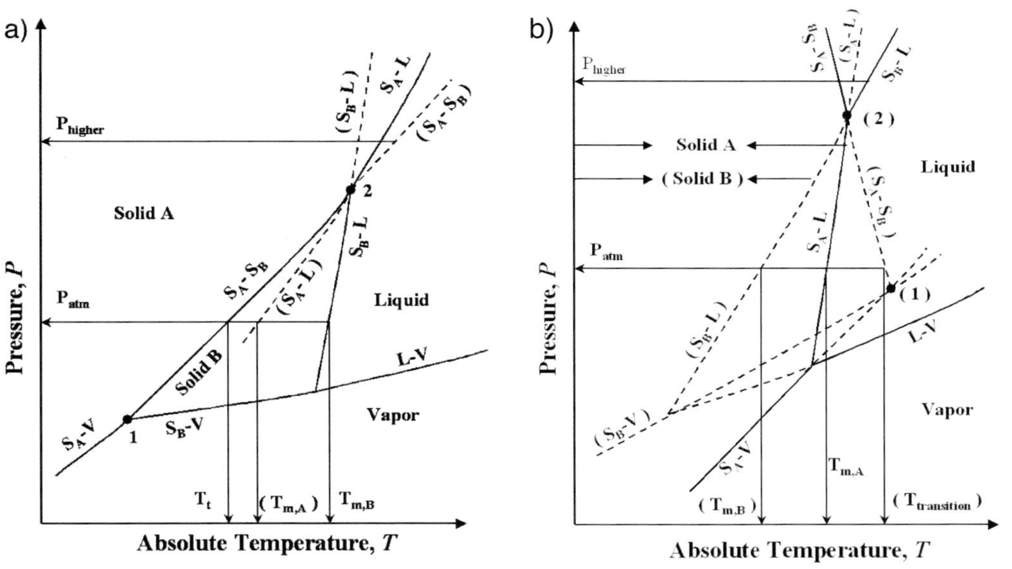

Fig. 2.4 Pressure–temperature plots of (a) an enantiotropic system and (b) a monotropic system. S, L and V refer to solid, liquid and vapor phase, respectively. Subscripts A and B refer to the polymorphs A and B, respectively. Solid lines denote stable phases while dashed lines denote metastable phases. Point 1 refers to the solid A↔solid B↔vapor triple point, while point 2 refers to the solid A↔solid B↔liquid triple point. Parentheses are used to denote features that involve a metastable phase. (Adapted from [33]).

In Fig. 2.4, S_A-S_B is the equilibrium curve between the two polymorphs in the absence of vapor. The curves drawn with dashed lines represent metastable phases. Point 1 refers to the solid A↔solid B↔vapor triple point, which is a three-phase transition point. Below point 1, polymorph A has a lower vapor pressure and is the solid phase in stable equilibrium with the vapor phase. Above point 1, polymorph B has a lower vapor pressure and is the solid phase in stable equilibrium with the vapor phase. Point 2 refers to the solid A↔solid B↔liquid triple point. It is the *condensed transition point*, because just below it polymorph B is in stable equilibrium with the liquid phase, while just above it polymorph A is in stable equilibrium with the liquid phase [34].

Fig. 2.4 (a) represents an enantiotropic system. The *ordinary transition point* is shown on the S_A-S_B curve as the intersection with the line corresponding to atmospheric pressure, P_{atm}. This intersection is a *real transition point*, because the corresponding temperature is less than the melting point of the two polymorphs. The affect of pressure on the transition temperature or on the curve S_A-S_B corresponds quantitatively with the Clapeyron equation (Eq. 15) [34, 35]. Because, the difference between the molar volumes of two polymorphs is usually very small, the slope of the S_A-S_B curve is very large. In other words, changing the pressure has negligible affect on the transition point. In Fig. 2.4, the slope of the S_A-S_B curve has been adapted to improve clarity. Figure 2.4 (b) represents a monotropic system. The transition point shown on the S_A-S_B curve is a *virtual transition point*, because the corresponding temperature is greater than the melting points of the two polymorphs.

2.4.3
Inversion of Polymorphic Behavior

The application of high pressure may cause the apparent relationship between the two polymorphs to change from monotropic to enantiotropic and vice versa. As shown in Fig. 2.4 (a), the line corresponding to P_{higher} (a higher pressure beyond point 2) will intersect the melting point curves for polymorphs A and B before it intersects the S_A-S_B equilibrium curve. Hence, the polymorphic system that was enantiotropic at atmospheric pressure appears to show monotropic behavior at higher pressure. Similarly, in Fig. 2.4 (b), the line corresponding to P_{higher}, a higher pressure beyond point 2, will intersect the S_A-S_B equilibrium curve before it intersects the melting point curves for polymorphs A and B. In this case, the polymorphic system appears to change from monotropic at atmospheric pressure to enantiotropic at a higher pressure. Thus, the terms enantiotropy and monotropy appear to be somewhat restricted in their application, and it is therefore necessary to specify the pressure and temperature under which the polymorphs are enantiotropic or monotropic [36]. Ceolin et al. have provide a detailed discussion on the influence of pressure on the enantiotropic and monotropic behavior of polymorphs [37]. However, importantly, the *true* thermodynamic definition of an enantiotropic system is a system that has a stable triple point, solid A↔solid B↔vapor (point 1 in Fig. 2.4 a). The true thermodynamic definition of a monotropic system is a system that has a metastable triple point, solid A ↔ solid B ↔ vapor (point 1 in Fig. 2.4b).

2.5 Rules to Predict Thermodynamic Relationships Between Polymorphs

Several rules have been developed to predict the relative thermodynamic stability of polymorphs and whether the relationship between the polymorphs is enantiotropic or monotropic. Tammann in 1926 [38] was the first to develop a number of these rules. Later, Burger and Ramberger in 1979 [23, 39], Yu in 1995 [30], and Grunenberg et al. in 1996 [3], extended the applications of these rules. We will now discuss each rule in detail.

2.5.1 Heat of Transition Rule

This rule states that if an endothermic phase transition is observed at a particular temperature (referred to as the experimental or kinetic transition temperature), the thermodynamic transition point lies below this temperature. Consequently, we can conclude that the two polymorphs are enantiotropically related. If an exothermic phase transition is observed at a particular temperature (experimental or kinetic transition temperature), there is no thermodynamic transition point below this temperature. This can occur either when the two polymorphs are monotropically related or when the two polymorphs are enantiotropically related and in addition have their thermodynamic transition temperature higher than the experimentally observed transition temperature.

This rule can be explained with reference to Fig. 2.2, by noting that all spontaneous phase transitions must result in a decrease in free energy. Application of this rule assumes first that the free energy isobars intersect at most once and, second, that the enthalpy isobars, H_A and H_B, of the two polymorphs do not intersect. The second assumption may be invalid when there are significant differences between the conformations of the molecules in the two polymorphs [23, 39, 40]. Thus, exceptions to the heat of transition rule may be expected with conformational polymorphism. Nevertheless, Burger and Ramberger found that the heat of transition rule makes correct predictions in 99% of cases [39].

2.5.2 Heat of Fusion Rule

The heat of fusion rule states that if the higher melting polymorph has the lower heat of fusion then the two polymorphs are enantiotropic, otherwise they are monotropic. Often, the rate of polymorphic transition is too slow to allow for an accurate measurement of the heat of transition, in which case the heat of fusion rule may be applied. This rule is based on the assumption that the heat of transition can be approximated by the difference between the heats of fusion of the polymorphs.

However, Burger and Ramberger [23] showed that the difference between the heats of fusion of the polymorphs ($\Delta H_{\mathrm{f,A}}$ and $\Delta H_{\mathrm{f,B}}$) is not exactly equal to the heat of transition between two polymorphs ($\Delta H_{\mathrm{B}\to\mathrm{A}}$). A more exact relationship is given by Eq. (16), which introduces the difference in heat capacity to account for the temperature dependence of the change in enthalpy.

$$\Delta H_{\mathrm{B}\to\mathrm{A}} = \Delta H_{\mathrm{f,A}} - \Delta H_{\mathrm{f,B}} + \int_{T_{\mathrm{f,A}}}^{T_{\mathrm{f,B}}} (C_{P,\mathrm{liquid}} - C_{P,\mathrm{A}})\mathrm{d}T \qquad (16)$$

where $C_{P,\mathrm{liquid}}$ and $C_{P,\mathrm{A}}$ are the molar heat capacities at constant pressure for the supercooled liquid and polymorph A, respectively. Burger and Ramberger further showed that the error in predictions based on the heat of fusion rule may originate from the heat capacity term in Eq. (16). Thus, exceptions to this rule may arise when the enthalpy curves of the two polymorphs diverge significantly, or when the difference between the melting points of the two polymorphs is larger than ~30 K [29].

2.5.3
Entropy of Fusion Rule

According to this rule, if the polymorph with higher melting point has a lower entropy of fusion, then the two polymorphs are enantiotropically related [41]. Conversely, if the polymorph with higher melting point has a higher entropy of fusion then the two polymorphs are monotropically related. The entropy of fusion of a polymorph (ΔS_{f}) can be obtained from its heat of fusion (ΔH_{f}) and melting point (T_{f}) as follows:

$$\Delta S_{\mathrm{f}} = \frac{\Delta H_{\mathrm{f}}}{T_{\mathrm{f}}} \qquad (17)$$

Yu [30] has provided a detailed thermodynamic analysis of this rule and has further extended its application.

2.5.4
Heat Capacity Rule

This rule states that, for a pair of polymorphs, the relationship is enantiotropic if the polymorph with the higher melting point also has the higher heat capacity at a given temperature. Conversely, the relationship is monotropic if the polymorph with the higher melting point has the lower heat capacity at a given temperature. Because different polymorphs have similar heat capacities and because the differences in heat capacities are difficult to measure accurately, this rule has limited applicability.

2.5.5
Density Rule

The density rule, which is based on Kitaigorodskii's principle of closest packing for molecular crystals [42], states that, for a non-hydrogen-bonded system at absolute zero, the most stable polymorph will have the highest density, because of stronger intermolecular van der Waals interactions. Thus, according to this rule, the crystal structure with most the efficient packing will also have the lowest free energy. However, in some cases, energetically favorable hydrogen bonds compensate for the loss of van der Waals energy and thus stabilize the polymorph with the lower density. For example, with resorcinol [13, 43], acetazolamide [44], and acetaminophen [45] the metastable polymorph has closer molecular packing and higher density resulting from hydrogen bonding.

2.5.6
Infrared Rule

This rule states that, for hydrogen-bonded polymorphs, the polymorph with the higher bond stretching frequency may be assumed to have the greater entropy. It assumes that the bond stretching vibrations are weakly coupled to those in the rest of the molecule. Burger and Ramberger have discussed exceptions to this rule [39].

2.6
Relative Thermodynamic Stabilities of Polymorphs

As mentioned above, knowledge of the polymorphic behavior (enantiotropic or monotropic) of a drug is important in the pharmaceutical industry to ensure that the final product contains the desired polymorph. The sudden appearance or disappearance of a polymorph can present a problem in process development. Similarly, serious pharmaceutical consequences arise if transformation occurs in a dosage form. If the metastable form of a monotropic system is desired, precautions must be taken to maintain appropriate conditions to avoid transformation from a metastable into a stable polymorph. No such precautions are necessary if the stable form of a monotropic system is desired. To obtain and to maintain the desired polymorph in an enantiotropic system, knowledge of the transition point is necessary. Bernstein et al. [46] have provided a detailed discussion of the above-mentioned possibilities for a dimorphic system.

From the discussion in Section 2.4, we note that the free energies of polymorphs may exhibit different dependences on temperature and/or pressure. From their analysis of the Cambridge Structural Database (CSD), Gavezotti et al. [47] found that, at room temperature, in most cases the free energy difference between polymorphs has the same sign as the enthalpy difference between them. Furthermore, the free energy difference between polymorphs is usually in the range of ~10 kJ mol^{-1} or less [47, 48], which is comparable to the average kinetic energy of a molecule at room temperature (= RT ~ 2.5 kJ mol^{-1}). Thus, changes in temperature and/or pressure may alter the relative thermodynamic stability of polymorphs and may induce a polymorphic transformation. Such changes in pressure and temperature often occur during pharmaceutical processing steps, such as tableting, grinding, and drying. According to the Ehrenfest system of classification of phase transitions [49], most solid–solid polymorphic transformations are first order transitions, i.e., the first order derivative of free energy with respect to temperature and/or volume is discontinuous at the transition point. The mechanism of solid–solid polymorphic transformation most likely involves an intermediate disordered state [1, 11, 13]. Morris et al. [50] have proposed the use of pressure–temperature diagrams, such as that shown in Fig. 2.4, to explain phase transitions that occur during tableting.

2.7 Crystallization of Polymorphs

Various methods are available to crystallize different polymorphs of a compound, such as cooling from the melt, sublimation, recrystallization from single or mixed solvents, altering the pH of the solution, as well as the presence of or the addition of tailor-made additives. In the pharmaceutical industry, different polymorphs are usually prepared by recrystallization from solutions, employing various solvents and different experimental conditions, such as temperature, initial supersaturation, rate of desupersaturation, and rate of agitation. Crystallization is a multi-step process, which includes nucleation, crystal growth and Ostwald ripening of the crystals [51]. Nucleation and crystal growth steps together determine the nature of polymorph that finally crystallizes from the solution. However, nucleation is the most critical step in the crystallization of individual polymorphs [1].

2.7.1 Nucleation of Polymorphs

Nucleation is the formation of minute nuclei (embryos or seeds) in a supersaturated solution, which can act as centers of crystallization [51]. Nucleation can be classified into primary nucleation, in which no crystals are initially present in the solution, and secondary nucleation, in which crystals of the solute are already present or are deliberately added to the solution as seeds. Primary nucleation can be further classified into homogeneous nucleation, which occurs spontaneously in bulk solutions, and heterogeneous nucleation, which occurs at interfaces or surfaces and may be induced by foreign particles. Homogeneous nucleation is practically feasible only in a small volume of solution by the microdroplet levitation technique [52, 53] and is generally not observed in solutions of volume greater than 100 μL. In practice, heterogeneous nucleation and/or secondary nucleation are commonly encountered.

Reduction in free energy is the thermodynamic driving force for homogeneous nucleation. The classical nucleation model is based on the work of Gibbs [54], Becker and Döring [55], and Volmer [56]. The Volmer model assumes that the viability of clusters (pre-nucleation molecular assemblies with mean radius = r) depends on attainment of the critical cluster size (with mean radius = r_c). The total free energy of a cluster (ΔG_{Total}) is the algebraic sum of a volume free energy term (ΔG_{Volume}) that favors aggregation of molecules and a surface free energy term ($\Delta G_{\text{Surface}}$) that favors the dissolution of molecular clusters. Thus,

$$\Delta G_{\text{Total}} = \Delta G_{\text{Surface}} + \Delta G_{\text{Volume}} = 4\pi r^2 \gamma + \left(\frac{-4\pi r^3 kT \ln \sigma}{3\nu} \right) \tag{18}$$

$$\Delta G_c^* = \left[\frac{16\pi \nu^2}{3(kT \ln \sigma)^2} \right] \tag{19}$$

where r is the mean radius of the cluster, k is Boltzmann's constant, T is absolute temperature, γ is the interfacial free energy between the nucleus and the supersaturated solution, σ is the supersaturation ratio, defined as the ratio of solute concentration in the supersaturated solution to that in the saturated solution, ν is the molecular volume, and ΔG_c^* is the free energy barrier to nucleation. The nucleation rate can be calculated from ΔG_c^* as follows:

$$J = A_n \exp\left(\frac{-\Delta G_c^*}{kT} \right) = A_n \exp\left(\frac{-16\gamma^3 \nu^2}{3k^3 T^3 (\ln \sigma)^2} \right) \tag{20}$$

where A_n is the pre-exponential factor. Furthermore, γ is given by Eq. (21) (proposed by Mersmann [57]).

$$\gamma = 0.414\ kT (c^s N_A)^{2/3} \ln(c^s / c^*) \tag{21}$$

where c^s is the molar density of the polymorph, c^* is the solute concentration, and N_A is Avogadro's number. The first evidence for the existence of clusters in supersaturated solutions came from the work of Mullin and Leci [58]. Later, various diffusion and spectroscopic studies on supersaturated solutions confirmed the existence of clusters [59–61]. In the initial stage of cluster formation ($r < r_c$), the surface free energy term dominates over the volume free energy term and the cluster has a tendency to dissolve. However, supersaturation provides the necessary driving force to overcome the disruptive surface free energy and promotes cluster growth. Eventually, the cluster attains the critical size ($r = r_c$) at which the surface term and volume term exactly balance. At this point the total free energy of the cluster attains a maximum, which corresponds to the activation free energy of nucleation (Fig. 2.5). At this maximum, the cluster is a critical cluster. Thus, supersaturation is required to overcome the free energy barrier to nucleation. After this stage ($r > r_c$), the cluster becomes viable and is termed a nucleus, which eventually grows

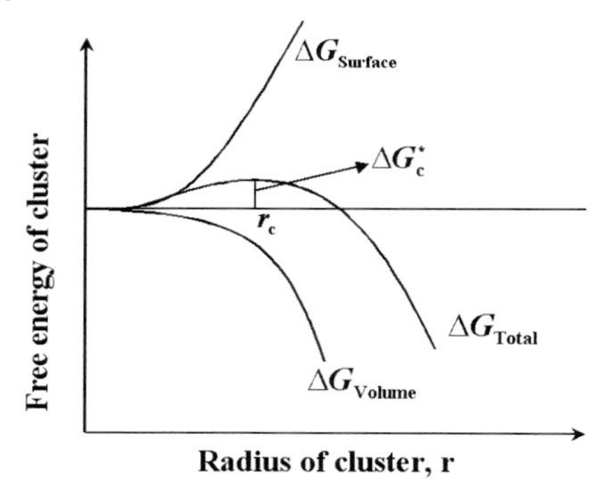

Fig. 2.5 Plot of Gibbs free energy change of clusters during molecular aggregation that leads to nucleation. $\Delta G_{Surface}$ is the surface free energy term, ΔG_{Volume} is the volume free energy term, ΔG_{Total} is the total free energy of the cluster, ΔG_c^* is the activation free energy of the critical cluster, r is the mean radius of the cluster, and r_c is the mean radius of the critical cluster.

into a crystal. If a cluster is smaller than the critical size required for a nucleus, the cluster dissolves and does not yield a crystal.

Let us reconsider the above dimorphic system with two polymorphs A (more stable) and B. Etter [62] proposed that, during nucleation, different clusters compete for molecules (Fig. 2.6 a). The free energy barrier of cluster formation will determine the concentration of each type of cluster. Furthermore, there will be a characteristic value of r_c and ΔG_c^* corresponding to each cluster. The polymorph that first crystallizes will correspond to the cluster with the lowest free energy barrier, which in this case is the metastable polymorph B because $\Delta G_{c,B}^* < \Delta G_{c,A}^*$ (Fig. 2.6 b). The polymorph that nucleates first comes from the

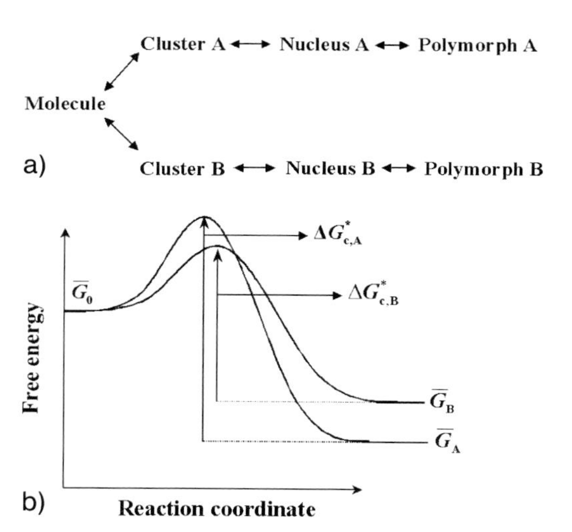

Fig. 2.6 (a) Clusters competing for molecules during crystallization of polymorphs A and B. (Adapted from [62]). (b) Free energy barrier associated with crystallization of polymorphs A and B. (Adapted from [29]) $\Delta G_{c,A}^*$ and $\Delta G_{c,B}^*$ are the activation free energies associated with nucleation of polymorphs A and B, respectively. $\overline{G}_A$ and $\overline{G}_B$ are the partial molar free energies of polymorphs A and B, respectively, and $\overline{G}_0$ is the partial molar free energy of the solute in the supersaturated solution.

cluster that exhibits the fastest growth rate as a result of its lowest free energy barrier to nucleation. Although the stable polymorph A may have the greater thermodynamic drive to crystallize, polymorph B may nucleate first due to its higher nucleation rate. However, the nature of polymorph that eventually crystallizes is determined by the combination of the relative nucleation rates and the relative crystal growth rates of the polymorphs [46].

2.8 Introduction to Solvates and Hydrates

Often, crystallization from solution yields crystals of a molecular adduct, which contains two or more chemical components together in the same crystalline lattice. Solvates and hydrates are common molecular adducts. Haleblian [5] defined a solvate as a crystalline molecular compound in which molecules of the solvent of crystallization are incorporated into the host lattice, consisting of unsolvated molecules. A hydrate is a special case of a solvate, when the incorporated solvent is water. The presence of solvent molecules in the crystal lattice influences the intermolecular interactions and confers unique physical properties to each solvate. Therefore, a solvate has its own characteristic values of internal energy, enthalpy, entropy, Gibbs free energy, and thermodynamic activity. Consequently, the solubility and dissolution rate of a solvate differ from those of the corresponding unsolvated phase and can result in differences in bioavailability of drugs. In general, a solvate will have a lower solubility in the solvent, from which it crystallizes as a solvate, than the unsolvated phase(s) [63]. Subsequent sections will focus on hydrates, because they are more common than solvates of organic solvents [64]. Nevertheless, the concepts developed below are applicable to any solvate system.

2.8.1 Thermodynamics of Hydrates

On the basis of the water uptake behavior at different water activities, hydrates can be classified as either stoichiometric or non-stoichiometric. Stoichiometric hydrates for which the mole ratio of water/host is constant (Fig. 2.7) have a defined stoichiometry over a range of water activities. Examples of important stoichiometric pharmaceutical hydrates include ampicilline trihydrate [65] and theophylline monohydrate [66]. However, for non-stoichiometric hydrates (Fig. 2.8, below), the mole ratio, water/host, may vary continuously as a function of water activity.

Figure 2.7 shows a hydrate system that forms two stoichiometric hydrates with m and n moles of water ($n > m$), respectively. Formation of m-hydrate from the anhydrous phase can be expressed as:

$$\mathrm{D(solid)} + m\mathrm{H_2O} \underset{}{\overset{K_{\mathrm{h},m}}{\rightleftharpoons}} \mathrm{D.mH_2O(solid)}$$

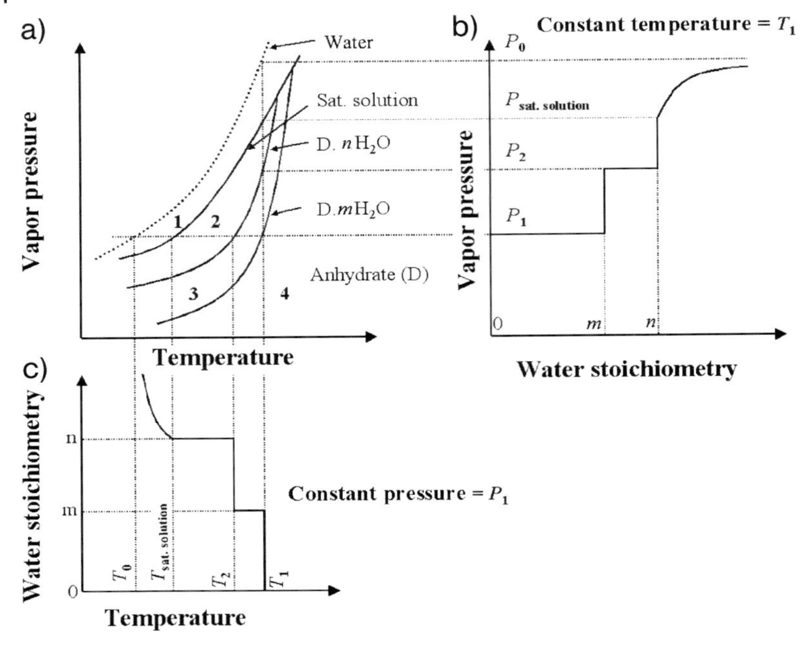

Fig. 2.7 (a) Vapor pressure of water versus temperature diagram of a hypothetical drug D, forming two stoichiometric hydrates, along with the vapor pressure–temperature curve for pure water. Regions labeled 1–4 represents the range of pressure–temperature values for which the solution of D in water, D.nH_2O, D.mH_2O, and D, respectively, are the phases that can exist under stable equilibrium. (b) Dependence of water stoichiometry on vapor pressure at constant temperature, T_1. (c) Dependence of water stoichiometry on temperature at constant pressure, P_1. (Adapted from [67]).

From the law of mass action the equilibrium constant, $K_{h,m}$, for above reaction can be expressed by Eq. (22) (proposed by Shefter and Higuchi [63], stated also by Grant and Higuchi [68]).

$$K_{h,m} = \frac{a[D.mH_2O(solid)]_{eq}}{a[D(solid)]_{eq}(a[H_2O]_{eq,m})^m} \tag{22}$$

Formation of the n-hydrate can be expressed as:

$$D.mH_2O(solid) + (n-m)H_2O \xrightleftharpoons{K_{h,n}} D.nH_2O(solid)$$

The equilibrium constant, $K_{h,n}$, for the reaction can be expressed by Eq. (23).

$$K_{h,n} = \frac{a[D.nH_2O(solid)]_{eq}}{a[D.mH_2O(solid)]_{eq}(a[H_2O]_{eq,n})^{n-m}} \tag{23}$$

Where $a[\mathrm{D}.m\mathrm{H_2O(solid)}]_{\mathrm{eq}}$ and $a[\mathrm{D}.n\mathrm{H_2O(solid)}]_{\mathrm{eq}}$ are the equilibrium activities of the corresponding crystalline hydrates, $a[\mathrm{D(solid)}]_{\mathrm{eq}}$ is the equilibrium activity of the anhydrate crystal, while $a[\mathrm{H_2O}]_{\mathrm{eq},m}$ and $a[\mathrm{H_2O}]_{\mathrm{eq},n}$ are the equilibrium activities of water. We next assume constant activity of unity for the three solid crystalline phases, i.e., for D, D.mH$_2$O and D.nH$_2$O. This assumption will hold for pure crystalline phases with negligible crystal defects. Using this assumption the standard Gibbs energy of formation of D.mH$_2$O(solid) can be expressed in terms of the equilibrium constant as follows (stated by Carstensen [69]):

$$\Delta G^{\ominus}_{\mathrm{D}\to\mathrm{D}.m\mathrm{H_2O}} = -RT\ln K_{\mathrm{h},m} = -RT\ln(a[\mathrm{H_2O}]_{\mathrm{eq},m})^{-m} \tag{24}$$

Similarly, the standard Gibbs energy of formation of D.nH$_2$O(soldi) can be expressed by Eq. (25).

$$\Delta G^{\ominus}_{\mathrm{D}.m\mathrm{H_2O}\to\mathrm{D}.n\mathrm{H_2O}} = -RT\ln K_{\mathrm{h},n} = -RT\ln(a[\mathrm{H_2O}]_{\mathrm{eq},n})^{-(n-m)} \tag{25}$$

Under a given set of conditions other than the standard conditions, the change in free energy for formation of D.mH$_2$O ($\Delta G_{\mathrm{D.}\to\mathrm{D}.m\mathrm{H_2O}}$) is given by

$$\Delta G_{\mathrm{D}\to\mathrm{D}.m\mathrm{H_2O}} = \Delta G^{\ominus}_{\mathrm{D}\to\mathrm{D}.m\mathrm{H_2O}} + RT\ln\left\{\frac{a[\mathrm{D}.m\mathrm{H_2O(solid)}]}{a[\mathrm{D(solid)}](a[\mathrm{H_2O}])^{m}}\right\} \tag{26}$$

D.mH$_2$O(solid) is obtained spontaneously when its formation results in the reduction of free energy, i.e., when $\Delta G_{\mathrm{D}\to\mathrm{D}.m\mathrm{H_2O}}$ is less than zero:

$$\Delta G_{\mathrm{D}\to\mathrm{D}.m\mathrm{H_2O}} = -RT\ln K_{\mathrm{h},m} + RT\ln(a[\mathrm{H_2O}])^{-m} < 0 \tag{27}$$

$$a[\mathrm{H_2O}] > (K_{\mathrm{h},m})^{-1/m} \tag{28}$$

Similarly, under a given set of conditions other than the standard conditions, the change in free energy for formation of D.nH$_2$O(solid) ($\Delta G_{\mathrm{D}.m\mathrm{H_2O}\to\mathrm{D}.n\mathrm{H_2O}}$) is given by Eq. (29).

$$\Delta G_{\mathrm{D}.m\mathrm{H_2O}\to\mathrm{D}.n\mathrm{H_2O}} = \Delta G^{\ominus}_{\mathrm{D}.m\mathrm{H_2O}\to\mathrm{D}.n\mathrm{H_2O}} + RT\ln\left\{\frac{a[\mathrm{D}.n\mathrm{H_2O(solid)}]}{a[\mathrm{D}.m\mathrm{H_2O(solid)}](a[\mathrm{H_2O}])^{n-m}}\right\} \tag{29}$$

D.nH$_2$O(solid) is obtained spontaneously when its formation results in the reduction of free energy, i.e., when $\Delta G_{\mathrm{D}.m\mathrm{H_2O}\to\mathrm{D}.n\mathrm{H_2O}}$ is less than zero:

$$\Delta G_{\mathrm{D}.m\mathrm{H_2O}\to\mathrm{D}.n\mathrm{H_2O}} = -RT\ln K_{\mathrm{h},n} + RT\ln(a[\mathrm{H_2O}])^{-(n-m)} < 0 \tag{30}$$

$$a[\mathrm{H_2O}] > (K_{\mathrm{h},n})^{-1/(n-m)} \tag{31}$$

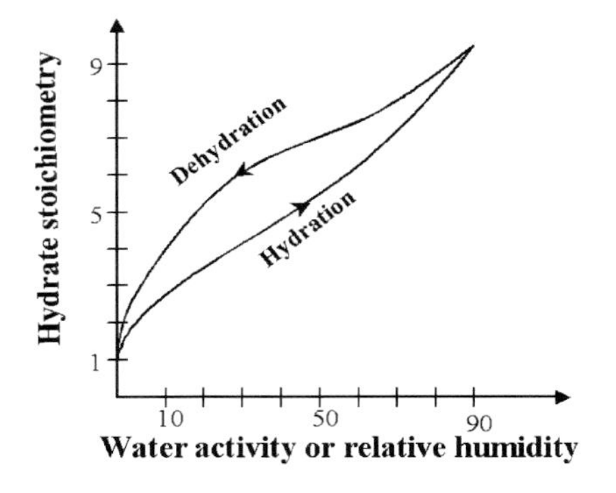

Fig. 2.8 Water uptake behavior for a non-stoichiometric hydrate (in this case cromolyn sodium). (Adapted from [70]). Dehydration curve refers to change in water stoichiometry with decreasing water activity, while the hydration curve refers to change in water stoichiometry with increasing water activity.

Therefore, D.mH$_2$O is the thermodynamically stable phase when water activity is greater than $(K_{h,m})^{-1/m}$, while D.nH$_2$O is the thermodynamically stable phase when water activity is greater than $(K_{h,n})^{-1/(n-m)}$.

The water activity ($a[H_2O]$) is equal to the relative humidity (r), defined as the ratio of vapor pressure of water (P) to the saturated vapor pressure of pure water (P_0) at the given temperature. Eqs. (28) and (31) can be used to calculate, at a given temperature, the range of relative humidities under which the given hydrate is stable, corresponding to the values of P_2 and P_1 in Fig. 2.7 (b). Similarly, Eqs. (28) and (31) can also be used to predict the temperature range, at a given water activity, within which the hydrate is stable, corresponding to T_2 and T_1 in Fig. 2.7 (c).

In the classic example of cromolyn sodium, a non-stoichiometric hydrate (Fig. 2.8), Cox et al. [70] showed that, with increasing (or decreasing) relative humidity, crystals of cromolyn sodium absorb (or lose) water by continuous adjustment of the crystal lattice, as shown by continuous changes in the powder X-ray diffraction patterns. The formation of non-stoichiometric hydrates may be thought of as water entering (or leaving) the crystal of pure drug to give a new solid phase, the structure of which was later solved and appears to be variable, depending on the water content [71]. In crystals of cromolyn sodium, one sodium ion has a fixed site, while the second sodium ion is disordered over three sites. The crystal lattice expands (or contracts) as water molecules enter (or leave) the crystal.

2.9 Summary

Each solid phase (such as a polymorph, solvate or hydrate) of a chemical substance has a characteristic value for each of its physicochemical properties, including thermodynamic values. Thermodynamics enables us to understand and to predict the influence of environmental conditions, such as temperature, pressure, and relative humidity, on the nature of each solid phase under equilibrium conditions, on the tendency of one phase to transform into another, and hence on the relative stability of the solid phases. Such knowledge is essential for determining the optimum conditions for crystallization and storage of the solid phase desired.

References

1 D. J. W. Grant, Theory and origin of polymorphism, in: *Polymorphism in Pharmaceutical Solids*, Marcel Dekker, Inc., New York, pp. 1–33, **1999**.

2 S. Datta, D. J. W. Grant, *Nat. Rev., Drug Discovery*, **2004**, 3, 42–57.

3 A. Grunenberg, J. O. Henck, H. W. Siesler, *Int. J. Pharm.*, **1996**, 129, 147–158.

4 T. L. Threlfall, *Analyst*, **1995**, 120, 2435–2460.

5 J. K. Haleblian, *J. Pharm. Sci.*, **1975**, 64, 1269–1288.

6 J. Bernstein, *NATO Sci. Ser., II: Mathematics, Phys. Chem.*, **2002**, 68, 244–249.

7 C.-H. Gu, V. Young Jr, D. J. W. Grant, *J. Pharm. Sci.*, **2001**, 90, 1878–1890.

8 G. G. Z. Zhang, C.-H. Gu, M. T. Zell, R. T. Burkhardt, E. J. Munson, D. J. W. Grant, *J. Pharm. Sci.*, **2002**, 91, 1089–1100.

9 D. D. Tiffani, G. E. Peck, J. G. Stowell, K. R. Morris, S. R. Byrn, *Pharm. Res.*, **2004**, 21, 860–866.

10 M. A. Moustafa, S. A. Khalil, A. R. Ebian, M. M. Motawi, *J. Pharm. Pharmacol.*, **1972**, 24, 921–926.

11 M. Otsuka, M. Matsumoto, S. Higuchi, K. Otsuka, N. Kaneniwa, *J. Pharm. Sci.*, **1995**, 84, 614–618.

12 J. Bernstein, A. T. Hagler, *J. Am. Chem. Soc.*, **1978**, 100, 673–681.

13 S. R. Byrn, R. R. Pfeiffer, J. G. Stowell, *Solid-state Chemistry of Drugs*, SSCI-Inc., West Lafayette, IN, **1999**.

14 E. J. Graeber, B. Morosin, *Acta Crystallogr., Sect. B: Struct. Crystallogr. Crystal Chem.*, **1973**, B30, 310–317.

15 P. Prusiner, M. Sundaralingam, *Acta Crystallogr., Sect. B: Struct. Crystallogr. Crystal Chem.*, **1976**, B32, 419–426.

16 G. U. Kulkarni, P. Kumardas, C. N. R. Rao, *Chem. Mater.*, **1998**, 10, 3498–3505.

17 M. Perrin, P. Michel, *Acta Crystallogr., Sect. B: Struct. Crystallogr. Crystal Chem.*, **1973**, B29, 258–263.

18 D. Singh, P. V. Marshall, L. Shields, P. York, *J. Pharm. Sci.*, **1998**, 87, 655–662.

19 N. Blagden, R. J. Davey, L. Lieberman, R. Williams, R. Payne, R. Roberts, R. Rowe, R. Docherty, *J. Chem. Soc., Faraday Trans.*, **1998**, 94.

20 S. R. Vippagunta, H. G. Brittain, D. J. W. Grant, *Adv. Drug Deliv. Rev.*, **2001**, 48, 3–26.

21 T. Threlfall, *Org. Process Res. Develop.*, **2003**, 7, 1017–1027.

22 M. J. Buerger, Crystallographic aspects of phase transformations, in: *Phase Transformation in Solids*, W. A. Weyl (ed.), John Wiley and Sons, New York, pp. 183–211, **1951**.

23 A. Burger, R. Ramberger, *Mikrochim. Acta*, **1979**, 2, 259–271.

24 E. F. Westrum, Jr., J. P. McCullough, Thermodynamics of crystals, in: *Physics and Chemistry of the Organic Solid State*, A. Weissberger (ed.), John Wiley, New York, pp. 3–178, **1963**.

25 A. Einstein, *Ann. Physik*, **1907**, 22, 180–190.

26 P. Atkins, J. Paula, *Physical Chemistry*, W.H. Freeman and Company, New York, **2002**.

27 O. Lehmann, *Molekularphysik*, Engelmann, Liepzig, **1888**.

28 M. J. Buerger, M. C. Bloom, *Z. Kristallogr.*, **1937**, 96.

29 J. Bernstein, *Polymorphism in Molecular Crystals*, Clarendon Press, Oxford, **2002**.

30 L. Yu, *J. Pharm. Sci.*, **1995**, 84, 966–974.

31 C.-H. Gu, D. J. W. Grant, *J. Pharm. Sci.*, **2001**, 90, 1277–1287.

32 J. E. Ricci, *The Phase Rule and Heterogeneous Equilibrium*, D. Van Nostrand Company, Inc., New York, **1951**.

33 W. C. McCrone, Polymorphism, in: *Physics and Chemistry of the Organic Solid State*, M. M. Labes (ed.), Interscience Publisher, New York, pp. 726–767, **1965**.

34 H.G. Brittain, Application of the phase rule to the characterization of polymorphic systems, in: *Polymorphism in Pharmaceutical Solids*, H.G. Brittain (ed.), Marcel Dekker, Inc., New York, **1999**.

35 U.J. Griesser, M. Szelagiewicz, U.C. Hofmeier, C. Pitt, S. Cianferani, *J. Thermal Anal. Calorimetry*, **1999**, 57, 45–60.

36 S.T. Bowden, *The Phase Rule and Phase Reactions*, Macmillan and Co., London, **1938**.

37 R. Ceolin, S. Toscani, J. Dugue, *J. Solid State Chem.*, **1993**, 102, 465–479.

38 G.H.J.A. Tammann, *The States of Aggregation (translation by Mehl, F. F.)*, Constable and Company, London, **1926**.

39 A. Burger, R. Ramberger, *Mikrochim. Acta*, **1979**, 2, 273–316.

40 D.D. Wirth, G.A. Stephenson, *Org. Process Res. Develop.*, **1997**, 1, 55–60.

41 A. Burger, *Pharmacy Int.*, **1982**, 3, 158–163.

42 A.I. Kitaigorodskii, *Molecular Crystals and Molecules*, Academic Press, New York, **1973**.

43 J.M. Robertson, A.R. Ubbelohde, *Proc. Royal Soc. London Ser. A.*, **1938**, 167, 136–147.

44 U.J. Griesser, A. Burger, K. Mereiter, *J. Pharm. Sci.*, **1997**, 86, 352–358.

45 M. Haisa, S. Kashino, H. Maeda, *Acta Crystallogr., Sect. B: Struct. Crystallogr. Crystal Chem.*, **1974**, 30, 2510–2512.

46 J. Bernstein, R.J. Davey, J.-O. Henck, *Angew. Chem., Int. Ed.*, **1999**, 38, 3441–3461.

47 A. Gavezzotti, G. Filippini, *J. Am. Chem. Soc.*, **1995**, 117, 12299–12305.

48 J.D. Dunitz, *Acta Crystallogr., Sect. B: Struct. Crystallogr. Crystal Chem.*, **1995**, 51, 619–631.

49 P. Ehrenfest, *Proc. Acad. Sci., Amsterdam*, **1933**, 36, 153–157.

50 K.R. Morris, U.J. Griesser, C.J. Eckhardt, J.G. Stowell, *Adv. Drug Deliv. Rev.*, **2001**, 48, 91–114.

51 J.B. Mullin, *Crystallization*, Butterworth-Heinemann, Boston, MA, **2001**.

52 P. Stockel, H. Vortisch, T. Leisner, H. Baugmartel, *J. Mol. Liq.*, **2002**, 96/97, 153–175.

53 S. Santesson, J. Johansson, L.S. Taylor, I. Levander, S. Fox, M. Sepaniak, S. Nilsson, *Anal. Chem.*, **2003**, 75, 2177–2180.

54 J.W. Gibbs, *The Scientific Papers of J.W. Gibbs: On the Equilibrium of Heterogeneous Substances*, Dover Publications, Inc., New York, **1961**.

55 R. Becker, W. Döring, *Ann. Physik (Berlin)*, **1935**, 24, 719–752.

56 M. Volmer, *Kinetik der Phasenbildung*, Steinkopf, Leipzig, **1939**.

57 A. Mersmann, *J. Crystal Growth*, **1990**, 102, 841–847.

58 J.W. Mullin, C.L. Leci, *Philosophical Mag. A: Phys. Condensed Matter: Struct., Defects Mech. Prop.*, **1969**, 19, 1075–1077.

59 M.A. Larson, J. Garside, *Chem. Eng. Sci.*, **1986**, 41, 1285–1289.

60 M.K. Cerreta, K.A. Berglund, *J. Crystal Growth*, **1987**, 84, 577–588.

61 A.S. Myerson, P.Y. Lo, *J. Crystal Growth*, **1991**, 110, 26–33.

62 M.C. Etter, *J. Phys. Chem.*, **1991**, 95, 4601–4610.

63 E. Shefter, T. Higuchi, *J. Pharm. Sci.*, **1963**, 52, 781–791.

64 J.A.R.P. Sarma, G.R. Desiraju, Crystal engineering: design and application of functional solids, M. Zaworotko (ed.), Kluwer, Norwell, MA, pp. 325–340, **1999**.

65 H. Zhu, D.J.W. Grant, *Int. J. Pharm.*, **1996**, 139, 33–43.

66 H. Zhu, C. Yuen, D.J.W. Grant, *Int. J. Pharm.*, **1996**, 135, 151–160.

67 S. Glasstone, *Textbook of Physical Chemistry*, D. Van Nostrand Company, New York, **1946**.

68 D.J.W. Grant, T. Higuchi, *Solubility Behavior of Organic Compounds*, John Wiley, New York, **1990**.

69 J.T. Carstensen, *Advanced Pharmaceutical Solids*, Marcel Dekker, Inc., New York, **2001**.

70 J.S.G. Cox, G.D. Woodard, W.C. McCrone, *J. Pharm. Sci.*, **1971**, 60, 1458–1465.

71 L.R. Chen, V.G. Young, Jr., D. Lechuga-Ballesteros, D.J.W. Grant, *J. Pharm. Sci.*, **1999**, 88, 1191–1200.

3
Characterization of Polymorphic Systems Using Thermal Analysis

Duncan Q. M. Craig

3.1
Introduction – Scope of the Chapter

Thermal methods, and in particular differential scanning calorimetry (DSC), have been one of the mainstay methods for the study of polymorphs, providing a means of both identification and characterization. Numerous excellent texts are available that outline the uses of these methods in the context of polymorphism, including those of Brittain [1, 2], Ford and Timmins [3] and Giron [4] and the interested reader is directed to these for further information on aspects not covered here. Indeed, it is important to define from the outset the scope of the present chapter in the context of both the available literature and the objectives of the book as a whole. The field of thermal analysis and the level of understanding of polymorphism have both grown so rapidly in recent years that a single chapter covering all the studies overlapping the two areas would be prohibitively lengthy. It is, for example, worth considering the number of new thermal techniques that have been either developed *de novo* or only recently introduced into the pharmaceutical arena over the last ten or so years – these include modulated temperature DSC, hyper-DSC and microthermal analysis. Similarly, a range of interfaced techniques such as Raman-DSC have also been described recently. On this basis, therefore, the current text is intended to introduce the reader both to "traditional" thermal methods and also to highlight some of these newer approaches that are only beginning to be used for the study of polymorphism but which show considerable potential.

Initially, the principles of DSC and the use of this method for the detection of polymorphic drugs, excipients and dosage forms will be outlined. DSC is and will probably remain the most commonly used method for initial polymorph screening and hence familiarity does not diminish its importance to the field. Nevertheless, in several studies, DSC data has been considered in non-conventional ways and some of these are highlighted to demonstrate that, even with this traditional technique, several new avenues may be usefully explored. Secondly, the use of DSC in conjunction with other techniques is considered, either

Polymorphism: in the Pharmaceutical Industry. Edited by Rolf Hilfiker

ISBN: 3-527-31146-7

in the context of utilizing more than one approach in a complementary manner or in the sense of using techniques that have been physically interfaced. It is fairly unusual for DSC to be used in complete isolation in any case, but the emphasis here is on non-routine combinations whereby complementarity is particularly well exemplified. Finally, the use of a number of other techniques, ranging from the well known (thermogravimetric analysis, thermal microscopy) to emerging methods (microthermal analysis, thermally stimulated current), will be considered. In this manner it is hoped that the chapter will direct the reader towards some of the more interesting recent developments in the field as well as exemplifying the more standard approaches to thermal characterization of polymorphism.

3.2 Use of Differential Scanning Calorimetry for the Characterization of Polymorphs

3.2.1 Principles of DSC in the Context of Polymorphism

Differential scanning calorimetry (DSC) was introduced in the 1960s as a development of the existing technique of differential thermal analysis (DTA). This latter technique involved monitoring the temperature difference between a sample and a reference during a heating or cooling cycle, the principle being that a thermal event such as melting or crystallization will result in a temperature difference between the two as heat is either absorbed (in the case of melting) or released (with crystallization) by the sample. Unfortunately, the technique does not lend itself well to quantitative analysis and has now been effectively entirely superseded (at least in the pharmaceutical field) by DSC. The original concept of DSC (which is still used in Perkin Elmer instruments) is one known as power compensation, whereby the sample and reference are maintained at the same temperature and the heat flow required to keep the two at thermal equilibrium is measured, thereby allowing both the temperature and the energy associated with a thermal event to be easily assessed. An alternative but closely related approach is heat flux DSC whereby the heat differential between the two samples is measured as a function of temperature (i.e., it has strong similarities to DTA in terms of basic principle). The energy associated with the transition is calculated via Eq. (1):

$$dQ/dt = \Delta T/R \tag{1}$$

where dQ/dt is the heat flow, ΔT is temperature difference between the furnace and the crucible and R is the thermal resistance of the heat path between the furnace and the crucible.

These two modes have formed the basis of all DSCs for several decades, with improvements in cell design and software being the main innovations involved, to the extent that some would argue that the baseline sensitivity of modern DSCs as well as the sophistication of the data manipulation are now as optimal as the basic design will allow. There have, inevitably, been more radical innovations, particularly in the last 10 or so years, some of which (modulated temperature DSC, hyper-DSC) are covered below. A notable further new approach to the DSC concept has been the introduction of Tzero DSC, whereby a third thermocouple has been introduced to compensate for many of the factors that cause loss of baseline sensitivity. However, the basic usage of this method in terms of the types of experiment that may be performed remains similar to conventional techniques.

Irrespective of the type of DSC instrument used, the type of information that may be obtained is uniform. In brief, DSC raw data shows heat flow (power in mW) plotted against temperature (usually presented in °C, although clearly more correctly in K), the former referring to the heat flux difference between the sample and reference. Consideration of the basic parameters associated with measurement indicates that the power (J s^{-1}) at a constant scanning rate (K s^{-1}) will be related to the heat capacity via

$$dQ/dt = Cp dT/dt \tag{2}$$

where dQ/dt is the heat flow and dT/dt is the heating rate. In effect, therefore, the measured intrinsic quantity associated with the material is the heat capacity; indeed in the thermal analysis literature it is usual best practice to present the DSC data as heat capacity rather than power against temperature, although this has not been widely adopted within the pharmaceutical sciences as yet. It is, however, helpful to consider that the baseline is directly related to the heat capacity of the system. As the sample undergoes a thermal event it is effectively altering the total heat capacity of the system due to the latent heat associated with the melting, crystallization etc., this being seen as a peak or, in the case of a glass transition, a shift in baseline. A further consideration of an entirely practical nature is that, unfortunately, some instruments show endotherms as upward peaks and crystallization exotherms as downward curves while others use the opposite convention. This is largely alleviated by the use of a clear scale on the y-axis, but nevertheless lends some confusion when comparing different studies.

Integration of Eq. (2) indicates that the area under the curve for a thermal event will be proportional to the energy involved in the process; hence, by suitable calibration with standards with known melting points, heats of fusion and heat capacities, it is possible to obtain an extremely thorough characterization of a range of samples. The thermal response most relevant to the current context is melting, as polymorphs may exhibit different melting points (discussed in more detail below). Consequently, it is worth considering the processes involved in this response. Melting is a first-order thermodynamic process, i.e., it depends solely on the temperature under given pressure conditions. Therefore, theoretically, the sample should pass from the solid to the liquid state effectively instantaneously as soon as the melting point is reached and the melting endotherm should take place over an infinitely small temperature range. In practice, several factors result in the melting peak being observed over a finite temperature range, including the response time of the instrument, the possibility of thermal lags within the pan (these are minimized by keeping the sample size as small as is practicable) and the possibility of crystal imperfections or contaminants (or, with polymers, molecular weight inhomogeneity) that serve to broaden the range of bond strengths within the sample and hence the melting range. Nevertheless, materials such as pure metals show very sharp melting peaks, with polymers being at the opposite extreme for the reasons mentioned above. As most low molecular weight drugs show relatively sharp peaks, differentiation between polymorphs on the basis of subtle differences in melting point is often possible. However, several considerations need to be taken into account in this respect. Firstly, it is common practice (within the pharmaceutical sciences) to use the peak temperature as the melting point for a sample. This is not strictly accurate as the melting point is more correctly identified as the onset of the melting peak, with the peak representing the temperature corresponding to the most rapid measured heat change rather than the thermodynamic melting point as such. Secondly, there will inevitably be some variance in measured melting points between runs; hence there are many instances (which admittedly do not tend to get reported in the literature) whereby DSC is not able to differentiate between polymorphs within the noise of repeat measurements, the point being that, unfortunately, the absence of a difference in measured melting point does not necessarily indicate the absence of polymorphic effects.

While polymorph identification may be performed by simple melting point studies, in practice such studies are not always so straightforward, particularly when one considers that one of the fundamental weaknesses of DSC is that it is an invasive technique; a heating or cooling signal is being applied to the sample, which may cause alterations in structure over and above the intended transition of interest. This is particularly pertinent for polymorphic studies as metastable forms may be transformed during the heating run. Indeed, one of the more interesting aspects of using DSC in this context is the possibility of monitoring transformations between forms via melt recrystallization, whereby the melting of a metastable form is accompanied by the simultaneous (or partially simultaneous) transformation into a second, more stable form. This may be envisaged as a clear three-stage process (melt, recrystallization, melt) or as a subtle shoulder on a melting curve. The scanning rate may be of particular importance in resolving these processes, with slower rates allowing a greater time period for the transformation to manifest. An example of such a staged transformation is shown in Fig. 3.1 for the interconversion of polymorphs of the quinolone antibiotic premafloxacin [5]. Form I shows a series of transformations on heating; more specifically, two melt/recrystallizations were observed, ascribed to conversions into Forms II and III, respectively, followed by melting of Form III. While informative, such simultaneous processes render the calculation of the heats of fusion and crystallization of the individual forms difficult.

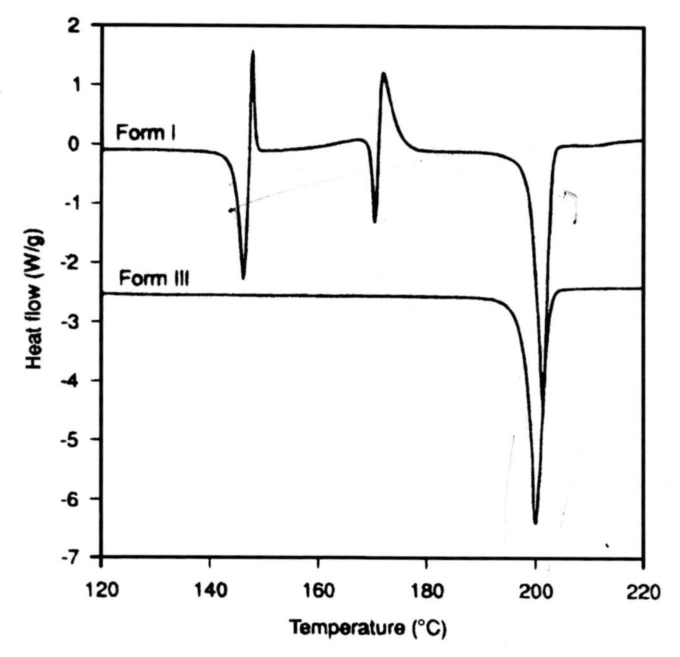

Fig. 3.1 DSC curves of Form I and Form III premafloxacin samples [5].

In addition to melt recrystallization responses, polymorphic forms may undergo transitions at temperatures below the melting point of either form, which may be seen as small endotherms or exotherms in the DSC curve; an example is shown in Fig. 3.2 for flunisolide, which undergoes a series of solid-state transformations between forms on heating [6]. A related concept is the question of whether polymorphic forms are monotropic (i.e., only a single stable form exists irrespective of temperature) or enantiotropic (one or other is stable above or below a specific temperature). This is not routinely addressed within the pharmaceutical industry at present, but nevertheless DSC does allow some insights into which of the two types of system may be present. A very considerable contribution in this respect was made by the group of Burger [7, 8] who developed a series of rules for differentiating between the two (Table 3.1). The rationale behind these rules is beyond the scope of the current text but the interested reader is referred to the original work or to the review by Giron [4] (see also Chapter 2). Briefly, however, the rules attempt to differentiate between enantiotropic and monotropic transitions via a series of observations based on the free-energy phase diagram for the two types of system. The observed solid-state transitions between forms may allow differentiation between enantiotropy and monotropy as well as identification of transition temperatures.

In summary, therefore, DSC may be used in the first instance to differentiate between polymorphs on the basis of melting point but may also be used to

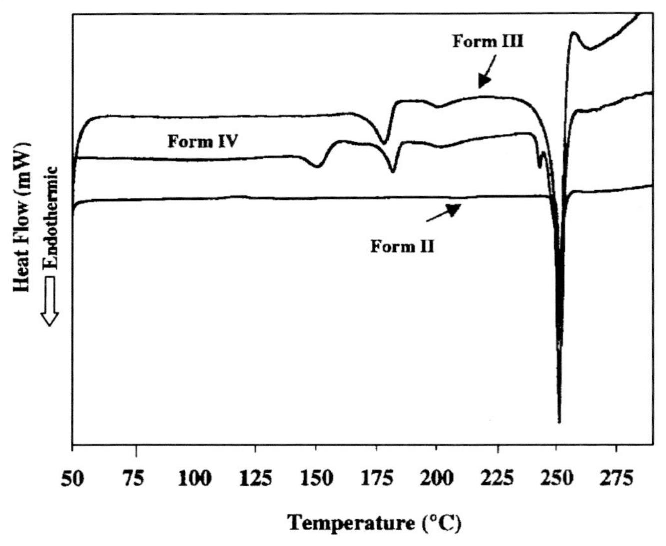

Fig. 3.2 DSC curves for flunisolide anhydrous polymorphic forms II, III and IV [6].

Table 3.1 Summary of means of differentiating between monotropic and enantiotropic polymorphism, with I being the higher melting form. (Adapted from Burger [8] through Giron [4].)

Enantiotropy	Monotropy
I stable > transition	I always stable
II stable < transition	II not stable at any temperature
Transition reversible	Transition irreversible
Solubility I higher < transition	Solubility I always lower than II
Transition II to I endothermic	Transition II to I exothermic
$\Delta H_F^I < \Delta H_F^{II}$	$\Delta H_F^I > \Delta H_F^{II}$
IR peak I before II	IR peak I after II
Density I < density II	Density I > density II

study the transformation behavior of metastable systems and to probe the nature of the interrelationship between the polymorphic forms. As demonstrated below, analysis of DSC data may take further, more sophisticated forms but the basic means of analysis outlined above have nevertheless served the pharmaceutical community effectively for several years.

3.2.2 Examples of the Uses of DSC: Characterization of Drugs, Excipients and Dosage Forms

There are a plethora of studies whereby different drug polymorphic forms have been identified and studied, many of which have been outlined [1–4]. A recent example of one such thorough study on a complex system is that of Dong et al. [9], whereby seven anhydrous polymorphs of neotame were prepared and characterized using DSC, solid-state NMR, XRD, FTIR, SEM, dynamic vapor sorption and density measurements, exemplifying the advisability of a multiple approach to studying polymorphism that will be expanded upon in a later section. Figure 3.3 shows the DSC data for these forms. Briefly, Forms A, D, F and G showed single melting endotherms, with no evidence of phase transitions

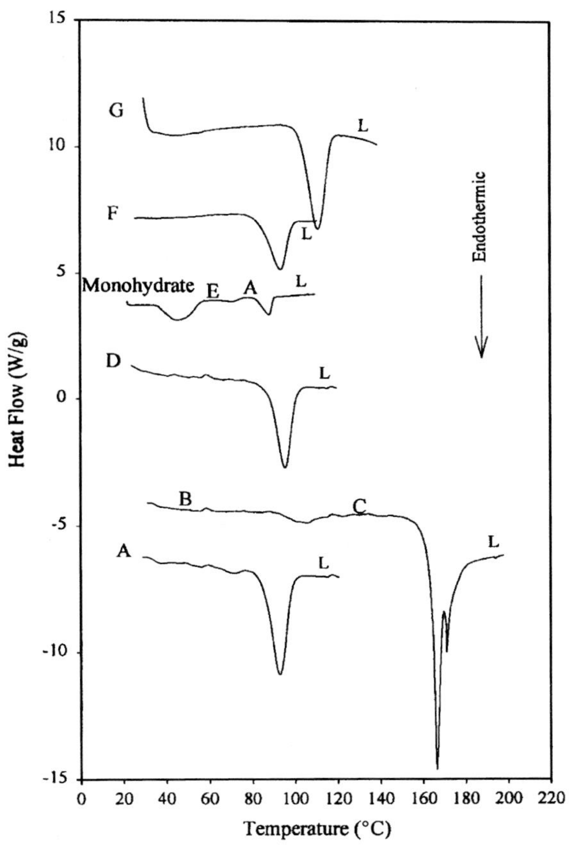

Fig. 3.3 DSC curves of neotame anhydrate polymorphs A–G. Each curve is labeled with the corresponding polymorph while the liquid state is labeled L [9].

prior to melting. However, Form B underwent an endothermic transition at 105 °C to Form C, which then melts at 166 °C followed by decomposition. Form E was produced by dehydration of the monohydrate and then converts into Form A. The authors suggested that, with reference to Burgers rules, Forms A and E and Forms B and C are enantiotropically related. The authors were able to correlate the DSC data with the other techniques to provide a series of conclusions linking the density of the polymorphs to the conditions in which they are formed and the unit cell structures involved.

A major concern in industry, which has received limited attention in the published literature, is the quantitation of mixtures of polymorphs of drugs. Clearly only one polymorph may be stable at any one temperature but nevertheless the kinetic stability of the metastable form may be such that it persists for a considerable period, and hence mixes are a very realistic possibility. An interesting paper by Vitez [10] directly addresses this issue. While the author was not able to name the drugs under study for reasons of confidentiality, he demonstrated the possibilities and, just as importantly, the limitations of using DSC to provide a quantitative analysis of polymorph content. Briefly, the author studied a series of systems whereby distinct polymorphs had been identified and compared mixtures of the two to the pure forms. In general, the approach is highly dependent on the extent to which the forms may be distinguished by thermal analysis, such that even drugs with a melting point difference of 10 °C could not be quantitatively characterized in mixes. This is exemplified by the study shown in Fig. 3.4, where two forms of a drug, for which the melting points of the two are distinct (peak melts at 104 and 114 °C), were mixed in known proportions. The composite DSC curve, however, did not show a baseline differentiation between the two forms, indicating that quantitative analysis was not possible. The author also explored the possibility of using different scanning speeds, as lower speeds increase resolution at the expense of sensitivity. However, this was unsuccessful in the case cited. While being a "negative result", the study demonstrates several important principles, not least of which being that the melting behavior of a composite system will not necessarily (or even probably) represent the sum of the melting behavior of the individual components, in part because the presence of one material in the molten state will influence the subsequent behavior of the higher melting component.

One aspect of the use of DSC for polymorphic studies that has received particular attention recently in the literature has been the study of the effects of processing on polymorphic form. Not only may traditional methods of processing such as granulation and tableting cause conversions between polymorphs but, in addition, several more recently introduced techniques such as supercritical fluid technology may alter the crystal form of a drug. An example of the latter is provided by Gosselin et al. [11], whereby the effects of rapid expansion of supercritical carbon dioxide solutions (RESS) were used to generate carbamazepine microparticles. The authors reported that the polymorph formed could be manipulated by the conditions used for the RESS process, with combinations of all four polymorphs and a varying degree of crystallinity being found to depend

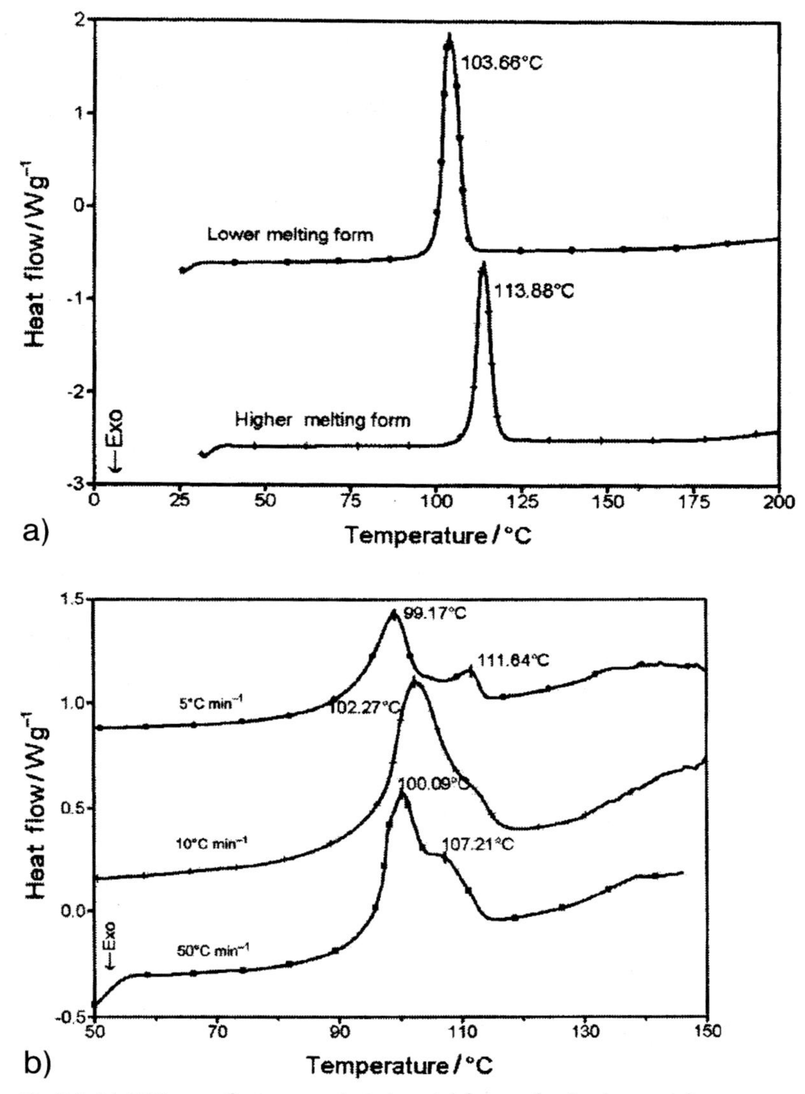

Fig. 3.4 (a) DSC curve for two unsolvated crystal forms of a developmental drug. (b) DSC curves for ca. 50% mixes of the two polymorphic forms run at a range of scanning speeds [10].

on the pressure used, extraction temperature and expansion temperature. Interestingly, in relation to the previous study described here, the authors used XRPD rather than DSC to determine the proportion of polymorphs present. As discussed later, XRD is routinely used in conjunction with DSC to study drug polymorphism and the relative merits of the two approaches continue to be discussed. In terms of determining proportions in mixtures, the evidence points towards XRPD being the technique of choice in many cases due to clear baseline differentiation between peaks, which, as Vitez [10] has shown, is often or even usually not possible with DSC. However, recent developments in thermoanalytical instrumentation such as high speed DSC, discussed later, may have a major influence on thinking in this regard.

Another example of the effects of supercritical fluid technology on crystal form is provided by Velaga et al. [12], who examined the crystallization of budesonide and flunisolide from acetone and methanol solutions using supercritical carbon dioxide as the crystallization medium. Again the authors reported a dependence of crystal form on the conditions used, assessed using DSC and XRPD, with two new polymorphs of flunisolide being reported using this method of preparation.

It has long been recognized that more traditional processing methods may also influence the polymorphic form of drugs. For example, Oguchi et al. [13] studied the effect of grinding different polymorphs of chenodeoxycholic acid to the amorphous state and monitoring the subsequent crystal forms of the recrystallized material on heating in DSC. Interestingly, the authors noted that the morphological history of the sample had a profound effect on the recrystallization behavior despite the ground material clearly showing an amorphous halo in XRPD prior to recrystallization. More specifically the authors demonstrated that amorphous material prepared from Form I recrystallized at 119.8 °C (which corresponded to the melting point of Form III), while amorphous material prepared by grinding Form III recrystallized at 147.3 °C, with a melting peak at ca. 165 °C seen in both cases, corresponding to the melting of Form I (Fig. 3.5). Variable-temperature XRD indicated that, at the temperatures corresponding to the exotherms, the samples converted into Form I, which subsequently melted at ca. 165 °C. The authors suggested that the difference in behavior could be related to the persistence of crystalline nuclei within the sample, as suggested earlier by Yonemochi et al. [14].

In addition to grinding, compression may also lead to polymorphic change [15]. Panchagnula et al. [16] have studied the polymorphic behavior of mefenamic acid, noting that the drug exists in Form I in commercial samples but converts into Form II on heating. Compression in an intrinsic dissolution test apparatus similarly caused conversion into Form II. However, a difficulty associated with studying the effects of compaction is a practical one in that tablets, unless unrealistically small, will not fit into a DSC pan, leading to the necessity of scraping samples or breaking the tablet, neither of which is optimal. However, we discuss an alternative approach below.

A further processing issue associated with polymorphism is the change in crystal form induced by freeze drying. The usual expectation for a freeze-dried drug is that a material will be in the glassy state (or in a glassy solution) in the final product. However, many materials (particularly some commonly used excipients) may crystallize during the freezing and/or drying process – hence it is necessary to understand how the process may alter the crystal structure. This is

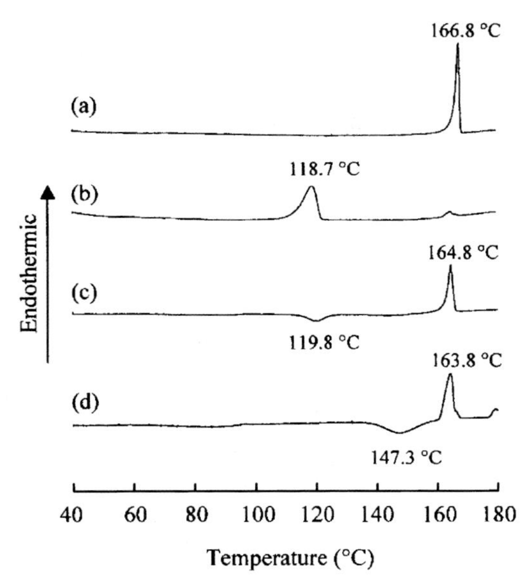

Fig. 3.5 DSC curves of intact and ground chenodeoxycholic acid: (a) Form I crystals, (b) Form III crystals, (c) ground Form I, and (d) ground Form III [13].

particularly pertinent to mannitol, which may remain amorphous or recrystallize, depending on the formulation and parameters used. Telang et al. [17] studied the effects of the addition of a range of salts on the polymorphic form of mannitol produced by evaporation from an aqueous solution. They noted that the presence of NaCl influenced the polymorphic form of the mannitol; interestingly, the form generated (α or δ) depended on the sample size of the generated batch. The authors interpreted this in terms of the kinetics of the nucleation and growth process, with the increased time of crystallization for the larger sample favoring the α polymorph. The frozen systems themselves may also exhibit polymorphism; Chongprasert et al. [18] demonstrated that glycine can be found in the α, β or γ form in frozen aqueous systems as well as the freeze-dried product, depending on the conditions used. In particular, for quench cooled solutions the authors were able to distinguish four separate endotherms immediately below the melting of ice by using slow heating rates. These were attributed to the melting of a β-glycine/ice eutectic, dissolution of α crystals, and the melting of the γ glycine/ice eutectic (the fourth could not be resolved sufficiently for interpretation).

DSC has also been extensively used for the characterization of multicomponent dosage forms. The technique is a staple approach in characterizing solid dispersions of drugs in polymers and other carriers. For example, Zajc et al. [19] examined the thermal behavior of nifedipine in mannitol dispersions prepared by the hot melt method. In this study a range of techniques in addition to DSC were used, including FTIR, to characterize the dispersions. Notably, FTIR was able to demonstrate that the nifedipine was present in a metastable polymorph after processing, which was not detected using DSC. Indeed, the authors point out that in an earlier study Burger et al. [20] reported that mannitol polymorphs I and II showed very small differences in DSC profile (mps 166.5 and 166 °C, heats of fusion 53.5 and 52.1 kJ mol^{-1}, respectively). This, therefore, demonstrates that irrespective of the high usefulness of DSC in detecting polymorphs in many studies, it is always advisable to use supporting techniques before drawing definitive conclusions.

As well as the polymorphic form of the drug, it is also important to consider the possibility of the excipients exhibiting polymorphism. This is particularly true of polymers such as poly(ethylene glycol), which can exist in extended or folded chain forms [21], but is also an essential consideration in the context of lipids. Sato [22] has reviewed the crystallization behavior of fats and lipids, while several studies have examined the melting behavior of pharmaceutical lipids. For example, Hamdani et al. [23] studied the derivatized lipids Compritol and Precirol, which may be used for melt pelletization, demonstrating the complex polymorphic behavior and processing sensitivity of these systems. Indeed, the fact that pharmaceutically relevant lipids are almost invariably chemically as well as physically complex renders their characterization difficult. One means of presenting such data that may be of use in certain circumstances is to employ solid fat content diagrams, whereby the percentage in the solid state is calculated from the DSC data as a function of temperature (easily calculated from the area under the curve representing the solid phase at a given temperature). This is used widely in food science and has been suggested for use in pharmaceutical systems; an example is presented in Fig. 3.6, showing the raw DSC data and corresponding SFC diagram for Gelucire 50/13 on storage for up to 180 days [24], from which it is possible to predict the solid and liquid content at any temperature, such as body temperature, and to visualize how the drug alters this distribution. The interested reader is also referred to the excellent paper by Brubach et al. [25], which reports the use of a novel temperature-dependent XRD approach to delineate the chemical species responsible for individual DSC peaks in Gelucires (Section 3.3.3).

3.2.3
Further Uses of DSC

Several other studies have attempted to use DSC in a manner that is not standard within the pharmaceutical sciences but which may hold considerable promise as a means of extracting additional information. One such approach is the application of curve fitting techniques to characterize the DSC data. There is some controversy regarding this approach as the physical significance of the fitted curves is sometimes open to question, as is the generalizability of the available software packages across the different DSC models (each with different furnace time constants). Nevertheless the approach is undoubtedly interesting

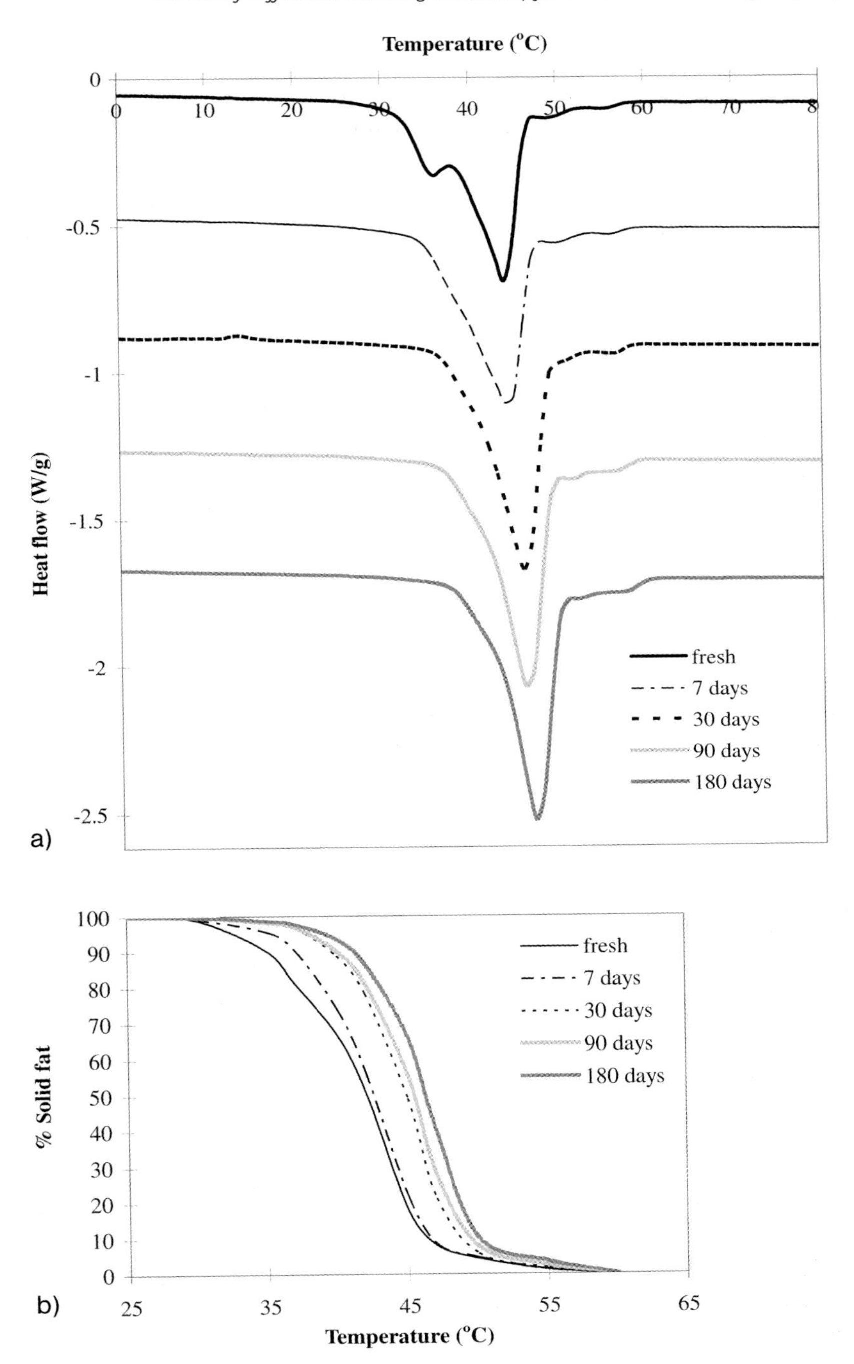

Fig. 3.6 (a) DSC curves and (b) solid fat content of Gelucire 50/13 after storage at 37 °C [24].

and arguably underexploited. The method is exemplified by the work of Leitao et al. [26] who studied terfenadine polymorphs recrystallized using various solvent evaporation protocols, enabling systems (and hence DSC curves) to be obtained for a range of mixed polymorphic systems. The authors developed an empirical yet flexible approach based on the curve for the pure (higher melting) polymorph. A fitting expression was derived for this peak and the mixed systems then fitted using Eq. (3):

$$t_{ons} = x_{inf}\left[\frac{\partial\eta(x_{inf})}{\partial x}\right]^{-1}\eta(x_{inf}) \tag{3}$$

whereby t_{ons} is the onset temperature, x_{inf} is the temperature of the inflection and $\delta\eta(x_{inf})$ is the expected heat flux at this temperature. The parameter $\eta(x)$ incorporates a variable factor that allows for curve asymmetry. Curve fitting using this approach was compared to the Gaussian method, with the authors concluding that the Gaussian approach was inadequate in terms of both fitting ability and physical interpretation of the data. Figure 3.7 gives examples of the use of the approach for a range of curves. The authors proposed that the fitted curves corresponded to different polymorphs of the drug.

Several studies have applied thermodynamic models to DSC data to extract more fundamental information concerning the systems under study. Yoshihashi et al. [27] attempted to directly correlate DSC data for different polymorphs with intrinsic dissolution rate. In essence these authors calculated the theoretical heat of fusion at the temperature of the dissolution test, using an approach suggested previously [28], thereby obtaining a measure of the bond strength at the temperature of the study (they made several interesting refinements to the ear-

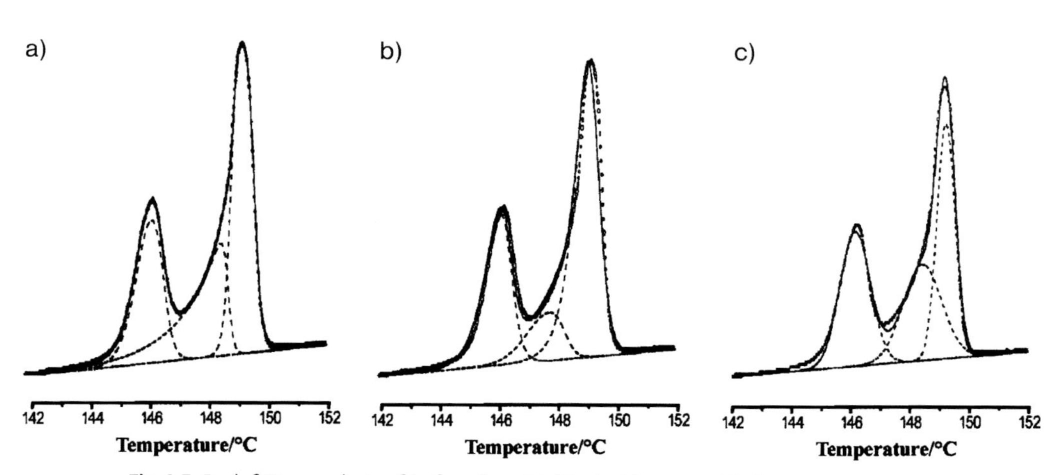

Fig. 3.7 Peak fitting analysis of terfenadine: (a) fitted with an empirical approach using variable parameters, (b) fitted with an empirical approach using non-variable parameters, and (c) Gaussian fit [26].

lier model, including a correction for partial and full amorphicity). In brief, the authors used the assumption that the entropy of melting remains essentially constant irrespective of temperature to derive the relationship

$$\Delta H_x = \Delta H_m \cdot \frac{T_x}{T_m} - \Delta H_{cry} \tag{4}$$

where ΔH_x is the heat of fusion at temperature T_x, ΔH_m is the heat of fusion at the melting point T_m, ΔH_{cry} is the heat of crystallization. The authors were then able to demonstrate a linear relationship between the intrinsic dissolution rate (at 25 °C) and the heat of fusion at 25 °C, which appeared to hold for different polymorphs and the glassy form of the drug (Fig. 3.8). The authors suggested that not only could this approach be used to predict the effect of physical composition on the dissolution behavior but could also be used the other way round, i.e., to predict polymorph mix ratio or degree of crystallinity.

Numerous studies have involved thermodynamic consideration of polymorphic transitions, both within the pharmaceutical and more general literature. Examples of the latter include the study by Naoki et al. [29], who derived the pressure and temperature phase diagram for hydroquinone, which exists in three polymorphic forms, α, β and γ. The authors reported that the α-form transforms into a new phase (δ) at elevated pressures. Interestingly, these authors used DTA to determine the phase boundaries for the system due to the possibility of using large sample sizes ($\sim$1 g) under high pressure (up to 100 MPa), which is difficult to perform using conventional DSC. Given the interest in pressure-induced polymorphic changes within the pharmaceutical sciences, the analysis described by these authors is worth noting. Studies that are more directly associated with pharmaceutical systems include that of Yu [30], who produced an interesting paper describing an approach to delineation of the thermodynamic parameters associated with polymorph transitions, with a view to allowing extrapolation of the free energy difference to storage temperature. Considering the process

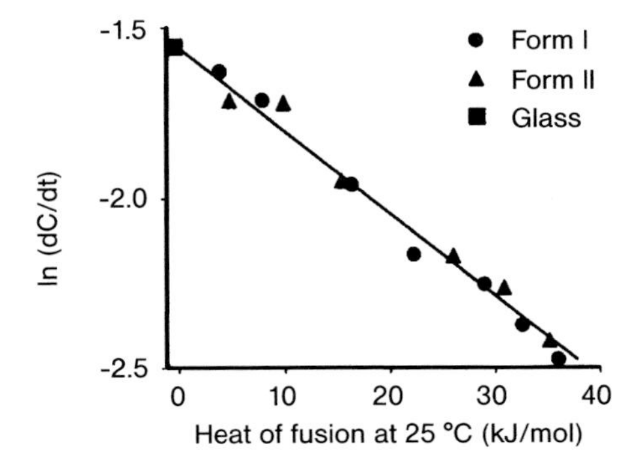

Fig. 3.8 Relationship between heat of fusion and initial dissolution rate of terfenadine for two polymorphs and the amorphous form [27].

Form II $(T_{\mathrm{MII}}) \rightarrow$ Form I (T_{MII})

with Form II being the lower melting form, the author derived expressions for the free energy and enthalpies of the process:

$$\Delta G_0 = \Delta H_{\mathrm{MI}}(T_{\mathrm{MII}}/T_{\mathrm{MI}} - 1) + (C_{p.\mathrm{L}} - C_{p,\mathrm{I}})[T_{\mathrm{MI}} - T_{\mathrm{MII}} - T_{\mathrm{MII}} \ln(T_{\mathrm{MI}}/T_{\mathrm{MII}})] \tag{5}$$

$$\Delta H_0 = \Delta H_{\mathrm{MII}} - \Delta H_{\mathrm{MI}} + (C_{p.\mathrm{L}} - C_{p,\mathrm{I}})(T_{\mathrm{MI}} - T_{\mathrm{MII}}) \tag{6}$$

$$\Delta S_0 = \Delta H_{\mathrm{MII}}/(T_{\mathrm{MII}} - \Delta H_{\mathrm{MI}}/T_{\mathrm{MI}} + (C_{p.\mathrm{L}} - C_{\mathrm{p,I}}) \ln(T_{\mathrm{MI}}/T_{\mathrm{MII}})] \tag{7}$$

where ΔG_0, ΔH_0 and ΔS_0 are the standard free energy, enthalpy and entropy of the process; ΔH, T and C_p are the enthalpies, temperatures and heat capacities (at constant pressure), with subscripts referring to the melting point (M) and physical form (I or II or liquid, L). The free energy at temperature T is given by Eq. (8).

$$\Delta G(T) = \Delta G_0 - \Delta S_0(T - T_{\mathrm{MII}}) \tag{8}$$

The analysis allows both the calculation of the free energy at any temperature and the estimation of the temperature at which the free energy difference is zero, i.e., the transition temperature. Yu [30] then applied this method to 96 pairs of polymorphs and noted a high level of agreement with experimentally derived values of the transition temperature. This model was used by Caira et al. [31] to characterize the free energy differences between two forms of tolbutamide, which they then related to XRD data to interpret the thermodynamic differences to fundamental crystal structure.

3.3 Combined Approaches

3.3.1 Multi-instrument Approaches

As mentioned earlier, DSC is rarely used in isolation when studying polymorphism and the usual situation is for several methods, particularly spectroscopic approaches, to be employed in combination. Of these, by far the most common is XRD, which provides an excellent means of linking thermal behavior with crystal structure. To this effect a brief subsection is given whereby some studies that have gone into the crystallographic aspects in particular detail are highlighted. The reader is referred to an interesting review by Giron [32] on combined approaches; this paper is discussed in more detail below. Virtually all the work discussed thus far has involved using DSC in combination with at least one other technique, with this text highlighting the thermal aspects of the study. Here, the emphasis is more on exemplifying how combined approaches may strengthen the conclusions inferred from any single technique, particularly DSC.

Grcman et al. [33] used a combination of DSC, thermogravimetric analysis, temperature-resolved XRD, hot stage microscopy, solubility measurements, FTIR and SEM to study the polymorphic transformations of doxazosin mesylate. This drug may exist in seven polymorphic forms as well as a relatively stable amorphous form, hence the propensity for complexity is considerable. Interestingly, only five of these forms were recognized previously, with the additional two being identified for the first time in this study. DSC may provide evidence for the presence of additional forms in terms of anomalous melting behavior but is not in itself unequivocal. However, the authors confirmed the presence of these forms using FTIR and XRD studies, thereby demonstrating that DSC is extremely useful for indicating the possibility of new form generation and for studying the kinetics of transformation, but for details of molecular structure and for unequivocal confirmation of the presence of additional forms it is necessary to use spectroscopic techniques. Similarly, Giordano et al. [34] studied the polymorphic forms of *rac*-5,6-diisobutyryloxy-2-methylamino-1,2,3,4-tetrahydronaphthalene hydrochloride (CHF 1035). These authors used Raman and carbon-13 nuclear magnetic resonance spectroscopies, powder XRD (temperature controlled) and a range of thermal methods. The authors identified three polymorphs for this material but, interestingly, inferred from the XRD studies that Form I underwent reversible structural rearrangements at ca. 60 and 75 °C. Corresponding DSC studies showed two small endothermic events at these temperatures (Fig. 3.9 A). The authors suggested that these represented conformational shifts with low energy barriers and were able to incorporate these forms into the suggested interconversion scheme (Fig. 3.9 B).

Schmidt et al. also used spectroscopic and thermal methods to study polymorphic conversions of prilocaine hydrochloride [35]. Several interesting points arise from this investigation. Firstly, the authors used FT Raman, XRD and FTIR spectroscopy to study the molecular conformation of the drug in the various polymorphs. Secondly, they suggest that, despite using a wide range of processes for polymorph generation, one polymorph (Form II) could only be obtained via desolvation of the dioxane solvate. Finally, the authors used Burgers rules to suggest that the Form II was thermodynamically unstable, on the basis of both the lower melting point and heat of fusion, but reported that it was kinetically stable for at least a year, exemplifying the important point that thermodynamic instability does not necessarily equate to kinetic instability within the lifetime of a typical pharmaceutical product. Zhang et al. [36] performed a study on sulfamerazine polymorphs that had previously proved to be controversial in terms of identifying a reliable preparation and characterization protocol. Several aspects of this study merit attention, including the successful development of a method of bulk producing the metastable polymorph II, the thermodynamic and kinetic characterization of the polymorphs and their interconversion, and the study of the effects of pro-

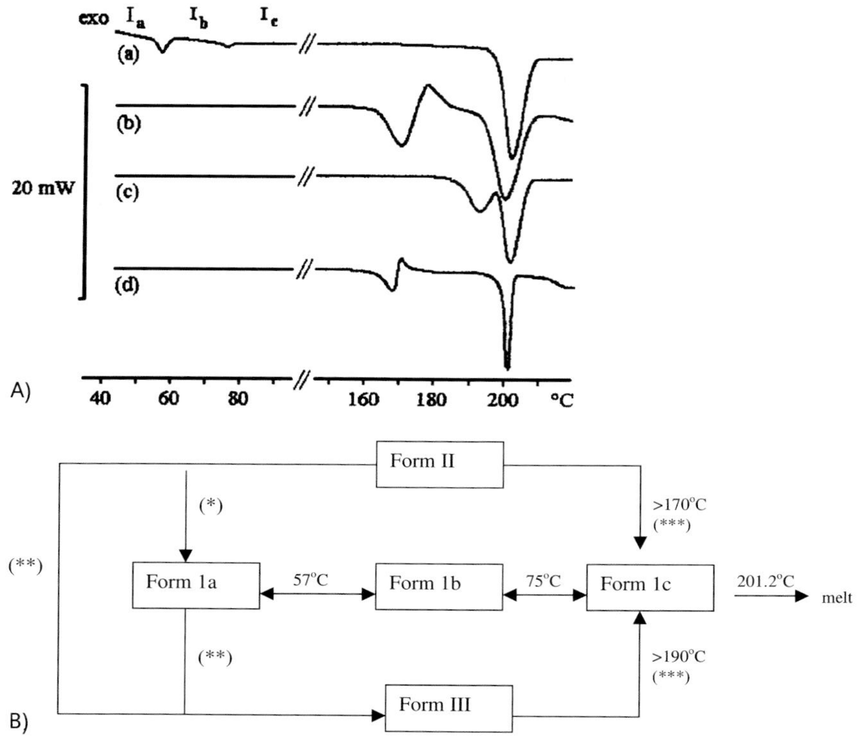

Fig. 3.9 (A) Thermal behavior of CHF 1035 polymorphs: (a) Form I (showing conformationally distinct forms), (b) Form II, (c) Form III, and (d) a 1:1 mixture of enantiomers. (B) Interconversion scheme of CHF 1035. Key: (*) by suspension with Form Ia seeds in acetone or methyl ethyl ketone at room temperature; (**) by suspension with Form III seeds in acetone or methyl ethyl ketone at room temperature; (***) melt-recrystallization event.

cessing on the polymorphic form. In the present context, it is worth highlighting that the authors used variable-temperature NMR (Fig. 3.10) and XRD in conjunction to monitor the changes in polymorphic form as a function of temperature and thermal history, with both demonstrating the enantiotropic interconversion of Form II into I on raising the temperature but the non-immediate reversion on cooling back to 25 °C, again demonstrating the importance of considering both thermodynamics and kinetics.

A particularly interesting study in terms of combining DSC with vibrational spectroscopic (FTIR) studies is that of Wouters et al. [37]. This was not a pharmaceutical study in that it involved polymorphic characterization of a lipid, *N*-stearoylethanolamine (NSEA), but the principles are, of course, still relevant and transferable to the pharmaceutical arena. These authors used DSC to study the transitions of this material on heating (Fig. 3.11 a). The peak at 102 °C was ascribed to the melting of the material while the two minor transitions were

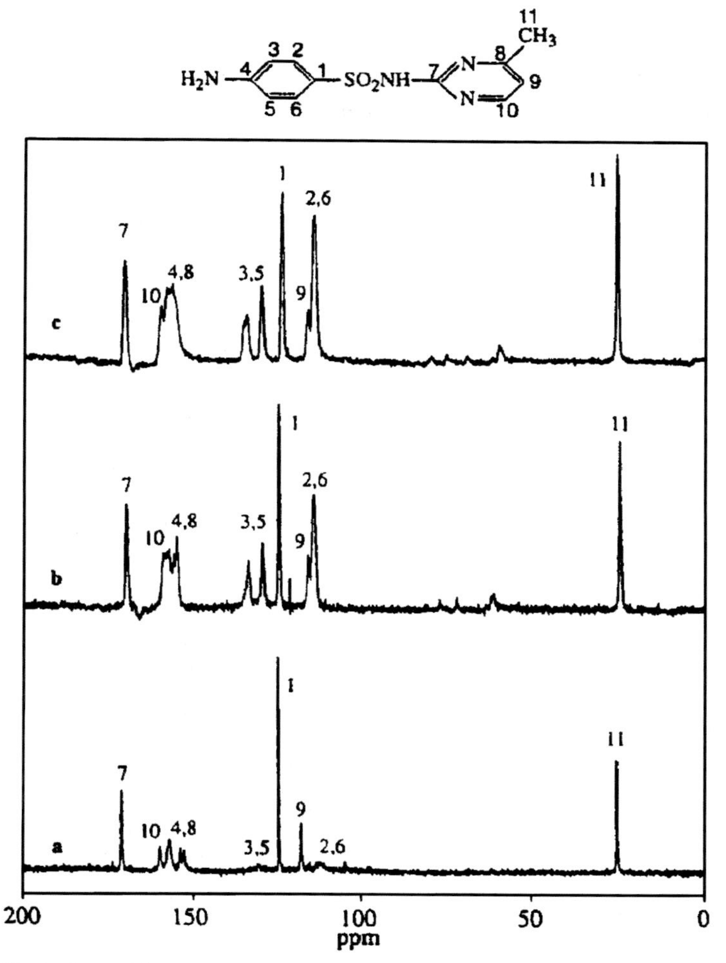

Fig. 3.10 Variable temperature carbon-13 nuclear magnetic resonance spectra of sulfamerazine; (a) bottom spectrum; Form II at 25 °C, (b) spectra at 200 °C, showing conversion into Form I, and (c) at 25 °C after heating to 200 °C, showing persistence of Form I [36].

ascribed to polymorphic transformations (I into II into III), as previously suggested by Ramakrishnan et al. [38]. The authors then used variable-temperature XRD and FTIR to study these forms. Based on the combined spectroscopic and thermal data the authors developed a hypothesis with regard to the structure of the polymorphs in terms of molecular conformation and mobility, as outlined in Fig. 3.11 (b). In brief the authors suggested that the geometry of the polar heads is not radically altered during initial heating as these groups are held in place by extensive hydrogen bonding, but the hydrocarbon chains undergo a

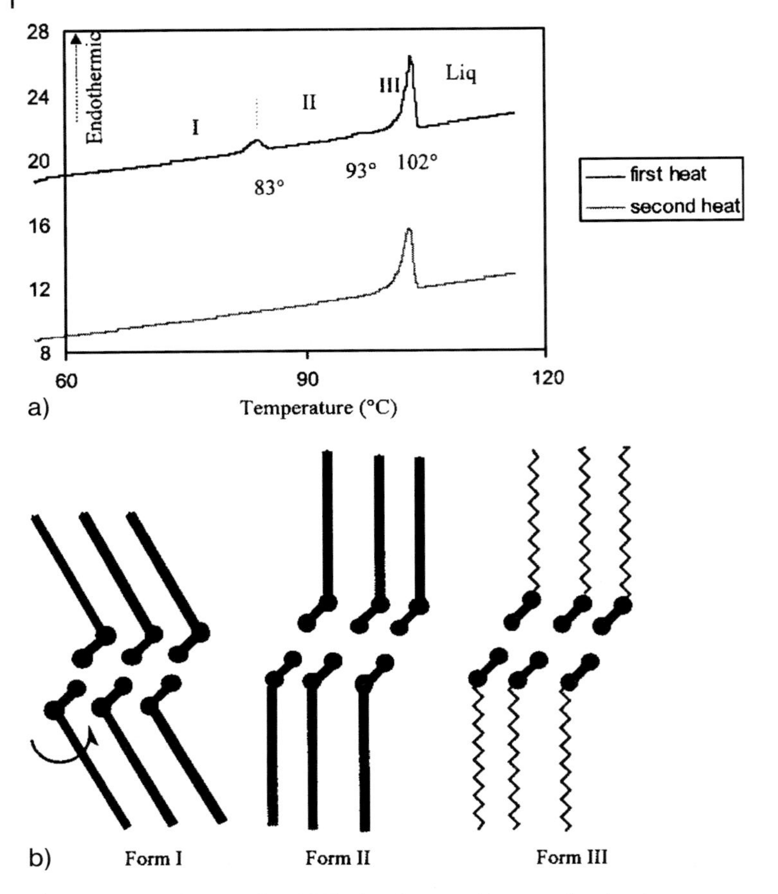

Fig. 3.11 (a) DSC curves for NSEA, showing transitions on first heating between Forms I, II and III. (b) Proposed schematic for conformation change of NSEA on heating (see text for explanation) [37].

change in tilt angle between Forms I and II, which is reflected by the spacings detected in the small-angle region of the XRD. The FTIR data enabled the authors to suggest that the CH_2–CO function of the molecule (i.e., between the polar head group and the hydrocarbon chain, as indicated in Fig. 3.11 (b) undergoes the key rearrangement. Furthermore, they proposed that the Form II into III transformation actually represents a destabilization of the lipid packing, leading to a liquid crystalline structure whereby the lateral chains are free to rotate. The study is of interest over and above the specific findings in that it represents an excellent example of how DSC, XRD and FTIR may be used in conjunction, with the combination providing information in terms of transition identification and thermodynamic characterization, structural change and molecular mobility that would not be possible in the absence of any of the three.

3.3.2
Thermal and Crystallographic Studies

While the above section deals with multiple approaches that include XRD, the combination of thermal and crystallographic studies is so central to the effective characterization of polymorphism that this partnership is highlighted in more detail here. In particular, several studies have involved a strong emphasis and level of detail in linking the thermal behavior associated with the transformations with specific structural rearrangements, thereby providing a strong contribution to our understanding of the fundamental mechanisms involved.

The study of Pfeffer-Hennig et al. [39] on the polymorphic forms of a new experimental drug (API-CG3) provides an example. The authors used DSC to identify two enantiotropic forms (named Mod A and B) with a transition temperature ca. 50 °C. However, they also used a combination of single-crystal XRD (for Mod A) and synchrotron powder data (for Mod B as this formed only very small crystallites; see below) to understand the molecular packing of the two polymorphs. The two molecular solid-state orientations were found to be very similar, with the main difference lying in the orientation of the terminal ethyl group, in turn leading to differences in row-to-row packing orientation. The authors were then able to interpret the solid-state transformation in terms of an anticlockwise rotation of molecules within the row structures, thereby providing a specific molecular interpretation for the behavior of the polymorphs. Vogt et al. [40] studied the polymorphic forms of tranilast. Interestingly, the forms exhibited very similar thermal properties but could be distinguished and characterized by single-crystal XRD. However, such analysis was not possible for all polymorphs as suitable samples could not be prepared, necessitating the use of additional spectroscopic techniques such as FTIR, FT-Raman and solid-state NMR, thereby providing another example of how multi-instrument approaches allow a more complete characterization. The XRD studies are nevertheless of particular interest as the authors provided a detailed analysis of the formation of a hydrogen-bonded network in the solid state, and suggested that the relative stabilities of the polymorphic forms could be related to the strength of the intermolecular hydrogen bonding. A further example of a similarly thorough characterization is that of Caira et al. [41], who used powder and single-crystal XRD data in conjunction with a range of thermal and spectroscopic techniques to examine polymorphs and solvates of fluconazole.

While XRD is undoubtedly a pivotal method for structural elucidation, the issue of sample availability for single-crystal studies recurs in several studies where XRD provides a major focus of the work, with one alternative being to use synchrotron sources to achieve high quality powder data. For example, Nunes et al. [42] used high-resolution synchrotron XRD powder diffraction data to identify the structure of mannitol hydrate. Similarly, Edwards et al. [43] used time-resolved X-ray scattering using synchrotron radiation to study polymorphic transitions in carbamazepine. This technique allows simultaneous wide and small angle scattering so as to monitor the α-into-γ polymorphic transition iden-

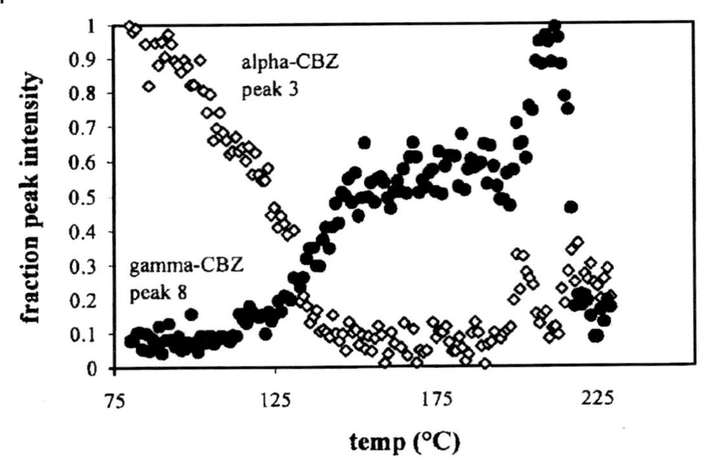

Fig. 3.12 Fractional intensity of characteristic WAXS peaks of α (d=6.6 Å) and γ (d=7.2 Å) carbamazepine heated at 5 °C min^{-1} [44].

tified using DSC. These authors were able to monitor the proportionality of the transition at a range of scanning speeds (Fig. 3.12), thereby allowing the possibility of both structural and kinetic evaluation of the mixed system which, as discussed previously [10], is difficult to achieve using DSC.

XRD may also be used to construct phase diagrams for polymorphic transformations. Espeau et al. [44] studied two polymorphs of paracetamol (the monoclinic Form I and orthorhombic Form II) as a function of pressure and temperature, using variable-temperature XRD in conjunction with DSC to construct the pressure–temperature phase diagram and to understand the enthalpic and volume changes associated with the transition. Interestingly, the authors suggested that the stability hierarchy of the forms may be inferred at any temperature from their sublimation curves.

Overall, therefore, XRD provides information that complements and in many cases exceeds that obtained using DSC, particularly if variable temperature and/or synchrotron sources are used. Indeed, DSC is not capable of providing direct structural and molecular information, while the traditional strengths of the calorimetric approach (e.g., transition identification) may be matched in some instances if variable temperatures and scanning speeds are used in XRD studies. Nevertheless, DSC has distinct advantages, including the ease, simplicity and rapidity of the measurements and the fact that it provides direct information on thermodynamic parameters associated with the transitions, while XRD information is more limited if powder techniques must be used, which is often the case if single crystals may not be easily prepared and (as is almost invariably the case) access to synchrotron sources is limited. That said, the question of whether DSC is "better" or "worse" than XRD is largely academic as using the two in conjunction clearly provides the possibility of structural and thermodynamic characterization. When these two are used with vibrational spectroscopy as well the possibility of measuring molecular mobility arises. On that basis, the possibility of being able to resolve the thermodynamics, structure and molecular mobility must surely provide as sound a basis as is possible for predicting the most important yet poorly understood aspect as far as the pharmaceutical field is concerned, the kinetics of the transformations.

3.3.3 Interfaced Techniques

In addition to approaches whereby more than one technique is used to study the same sample, there are also examples of techniques whereby two methods are directly interfaced. In the context of the present chapter, this is considered in terms of DSC being interfaced with other methods, although the subject of coupled techniques is a broad one in its own right. Giron has reviewed the use of coupled techniques, both those involving thermal analysis within the pharmaceutical sciences in a general sense [45] and several approaches that are particularly useful for solvates rather than polymorphs (in particular techniques involving coupling to thermogravimetric analysis) [46]. Here, several techniques used specifically for polymorphism are highlighted.

Sprunt et al. [47] have described the use of DSC coupled with FT-Raman (named by the authors as simultaneous FT-Raman-DSC or SRD [48]) as a means of studying polymorphism in *sn*-1,3-distearoyl-2-oleoglycerol (SOS). In essence the technique involves using a fiber optic probe to obtain the Raman spectra of the sample while in the DSC pan (obviously the sample has to be unsealed for this to take place, with some compromise of baseline stability). The authors obtained isothermal Raman spectra for five polymorphic forms at temperatures approximately 10 °C below the melting points (Fig. 3.13), from which they derived structural information that agreed with the existing knowledge base regarding these techniques.

Allais et al. [49] have used XRD coupled to DSC to study another lipid, trilaurin, mixed with 4% cholesterol. The technique involves coupling a time-resolved synchrotron XRD that, by coupling to a DSC, allows measurement of structural change as a function of temperature using heating rates of ca. 1 °C min^{-1}. The technique appears to be particularly useful for systems such as lipids, whereby complex melting peaks are apparent, allowing identification of the structural component responsible for a particular peak. This is demonstrated for the example of the β' to β_2 to β_1 transition (Fig. 3.14). The DSC data shows two clear transitions at 28 °C (small exothermic) and 45 °C (large endothermic). The SAXS data (where q is equal to $2\pi/d$ with d being the repeat distance) shows a peak at 0.193 $Å^{-1}$, corresponding to the β' form up to 28 °C, which is then replaced by a peak at 0.199 $Å^{-1}$ that corresponds to the β_2 form; hence the technique has allowed interpretation of the exotherm as the β' to β_2 transition. However, it is also possible to discern the β_2 to β_1 transition in the WAXS signal by the appearance of characteristic peaks at 1.60 and 1.65 $Å^{-1}$ corresponding to β_1, in

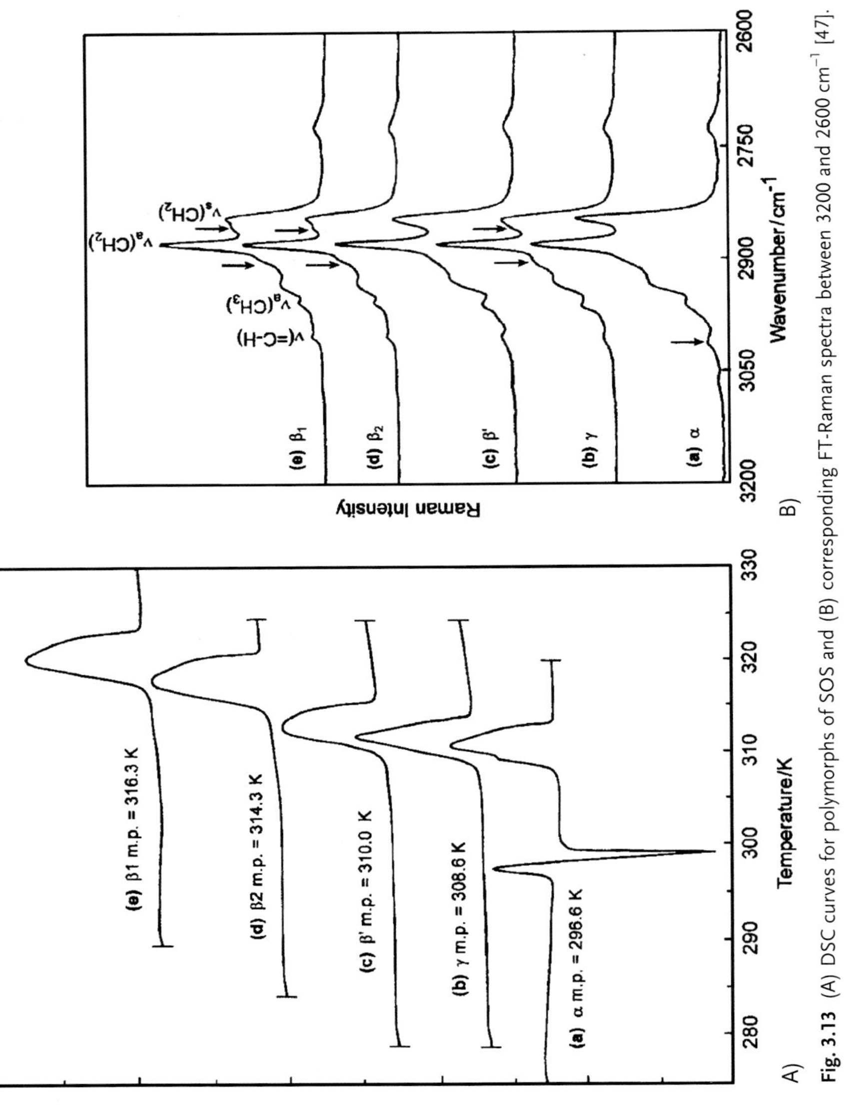

Fig. 3.13 (A) DSC curves for polymorphs of SOS and (B) corresponding FT-Raman spectra between 3200 and 2600 cm^{-1} [47].

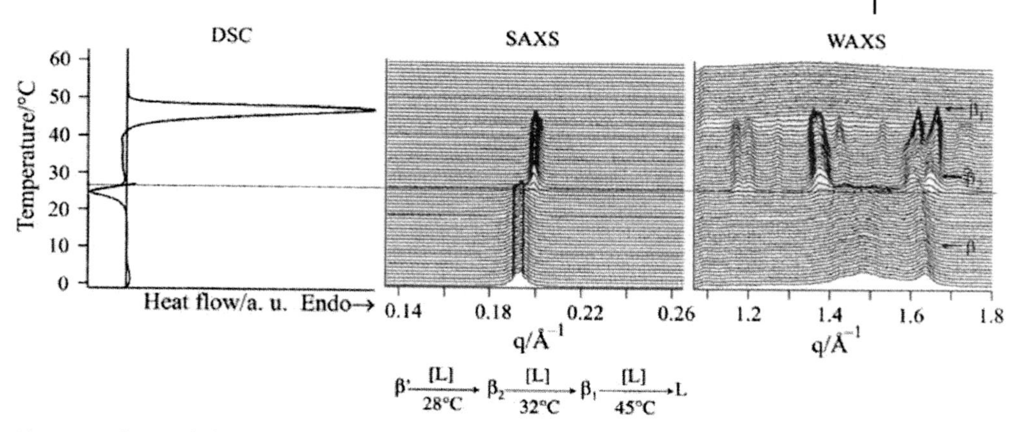

Fig. 3.14 Trilaurin/cholesterol (96/4) studied by calorimetry interfaced with XRD (see text for explanation) [49].

contrast to the peaks at 1.62 and 1.665 $Å^{-1}$ corresponding to the β_2 peaks. This transformation is not easily discernible by DSC but takes place on the approach to the endothermic melting point which, as can now be seen, corresponds to the β_1 rather than the β_2 to liquid transition. This elegant technique has also been used to characterize the complex melting behavior of the complex polyglycolyzed triglyceride mix Gelucire 50/13 [25, 50]. In these studies the group also interfaced the system to infrared spectroscopy, thereby also allowing molecular mobility to be simultaneously assessed.

3.4 Additional Thermal Methods for the Study of Polymorphism

3.4.1 Thermogravimetric Analysis

While thermogravimetric analysis is not, strictly speaking, used for the study of polymorphs on the basis that polymorphic transitions do not involve a change in mass, this chapter would be incomplete without some mention of the technique due to its pivotal role in characterizing solvates of drugs and excipients. Thermogravimetric analysis (TGA) involves the measurement of mass as a function of temperature, thereby allowing changes in mass as material is evolved to be identified and quantified. As such it plays a central role in determining the stoichiometry of solvates (easily calculated from the weight loss per known amount of original sample) and may also give an indication of the type of binding involved, in the first instance on the basis of the temperature at which the solvent is lost. It is also possible to calculate the kinetics and from that the activation energy of the loss process by, for example, running the experiment at a range of scanning speeds. It can also be used to examine thermal degradation, although this is not an application that has been widely used within the pharmaceutical field as it is not advisable to extrapolate such data to room temperature, and hence is of relevance only if one is heating a drug to very elevated temperatures. A key weakness of the technique is that it does not tell the operator the nature of the substance being lost, only the mass, and a classic mistake it to make unsupported assumptions regarding the nature of the materials lost. On this basis there has been considerable interest in coupling the method to analytical approaches such as mass spectroscopy so as to allow chemical identification of the lost material. For more information on this technique the reader is referred to any of the texts mentioned previously that review the use of thermal and related methods in a more general context [1–4, 45, 46].

3.4.2
Thermal Microscopy

Thermal microscopy, often referred to as hot stage microscopy, is a very widely used technique for studying polymorphism [51–54]. While being semi-quantitative in nature, the method often provides invaluable insights into the thermal transitions undergone by the polymorphs. In essence the technique involves viewing the sample using some form of optical microscopy while also employing a hot stage to control the temperature of the sample. The technique may be quantitative both in terms of identifying the temperature of the transition and also in terms of measuring light intensity as a function of temperature. In terms of polymorphic transitions, the information derived can be broadly categorized into identifying changes in crystal habit associated with the polymorphic change and identifying changes in light transmission or reflection as-

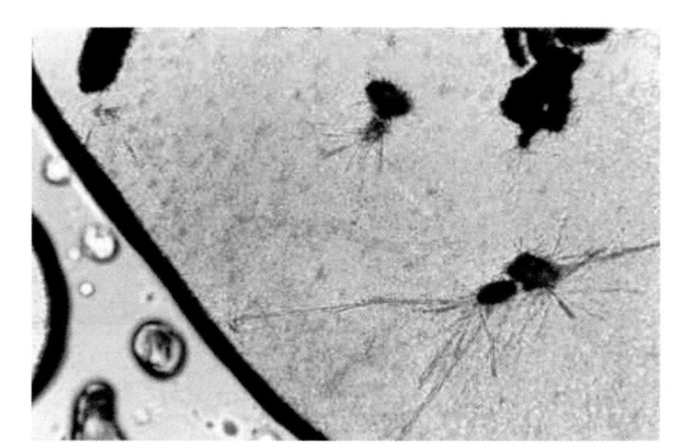

Fig. 3.15 Hot stage differential interference contrast images of 5% caffeine in Gelucire 50/13 at 50 °C during heating. Note the extensive formation of needle-shaped crystals of caffeine [24].

sociated with the transition. An example of the use of thermal microscopy for the study of crystal habit is provided by Childs et al. [55] who studied the polymorphic behavior of metformin hydrochloride. These authors showed the generation of needle-like crystals of Form A at the expense of the metastable Form B on incubation at 80 °C. Similarly, Khan and Craig [24] indicated that caffeine changed polymorphic form the stable Form II to the metastable enantiotrope Form I on dispersion in Gelucire 50/13 (Fig. 3.15); this can be seen by the emergence of needle shaped crystals indicative of Form I.

In addition, Raman and IR microscopy may be used with a temperature controlled stage to allow spectral information to be obtained. Examples of this include a study by Szelagiewicz et al. [56], who identified the polymorphic forms of paracetamol and lufenuron by obtaining the Raman spectra as a function of temperature.

3.4.3
Heat of Solution Studies

An elegant and arguably under-utilized method of detecting and characterizing polymorphic transitions is to measure the heat of solution. Briefly, the method involves measuring the temperature change in a known volume of solvent on dissolving a solid, from which the enthalpy of dissolution may be easily calculated. There are two key strengths to this approach. Firstly, several studies have demonstrated that the heat of solution (as the enthalpy measurement is usually referred to) is extremely sensitive to crystal form and hence may be used as a means of detecting the presence of polymorphs. This is exemplified by a study by Souillac et al. [57] whereby the heat of solution of cimetidine was studied using surfactant systems as solvents, these being chosen due to the poor wettability of the drug in aqueous systems. The authors stress that the melting points of the polymorphs are very similar (hence DSC is not an appropriate tool for studying this particular drug). They examined a range of batches of cimetidine, detecting significant differences between their heats of solution, which were ascribed to differences in polymorphic profile, and confirmed using spectroscopic techniques. The second strength of this approach lies in the possibility of utilizing the data obtained to define the thermodynamic parameters associated with different crystal forms. An early attempt to achieve this by Craig and Newton [28] involved studying the heat of solution of poly(ethylene glycols) prepared using different thermal histories, which led to different chain folding and degree of crystallinity profiles. The authors found remarkably large differences in the overall heats of solution for different samples of the same polymer and developed a model whereby the heat of solution was considered to represent the sum of the heat of fusion at the temperature of the experiment and the heat of molecular interaction with the solvent. By using a corrected value of the heat of fusion obtained using DSC they were able to estimate the enthalpy of interaction with the solvent. A more sophisticated model has been developed by Gu and Grant [58] whereby the transition temperature of, and free energy differences between, enantiotropic polymorphs is estimated from their heats of solution. The authors found the model to be applicable to a wide range of drugs under study, thereby allowing a key parameter to be estimated using a simple series of measurements.

3.4.4
Modulated Temperature DSC

Modulated temperature DSC (MTDSC) is a software modification of conventional DSC in that a perturbation is superimposed on the conventional linear signal. With the TA Instruments model this perturbation takes the form of a sine wave, although with other instruments a step wave or saw tooth wave is also found. Irrespective of the method used, the key aspect is that the response to the modulation gives the heat capacity of the sample as a function of temperature. Consequently, by measuring the Fourier Transform average of the overall signal (the total heat flow) and subtracting the heat capacity signal (the reversing heat flow) it is possible to obtain the kinetic component associated with non-reversing responses such as crystallization (the non-reversing heat flow). The instrument therefore gives three signals in a single heating run, allowing the heat capacity, kinetic and total heat flow components to be delineated. For more information the interested reader is referred to several available texts, e.g., [59–61].

The method has been of undoubted use for the study of amorphous systems but is more controversial for the study of polymorphism. This is because the melting response is such that the sample is not able to maintain steady state with the modulated signal. In simple terms, the success of the deconvolution process is dependent on the assumption that the modulation is rapid in relation to the underlying process, thus reflecting the heat capacity of the sample at that temperature and not the process itself. Melting is, however, a highly energetic process that occurs over a narrow temperature range for low molecular weight samples, hence in practice signals are seen in both the reversing and non-reversing components, which are difficult to interpret in terms of the fundamental properties of the sample. This has led to the common practice of not using MTDSC to study melting and hence polymorphic transitions. Nevertheless, MTDSC has been of use and interest in several studies of polymorphic forms and transitions. Keymolen et al. [62] used MTDSC to characterize the glassy behavior of nifedipine in the context of the recrystallization of this drug into different polymorphic forms, while Six et al. [63] studied hot-stage extrudates of itraconazole in Eudragit E100, using MTDSC to characterize the glassy and liquid crystalline phases of the drug prior to recrystallization into different polymorphs. However, while the application of the technique to the melting process remains limited, there is a growing interest in using it to study the transformation process itself. Kawakami and Ida [64] studied the endothermic transformation of frusemide and tolbutamide, indicating that the transition could be deconvoluted into the reversing and non-reversing signals (Fig. 3.16). Interestingly,

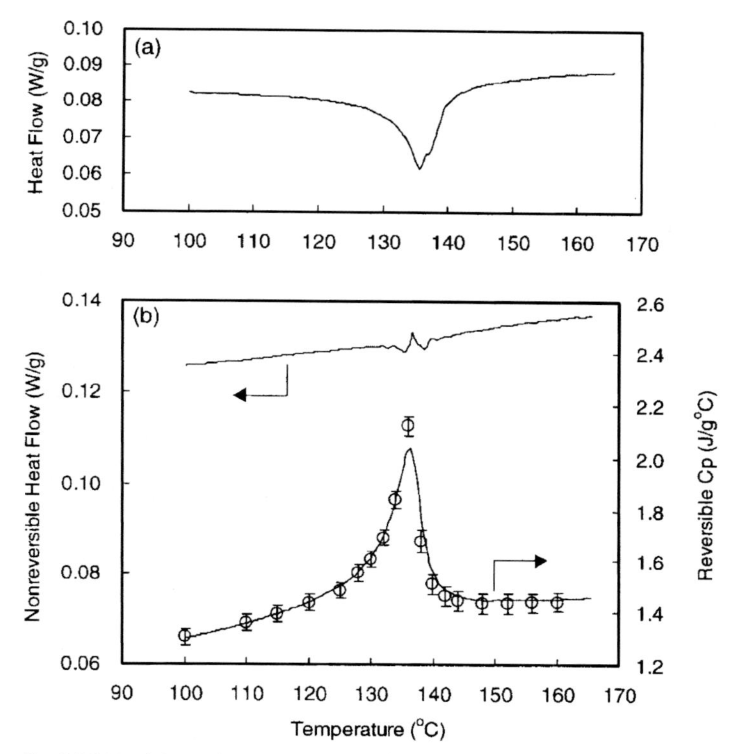

Fig. 3.16 Total (a) and non-reversing and reversing (b) heat flows for frusemide. Open circles represent quasi-isothermal measurements (see text for explanation) [67].

the response for frusemide was dominated by the reversing signal, indicating a second-order process whereby instead of there being an enthalpy "jump" between the polymorphs the enthalpy and free energy are continuous. The authors in turn suggested that this may lead to a kinetically reversible transition. In contrast, sulfamerazine showed a transition that was dominated by the non-reversing signal, which indicates a first-order response and kinetic non-reversibility. This is an important paper as it suggests that MTDSC may be used to determine whether a sample will switch between enantiotropic polymorphs effectively immediately on going through the transition or not; further work in this area may allow one to predict the rate at which the metastable polymorph will transfer to the stable one.

A related study that is, at the time of writing, only available in abstract form is that of Manduva et al. [65] who studied the enantiotropic transition of caffeine. These authors studied the effects of heating rate on the transition, showing not only a clear rate dependence but also that there is a kinetic component to the transition. They also used quasi-isothermal studies whereby the sample is modulated

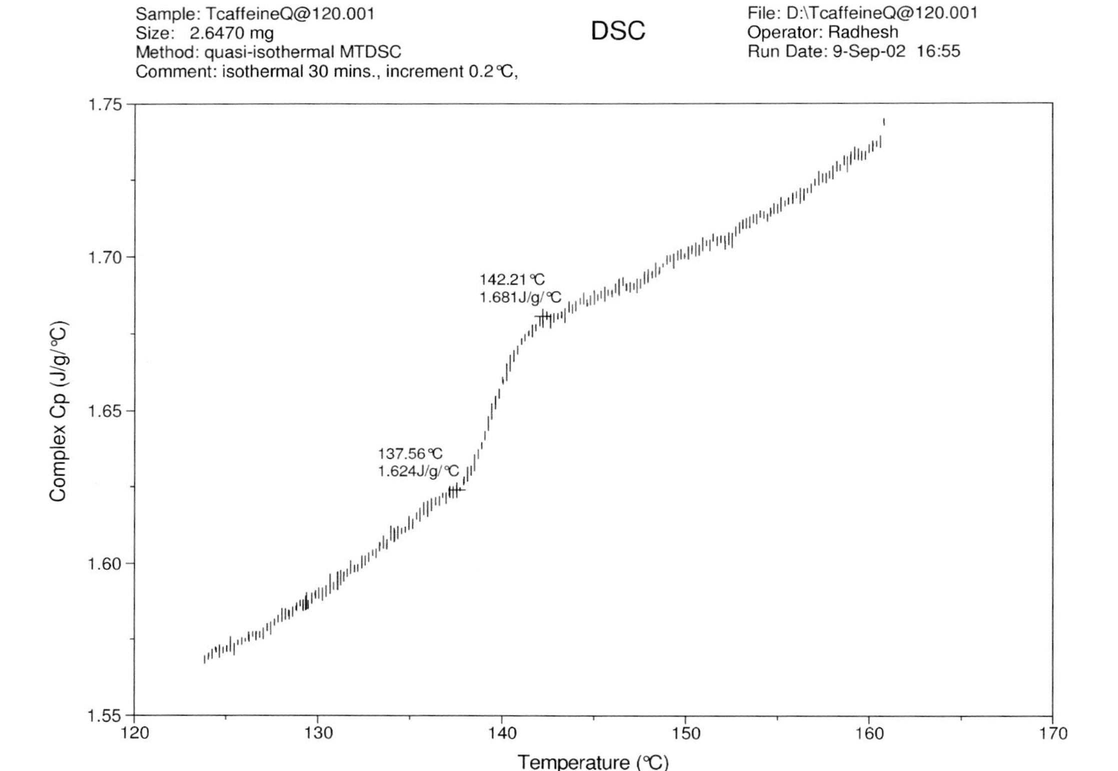

Fig. 3.17 Quasi-isothermal MTDSC studies on caffeine in the region of the enantiotropic transition (see text for explanation) [68].

at a particular temperature, which is then incrementally ramped up, leading to a series of heat capacities (Kawakami and Ida [64] used the same method in part of their study). The authors argued that this process would eradicate the kinetic effect, as the sample is being heated at an infinitesimally slow rate, thereby allowing the non-kinetic value of the transition to be obtained (Fig. 3.17). Both these studies give a strong indication that while conventional MTDSC may be of limited use in the study of polymorphic melts, there is a potentially very exciting role for the technique in studying the associated transitions themselves.

3.4.5 High-speed DSC

There has been considerable recent interest in high-speed DSC techniques, whereby heating rates of several hundred $^\circ C\ min^{-1}$ are possible, the most widely studied method currently being Hyper-DSC™. The method has been consider-

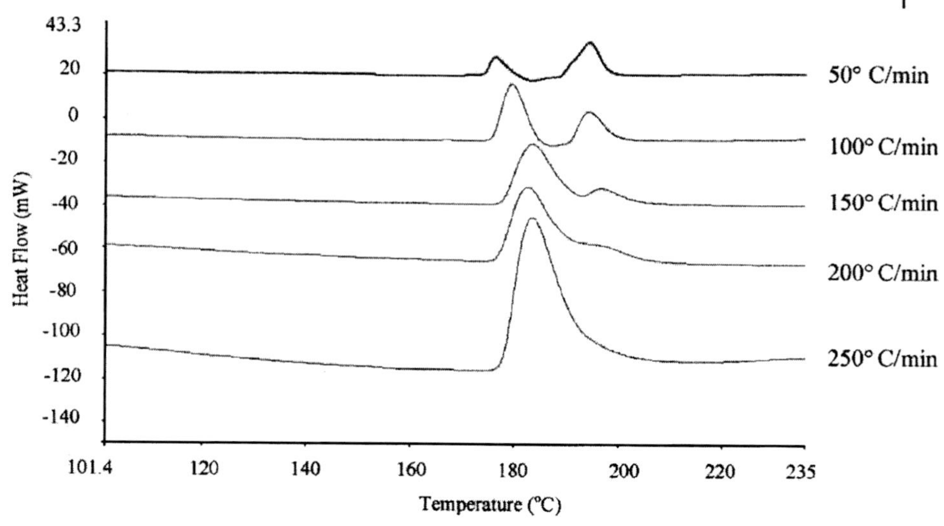

Fig. 3.18 High-speed DSC curves for carbamazepine Form III run at a range of speeds, with the melting of the metastable form being visualized in isolation at high scan rates [70].

ed in the context of polymorphism in particular due to the possibility of using high scanning speeds to allow metastable forms to be studied, the idea being that the conversion into the more stable form, which would occur during the timescale of the experiment for conventional DSC, would not take place, thereby allowing resolution between polymorph peaks and analysis of the individual endotherms. This is exemplified by the study of McGregor et al. [66] whereby the thermal properties of carbamazepine polymorphs were examined. The authors were able to isolate the endotherm of the metastable Form III polymorph (Fig. 3.18) and were also able to characterize mixes of the two polymorphs such that quantitation of the mix was possible. At the time of writing, only a limited number of studies have used this approach, but it is strongly anticipated that this method will become widely used for the study of polymorphic forms and transitions.

3.4.6 Microthermal Analysis

Microthermal analysis represents an interface technique whereby the tip of an atomic force microscope is replaced by a small thermistor, thereby allowing thermal analysis to be performed on specific regions (currently of the order of a few square microns) of complex samples [67, 68]. The method is generally used in one of three modes. Firstly, localized thermal analysis (LTA), whereby the physical position of the tip is measured as a function of temperature – hence when the sample undergoes a transition such as melting the tip penetrates into the sample and the temperature at which this movement takes place is recorded. Secondly, the tip may be rastered over the surface and the thermal conductivity measured, allowing an alternative form of mapping. Thirdly, the differential heat between the tip and a remote sensor may be measured as a specific region is heated, thereby allowing a form of highly localized DSC. The method has attracted interest as a means of characterizing complex dosage forms and has also been explored as a way of characterizing polymorphism. This is exemplified by a study by Sanders et al. [69] wherein cimetidine polymorphs were assessed on a compressed surface. A particularly interesting aspect of this was the use of differential heat flow studies on the tablet, whereby the melting of a cimetidine polymorph was recorded. The authors also used thermal conductivity measurements to map the surface, suggesting that it is possible to locate regions of different polymorphic forms on the basis of differential water uptake. Murphy et al. [70] demonstrated that it is possible to differentiate between indomethacin polymorphs (α and γ) using LTA measurements, while Manduva et al. [71] produced an intriguing result while studying caffeine polymorphs in that the enantiotropic transition could be identified using LTA studies (Fig. 3.19); this is suggested to be due to the structural rearrangement and volume change associated with the transition. The authors used this measurement to differentiate between Form II (which shows the transition on heating) and Form I (which does not) generated on compression.

As this technique becomes more sophisticated, both in terms of resolution and the type of measurement possible, the thermal mapping of polymorphic forms will, probably, become increasingly prevalent.

3.4.7 Thermally Stimulated Current

Thermally stimulated current (TSC) is an analytical technique that, while being in use for several years in the polymer and related fields, has only recently begun to be explored as a pharmaceutical tool but is nevertheless beginning to generate considerable interest. In brief the technique involves the application of an electric field to a sample at elevated temperatures, nominally above the transition of interest, followed by rapid cooling so as to "freeze in" the polarization. The sample is then reheated and, as the sample undergoes a thermal rearrangement, a current (generated by the movement of dipoles) is recorded as a function of temperature. A key strength of the technique is the great sensitivity to transitions such as T_g and the ability to calculate relaxation times, fragility and a range of other parameters with comparative ease. The method has been explored recently as a possible tool for the study of polymorphism, with Shmeis et al. [72] demonstrating that the method may easily detect the transition between two polymorphs of an experimental drug (LAU254). In addition, Boutonnet-Fagegaltier et al. [73] measured the molecular mobility of two polymorphs of irbesartan, showing in the first instance that the two forms (A and B) have TSC profiles that are clearly distinguishable (Fig. 3.20) and also that the tem-

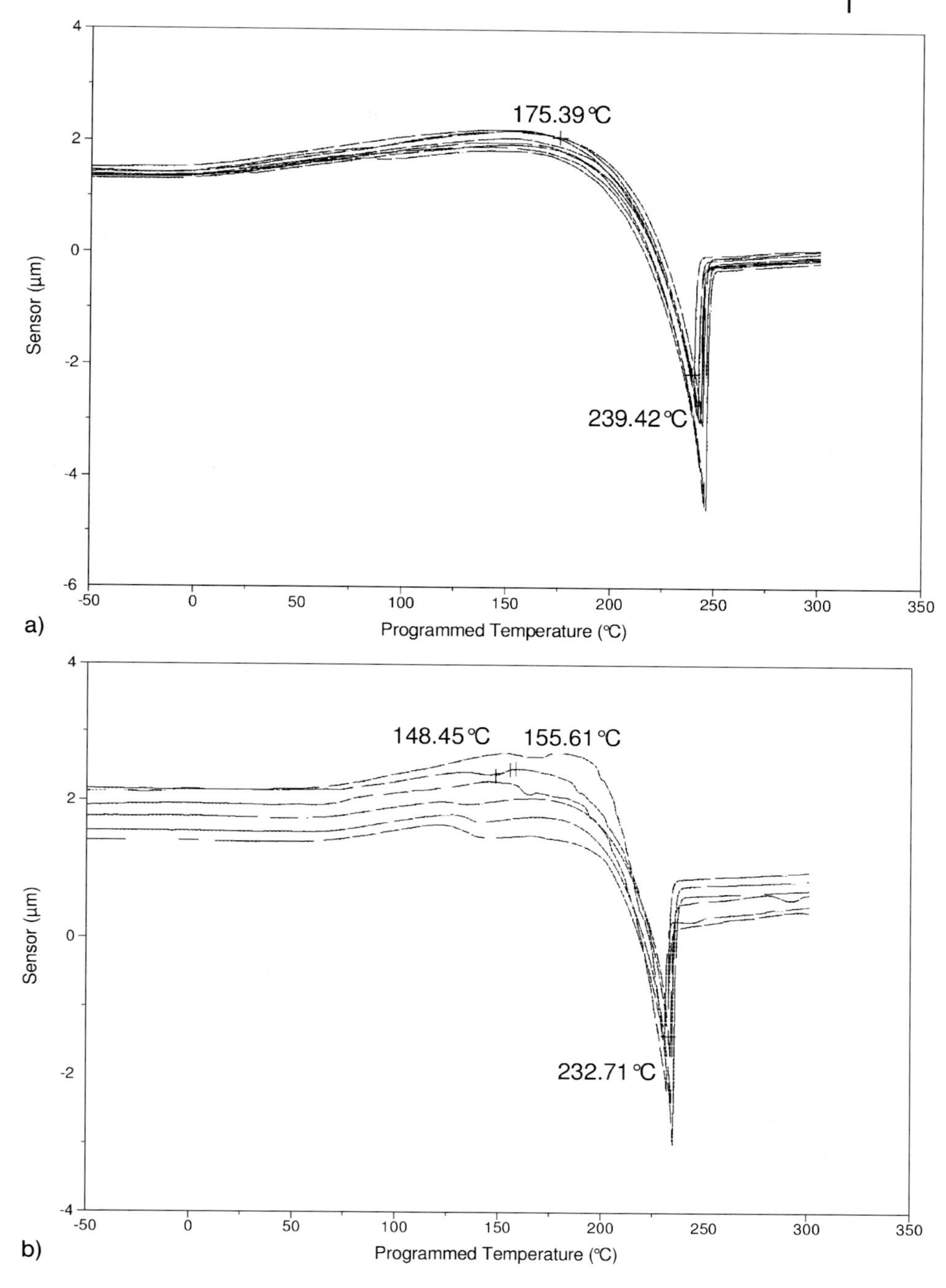

Fig. 3.19 Localized thermal analysis results for compacts of (a) Form I and (b) Form II caffeine, showing a discontinuity corresponding to the transition between Forms II and I polymorphs prior to the melting of Form I [71].

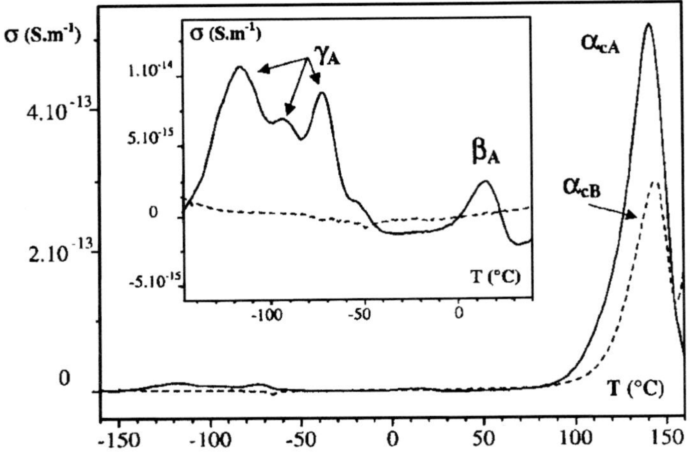

Fig. 3.20 TSC spectra of (A), solid line, and (B), dashed line, crystalline forms of irbesartan obtained after polarization of the sample at 150 °C. Inset: enlargement of TSC response in the low temperature range [73].

perature dependence of the relaxation time may be assessed for the two, thereby allowing a thorough characterization of the mobility profile of the polymorphs.

3.5 Conclusions

The chapter has highlighted the possibilities and limitations of thermal methods for the study of polymorphism, both in terms of established and emerging techniques. DSC remains a key tool for the study of this issue, both in terms of identification and characterization, with there being several possibilities for more sophisticated analysis of the thermodynamic parameters associated with polymorphic transitions. However, the use of DSC in isolation has limitations and combination with crystallographic and vibrational spectroscopic techniques is almost certainly advisable in most cases. In terms of emerging techniques, the use of interfaced systems is certainly of interest, while recently introduced methods such as MTDSC (for studying the polymorphic transitions), high-speed DSC (for isolating metastable polymorphs) and microthermal analysis (for *in situ* characterization) are examples of interesting new approaches that are able to study polymorphic systems in ways not previously possible. In particular, we are now able to not just identify the presence of polymorphism but we are also starting, using thermal methods, to understand the thermodynamics and kinetics of the transformation, the molecular mobility of the polymorphs, the quantity of polymorphs present in mixes and the spatial location of the polymorphs in complex systems. Consequently, the level of information that may be derived from appropriate and imaginative use of thermal methods continues to rise and it is envisaged that this family of methods will remain a key tool in the study of pharmaceutical polymorphism for many years.

Acknowledgement

The author would like to thank Ms Ljiljana Dakic for her invaluable help in compiling this chapter.

References

1 H.G. Brittain, *Physical Characterization of Pharmaceutical Solids*, Marcel Dekker, New York, **1995**.
2 H.G. Brittain, *Polymorphism in Pharmaceutical Solids*, Marcel Dekker, New York, **1999**.
3 J.L. Ford, P. Timmins, *Pharmaceutical Thermal Analysis*, Ellis Horwood, Chichester, **1989**.
4 D. Giron, *Thermochim. Acta* **1995**, *248*, 1–59.
5 W.C. Schinzer, M.S. Bergren, D.S. Aldrich, R.S. Chao, M.J. Dunn, A. Jeganathan, L.M. Madden, *J. Pharm. Sci.*, **1997**, *86*, 1426–1431.
6 S.P. Velaga, R. Berger, J. Carlfors, *Pharm. Res.*, **2002**, *19*, 1564–1571.
7 A. Burger, *Sci. Pharm.* **1977**, *45*, 269.
8 A. Burger, *Acta Pharm. Technol.* **1982**, *28*, 1–20.
9 Z. Dong, B.E. Padden, J. S. Salsbury, E.J. Munson, S.A. Schroeder, I. Prakash, D.J.W. Grant, *Pharm. Res.* **2002**, *19*, 330–336.
10 I.M. Vitez, *J. Therm. Anal. Calorim.* **2004**, *78*, 33–45.
11 P.M. Gosselin, R. Thibert, M. Preda, J.N. McMullen, *Int. J. Pharm.* **2003**, *252*, 225–233.
12 S.P. Velaga, R. Berger, J. Carlfors, *Pharm. Res.* **2002**, *19*, 1564–1571.
13 T. Oguchi, N. Sasaki, T. Hara, Y. Tozuka, K. Yamamoto, *Int. J. Pharm.*, **2003**, *253*, 81–88.
14 E. Yonemochi, Y. Inoue, G. Buckton, A. Moffat, T. Oguchi, K. Yamamoto, *Pharm. Res.*, **1999**, *16*, 835–840.
15 K.R. Morris, U.J. Griesser, C.J. Eckhardt, J.G. Stowell, *Adv. Drug Deliv. Rev.*, **2001**, *48*, 91–114.
16 R. Panchagnula, P. Sundaramurthy, O. Pillai, S. Agrawal, Y. Raj, *J. Pharm. Sci.*, **2004**, *93*, 1019–1029.
17 C. Telang, R. Suryanarayanan, L. Yu, *Pharm. Res.*, **2003**, *20*, 1939–1945.
18 S. Chongprasert, S.A. Knopp, S.L. Nail, *J. Pharm. Sci.*, **2001**, *90*. 1720–1728.
19 N. Zajc, A. Obreza, M. Bele, S. Srcic, *Int. J. Pharm.*, **2005**, *291*, 51–58.
20 A. Burger, J.O. Henck, S. Hetz, J.M. Rollinger, A.A. Weissnicht, H. Stottner, *J. Pharm. Sci.*, **2000**, *89*, 457–468.
21 C.P. Buckley, A.J. Kovacs, *Coll. Polym. Sci.*, **1976**, *254*, 695–715.
22 K. Sato, *Chem. Eng. Sci.* **2001**, *56*, 2255–2265.
23 J. Hamdani, A.J. Moes, K. Amighi, *Int. J. Pharm.*, **2003**, *260*, 47–57.
24 N. Khan, D.Q.M. Craig, *J. Pharm. Sci.*, **2004**, *93*, 2962–2971.
25 J.B. Brubach, M. Ollivon, V. Jannin, B. Mahler, *Bull. Tech. Gattefosse*, **2004**, *97*, 41–50.
26 M.L.P. Leitao, J. Canotilho, M.S.C. Cruz, J.C. Pereira, A.T. Sousa, J.S. Redinha, *J. Therm. Anal. Calorim.*, **2002**, *68*, 397–412.
27 Y. Yoshihashi, H. Kitano, E. Yonemochi, K. Terada, *Int. J. Pharm.*, **2004**, *204*, 1–6.
28 D.Q.M. Craig, J.M. Newton, *Int. J. Pharm.*, **1991**, *74*, 43–48.
29 M. Naoki, T. Yoshizawa, N. Fukushima, M. Ogiso, M. Yoshino, *J. Phys. Chem. B*, **1999**, *103*, 6309–6313.
30 L. Yu, *J. Pharm. Sci.* **1995**, *84*, 966.
31 M.R. Caira, S. A. Bourne, C.L. Oliver *J. Therm. Anal. Calorim.*, **2004**, *77*, 597–605.
32 D. Giron, *J. Therm. Anal. Calorim.*, **2001**, *64*, 37–60.

33 M. Grcman, F. Vrecer, A. Meden, *J. Therm. Anal. Calorim.*, **2002**, *68*, 373–387.

34 F. Giordano, A. Rossi, J. R. Moyano, A. Gazzaniga, V. Massarotti, M. Bini, D. Capsoni, T. Peveri, E. Redenti, L. Carima, M. D. Alberi, M. Zanol, *J. Pharm. Sci.*, **2001**, *90*, 1154–1163.

35 A. C. Schmidt, V. Niederwanger, U. J. Griesser, *J. Therm. Anal. Calorim.*, **2004**, *77*, 639–652.

36 G. G. Z. Zhang, C. Gu, M. T. Zell, R. T. Burkhardt, E. J. Munson, D. J. W. Grant, *J. Pharm. Sci.*, **2002**, *91*, 1089–1100.

37 J. Wouters, S. Vandevoorde, C. Culot, F. Docquir, D. M. Lambert, *Chem. Phys. Lipids*, **2002**, *119*, 13–21.

38 M. Ramakrishnan, V. Sheeba, S. S. Koormath, M. J. Swamy, *Biochim. Biophys. Acta* **1997**, *1329*, 302–310.

39 S. Pfeffer-Hennig, P. Piechon, M. Bellus, C. Goldbronn, E. Tedesco, *J. Therm. Anal. Calorim.*, **2004**, *77*, 663–679.

40 F. G. Vogt, D. E. Cohen, J. D. Bowman, G. P. Spoors, G. E. Zuber, G. A. Trescher, P. C. Dell'Orco, L. M. Katrincic, C. W. Debrosse, R. C. Haltiwanger, *J. Pharm. Sci.*, **2005**, *94*, 651–665.

41 M. R. Caira, K. A. Alkhamis, R. M. Obaidat, *J. Pharm. Sci.*, **2004**, *93*, 601–611.

42 C. Nunes, R. Suryanarayanan, C. E. Botez, P. W. Stephens, *J. Pharm. Sci.*, **2004**, *93*, 2800–2809.

43 A. D. Edwards, B. Y. Shekunov, R. T. Forbes, J. G. Grossman, P. York, *J. Pharm. Sci.*, **2001**, *90*, 1106–1114.

44 P. Espeau, R. Ceolin, J.-L. Tamarit, M.-A. Perrin, J.-P. Gauchi, F. Leveiller, *J. Pharm. Sci.*, **2005**, *94*, 524–539.

45 D. Giron, *J. Therm. Anal. Calorim.*, **2002**, *68*, 335–357.

46 D. Giron, *J. Therm. Anal. Calorim.*, **2001**, *64*, 37–60.

47 J. C. Sprunt, U. A. Jayasooriya, R. H. Wilson, *Phys. Chem. Chem. Phys.*, **2000**, *2*, 4299–4305.

48 J. C. Sprunt, U. A. Jayasooriya, *Appl. Spectrosc.*, **1997**, *51*, 1410.

49 C. Allais, G. Keller, P. Lesieur, M. Ollivon, F. Artzner, *J. Therm. Anal. Calorim.*, **2003**, *74*, 723–728.

50 J. B. Brubach, M. Ollivon, V. Jannin, B. Mahler, C. Bourgaux, P. Lesieur, P. Roy, *J. Phys. Chem. B*, **2004**, *108*, 17721–17729.

51 M. Kuhnert-Brandstatter, *Thermomicroscopy in the Analysis of Pharmaceuticals*, International Series of Monographs on Analytical Chemistry, Vol. 45, Permagon Press, Oxford, **1971**.

52 M. Kuhnert-Brandstatter, *Thermomicroscopy of Organic Compounds*, in Comprehensive Analytical Chemistry, Vol. 16, G. Svehla (ed.), Elsevier, Amsterdam, **1982**.

53 I. M. Vitez, A. W. Newman, M. Davidovich, C. Kiesnowski, *Thermochim. Acta*, **1998**, *324*, 187.

54 S. L. Childs, L. J. Chyall, J. T. Dunlap, D. A. Coates, B. C. Stahly, G. P. Stahly, *Cryst. Growth Design*, **2004**, *4*, 441–449.

55 N. E. Variankaval, K. I. Jacob, S. M. Dinh, *J. Cryst. Growth*, **2000**, *217*, 320–331.

56 M. Szelagiewicz, C. Marcolli, S. Cianferani, A. P. Hard, A. Vit, A. Burkhard, M. von Raumer, U. C. Hofmeier, A. Zilian, E. Francotte, R. Schenker, *J. Therm. Anal. Calorim.*, **1999**, *57*, 23–43.

57 P. O. Souillac, P. Dave, J. H. Rytting, *Int. J. Pharm.*, **2002**, *231*, 185–196.

58 C.-H. Gu, D. J. W. Grant, *J. Pharm. Sci.*, **2001**, *90*, 1277–1287.

59 M. Reading, *Thermochim. Acta*, **1994**, *238*, 295–307.

60 M. Reading, D. Elliot, V. L. Hill, *J. Therm. Anal.*, **1993**, *40*, 949–955.

61 N. Coleman, D. Q. M. Craig, *Int. J. Pharm.*, **1996**, *135*, 13–29.

62 B. Keymolen, J. L. Ford, M. W. Powell, A. R. Rajabi-Siahboomi, *Thermochim. Acta*, **2003**, *397*, 103–117.

63 K. Six, C. Leuner, J. Dressman, G. Verreck, J. Peeters, N. Blaton, P. Augustjins, R. Kinget, G. Van den Mooter, *J. Therm. Anal. Calorim.*, **2002**, *68*, 591–601.

64 K. Kawakami, Y. Ida, *Thermochim. Acta*, **2005**, *427*, 93–99.

65 R. Manduva, D. Q. M. Craig, V. L. Kett, D. A. Adkin, *Am. Assoc. Pharmaceutical Sci.*, **2003**, *5*, T2267.

66 C. McGregor, M. H. Saunders, G. Buckton, R. D. Saklatvala, *Thermochim. Acta*, **2004**, *417*, 231–237.

67 A. Hammiche, D. M. Price, E. Dupas, G. Mills, A. Kulik, M. Reading, J. M. R. Weaver, H. M. Pollack, *J. Microscopy-Oxford*, **2000**, *199*, 180–190.

68 D. Q. M. Craig, V. L. Kett, C. S. Andrews, P. G. Royall, *J. Pharm. Sci.*, **2002**, *91*, 1205–1213.

69 G. H. W. Sanders, C. J. Roberts, A. Danesh, A. J. Murray, D. M. Price, M. C. Davies, S. J. B. Tendler, M. J. Wilkins, *J. Microsc.-Oxford*, **2000**, *198*, 77–81.

70 J. R. Murphy, C. S. Andrews, D. Q. M. Craig, *Pharm. Res.*, **2003**, *20*, 500–507.

71 R. Manduva, V. L. Kett, D. A. Adkin, D. Q. M. Craig, *J. Pharm. Pharmacol.*, **2004**, *56*, 65S.

72 R. A. Shmeis, S. L. Krill, *Thermochim. Acta*, **2005**, *427*, 61–68.

73 N. Boutonnet-Fagegaltier, J. Menegotto, A. Lamure, H. Duplaa, A. Caron, C. Lacabanne, M. Bauer, *J. Pharm. Sci.*, **2002**, *91*, 1548–1560.

4
Solid-state NMR Spectroscopy

Joseph W. Lubach and Eric J. Munson

4.1
Introduction

Nuclear magnetic resonance (NMR) spectroscopy is one of the most important analytical techniques for the determination of molecular structure in solution. High-resolution NMR spectroscopy of pharmaceutical systems has mostly been applied to solutions. The development of magic-angle spinning in combination with cross polarization has made routine high-resolution analysis of solids a reality [1–7], and the technique has become a very powerful tool for investigations of solid-state chemistry. In particular, solid-state NMR spectroscopy provides a wealth of information about the physical and chemical state of pharmaceutical solids, especially within a formulated product.

Pharmaceutical solids have traditionally been characterized by thermal, diffractional, spectroscopic methods, and various forms of microscopy. Solid-state NMR spectroscopy is now used extensively in pharmaceutical analysis, from polymorph identification all the way to final dosage form characterization [8–15]. It can characterize bulk drugs by identifying and quantitating polymorphs and mixtures of polymorphs (or hydrates/solvates), as well as mixtures of crystalline and amorphous forms. Drugs and/or excipients can be analyzed within a formulated product without sample preparation, making it a non-invasive technique. In addition, the non-destructive nature of solid-state NMR analysis means that the exact same material can be analyzed using other analytical techniques. For example, entire tablets can be put into a spectrometer and analyzed *as is*, so the entire state of a formulation can be analyzed without the risks of grinding or heating. Another advantage is that the entire sample is simultaneously analyzed, rather than just a small part or the surface. The excipients will almost always have different chemical shifts than the active pharmaceutical ingredient (API), or have no peaks at all (such as talc in a ^{13}C experiment), making it a very selective technique. For some samples, carbon-containing excipients can be selectively eliminated from the spectrum using NMR relaxation properties, resulting in a spectrum of only the analyte of interest, as is shown later in the chapter.

Polymorphism: in the Pharmaceutical Industry. Edited by Rolf Hilfiker

ISBN: 3-527-31146-7

4.1.1
Basics of Solid-state NMR

The basic principles of both solution and solid-state NMR spectroscopy can be found in several excellent references [16–19]. However, a brief introduction to the most common methods used to acquire solid-state NMR spectra of pharmaceuticals is given here. Most solid-state NMR spectra of pharmaceutical solids are acquired using cross polarization with magic-angle spinning (CPMAS), most often with ^{13}C detection. CPMAS is a double resonance technique with cross polarization (CP) facilitating magnetization transfer from abundant spins (usually ^{1}H) to dilute spins (e.g., ^{13}C, ^{15}N) [4–6, 20]. ^{1}H to ^{13}C CP results in a four-fold increase in sensitivity relative to Bloch decay (single-pulse) ^{13}C experiments, and the time period between each acquisition is usually much shorter because the ^{1}H spin–lattice relaxation time (T_1) determines the repeat rate. Magic-angle spinning (MAS) serves to reduce or eliminate the inhomogeneous broadening effects of chemical shift anisotropy (CSA), such that only the isotropic chemical shift is observed [7, 21]. These two techniques, coupled with high-power ^{1}H decoupling, have enabled acquisition of solid-state NMR spectra that approach the resolution attainable in solution NMR experiments [1–3].

4.2
Applications

As mentioned previously, solid-state NMR spectroscopy can provide a wealth of information about a particular pharmaceutical formulation. Much of the information is based upon the fact that both structure and dynamics can be studied. The non-destructive and non-invasive aspects were discussed in the introduction. Other important information that solid-state NMR spectroscopy can provide to characterize pharmaceutical solids is highlighted here. First, the ability to identify different forms of drugs and excipients is shown. Second, the difference in relaxation times between crystalline drugs and amorphous excipients is used to selectively remove one component from the spectrum. Third, the ability to study complex amorphous systems consisting of both drugs and excipients is illustrated. Finally, an extensive discussion of the ability of solid-state NMR spectroscopy to quantify different forms is given because of the relevance of form identification and quantitation to the pharmaceutical field.

4.2.1
Identification

Solid-state NMR spectroscopy is well-suited for the study of polymorphism because small changes in conformation and/or local electronic structure cause observable differences in chemical shift. Different crystal forms can have relatively large changes in the chemical shift of some resonances. There are many examples of different crystalline forms giving unique NMR spectra [8, 22–25]. Padden et al. have compared solid-state NMR spectroscopy with powder X-ray diffraction (PXRD) in an investigation of neotame polymorphs. Following the conversion of a mixture of anhydrous forms over time, they found that PXRD could not detect a change in the patterns during the conversion process, whereas up to seven forms were detected in ^{13}C NMR spectra [25]. Byrn and coworkers have published several papers investigating polymorphism in drug compounds [22, 23, 26]. In a ^{13}C study of benoxaprofen, nabilone, and cefazolin, they found that CPMAS NMR spectra of the different crystal forms of the drugs were indeed different [22]. They also performed preliminary studies on pharmaceutical granulations of these drugs and found that the spectrum of the API could be discerned in the formulation. Five different crystal forms of prednisolone *tert*-butylacetate were investigated in another study, and again each solid form had a unique ^{13}C NMR spectrum [23]. Four of the forms were solvates, while one was anhydrous. The authors were able to relate changes in crystal packing to oxygen reactivity, as one of the forms exhibited a channel down the hexagonal axis of the crystal lattice. This channel allowed oxygen to penetrate into the lattice and react with the drug.

Work by Chang et al. [27] illustrates a case in which solid-state NMR was used to disprove the presence of aspirin polymorphic forms. The authors looked at both bulk aspirin in different crystal habits, as well as numerous commercial tableted dosage forms. The ^{13}C CPMAS NMR spectra of the different habit forms were identical, showing that the change in macroscopic morphology was not due to a change in crystal packing. No interactions between the aspirin and excipients were observed in the spectra of the dosage forms. However, when dissolved and lyophilized the aspirin showed a strong interaction with the buffer components of the formulation.

Perhaps the aspect of solid-state NMR spectroscopy of greatest value to pharmaceutical scientists is the unique ability to study formulations without the need to modify the sample. Products need to be characterized at all stages of development, all the way through to the final formulation and product life cycle. Each ingredient may exist in a number of solid forms, which can render detailed analysis difficult. Also, with other techniques it may not always be clear if the signals are from the API or excipients, which is rarely the case with NMR spectroscopy. It is critical to know the state of each component in the formulation, especially during stability studies where physical transformations may have marked effects on the integrity of the study. Solid-state stability studies are regularly carried out at elevated temperatures and relative humidities, both of which can facilitate changes in the solid state such as polymorphic interconversions, hydrate formation, or recrystallization of amorphous APIs or excipients. Solid-state NMR spectroscopy can simultaneously look at both the API and excipients without the need for sample preparation or physical separations. Within a given formulation it is possible to identify and even quantify which form(s) is (are) present, even at very low API loadings. Although sensitivity is ultimately limited by the amount of instrument time allotted to a given sample, identification of a polymorph in a formulation should be relatively straightforward at API loadings down to 0.25% (w/w) for most crystalline compounds. Lubach and coworkers [24] used the example of bupivacaine solid forms formulated in lipospheres and a protein matrix. The authors were able to identify the form of bupivacaine present in the formulation following processing, as well as identify changes in the matrix, illustrating the importance of being able to characterize both API and excipients simultaneously without disturbing the nature of the sample.

4.2.2 Selectivity

One may even utilize the NMR properties of the components of a given sample to select out unwanted parts of the spectrum. For instance, crystalline prednisolone has a relatively slow magnetization decay during cross polarization (^1H $T_{1\rho}$), meaning that a spectrum acquired at a long contact time, 15 ms for example, will give almost the same signal strength as a spectrum acquired at the optimal contact time (2 ms). Sulfobutylether-β-cyclodextrin (Captisol®) has a much more rapid ^1H $T_{1\rho}$, such that a spectrum acquired using a 15-ms contact time will essentially give no signal (Fig. 4.1). This property can be exploited when analyzing a mixture of an API and excipient such as prednisolone and Captisol®. Figure 4.2 shows the CPMAS NMR spectra of bulk crystalline prednisolone acquired with a 2-ms contact time (A), a 7.7% dispersion of prednisolone in Captisol® acquired with a 2-ms contact time (B), and the same mixture acquired with a 15-ms contact time (C) [28]. The large, broad peaks in the middle spectrum are from the cyclodextrin, and it is easily seen that the spectrum acquired at a contact time of 15 ms contains only prednisolone peaks. The Captisol® also has a longer ^1H T_1 than the prednisolone, so some of the excipient signal is already attenuated by using a recycle delay much shorter than 5 times

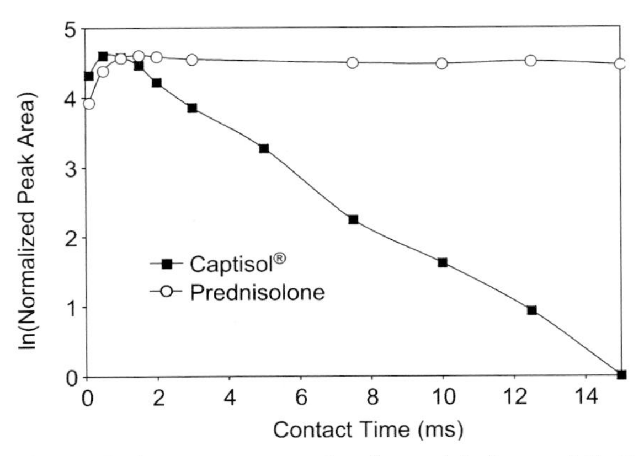

Fig. 4.1 Peak area vs. contact time for prednisolone and Captisol® in a 7.7% (w/w) dispersion, showing the differences in cross polarization kinetics between the API (prednisolone) and excipient (Captisol®).

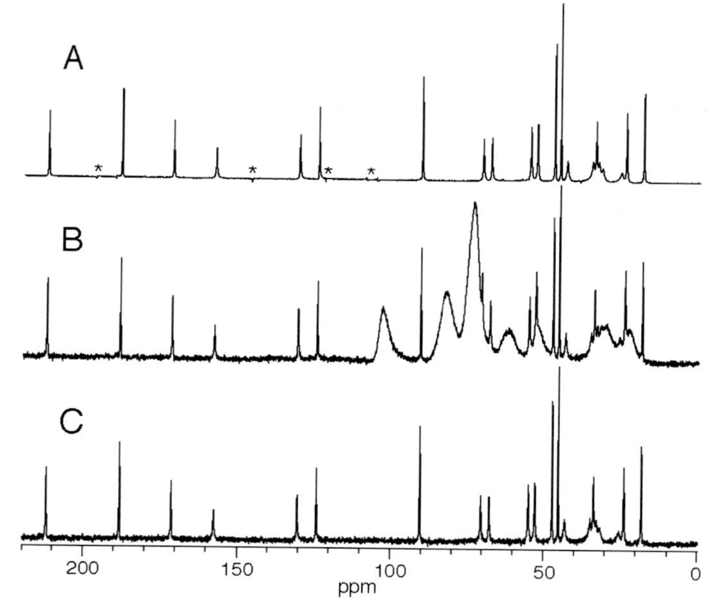

Fig. 4.2 ^{13}C CPMAS NMR spectra of crystalline prednisolone, 2-ms contact time (A) and a 7.7% (w/w) dispersion of prednisolone in Captisol®, contact times of 2 ms (B) and 15 ms (C). Magic-angle spinning rate was 5.0 kHz. (*) Denotes spinning sidebands not perfectly suppressed by the TOSS [43] pulse sequence.

the ^{1}H T_1. This is another NMR property that can be exploited if the properties of the given sample are amenable to this technique.

4.2.3 Mobility and Dynamics

It is sometimes desirable to prepare formulations using amorphous materials, especially with protein formulations, where lyoprotectants, cryoprotectants, and stabilizers do little good if they crystallize during formulation or storage. Amorphous materials are typically characterized by their glass transition temperature (T_g), which represents a temperature at which a dramatic change in molecular mobility occurs, as the glass moves into a rubbery state. Above the T_g amorphous materials are much more likely to crystallize. The inherent disorder, heterogeneity, and propensity to crystallize of the amorphous phase make it notoriously difficult to characterize and understand. Using NMR, one can obtain a wealth of information about the amorphous phase, despite significant losses in resolution relative to crystalline materials. One source of information accessible to solid-state NMR is molecular mobility, which can be probed using relaxation measurements, as well as variable temperature and dynamics experiments.

Yoshioka and coworkers [29–31] have examined kHz- and MHz-order molecular motions as a function of temperature and water content of proteins formulated in amorphous dextran matrices, using T_1, $T_{1\rho}$, and T_2 relaxation time measurements. Aso and coworkers [32] studied the effects of increasing water content on mobility of lyophilized sucrose and poly(vinylpyrrolidone) (PVP) formulations. They measured ^{13}C T_1 and $T_{1\rho}$ relaxation times to investigate motions of specific nuclei and functional groups. PVP was found to inhibit sucrose crystallization due to its effect on sucrose mobility, which was attributed to hydrogen bonding between sucrose hydroxyl groups and the PVP side-chain carbonyl. Information about local structure short-range order can also be reflected in lineshape and linewidth changes of amorphous compounds. Chemical shift differences may also occur when specific interactions are taking place between a drug and the amorphous matrix it is formulated in, and these interactions can be localized to certain moieties of the molecules involved, given that the resonances can be assigned properly. Examples of using solid-state NMR to study drug–matrix interactions can be found in [24, 33–36].

4.2.4 Quantitation of Forms

NMR spectroscopy is a quantitative technique by nature, with the observed signal corresponding directly to the relative number of like nuclei present in a given sample. Solid-state NMR is well suited to quantitate the various solid forms a compound may exist in, as the chemical shifts for peaks of different polymorphs or solvatomorphs (i.e., hydrates and solvates) can vary up to 10 ppm. Quantitation of the amounts of crystalline and amorphous material is also possible, as amorphous peaks are much broader than crystalline peaks, simply due to the disorder inherent in the amorphous phase. Chemical shifts of peaks of amorphous material can also vary markedly from corresponding crystalline peaks. Early literature suggested that cross polarization experiments were not quantitative unless the experiment contact time was optimized to give signals equaling molar ratios, or if internal standards were used to create a standard curve, much like doing quantitation with other analytical techniques. It was concluded that CPMAS NMR spectroscopy was still a viable and convenient technique to use for quantitation [37, 38]. Harris [39] provided an early account of using solid-state NMR spectroscopy for quantitative analysis, although quantitative studies in the literature are few. Suryanarayanan and Wiedmann [40] demonstrated the use of a standard curve and an internal standard (glycine) to quantitate mixtures of carbamazapine anhydrate and dihydrate. The authors of this study also employed complete saturation of the anhydrous form, which would have required a 2000-s recycle delay to be quantitative. The use of glycine carbonyl as an internal standard enabled them to use a 10-s recycle delay and complete the experiments in a much shorter time than would otherwise have been possible. Gao [41] investigated mixtures of delavirdine mesylate solid forms and was able to obtain quantitative results with a limit of detection around 2–3%. The author concluded that solid-state NMR is applicable to quantitation for both bulk drugs as well as formulations, but a complete characterization of relaxation and cross polarization dynamics is critical to any quantitative study.

One of the advantages of CPMAS NMR spectroscopy is the ability to quantitate forms *without the need to prepare a standard curve*. Several factors need to be considered to ensure that quantitative data is acquired. First of all, a recycle delay of at least 5 times the longest T_1 of the mixture must be utilized to ensure that the magnetization has returned to its full equilibrium value. If the recycle delay is too short, peaks in the resulting spectrum will be attenuated, and inaccurate quantitation data will result. In theory, shorter recycle delays could be used if a standard curve was prepared or the precise amount of saturation was known so that the correct peak areas could be calculated. Also of primary concern when performing any CPMAS quantitation experiment is the CP dynamics, or the rate at which magnetization is transferred from ^1H to ^{13}C (T_{CH}) and the rate of ^1H magnetization decay in the rotating frame ($T_{1\rho}$). Different solid forms of the same compound often have different T_{CH} and ^1H $T_{1\rho}$, and this affects the signal intensity achieved at a given experimental contact time. The contact time is the length of time both ^1H and ^{13}C pulses are simultaneously applied to induce cross polarization, and is generally on the order of 1–5 ms. Quantitation could also be performed using Bloch decay experiments, although the experiment time required relative to CP experiments limits their utility.

The first part of any quantitation experiment is to assign peaks to their respective solid form. The same nucleus (e.g., quaternary carbon in a phenyl ring) should be chosen for each form. These peaks must have sufficient resolution either to integrate the peak areas or perform deconvolution to calculate peak areas. If the CP dynamics of both (or all) components in a mixture are identical, quantitation can be performed simply by integrating the peak areas of the selected resonances of each respective form. However, if one polymorph has an optimal contact time of 2 ms, and the other polymorph has an optimal contact time of 4 ms, the results will not be quantitative. Acquiring a spectrum of a mixture of these two polymorphs using a 2-ms contact time would give an attenuated peak area of the second polymorph relative to that of the first polymorph.

The work of Offerdahl et al. [42] provides an example of quantitation employing CP experiments, using neotame, *N*-(3,3-dimethylbutyl)-L-aspartyl-L-phenylalanine methyl ester, an artificial sweetener. Neotame Forms A and G, and crystalline Form G and amorphous neotame were successfully quantitated using solid-state NMR spectroscopy. Figure 4.3 shows the ^{13}C CPMAS NMR spectrum of a 1 : 1 (wt/wt) mixture of neotame Forms A and G. Two well-resolved aromatic resonances representing the same carbon in each form were selected (these are expanded in the figure). The peak at 138.6 ppm is from Form G, while the peak at 135.6 ppm is due to Form A. This can be easily determined by comparing with spectra of the pure forms. This particular spectrum was ac-

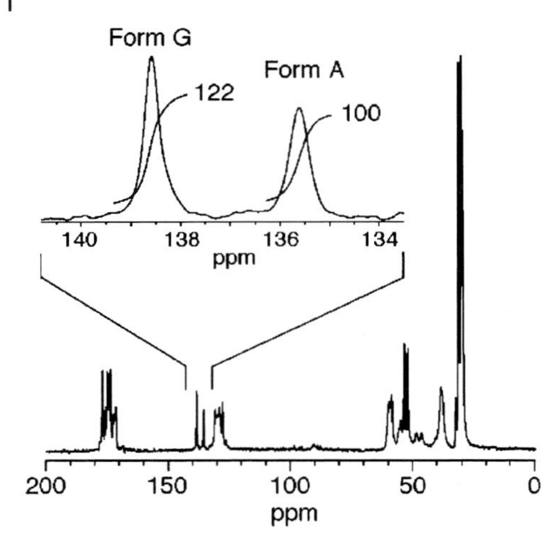

Fig. 4.3 ^{13}C CPMAS NMR spectrum of neotame containing 50% polymorphic Form A and 50% polymorphic Form G, with expansion from ~134 to 140 ppm to detail the peaks used for quantitation. The integrated relative peak areas are given.

quired using a CP contact time of 2 ms, and as seen in the spectrum the integrated relative peak areas for each form are not identical, indicating a difference in the CP dynamics of each form. Form A has a relative peak area of 100, while Form G has a relative peak area of 122. Figure 4.4 shows the CP dynamics profiles for both forms from 0 to 30 ms. This shows that the rate of magnetization decay ($T_{1\rho}$) is much more rapid in Form A than in Form G, such that if a spectrum of the same 1:1 mixture shown in Fig. 4.3 was acquired using a 30-ms contact time, only Form G would be present in the spectrum.

The issue of CP dynamics is corrected for by plotting ln(relative peak area) versus contact time, at contact times $>5T_{CH}$, where the magnetization profile is in linear decay, with the assumption that ^1H $T_{1\rho}$ is monoexponential. This plot is then extrapolated back to 0-ms contact time, where the y-intercept represents the theoretical case of instantaneous magnetization transfer from ^1H to ^{13}C. The antilog of the y-intercept of both forms represents the relative weight percent of each form as measured by NMR. For example, the sample containing 30% Form A and 70% Form G gave y-intercepts of 4.8991 and 5.7772, respectively, in a multiple-contact time experiment. The relative abundance of each form is then given by $e^{4.8991}$ for Form A and $e^{5.7772}$ for Form G. This calculation results in relative abundances of 134.2 and 322.9, respectively, for Forms A and G, which correspond to 29.4% Form A and 70.6% Form G. The measured weight percentages making up the actual sample were 29.71% Form A and 70.29% Form G – only a 0.3% difference from the NMR data. Table 4.1 summarizes the quantitation data for mixtures of neotame Forms A and G. The NMR data are accurate to within about 1%.

Quantitation of mixtures of crystalline and amorphous forms can be performed in much the same way as described above, and was carried out for neotame Form G and amorphous neotame. The disorder inherent in amorphous material results in very broad peaks with Gaussian lineshapes, which in turn re-

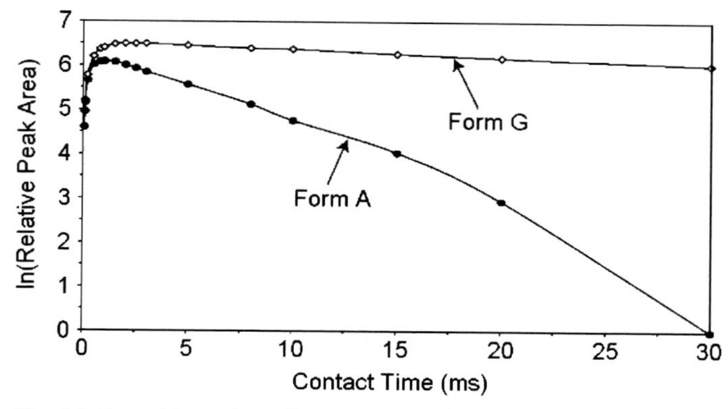

Fig. 4.4 Natural logarithm of relative peak area vs. contact time for a 1:1 mixture of neotame Forms A and G, showing large differences in cross polarization dynamics between the two polymorphic forms.

Table 4.1 Summary of data from quantitation experiments on neotame polymorphic mixtures containing Forms A and G.

% Form A	Form	Mass (g)	Wt.%	y-Intercept	Rel. area	NMR (%)	Diff (abs)
10	A	0.0478	12.62	7.2984	1478	13.7	1.1
	G	0.3309	87.38	9.1421	9340	86.3	
20	A	0.0759	21.20	4.9549	141.9	22.0	0.8
	G	0.2822	78.80	6.2208	503.1	78.0	
30	A	0.1034	29.71	4.8991	134.2	29.4	0.3
	G	0.2446	70.29	5.7772	322.9	70.6	
40	A	0.1273	39.12	5.0380	154.2	37.9	1.2
	G	0.1981	60.88	5.5337	253.1	62.1	
50	A	0.1961	49.73	4.8556	128.5	50.3	0.6
	G	0.1982	50.27	4.8449	127.1	49.7	
60	A	0.1954	61.56	4.8684	130.1	62.3	0.7
	G	0.1220	38.44	4.3667	78.8	37.7	
70	A	0.2413	70.29	4.8870	132.6	71.4	1.1
	G	0.1020	29.71	3.9714	53.1	28.6	
80	A	0.2966	80.88	4.8781	131.4	81.8	0.9
	G	0.0701	19.12	3.3723	29.1	18.2	
90	A	0.2961	88.41	7.1853	1320	89.0	0.6
	G	0.0388	11.59	5.0989	164	11.0	

duces both resolution and sensitivity. Nevertheless, CPMAS NMR is one of the best techniques for investigating and quantitating the amorphous phase. Figure 4.5 shows the spectrum of a 1 : 1 mixture of amorphous neotame and crystalline Form G. The broad peak at the base of the sharp crystalline peak at 138.6 ppm represents the amorphous material. Due to insufficient resolution between the two peaks used for quantitation, as will almost always be the case with crystalline and amorphous mixtures, deconvolution of the two peaks must be performed to calculate their respective peak areas. The calculated areas of the crystalline and amorphous peaks in the 1:1 mixture are shown in the inset of Fig. 4.5, with the amorphous peak normalized to 100, and the relative area of the Form G peak is 87.3. The fact that they are not equal indicates that the two solid phases have differing CP dynamics. This discrepancy is corrected for by doing a multiple-contact time experiment just as described above for the polymorphic mixture. An issue that can arise in quantitation involving amorphous material is the possibility of having amorphous impurities in the crystalline reference material. Because of the very broad nature of the amorphous peaks, at low levels they can disappear into the baseline and go undetected. Even highly crystalline substances may have some amorphous material present, and many forms of processing used in the pharmaceutical industry can induce amorphization of an otherwise crystalline sample.

In the quantitation experiments performed on crystalline and amorphous neotame mixtures, errors larger than expected were obtained, but the error was systematic, with the highest error occurring in the sample containing 20% amorphous, 80% Form G, decreasing down to the lowest error in the 90% amorphous sample. These results are summarized in Table 4.2. The error was due to amorphous impurities in the bulk Form G material that was previously assumed to be 100% crystalline. By plotting the amorphous weight % determined by NMR versus the calculated weight %, and extrapolating to 0% theoretical amorphous content, it was found that approximately 14% of the Form G standard used to prepare the amorphous/Form G mixtures was actually amorphous. To verify this, a slightly more complicated NMR experiment was performed on the standard material to probe the amorphous phase. This involved a pulse sequence that attenuated the crystalline peaks to a much greater extent

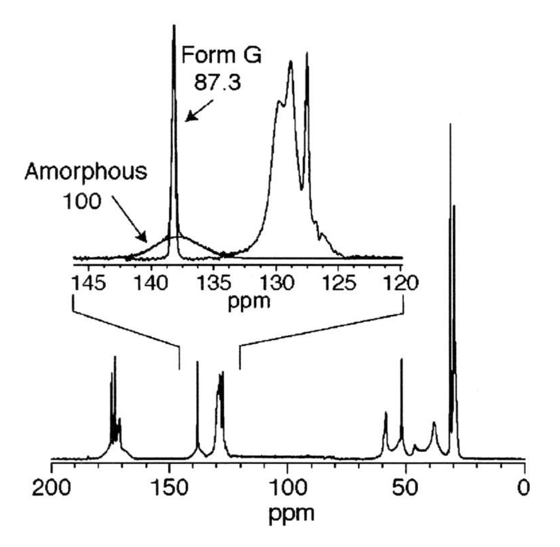

Fig. 4.5 ^{13}C CPMAS NMR spectrum of neotame containing 50% amorphous material and 50% polymorphic Form G, with expansion from ~120 to 145 ppm to detail the deconvoluted amorphous and crystalline peaks used for quantitation. The calculated relative peak areas are given.

Table 4.2 Summary of data from quantitation experiments on mixtures containing amorphous neotame and neotame Form G, before correction for amorphous impurities.

% Amorphous	Form	Mass (g)	Wt.%	Rel. area	NMR (%)	Diff (abs)
20	Amorphous	0.0966	20.07	119.3	32.1	12.0
	G	0.3847	79.93	252.8	67.9	
30	Amorphous	0.1381	26.93	137.2	36.2	9.2
	G	0.3748	73.07	242.2	63.8	
40	Amorphous	0.1845	41.08	124.5	47.7	6.6
	G	0.2646	58.92	136.6	52.3	
50	Amorphous	0.2701	51.08	121.4	56.3	5.3
	G	0.2587	48.92	94.1	43.7	
70	Amorphous	0.3220	67.10	121.5	69.8	2.7
	G	0.1579	32.90	52.5	30.2	
90	Amorphous	0.3788	47.48	131.5	91.5	1.5
	G	0.4190	52.52	12.2	8.5	

Table 4.3 Summary of data from quantitation experiments on mixtures containing amorphous neotame and neotame Form G, following correction for amorphous impurities.

% Amorphous	Form	Mass (g)	Wt.%	Rel. area	NMR (%)	Diff (abs)
20	Amorphous	0.1483	30.81	119.3	32.1	1.3
	G	0.3330	69.19	252.8	67.9	
30	Amorphous	0.1885	36.75	137.2	36.2	0.6
	G	0.3244	63.25	242.2	63.8	
40	Amorphous	0.2201	49.01	124.5	47.7	1.3
	G	0.2290	50.99	136.6	52.3	
50	Amorphous	0.3049	57.66	121.4	56.3	1.3
	G	0.2239	42.34	94.1	43.7	
70	Amorphous	0.3422	71.51	121.5	69.8	1.7
	G	0.1367	28.49	52.5	30.2	
90	Amorphous	0.3844	91.37	131.5	91.5	0.1
	G	0.0363	8.63	12.2	8.5	

than the amorphous peaks, simply based on the slow ^{1}H $T_{1\rho}$ relaxation properties of the crystalline form and relatively fast ^{1}H $T_{1\rho}$ of the amorphous component. A quantifiable net increase in the intensity of the amorphous peak relative to the crystalline peak resulted, effectively increasing the limit of detection for the amorphous material. This experiment gave a result of ~12% amorphous, which agreed well with the ~14% determined in the previous experiments. When correction factors were applied to account for the amorphous impurities, the accuracy of the quantitation data turned out to be quite good, again with absolute errors around 1% (Table 4.3).

4.3 Conclusions

Solid-state NMR spectroscopy is an extremely powerful tool at the disposal of the pharmaceutical scientist, and should be considered an integral part of pharmaceutical research and development. Its strengths lie in the fact that it is selective, non-destructive, and non-invasive, and that it is particularly adept at handling pharmaceutical formulations without altering properties of the formulation upon analysis. Quantitative analysis of mixtures of crystalline forms as well as mixtures of crystalline and amorphous forms without the need of a standard curve has been shown, along with other methods of quantitation. While no one technique is suited to fully characterize any solid-state system, solid-state NMR spectroscopy is an excellent complement to the well established classical instrumental methods used to characterize pharmaceutical solids.

References

1 J. Schaefer, E.O. Stejskal, *J. Am. Chem. Soc.* **1976**, *98*, 1031–1032.
2 J. Schaefer, E.O. Stejskal, R. Buchdahl, *Macromolecules* **1975**, *8*, 291–296.
3 E.O. Stejskal, J. Schaefer, J. S. Waugh, *J. Magn. Reson.* **1977**, *28*, 105–112.
4 A. Pines, M.G. Gibby, J.S. Waugh, *J. Chem. Phys.* **1972**, *56*, 1776–1777.
5 A. Pines, M.G. Gibby, J.S. Waugh, *J. Chem. Phys.* **1973**, *59*, 569–590.
6 S.R. Hartmann, E.L. Hahn, *Phys. Rev.* **1962**, *128*, 2042–2053.
7 E.R. Andrew, *Prog. Nucl. Magn. Reson. Spectrosc.* **1971**, *8*, 1–39.
8 H.G. Brittain, K.R. Morris, D.E. Bugay, A.B. Thakur, A.T.M. Serajuddin, *J. Pharm. Biomed. Anal.* **1993**, *11*, 1063–1069.
9 D.E. Bugay, *Pharm. Res.* **1993**, *10*, 317–327.
10 S.R. Byrn, R.R. Pfeiffer, G. Stephenson, D.J.W. Grant, W.B. Gleason, *Chem. Mater.* **1994**, *6*, 1148–1158.
11 P.A. Tishmack, D.E. Bugay, S.R. Byrn, *J. Pharm. Sci.* **2003**, *92*, 441–474.
12 S.M. Reutzel-Edens, J.K. Bush, *Am. Pharm. Rev.* **2002**, *5*, 112–115.
13 P.J. Saindon, N.S. Cauchon, P.A. Sutton, C.J. Chang, G.E. Peck, S.R. Byrn, *Pharm. Res.* **1993**, *10*, 197–203.
14 T. Mavromoustakos, I. Daliani, P. Zoumboulakis, A. Kolocouris, *Pharmakeutike* **2000**, *13*, 37–51.
15 M.R.M. Palermo de Aguiar, A.L. Gemal, R. Aguiar da Silva San Gil, *Quim. Nova* **1999**, *22*, 553–564.
16 R.K. Harris, *Nuclear Magnetic Resonance Spectroscopy: A Physicochemical View*, Longman Scientific & Technical, Harlow, **1986**.
17 M.H. Levitt, *Spin Dynamics: Basics of Nuclear Magnetic Resonance*, John Wiley & Sons, Chichester, **2001**.
18 K. Schmidt-Rohr, H.W. Spiess, *Multidimensional Solid-State NMR and Polymers*, Academic Press, London, **1994**.
19 C.P. Slichter, *Principles of Magnetic Resonance*, Springer-Verlag, Berlin, **1990**.
20 O.B. Peersen, X. Wu, S.O. Smith, *J. Magn. Reson. A* **1994**, *106*, 127–131.
21 E.R. Andrew, A. Bradbury, R.G. Eades, *Nature* **1959**, *183*, 1802–1803.
22 S.R. Byrn, G. Gray, R.R. Pfeiffer, J. Frye, *J. Pharm. Sci.* **1985**, *74*, 565–568.
23 S.R. Byrn, P.A. Sutton, B. Tobias, J. Frye, P. Main, *J. Am. Chem. Soc.* **1988**, *110*, 1609–1614.
24 J.W. Lubach, B.E. Padden, S.L. Winslow, J.S. Salsbury, D.B. Masters, E.M. Topp, E.J. Munson, *Anal. Bioanal. Chem.* **2004**, *378*, 1504–1510.

25 B.E. Padden, M.T. Zell, Z. Dong, S.A. Schroeder, D.J.W. Grant, E.J. Munson, *Anal. Chem.* **1999**, *71*, 3325–3331.

26 S.R. Byrn, B. Tobias, D. Kessler, J. Frye, P. Sutton, P. Saindon, J. Kozlowski, *Trans. Am. Cryst. Assoc.* **1989**, *24*, 41–54.

27 C.J. Chang, L.E. Diaz, F. Morin, D.M. Grant, *Magn. Reson. Chem.* **1986**, *24*, 768–771.

28 J.W. Lubach, S. Sotthivirat, V.J. Stella, E.J. Munson, unpublished data.

29 S. Yoshioka, Y. Aso, S. Kojima, S. Sakurai, T. Fujiwara, H. Akutsu, *Pharm. Res.* **1999**, *16*, 1621–1625.

30 S. Yoshioka, Y. Aso, S. Kojima, *Pharm. Res.* **1996**, *13*, 926–930.

31 S. Yoshioka, Y. Aso, S. Kojima, *J. Pharm. Sci.* **2002**, *91*, 2203–2210.

32 Y. Aso, S. Yoshioka, J. Zhang, G. Zografi, *Chem. Pharm. Bull.* **2002**, *50*, 822–826.

33 M.G. Vachon, J.G. Nairn, *Eur. J. Pharm. Biopharm.* **1998**, *45*, 9–21.

34 B.R. Rohrs, T.J. Thamann, P. Gao, D.J. Stelzer, M.S. Bergren, R.S. Chao, *Pharm. Res.* **1999**, *16*, 1850–1856.

35 S. Puttipipatkhachorn, J. Nunthanid, K. Yamamoto, G.E. Peck, *J. Controlled Release* **2001**, *75*, 143–153.

36 P. Mura, G.P. Bettinetti, A. Manderioli, M.T. Faucci, G. Bramanti, M. Sorrenti, *Int. J. Pharm.* **1998**, *166*, 189–203.

37 L.B. Alemany, D.M. Grant, R.J. Pugmire, T.D. Alger, K.W. Zilm, *J. Am. Chem. Soc.* **1983**, *105*, 2142–2147.

38 L.B. Alemany, D.M. Grant, R.J. Pugmire, T.D. Alger, K.W. Zilm, *J. Am. Chem. Soc.* **1983**, *105*, 2133–2141.

39 R.K. Harris, *Analyst* **1985**, *110*, 649–655.

40 R. Suryanarayanan, T.S. Wiedmann, *Pharm. Res.* **1990**, *7*, 184–187.

41 P. Gao, *Pharm. Res.* **1996**, *13*, 1095–1104.

42 T.J. Offerdahl, J.S. Salsbury, Z. Dong, D.J.W. Grant, S.A. Schroeder, I. Prakash, E.J. Munson, *J. Pharm. Sci.* **2005**, *94*, 2591–2605.

43 W.T. Dixon, J. Schaefer, M.D. Sefcik, E.O. Stejskal, R.A. McKay, *J. Magn. Reson.* **1982**, *49*, 341–345.

5
Vibrational Spectroscopic Methods in Pharmaceutical Solid-state Characterization

John M. Chalmers and Geoffrey Dent

5.1
Introduction

The vibrational spectroscopic methods covered in this chapter are the complementary techniques of infrared and Raman spectroscopy as well as terahertz spectroscopy. Both mid-infrared and Raman spectroscopy are well established as tools that provide spectra that can be considered as molecular structure fingerprints of an analyte; consequently, each has a prominent role in the solid-state characterization of an active pharmaceutical ingredient (API). The US Pharmacopoeia (USP) [1] states that:

"The infrared absorption spectrum of a substance, compared with that obtained concomitantly for the USP Reference Standard, provides perhaps the most conclusive evidence of the identity of the substance that can be realized from any single test."

Mid-infrared and/or Raman spectra are commonly featured in patent filings or litigation cases, alongside X-ray diffraction and thermal analysis data, as providing a definitive fingerprint of a particular pharmaceutical substance. In properly conducted measurements, a mid-infrared or Raman spectrum is characteristic of the solid state of the analyte, i.e., its stereochemistry; and, hence, its polymorphic form. This attribute has long been recognized, for instance, in 1955 Dickson et al. [2] reporting on a study of steroids commented:

"Infrared examination of several hundred steroids and steroid sapogenins encountered during synthetic work revealed that certain steroid samples, though known to be identical chemically, gave infrared absorption spectra that were different when examined in the solid state as either Nujol mulls or potassium bromide discs, but identical when examined as dilute solutions in either carbon disulphide or bromoform. The spectral differences were attributed to polymorphism; this was confirmed by X-ray diffraction."

Indeed it was this sensitivity to solid-state form that prompted Dickson et al. [2] to comment that for simple chemical structure interpretation of functional groups within a molecule that

Polymorphism: in the Pharmaceutical Industry. Edited by Rolf Hilfiker

ISBN: 3-527-31146-7

cipients. While most bands observed in the mid-infrared region are derived from intramolecular transitions of vibrational modes [6], low energy transitions occur in the far-infrared region. The low wavenumber (long-wavelength region) of the vibrational spectrum explores low-frequency motions in molecular systems; for organic molecules these include flexing or skeletal bending modes of entire individual molecules, intramolecular stretching modes involving heavy atoms, and intermolecular interactions of hydrogen-bonded molecules or weaker van der Waal's bonding between nearest neighbors [6]. Consequently, different solid forms can give rise to very distinct spectra in this region, and, very recently, with the introduction of new commercial instrumentation there has been a significant renewed interest in the potential of this region for polymorph characterization. Currently, although this is likely to be extended to higher wavenumber in the near future, the new instrumentation covers the approximate range 1.3 to 133.3 cm^{-1} (ca. 7.7 mm to 75 μm wavelength) or ca. 40 GHz to 4 THz; consequently, the technique is named Terahertz (THz) spectroscopy [7]. Near-infrared (NIR) spectroscopy also has a major role in analysis and material characterization within the pharmaceutical industry [8]. Although much of this use is as a quality control/quality assurance tool, it is widely used for chemical identification (ID) and qualification, raw material identification, and in at-line and on-line applications such as monitoring blend uniformity and granulation drying, and measuring quantitatively a wide variety of chemical and physical properties. Absorption bands in the near-infrared region, which covers the wavelength range 2500 to 780 nm, arise essentially from overtone and combination bands of fundamental X–H vibrations (e.g., OH, NH, CH) that occur in the mid-infrared region.

The primary focus of this book is *Polymorphism*; this will be discussed in this chapter within the context of the role of mid-infrared, Raman and THz spectroscopic characterization methods for solid-state pharmaceuticals. Any discussion of instrumentation and theory will be necessarily brief, since this is well covered in many other texts. Some limited consideration will be given to the different sampling and measurement techniques, as appropriate; this will be followed by a general discussion of the effects on vibrational spectra of hydration, solvation, crystalline form, conformation, tautomerism, hydrogen bonding, polymorphism, racemates, and formulation. The chapter will finish with some selected examples of the application of mid-infrared, near-infrared, Raman and THz spectroscopy in pharmaceutical solid-state characterization that illustrate particularly the preceding discussions.

5.2 Mid-infrared, Raman and THz Spectroscopy: Basic Comparison of Theory, Instrumentation and Sampling

5.2.1 Basic Theory

Infrared and Raman spectra are produced as a consequence of transitions between quantized vibrational energy states [9]. The transition energy difference between the ground state and the first excited state of most vibrational modes of organic molecules corresponds to energy of radiation in the mid-infrared region (4000 to 400 cm^{-1}).

In mid-infrared absorption spectroscopy, electromagnetic radiation in the infrared region passing through or incident upon a sample is selectively absorbed (attenuated in intensity) by fundamental molecular vibrations of the same frequency, providing that during the molecular vibration there is a change of dipole moment.

Raman spectra are produced as a consequence of inelastic scattering of monochromatic radiation. If monochromatic light is incident on a sample, then most of the scattered radiation is elastically scattered (Rayleigh scattering) in all directions; however, a small proportion (about 1 part in 10^4) is changed in frequency. The incident photons promote the potential energy of a molecule to a higher energy virtual state, most then relax to the ground state with the emission of photons unchanged in frequency (Rayleigh scattering). The inelastically scattered radiation, some of which is reduced in frequency (Stoke's bands) and some of which is raised in frequency (anti-Stoke's bands), constitutes the Raman spectrum of the sample. Since, at room temperature, most molecules are likely to reside in the ground vibrational state in practice, for spectral interpretation and comparative purposes, the Stoke's Raman spectrum is recorded and interrogated. In addition, while, strictly speaking, the position of a Raman band should be measured as a wavenumber shift, Δ cm^{-1}, from that of the excitation laser wavenumber, commonly it is simply expressed in the unit wavenumber, cm^{-1}. In contrast to mid-infrared spectroscopy, for a molecular vibration to be Raman active the vibration must be accompanied by a change of polarizability.

Figure 5.1 gives a schematic energy level diagram depicting the processes associated with near- and mid-infrared and Raman spectroscopy. More detailed explanations and descriptions of the theories underlying infrared [10–12] absorption and Raman [10, 11, 13, 14] scattering may be found in the references indicated.

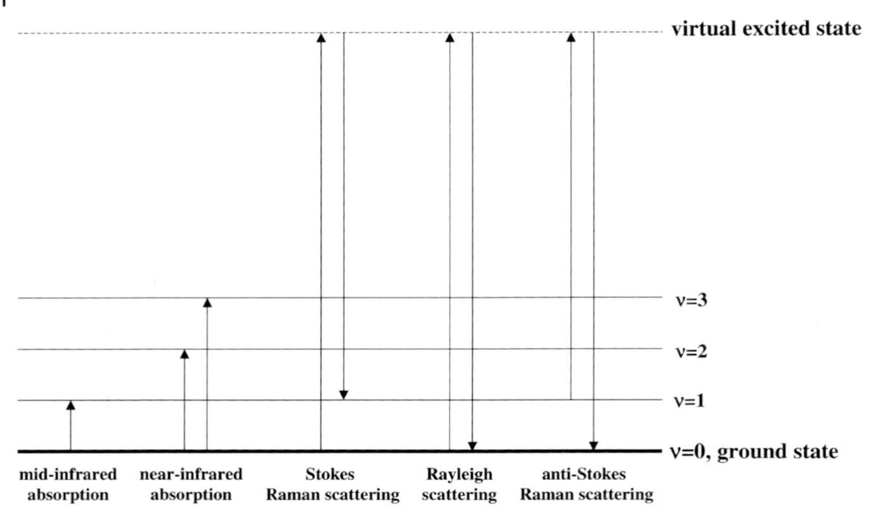

Fig. 5.1 Energy level diagram depicting the vibrational energy level transitions associated with infrared absorption and Raman and Rayleigh scattering processes.

Many fundamental vibrations of organic molecules give rise to mid-infrared absorption or Raman bands within relatively narrow wavenumber ranges, and many correlation tables exist in which these characteristic *group frequencies/wavenumbers* are listed, e.g., [15–18]. For instance, hydrocarbon stretches (νCH) normally occur within the range 3200 to 2800 cm^{-1}, with bands from vibrations of saturated species (e.g., $-CH_3$, $-CH_2$) lying below 3000 cm^{-1}, while those associated with unsaturated groups, such as aromatic CH and $=CH_2$ usually occur above 3000 cm^{-1}; the carbonyl stretch (νC=O) will typically be found within 1850 to 1550 cm^{-1}, with narrower regions within this range being more specific to type, such as ester, ketone, acid, etc. Consequently, in addition to pattern-recognition fingerprinting, both mid-infrared and Raman spectroscopy are widely used as qualitative tools in identifying the major functional groups present within materials and products related to the pharmaceutical industry.

The different selection rule that applies between the origin of infrared (change of dipole moment) and Raman (change of polarizability) bands means that the techniques are complementary. For instance, the stretching vibration of a carbonyl group, >C=O, is generally accompanied by a relatively large change in dipole moment and hence gives rise to an intense infrared absorption band, but the band is usually relatively much weaker in the counterpart Raman spectrum; whereas, the stretching vibration of a symmetrical group such as –C=C– generally gives rise to a relatively much more intense band in the Raman spectrum. For example, if one compares the pairs of ephedrine and pseudoephedrine spectra shown later as Fig. 5.5, one can see that the Raman spectra are dominated by the relatively very intense band near 1000 cm^{-1}, which is a characteristic of a mono-substituted aryl ring, whereas in the mid-infrared spectra in the region 3400 to 2700 cm^{-1}, the νOH and νNH (ν indicates a stretching vibration mode) are very prominent, but much weaker or barely seen in the Raman spectra. In general, bands within a Raman spectrum often appear sharper than many of those in its counterpart mid-infrared spectrum. Many bands in a mid-infrared spectrum are often broadened much more due to environmental effects (e.g., H-bonding). While mid-infrared and Raman spectroscopy are complementary in terms of the origin of bands, each technique has unique attributes as well as limitations. One of the often-quoted advantages of Raman spectroscopy (compared with mid-infrared spectroscopy) is its weak sensitivity to water, and hence a much easier opportunity to study aqueous based systems. However, the sensitivity of mid-infrared to bound water makes it particularly useful for discerning hydrated forms and studying inter- and intramolecular hydrogen bonding, as well as –OH groups. In 1960, Pimentel and McClellan [19] described infrared spectroscopy as providing "a definitive criterion for the detection of hydrogen bonds" and "direct evidence for the detection of the role of the proton in association". A distinct advantage for Raman spectroscopy is the low Raman scattering cross-section of silica, hence the capability of measuring directly solid-state samples contained within glass vessels, and the opportunity to use sampling probes coupled with optical fibers over several meters remote from the spectrometer [20, 21]. Many recent measurements made on solid-state pharmaceutical substance have employed an excitation laser lasing in the far-red of the visible region or the near-infrared region. This is to circumvent fluorescence emission which, when present, can swamp Raman emissions as it is much more intense, making it sometimes impossible to observe the Raman spectrum with a shorter wavelength visible excitation laser. There is, however, a loss of Raman intensity with increasing excitation laser wavelength, since the Raman signal varies with the fourth power of the excitation frequency. A common far-red laser in use today is a diode laser operating at 785 nm; commercial FT-Raman systems employ a Nd:YAG laser emitting in the near-infrared at 1064 nm. While one can sometimes also circumvent fluorescence by going to shorter wavelengths and use a UV-laser, there is a high likelihood that one will excite a resonance Raman spectrum [20] from an aromatic pharmaceutical. Such a spectrum will show preferential enhancement of some Raman bands and one will observe very significantly enhanced intensity bands that originate from the chromophore within the molecule giving rise to the UV electronic absorption. (Lasers operating in the UV region are also more expensive, and CCD detectors are not as sensitive [20].)

Neither infrared nor Raman spectroscopy, as normally practised, is an absolute technique, yet each, particularly infrared, after calibration or correlation and appropriate sampling can provide reproducible, precise quantitative methods.

5.2.2 Instrumentation Brief

While many variations in optical components and geometry exist in commercial instruments, the FT instrument *modus operandi* is usually described with reference to the basic Michelson interferometer. A schematic of this is shown in Fig. 5.2 (a).

Radiation from the source is essentially equally split by the beam-splitter; one beam is focused onto and returned by a fixed position mirror, the other beam is focused onto and returned by a moving mirror. On recombination, these two coherent beams interfere systematically according to their path-length difference. When their path-length difference is equivalent to one-half of one wavelength, then they interfere destructively; when their path-length difference is equivalent to an integral number of a wavelength they interfere constructively. For a monochromatic source, the resultant interferogram is a cosine wave; for a polychromatic source, the resultant interferogram has the form shown in Fig. 5.2 (b). A fast Fourier transform algorithm converts the interferogram, which is in the

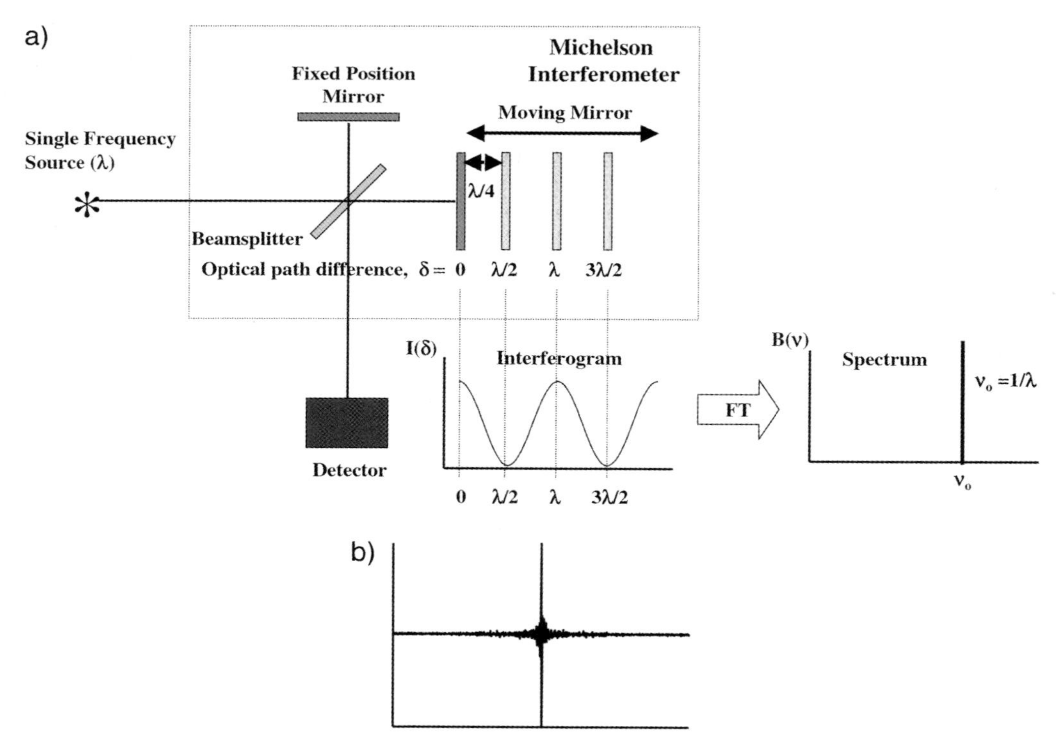

Fig. 5.2 (a) Schematic of a Michelson interferometer and the production of an interferogram from a monochromatic source; (b) example of an interferogram produced from a broad-band mid-infrared source.

FT-Raman measurements offer advantages over dispersive raman spectrometer.

Components of a modern raman spectrometer:

- Laser Source
- dispersive element (holographic grating spectrograph)
- CCD detector

Can combine thermomicroscopic measurement with FTIR microscopy + raman microscopy.

cording the spectrum of a standard calibrated pseudo-blackbody source, such as a tungsten lamp. In addition to its use in many cases for avoiding fluorescence interference, FT-Raman measurements offer advantages over those of a dispersive Raman spectrometer measurement of greater wavenumber precision, constant spectral resolution, and sampling of a greater sample volume [20], although this will be at the expense of reduced sensitivity. The primary components of a modern dispersive Raman spectrometer are a laser source, a dispersing element, such as a Czerny-Turner or holographic grating spectrograph, and a CCD detector (Fig. 5.3). Again, for comparing spectra between instruments, and particularly when comparing spectra recorded using different excitation wavelengths it is imperative that the spectra are corrected for the instrument response profile (Fig. 5.4).

Mid-infrared [24, 25], near-infrared [26, 27] and Raman [25, 28] (dispersive and FT-) spectrometer systems are commercially available combined with optical microscopes, and extensive use is made with such systems for characterizing spectroscopically and chemical imaging spatially resolved components within heterogeneous samples. Following visual inspection, a spectrum may be recorded from an area of a sample of interest; with a dispersive Raman spectrometer this may be as small as ca. 1 μm^2; in the mid-infrared region with a FTIR microscope the lower limit is ca. 10×10 μm. Thus, in some instances it is possible to obtain good quality spectra directly from single crystals. In addition to providing greater lateral spatial resolution, confocal Raman microspectroscopy can also enable some depth-resolved measurements [28]. Relative intensity, principal component or similar contour maps, axonometric plots, false-color images and the like may be recorded or generated from mapping, using single-point or linear-scan measurements, or more directly by imaging, using an array detector. Thermomicroscopy measurements may be combined with both FTIR microscopy [29–31] and Raman microscopy [31–33], such that, for example, an induced polymorphic transition and its consequent vibrational spectral characteristic

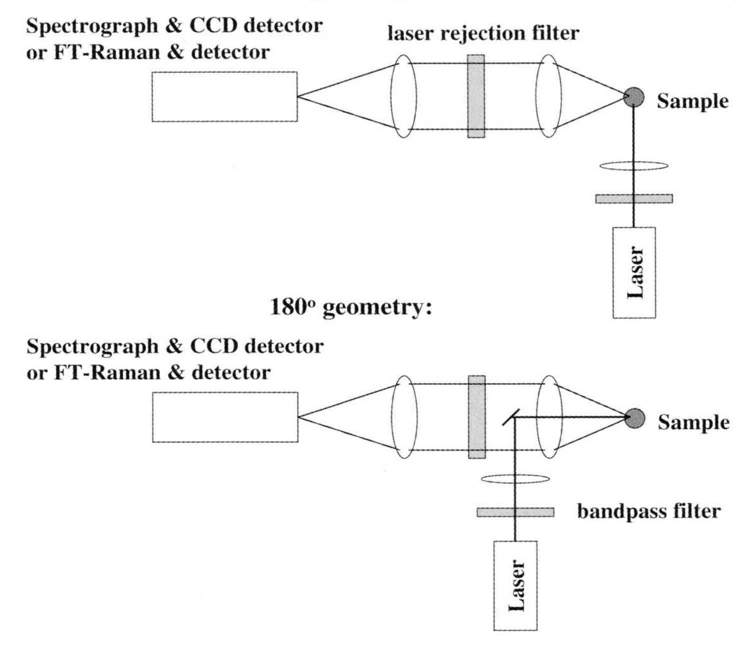

Fig. 5.3 Common Raman scattering and collection geometries.

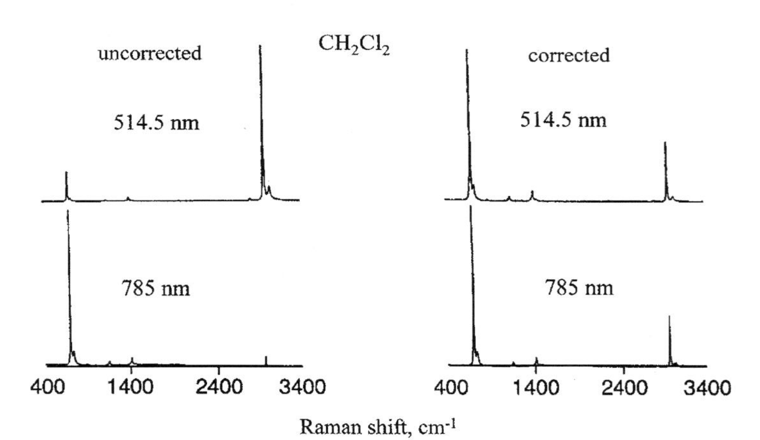

Fig. 5.4 Raman spectra of methylene chloride recorded with 514.5 and 785 nm excitation lasers. The spectra are shown before (as collected) and after correction for the instrument response function obtained with a standard tungsten source. (©2000, Copyright John Wiley & Sons Ltd., reproduced with permission from [20]).

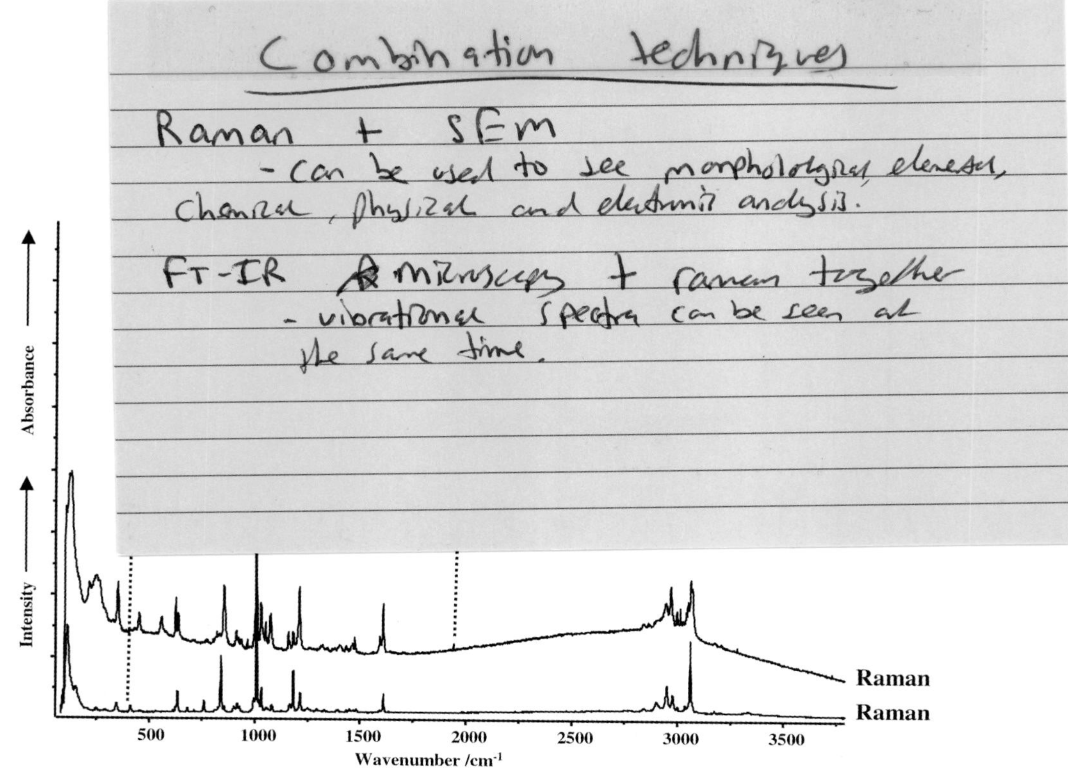

Fig. 5.5 Microprobe Raman and mid-infrared absorbance spectra recorded using a combined FTIR/Raman microscopy system, the Horiba Jobin Yvon LabRamIR^{TM}. The upper spectrum in each pair is that of pseudoephedrine; the lower spectrum in each pair is that of ephedrine. Spectra have been offset for clarity. (The strong band observed near 2300 cm^{-1} in the top spectrum is that of atmospheric CO_2 intrusion, and is not due to the ephedrine molecule). (Reproduced from [37] with permission of Horiba Jobin Yvon, Inc.).

changes may be studied simultaneously. Similarly, both techniques can be coupled with a variable temperature, variable humidity vapor pressure cell, so that one can follow the gain or loss of water molecules within a crystal structure [34]. A recent commercial innovation is a combined Raman spectroscopy–scanning electron microscopy (SEM) set-up [35]. The system allows for morphological, elemental, chemical, physical, and electronic analysis without moving the sample between techniques. Taking advantage of the attenuated total reflection (ATR) infrared sampling technique (Section 5.2.3.2), a combined FTIR microscopy and Raman microscopy instrument has been developed [36], whereby both vibrational spectra can be recorded from the same sample in a single instrument. An example is shown in Fig. 5.5, where the mid-infrared and Raman spectra are shown of ephedrine and pseudoephedrine, two molecules that only differ by the configuration at one chiral centre [37]. Although their chemical structures are very similar, their vibrational spectra are distinct, whether Raman or mid-infrared.

Recent advances in turnkey ultra-short pulsed lasers and semiconductor emitters and receivers have made working in the THz (far-infrared) region more straightforward [7]. An ultra-short pulsed laser generates a stream of near-infrared (800 nm) pulses at 80 MHz, with each pulse typically lasting about 70 fs.

The laser pulses are focused onto an Auston switch (a very fast semiconductor optical device). Electron–hole pairs are generated at the surface of the semiconductor. With a carefully aligned antenna arrangement on the semiconductor, the electron–hole pairs are accelerated by a DC electric field across the device. The resulting effect is to induce emission of short bursts of coherent terahertz radiation with each pulse. The light generated by this technique can then be collected using silicon lenses and utilized for spectroscopy or imaging.

5.2.3 Sampling

5.2.3.1 Raman Sampling [20, 23, 30]

It is easiest to start a section on vibrational spectroscopy sampling techniques with sampling for Raman measurements, since it is scattered radiation that is detected. Consequently, the measurement may be perceived as essentially non-destructive and requires little or no sample preparation, other than perhaps mounting the neat sample on a suitable support or in an appropriate container. Two common, 90° and 180°, scattering/collection geometries are employed (Fig. 5.3). For solid-state samples, typical powder holders are boiling-point/capillary tubes and glass vials, or the powder may be compacted into a suitable cup-shaped holder. When contained within a glass vessel, the incident laser beam is merely focused through the glass onto the solid analyte. This attribute means that it is often a simple matter to examine solid pharmaceutical ingredients or tablets within their containers. The laser may be focused directly or via a fiber-optic probe through such as an amber-colored USP glass vial [20]. Often an adequate Raman spectrum can be recorded from a tablet inside a transparent plastic container such as a blister pack, with minimal interference from Raman bands of the packaging material [38]. Many powders and crystals may simply be spread onto a microscope slide for a Raman microscopy study. In studies related to polymorphism/polymorphic issues, if a material is to be compacted, then one must be sure that this does not introduce any form transition, either full or partial. One must also be certain that the laser power at the sample is sufficiently low so as to not cause thermal damage or form change. This is particularly important for colored materials, which may absorb the radiation and warm up, dehydrate, or even degrade or char; and when using a Raman microscope, where the power density at the focus position could be high. Another key issue, particularly when examining individual crystals or crystalline material, is that of orientation. A laser beam is polarized and preferential enhancement of band intensities will occur if the direction of polarization of the laser is aligned with the direction of the molecular vibration, and band extinction or intensity reduction may be observed if the laser beam polarization is aligned perpendicular to the molecular vibration. Rotating a crystalline sample in the beam can highlight these effects, and for single/small crystal studies using a Raman microscope it is recommended that two spectra be recorded from the same sample that has been rotated through 90°. This effect can be averaged out and the potential for thermal damage/change reduced by using a rotating sample holder, which continuously refreshes the sample at the laser focus position. This clearly requires more readily available sample that it is not unduly fluorescent, and for a FT-Raman measurement that the speed of rotation is low enough (< 50 Hz) to prevent beats being seen across the recorded spectrum.

5.2.3.2 **Mid-infrared Sampling** [30, 39–41]

To record a mid-infrared spectrum appropriate for fingerprinting a solid-state pharmaceutical sample requires that either the mid-infrared radiation passes through and is absorbed by the sample, which has to be of the order of 10 μm thick or less if neat (or equivalent in a dispersion), or is reflected in some way from a sample. Consequently, there are various commonly used mid-infrared sampling techniques. Traditional transmission techniques are dispersions in a suitable mulling agent or as an alkali-halide disc, usually KBr. Sometimes, particularly for FTIR microscopy, a small amount of a solid sample may be thinned by compression in a cell. Reflection techniques include internal reflectance spectroscopy (or attenuated total reflectance, ATR, spectroscopy), specular reflectance, a reflection–absorption technique named transflectance, and diffuse reflectance. Photoacoustic spectroscopy is also an alternative mid-infrared solid sampling technique. Each of these techniques is discussed very briefly in subsequent sub-sections.

Since mechanical stress and heat may invoke a phase transformation, and polymorphs may also be interconverted by solvent-mediated processes, one must be critically aware of the effects any sampling procedure may induce [42]. With the potential for any mid-infrared sample preparation and/or presentation method to modify solid-state form, either due to grinding to reduce particle size or applied pressure, before composing this chapter we conducted a very limited survey of some experienced vibrational spectroscopists within several major pharmaceutical companies. Unsurprisingly, perhaps, there was no consensus and no single mid-infrared sampling technique was favored over all others for solid-state characterization. To some extent, their responses related to their position within the development stage of a drug substance. One spectroscopist working largely within the discovery stage clearly favored solution phase spectra, since these tended to be less complex and much more reproducible, and eliminate crystallinity and solid-state matrix effects. This is in line with the comments of Dickson et al. [2] (Section 5.1) and is a safe option if one is interested in the fundamental molecular structure relating to chemical group functionality. However, for distinguishing between polymorphs, solution spectra are, of course, excluded [43]. Another spectroscopist positioned further along the product chain and concerned with API release and testing favored Nujol mulls, with a definite recent trend towards increased use of the single-bounce ATR technique. For another, their default method was the single-bounce ATR approach, but would also use KBr discs for polymorphism studies, turning to Nujol mulls, if necessary, while one spectroscopist stated a preference for photoacoustic over diffuse reflectance. From this survey, one could conclude that for polymorphism studies photoacoustic FTIR spectroscopy is, however, comparatively rarely used. While some also increasingly use mid-infrared diffuse reflectance, some reservations were expressed about it as a preferred sampling technique because of a high potential for non-reproducibility of spectra, but the experimental sections of recent publications indicate that it has become a fairly common and, in appropriate circumstances, a satisfactorily used sampling technique. An attraction of the diffuse reflectance technique is that if the particle size of the analyte is appropriately small there is less likelihood of inducing crystal form transformation during sample preparation, compared with transmission techniques. In general, the single-bounce ATR method, with its simplicity of sampling, is becoming a prevalent sampling method and, along with diffuse reflectance, is supplanting traditional transmission methods of mulls and alkali-halide discs as a foremost sampling method; all, of those surveyed, of course, used Raman and FT-Raman to complement their mid-infrared measurements. In addition, as with any basic study of polymorphism, none simply relied on vibrational spectroscopic data alone; each used a range of techniques, including X-ray diffraction, thermal analysis and solid-state NMR spectroscopy.

Mid-infrared Sampling: Transmission Methods [39–41]

Mull Technique Here particles of a finely powdered solid are uniformly dispersed in an infrared transparent (or semi-transparent) liquid medium. Mineral oil (liquid paraffin, Nujol^{TM}) is the most widely used mulling agent, although the absorption bands due to the hydrocarbon will almost certainly obscure or interfere with details in the C–H stretching (3000–2800 cm^{-1}) and deformation (1500–1340 cm^{-1}) regions. The mull, which should have the consistency and appearance of a translucent paste, is sandwiched into a thin film between a pair of infrared transparent windows. The solid should be as finely ground as possible to minimize spectral artifacts arising from scatter, which superimposes a "sloping background" onto the spectrum, and anomalous dispersion, which leads to absorption bandshape distortion [44]. This is known as the Christiansen effect, and is a consequence of the anomalous dispersion in the refractive index.

Alkali-halide Disc Method Finely powdered dry potassium bromide will coalesce to form a clear disc with high transmission when pressed under high pressure in an evacuated die. In the alkali-halide disc sampling technique an intimate dilute mixture, ca. 200 : 1, of the finely ground solid sample analyte and dry powdered KBr is pressed into a self-supporting disc. As with the mulling technique, the particle size of the sample must be reduced to well below that of the shortest infrared wavelength being used, to reduce scattering loss and minimize absorption band distortions.

The KBr disc technique is a more aggressive sampling preparation than the Nujol mull method, but has advantages since KBr exhibits no absorption bands above 400 cm^{-1} (neglecting any adsorbed water or impurities) and is better adapted to micro preparations. However, generally, the technique is probably more prone to introducing changes in sample polymorphism, hydration state or crystallinity, tending to make it less reproducible and suitable for studying these effects. In some, albeit rare, circumstances ion exchange may occur between the analyte and alkali-halide.

Compression Cells Compression cells, fitted with mid-infrared transparent windows, are commonly used for reducing the thickness of a small amount of a solid sample for a FTIR microscopy investigation. Some cells employ diamond anvils both for compression and transmission windows.

Mid-infrared Sampling: Reflection Methods [39–41]

Internal reflection (attenuated total reflection, ATR) spectroscopy is a widely used sampling method for both qualitative and quantitative analyses of solids by mid-infrared spectroscopy. With the introduction in the 1990s of simple to operate, single-reflection ATR sampling accessories (see, for example, Fig. 5.6) the technique has become increasingly popular as a prime sampling technique for recording the mid-infrared spectra of solid-state pharmaceutical substances. The spectrum can be affected by changes of state if undue pressure is needed to ensure good ATR reflection element contact with powders. The external reflection technique of reflection–absorption (transflectance) at near normal angles of incidence is mainly used for recording spectra from thin absorbing samples deposited on reflective substrates. Useful specular reflection fingerprinting measurements are limited to examining optically thick samples. Diffuse reflection mid-infrared spectroscopy can provide a convenient means of examining many finely powdered or highly scattering solid samples, although spectral contrast and quality can be very influenced by particle size effects and specular reflection effects. Notwithstanding, diffuse reflection is a very important sampling technique, particularly for a wide range of near-infrared (NIR) applications.

Internal Reflection Spectroscopy Early mid-infrared internal reflection accessories were mostly used in the laboratory to study the surface layer characteristics of a continuous flat solid sample, such as a polymer film. However, with the introduction of both FTIR ATR microscopy and, particularly, modern single-reflection ("single-bounce") convenient operation ATR sampling accessories, this restriction on physical form has became largely irrelevant. Such sampling accessories are now commonly used in many analytical laboratories within the pharmaceutical industry.

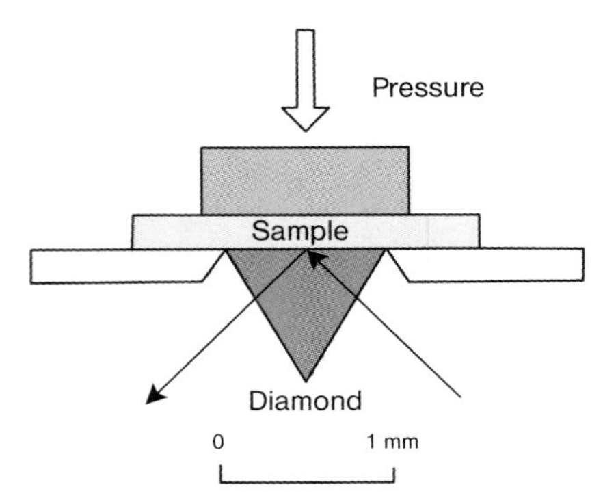

Fig. 5.6 Photograph of a single-bounce Golden Gate™ ATR accessory. The sample is mounted on top of the diamond internal reflection element and held in contact with the element by screwing down the pressure plate. On the right, a schematic of the sampling arrangement is shown together with a picture of the diamond element in its mount. (Reproduced by kind permission of Specac Ltd., Orpington, UK).

In mid-infrared ATR spectroscopy, a solid sample surface is held under low pressure in optical contact with the clean surface of an internal reflection element. The internal reflection element, commonly Type II diamond, Ge or ZnSe, is transparent (or semi-transparent) to mid-infrared radiation and has a higher refractive index than that of the sample. Infrared radiation passing through this prism at an incident angle greater than the critical angle will be totally internally reflected at the reflection element–sample surface boundary. This totally reflected beam will, however, have been attenuated by the mid-infrared absorption characteristics of the sample's surface layer, and hence when detected generates a mid-infrared absorption spectrum characteristic of the sample surface layer. The surface layer thickness interrogated by the beam depends on various parameters, including the refractive index of both the sample and reflection element, the angle of incidence, and the wavelength of the mid-infrared radiation, but will fall in the range of ca. 0.3 to 5 μm over the wavenumber range 4000 to 650 cm^{-1} (ca. wavelength range 2.5 to 15 μm). Owing to the wavelength dependency on the surface sampling depth probed, in an ATR spectrum the relative absorption band intensities will appear relatively more intense with decreasing wavenumber (increasing wavelength) than for a transmission spectrum.

Transflection An analytically useful fingerprint spectrum may be recorded by the near-normal incidence reflection–absorption (transflection) technique from a thin film (>100 nm) coated, or a sample spread, onto a flat reflective metal substrate, e.g., a gold mirror. With FTIR microscopy other common reflective supports are mid-infrared reflective glass slides, e.g., MirrIR^{TM} (Kevley Technologies, OH). In a FTIR microscope operating in the reflection mode the measurement is relatively straightforward. The infrared radiation beam at near normal angle of incidence makes a double-pass through the sample having been reflected back by the substrate. (NB: superimposed on the essentially double-path-length transmission spectrum will be a weaker specular reflectance spectrum that will mostly distort the stronger absorption bands, and in some circumstances may cause band inversion.)

Specular Reflection A pure specular reflection spectrum may be recorded directly from the surface of a flat, non-scattering, optically thick (opaque) sample. An analytically useful absorption index spectrum may be extracted from the recorded specular reflection spectrum by application of the Kramers-Kronig algorithm [39, 40].

Diffuse Reflection Diffuse reflection is a convenient method for recording a mid-infrared FTIR spectrum from a finely powdered sample that is dispersed in an excess of a dry powdered non-absorbing matrix, commonly KCl, and loaded into a sample cup. The KCl ensures that superimposed interferences from specular (front-surface) reflections from the analyte are minimized in the recorded diffuse reflection spectrum. In addition to the optical properties, the spectral contrast and quality of a diffuse reflectance spectrum depend strongly on a sample's physical (e.g., particle size, packing density) properties. In the mid-infrared region it is primarily a qualitative tool, although with care it can be used semi-quantitatively. Diffuse reflection is a very commonly used sampling technique with near-infrared (NIR) spectroscopy.

Photoacoustic Spectroscopy Mid-infrared spectra from solids having a wide range of physical forms may be recorded by photoacoustic FTIR spectroscopy, with minimal sample preparation. Essentially, the only requirement is that the sample be made to fit into the photoacoustic cell sample holder, although sample form will affect spectral contrast and intensity.

5.2.3.3 THz Spectroscopy Sample Presentation

High-density polyethylene (PE) and polytetrafluoroethylene (PTFE) powders will coalesce under pressure and can be formed into self-supporting discs in a manner analogous to a 13-mm diameter KBr disc used for a mid-infrared transmission measurement. Both PE and PTFE may be considered as essentially transparent in the THz region and, consequently, they are used as matrices for powder sample presentation for transmission measurements in this region. Because the relative intensity of bands in this region is generally much lower than in the mid-infrared region, discs contain a much higher proportion of analyte than for a typical KBr disc. A typical polymer-matrix disc for a transmission THz spectroscopy measurement may contain ca. 20% wt/wt of the drug/product. Another advantage is that, generally, there may often be no need to grind a sample since, because of the wavelengths of radiation used to interrogate a sample, the requirement for a fine particle size powder is much less.

5.3 Changes of State and Solid-state Effects on Infrared and Raman Spectra

5.3.1 Introduction

While this book focuses on polymorphism, and hence on solid-state issues, an appreciation of the changes occurring as one goes from the vapor state through the condensed states helps in building a picture of the effects and variations on band shapes and positions that occur in solid-state forms.

Many effects, both chemical and physical, can influence the behavior of materials and alter their vibrational spectra. In many basic interpretation tutorials of vibrational spectra that use simple molecules as examples there is an unstated assumption that the spectra are of pure materials in which the molecules are unaffected by their physical environment. However, the very name "vibrational" indicates that the spectra are very sensitive to the molecular environment and conformation. Whether the molecules are far apart (vapors), have limited freedom (liquids, solutions, some polymers) or are very restricted (solids, crystals) the spectra will be affected by the physical state. This must be considered when performing sample preparation. The most noticeable differences in spectra due to a change of state are between those in the vapor state, where rotational structure may be seen, in concentrated solutions, where bands may be broader and so less well resolved than in dilute solution, in polar solvents, where bands may broaden compared with non-polar solvents and cause a shift to lower frequency, and in the solid state, where either band broadening or band narrowing may be encountered, accompanied by small band shifts, often to lower wavenumber. APIs can be complex molecules and many solid-state effects can influence their spectra significantly. Here we use some relatively simple molecules to demonstrate these effects more clearly; more examples are discussed in Section 5.4.

5.3.2 Spectra of Gases, Liquids and Solutions

In the gas phase, molecules may be thought of as freely rotating, giving rise to rotational fine structure within the spectrum. (This can often be seen with intrusive atmospheric water vapor bands, from a poorly purged spectrometer, superimposed on a solid-state mid-infrared spectrum of a sample.) The number of molecular collisions increases with increasing pressure, such that the fine structure is lost, and only the contours of the rotational-vibrational bands are seen. In the liquid phase, the molecules press against either each other or the solvent molecules, preventing free rotation. The contour reduces to a single Lorentzian-shaped band. Whilst there may be some rotational and conformational mobility, the bands are generally much broader. Only in extreme cases will conformational effects lead to gross changes in the spectrum. In solids, molecular motion generally will be suppressed, the contour narrows, intensifies and becomes more Gaussian in shape [45]. In the ordered or crystalline state further band sharpening occurs, which in some cases may also be accompanied by band splitting.

Three-dimensional structures can also influence a vibrational spectrum. As a consequence of different packing, each polymorphic form will have a unique vibrational spectrum; however, the distinctions between spectra may be very slight, involving only small variations in some peak positions, shape and/or intensity [43, 46]. Clearly, it is important that one can distinguish between real relevant differences from those introduced by any sample preparation or presentation method!

5.3.3
Hydrogen Bonding

Infrared spectroscopy provides much information on hydrogen bonding. The –OH and –NH groups are those most commonly involved in intermolecular hydrogen bonding. Intermolecular hydrogen bonds are most clearly recognized; less easy is the intramolecular bond. In the condensed phase it is difficult to distinguish between inter- and intramolecular hydrogen bonding (Fig. 5.7). However, in weak solutions of inert solvents such as chloroethane the molecules may be separated far enough that intermolecular effects can be eliminated.

Electronic and steric factors strongly influence the formation of an intermolecular bond. Sterically hindered phenols, such as 2,6-di-t-butyl phenol, show a strong relatively much sharper free (non-H-bonded) OH band at ca. 3600 cm^{-1} in the solid state due to the inability of the OH group to physically bond with another molecule. The six-membered ring is a highly favored intramolecular bonded form, as shown, for example, by the simple molecule methyl salicylate. Its spectrum (Fig. 5.8) exhibits strong effects on both the OH and C=O stretching vibration bands, especially in the neat state. In the absence of hydrogen bonding, the OH stretching band would be expected to be centered in the range 3400–3300 cm^{-1} and the carbonyl band would be at ca. 1710 cm^{-1}. In the mid-infrared spectrum the effects of hydrogen bonding can be seen quite clearly in the significant shifts to lower wavenumber of both bands. The intramolecular hydrogen-bonded band is broader and weaker than the free OH to such an extent that in some compounds, such as hydroxy azobenzenes, it is not detectable in the infrared spectrum.

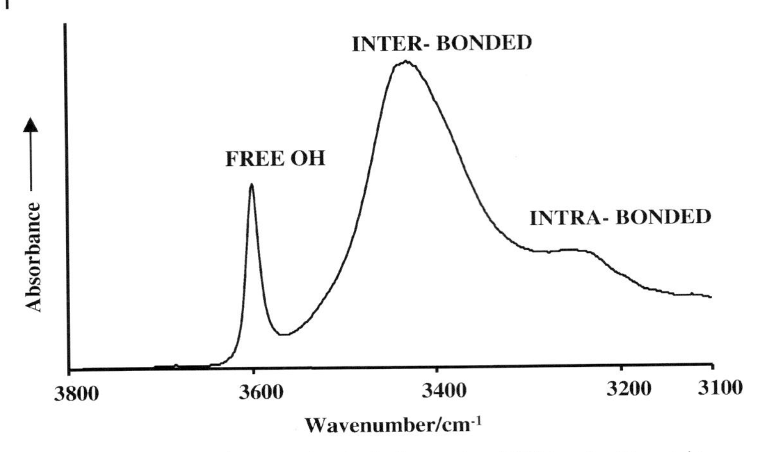

Fig. 5.7 Free, inter- and intramolecular hydrogen bond OH bands. (Copyright ©1997 the Royal Society of Chemistry, reproduced with permission from [40]).

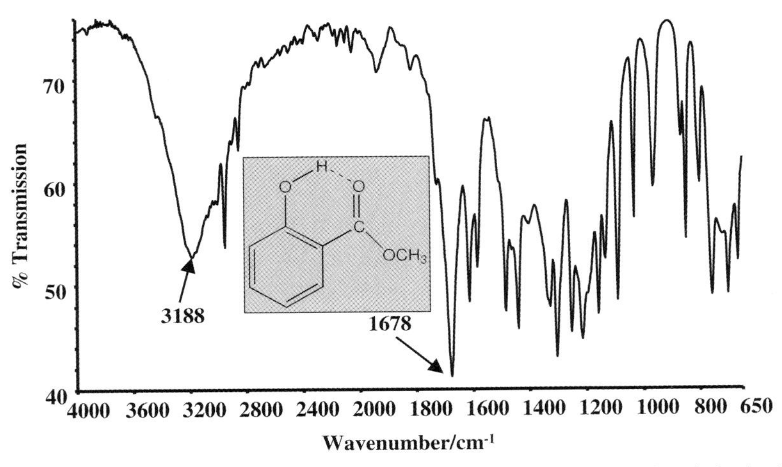

Fig. 5.8 Transmission mid-infrared spectrum of a capillary liquid film of methyl salicylate.

Free OH occurs at approximately 3500 cm^{-1} or above for most compounds. As a molecule forms a bond to become a dimer, a weaker broader band appears at lower wavenumber. An increase in concentration, in solution, causes this band to move to lower wavenumber, as trimerization and eventually polymerization through hydrogen bonds builds up. Many neat liquids and solids show strong intermolecular hydrogen bonding. Moving the molecules apart by successive dilution to weaker solutions usually, eventually, produces spectra showing the free OH or NH bands. Two exceptions are carboxylic acids and sterically hindered compounds. Carboxylic acids form very strong intermolecular hydrogen bonds. Many form strong dimers. A band at 960 cm^{-1} due to the dimer of-

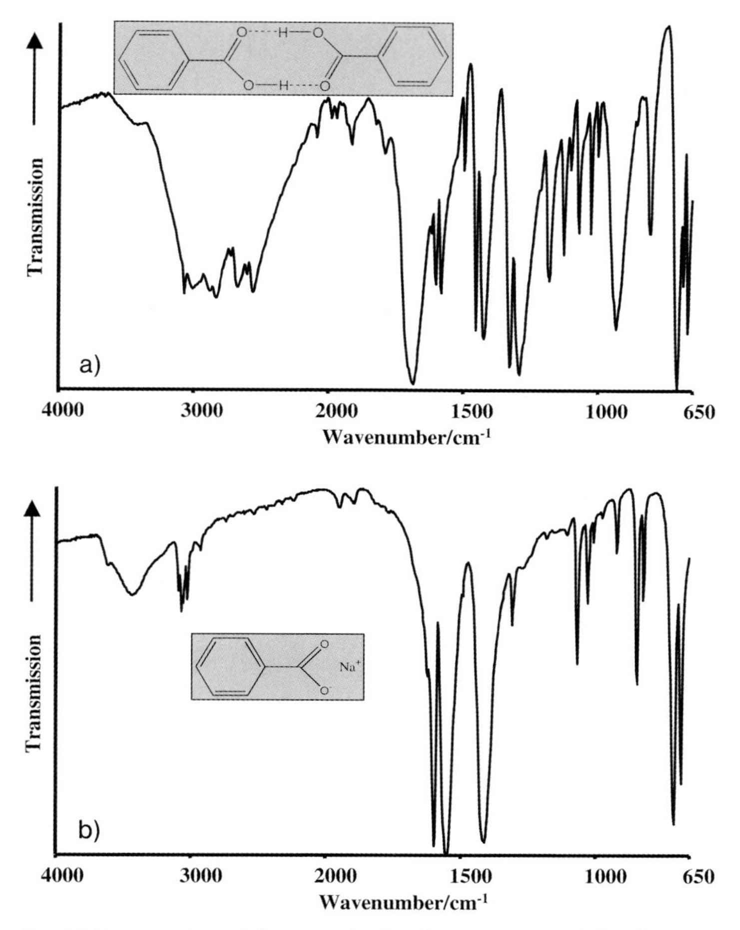

Fig. 5.9 Transmission solid-state mid-infrared spectra as KBr disks of (a) a carboxylic acid (benzoic acid) in the dimer form and (b) sodium benzoate.

ten appears in the mid-infrared spectrum. (Some of the strongest hydrogen bonds are formed in carboxylic acids; even in the vapor phase many carboxylic acids still occur in the dimer state.)

Carboxylic acids are often insoluble in aqueous systems, which can cause dissolution difficulties. In many APIs the free acid is converted into the salt form. The effect on the bands in the mid-infrared spectrum is dramatic. For example, in the benzoic acid spectrum shown in Fig. 5.9 (a) the carboxylic acid exhibits broad bands due to hydrogen bonding in the 3500 to 2000 cm^{-1} (νOH stretching) region. The strong carbonyl band occurs at 1710 cm^{-1}; the band at 960 cm^{-1} is associated with the H-bonded dimer ring. In forming the salt the hydrogen bonds disappear together with the relevant bands. The carbonyl moves to significantly lower wavenumber and forms a doublet with bands at ~1650 to 1530 and

~1400 cm^{-1} (Fig. 5.9b). The position of these bands depends on the acid and the size of the metal ion forming the salt.

5.3.4
Amine Salts (including Amino Acids)

Perhaps the ultimate in strong bonding effects on a spectrum is the hydrogen bonds formed by the creation of amine salts. Here, rather than specific bands due to $-NH_2$ or >NH groups, strong, broad bands occur due to the $-NH^+$ group. These bands occur much lower than the ~3400 cm^{-1} region where primary and secondary amine group bands appear. As an example, Fig. 5.10 compares the mid-infrared spectra of aniline with aniline hydrochloride. The amine salt bands spread over the range of ~3300 to 2000 cm^{-1}. In Fig. 5.10 (b), the two strong bands in the 2800 to 2300 cm^{-1} region are typical of aromatic amine salts, whilst that at ~2000 cm^{-1} is usually stronger in the spectra of aliphatic compounds. Bands indicative of primary amine salts and secondary amine salts appear at ~1550 cm^{-1}, which can be used to determine the type of amino group to which the salt is attached. However, this effect is usually only seen in relatively simple molecules and not at all in aromatics as the amino bands merge with the ring bands. At first glance, the broad bands in the 3500 to 2000 cm^{-1} region can easily be mistaken for a carboxylic acid; however, for an acid a band at ~1700 cm^{-1} is present (cf. Fig. 5.9 a).

In amino acid molecules both amine groups and carboxylic acid groups occur. If the groups are well separated in a molecule there will be no interaction and the bands due to both groups will be clear and distinct. However, if interaction

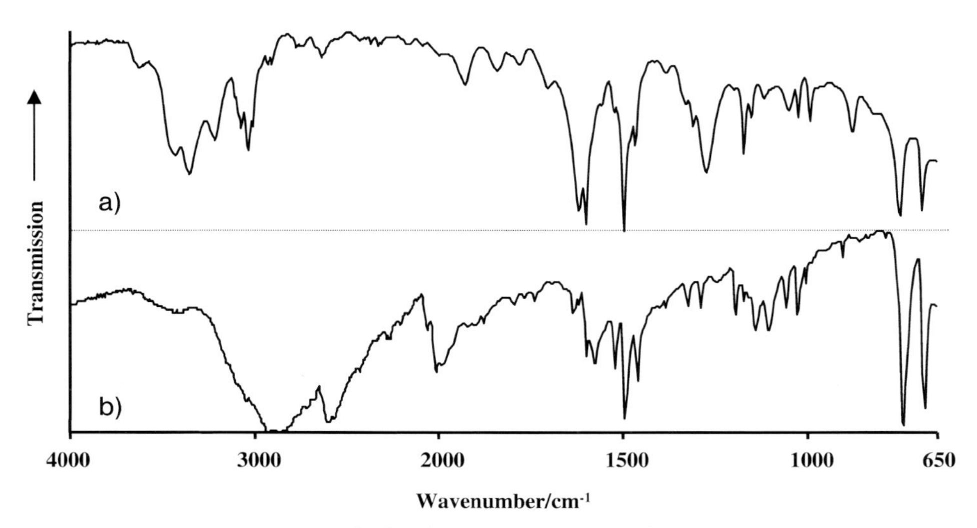

Fig. 5.10 Transmission mid-infrared transmission spectra of (a) aniline (liquid) and (b) aniline hydrochloride (solid, KBr disk).

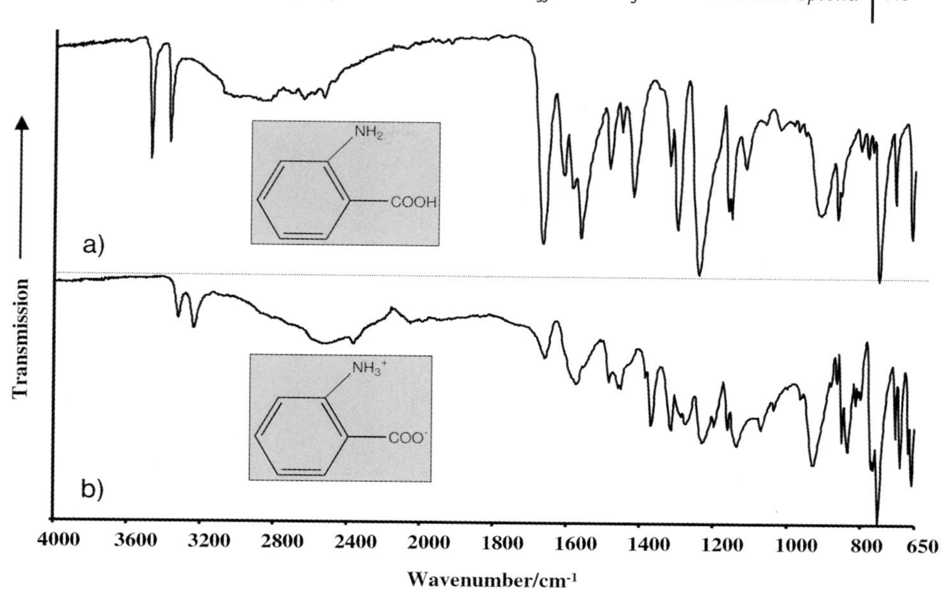

Fig. 5.11 ATR mid-infrared solid-state transmission spectra of solid anthranilic acid: (a) showing bands characteristic of free NH and carboxylic acid and (b) the zwitterion salt form.

occurs, both salts are formed. This may be due to intermolecular bonding or intramolecular bonding. In the latter case, the amine forms an internal salt with the acid group. This salt is usually referred to as a zwitterion. A molecule that shows these effects quite clearly is anthranilic acid (σ-aminobenzoic acid). In the solid state the acid exists in more than one form. In Fig. 5.11 (a) the mid-infrared spectra shows the form in which the carboxylic acid has formed the dimer leaving the amine groups free and unbonded. In another form, the zwitterion (Fig. 5.11 b), strong bonding occurs and the spectrum shows characteristic salt bands; also evident in the spectrum are some "free" NH bands but these are shifted due to intermolecular bonding.

5.3.5 Solids

In solids a potential hydrogen bond, occurring in the liquid phase, may be weakened or even excluded by the geometric constraints of the crystal lattice. The wavenumber position of such a band may rise. The most usual cause of band multiplication in the solid state is the presence of more than one independent molecule in the unit cell, each with its own combination of intermolecular and intramolecular bonding. In other cases co-operative motions between groups on adjacent molecules may afford coupled bands and hence band doubling in which one high wavenumber and one lower wavenumber band may appear. Another co-operative lattice motion is when all the molecules in the lattice move in unison. This gives rise to the very low frequency lattice or "phonon" modes, which can only be conveniently accessed by Raman spectroscopy or THz spectroscopy.

Specific solvation effects, which in the liquid state are usually no more than casual liaisons between the molecules, constantly being formed and broken, can lead in the solid state to a solvate in which molecules of the solvent become entrapped in the lattice, usually in a totally regular manner. If a molecule of acetone becomes trapped in the lattice, for example, it ought to be easily recognizable as such. Solvates do not have to be stoichiometric. Other solvates should also be visible in the spectrum, but often the bands from a molecule of water become lost in the bands due to strong OH or NH stretch hydrogen bonds in the sort of compounds in which hydration is common. It is not an absolute rule that hydrates only form from compounds with potential hydrogen bonding groups: the only rule is that a lattice should exist with a large enough hole to host a molecule of solvate. Therefore, hydrates are more common than other solvates. However, there is a greater chance of a stable hydrate forming if there are specific groups in a molecule with which the water can bond. Solvates are often mistakenly referred to as polymorphs [43, 46]. Figure 5.12 shows the mid-infrared spectrum of two hydrates, one organic and one inorganic.

The spectra of solids can therefore be more complicated than those of liquids or solutions. One further complication is the existence of polymorphism, which

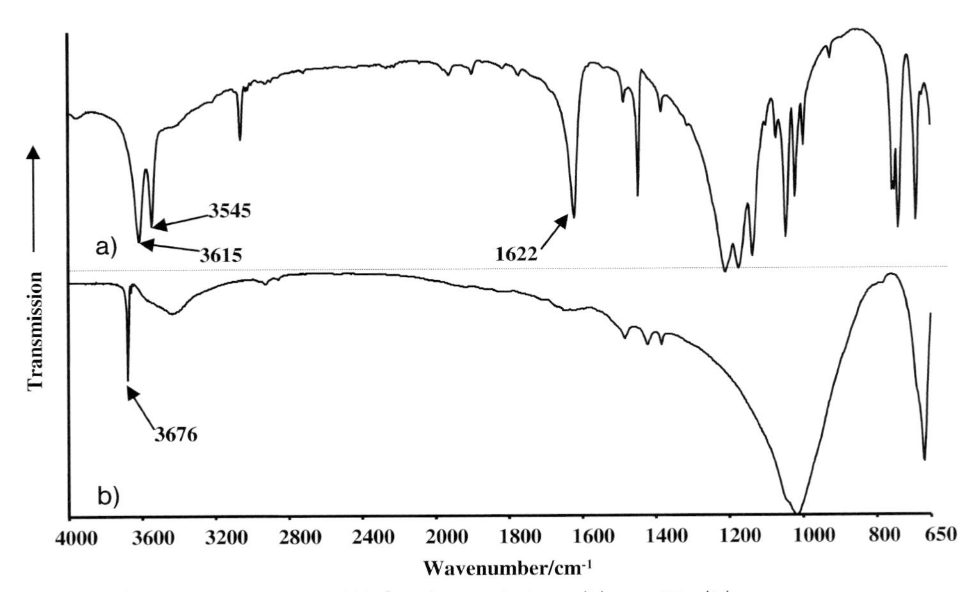

Fig. 5.12 Transmission mid-infrared transmission solid-state KBr disk spectra showing (a) hydrate water in an organic molecule and (b) hydrate water in an inorganic molecule.

is caused by different packing of the same molecule. In general, since only the packing varies, only small differences in the spectra might be expected. However, any spectral changes will likely reduce the chance of identification when comparing spectra with those in a spectral library, although good search programs should have been written so as to minimize the chance of missing spectra through small frequency and intensity changes. Sometimes, very considerable differences are seen between polymorphs, due to very different intermolecular bonding arrangements, a change in conformation of the molecule or a change in tautomer or ionization.

5.3.6 Polymorphism

The definition of polymorphism is a subject of much discussion in its own right [43]. For our purposes McCrone's definition [47] of a polymorph will suffice: "a solid crystalline phase of a given compound resulting from at least two crystalline arrangements of the molecules of that compound in the solid state". This excludes solvates and amorphous forms. There is discussion of multiple amorphous forms referred to as apolymorphs but we will restrict ourselves to the usual definition.

In the pharmaceutical industry the polymorphic form of a drug is very relevant as it affects the solubility and potentially the bioavailability of the drug. Much effort is demanded from regulatory authorities and patent attorneys to determine whether polymorphs exist. Apart from solubility, other manufacturing and tablet problems can arise. Products containing natural fats, waxes, soaps, sugars and polysaccharides can all change behavior when polymorphism occurs. Many spectroscopic and thermal techniques have been employed to study polymorphism. Vibrational spectra that can distinguish between two forms and quantitatively assess mixtures are vital tools.

As some polymorphic forms can be interconverted easily by physical means, the sample preparation for any analytical technique is very important. Forms can convert in different solvents, by grinding the solid when preparing as a halide disk or under pressure in an ATR study or in a compression cell. Mulling techniques are generally better than halide disk preparations, being gentler treatments. Diffuse reflection and sometimes photoacoustic techniques are employed for polymorphism examination, since they require minimal sample preparation, but for both procedures, as with mulls and alkali-halide discs sample preparations, particle size can significantly affect reproducibility, which is an important consideration for both qualitative and quantitative measurements. Raman spectroscopy, with minimal or no sample handling, is a favored technique, but can be limited in application with visible lasers due to fluorescence effects, even in colorless relatively pure pharmaceutical powders.

5.3.7
Enantiomers and Racemates

The bioavailability of a pharmaceutical drug substance may also be affected by the crystal form of an optical isomer. There are two principal types of crystalline racemates [48, 49]. Firstly conglomerates, which are a simple juxtaposition of crystals of two enantiomers (optical isomers). They are mixtures separable by physical means. Secondly, there are the much more common racemic compounds, whose crystals contain two enantiomers in equal numbers within the unit cell. The mid-infrared (and Raman) spectra of enantiomers in conglomerates are superimposable, since the arrangement of neighboring atoms in a pair of enantiomers is the same. The spectra are also indistinguishable from that of the conglomerate. While 90% of racemates occur as a racemic compound rather than a conglomerate [50], the production of pure enantiomers of drugs is of growing importance in the pharmaceutical industry. In racemic compounds the spectrum will differ significantly from that of the two enantiomers. A much-studied [51, 52] example is mandelic acid.

A cursory look at the mid-infrared spectra of Fig. 5.13 shows that the spectra of mandelic acid enantiomers are virtually identical (which they are) and differ little from the spectrum of the racemic compound. However, closer inspection shows that the OH and carbonyl bands are slightly shifted in the spectrum of the racemic compound and that there are significant differences in the 1500 to 1100 cm^{-1} fingerprint region, which reflect changes in the crystal structure. Indeed the crystal structure shows that the common carboxylic acid dimer is not

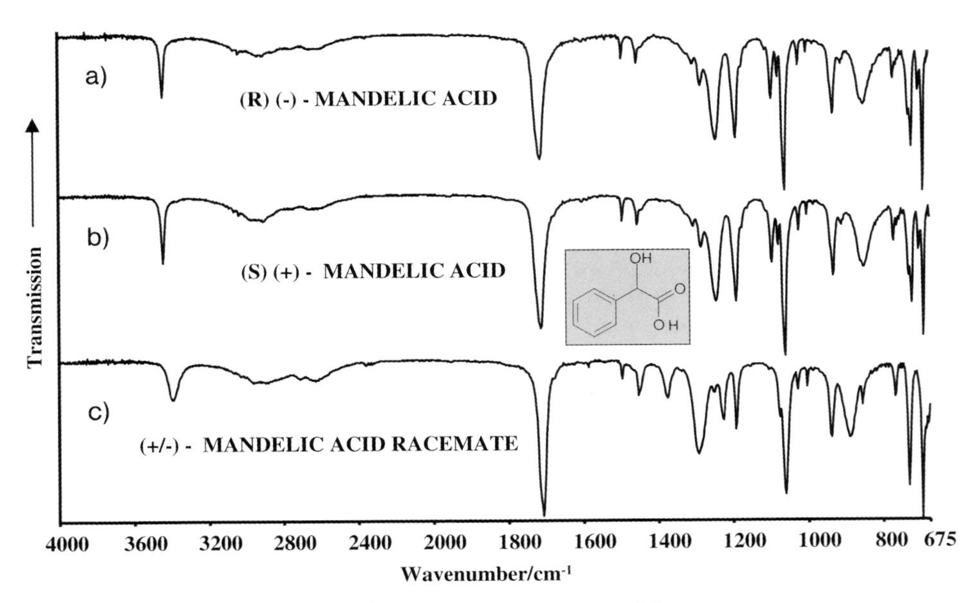

Fig. 5.13 ATR mid-infrared solid-state transmission spectra of the enantiomers (a) and (b), and racemate (c) of mandelic acid.

formed; this accounts for the shift in the broad band in the 900 cm^{-1} region. Consideration, therefore, must also be given to the appropriate use of other techniques, such as X-ray diffraction, to check whether racemates have been formed – the spectral differences observed in the fingerprint region could be interpreted as being due to polymorphism, or other crystal effects.

5.3.8 Tautomerism

The effects described so far have been largely due to molecular interaction and changes in physical state. However, structural changes can take place within the molecule, which also affect the spectra. The study of reactions by following the appearance or disappearance of a particular band associated with a specific chemical group is quite common. Rates of reaction and kinetic studies are also of particular importance in pharmaceutical plant design, and in process development. Another equilibrium state that can also be studied by vibrational spectroscopy is tautomerism. NMR is more sensitive for quantitative studies, but vibrational spectroscopy can easily and sensitively study the solid state [40].

The most frequently encountered tautomer is the keto-enol transform. The keto form is characterized readily in the mid-infrared by a strong >C=O group absorption band, usually in the region 1780 to 1700 cm^{-1}, while a broad –OH band in the region 3500 to 3200 cm^{-1} distinguishes the enol form.

5.3.9 Summary

Vibrational spectroscopy is most commonly thought of as a tool for identifying chemical groups or identifying materials. The fact that the whole molecule vibrates makes the spectrum sensitive to both the physical and chemical environment. As we have discussed, these factors need to be considered before attempting to interpret the spectra.

5.4 Examples and Applications

There are many examples in the patent literature of mid-infrared spectra being provided as part of a definitive portfolio of analytical data for a particular polymorphic form of a drug substance. Data from both vibrational spectroscopy techniques are also commonly debated and presented as evidence within litigation between manufacturers of pharmaceutical products. In addition, of course, mid-infrared spectra are used as one of several physicochemical measurements for supplying supporting documentation in new drug manufacturing submissions to Regulatory authorities. It would be easy to fill this section with comparative spectra showing how selected polymorphs of a particular molecule may be distinguished readily by their vibrational spectra. A few will be given, but we feel it much more useful to consider some experimental precautions that must be made, and how perhaps more can be extracted from spectra than just a simple form fingerprint. The application examples cited in this section will be mostly from relatively recent studies; more extensive references to studies of polymorphism using vibrational spectroscopy may be found in [43] and [46].

5.4.1 Polymorphism

As mentioned elsewhere in this book, the different molecular packing associated with different polymorphs of a drug substance can give rise to different physical properties that cause different behavior, such as dissolution rate, and consequently have a significant effect on a drug substance's bioavailability and its effective clinical performance. The study, characterization and understanding of the solid form (polymorphs, solvates and hydrates) are therefore of major importance within the pharmaceutical industry. Mid-infrared and Raman spectroscopy are both key tools in these areas, offering not only form fingerprinting but in some circumstances also insights into the crystalline arrangements of molecules.

A detailed investigation into different polymorphic modifications of the anticonvulsant drug carbamazepine (CBZ) was reported in 2000 [29]. A combination of techniques was used, including FTIR, hot-stage FTIR thermomicroscopy, XRPD, DSC and hot-stage microscopy. For the FTIR study, samples were examined both in transmission, as KBr discs, and by diffuse reflection. In the latter approach, the samples were ground gently with KBr and examined directly as a fine-powder mix, thereby avoiding extended mechanical grinding of the sample and any pressure necessary to form a disc. In this particular instance, however, no discernible differences were reported between corresponding spectra from the two mid-infrared sample preparation techniques. The "Experimental" section [29] stated that "the samples were ground gently with KBr and analyzed directly in the diffuse reflection mode, thus mechanically avoiding polymorphic transitions induced by extended grinding". More aggressive grinding might have led to an induced transformation, e.g., differences have been noted relating to a phase change between transmission spectra recorded from samples prepared as mulls when using preparations from more gentle hand-ground (agate pestle and mortar) and vibrator-ground (metal capsule with a steel ball) samples of Amphotericin B [53]. Here, the more aggressive vibrator-grinding promoted a transition from a crystalline to an amorphous phase. The FTIR spectra of Forms I, II and III of anhydrous CBZ are easily differentiated [29] (Fig. 5.14). Substantial differences are seen between the commercial form, Form III, and Form I. Of particular note is the lower position of the –NH stretching vibration band, appearing at 3464 cm^{-1} in Form III and 3473 cm^{-1} in Form II, indicating a stronger intermolecular hydrogen bond in Form III. Enantiotropic transformations between Forms III and I and between Forms II and I at elevated temperatures were also studied by hot-stage FTIR thermomicroscopy. In Form I, the

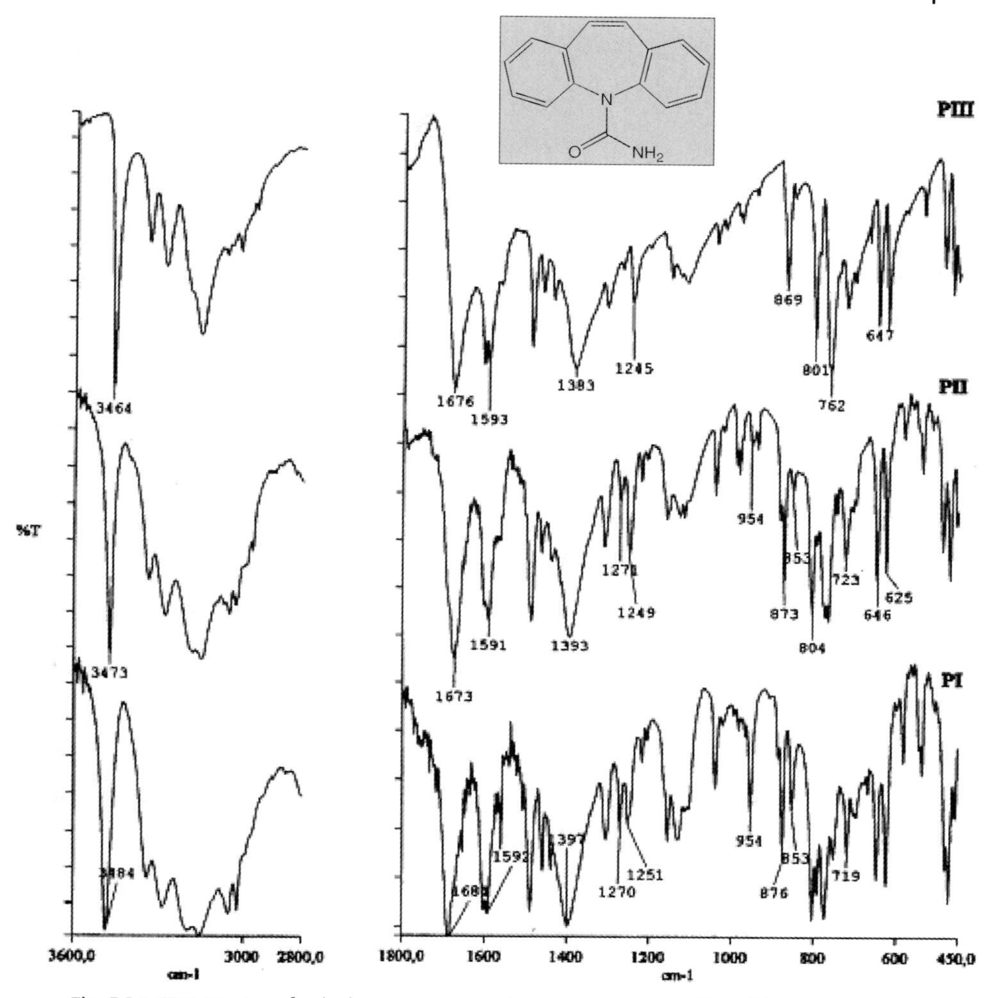

Fig. 5.14 FTIR spectra of anhydrous carbamazepine polymorphs; top: Form III; middle: Form II; bottom: Form I. (Reproduced from [29], Copyright ©2000, with permission from Elsevier).

–NH stretching vibration occurs at 3484 cm^{-1}, which in this case is in line with Burger's infrared rule as this, having the highest wavenumber band, i.e., the least strongly bonded –NH group, is the least stable form (strictly at zero degrees) [46]. A more recent study of CBZ reported the mid-infrared spectrum of a fourth anhydrous polymorph, Form IV [54]. FT-Raman spectroscopy has also been used to study the kinetics of the solid-state transformation between CBZ Form III and Form I [55]; spectra were collected *in situ* during isothermal heating from samples contained within an environmental chamber. Szelagiewicz et al. [32] used a dispersive Raman spectrometer in their thermomicrospec-

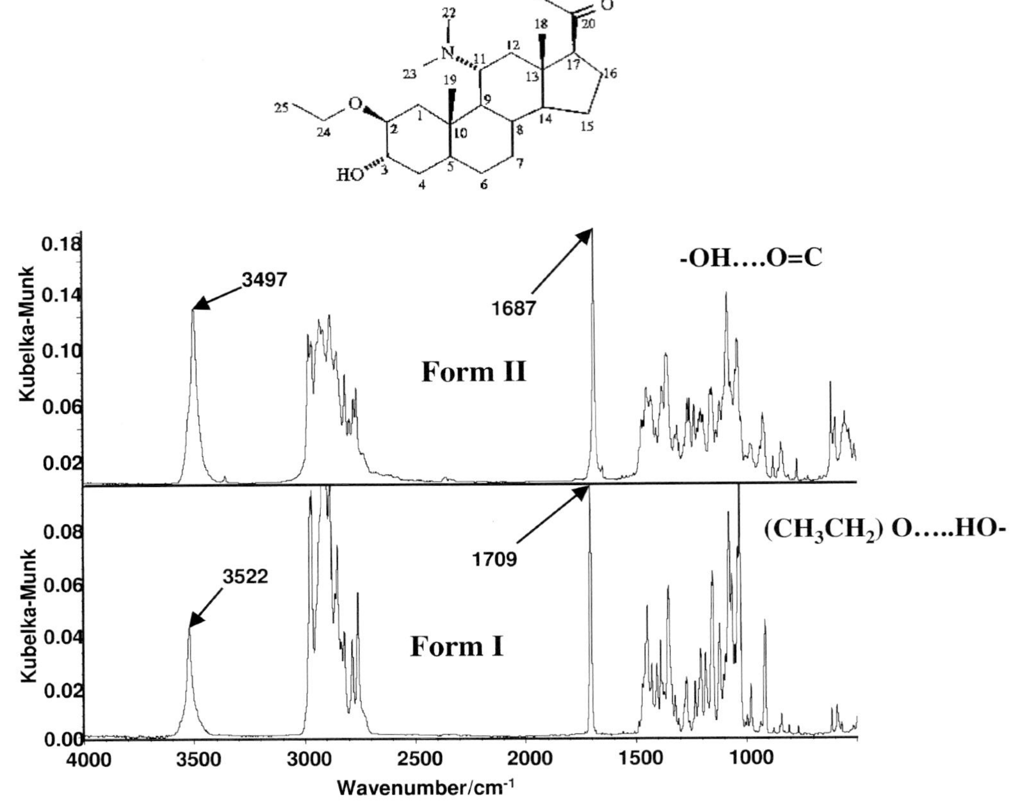

Fig. 5.15 Structure and mid-infrared diffuse reflection spectra of Minaxolone: Form I (lower) and Form II (upper). (Reproduced from [57], Copyright ©1998, with permission from Elsevier).

troscopy *in situ* studies for characterizing polymorphic forms of the drug substances paracetamol and the racemate lufenuron, while selecting to use FT-Raman to identify the polymorphic form in tablets. Both temperature-controlled measurements by FTIR microscopy and FT-Raman microscopy were employed, along with solid-state NMR and XRPD, in a multi-technique study of polymorphism in tedisamil dihydrochloride [33]. Mid-infrared spectroscopic examination of KBr discs of mefenamic acid have been used to measure the relative rate of conversion from polymorph I into polymorph II [56].

Diffuse reflection mid-infrared FTIR spectroscopy and FT-Raman spectroscopy, together with DSC, solid-state ^{13}C and ^{15}N NMR, have been used to investigate the steroidal anesthetic Minaxolone (Fig. 5.15) [57]. From X-ray crystallography, it was known that Minaxolone exists in two polymorphic forms, with the crystallographic asymmetric unit in each case being a single molecule, and that there was hydrogen bonding involving the hydroxyl group. The hydroxyl may hydrogen bond to one of two acceptors, either the C=O at C#20 or the oxygen that forms part of the ethoxy group linked to C#2. In the absence of hydrogen bonding, the keto group would be expected to give rise to a C=O stretching vibration at ca. 1710–1720 cm^{-1}. In the mid-infrared spectrum (see Fig. 5.15) of Minaxolone Form II the C=O stretching vibration occurs at 1687 cm^{-1}, whereas for Form I it is observed at 1709 cm^{-1}; in the Raman spectrum (Fig. 5.16) these bands occur at 1686 and 1702 cm^{-1}, respectively. The –OH stretching vibration wavenumber observed in the mid-infrared spectrum is lowered in both forms, occurring at 3522 and 3497 cm^{-1} for Forms I and II, respectively; this band might have been expected at ca. 3620 cm^{-1} in the absence of hydrogen bonding. The stronger hydrogen bonding observed for Form II is consistent with the "head-to-tail" linkages forming a chain structure involving hydrogen bonding between the –C=O and –OH groups of adjacent molecules. Form I, which involves hydrogen bonding between the –OH and oxygen of the ethoxy group, exists as a "head-to-head" ladder-like structure. In the Raman spectrum of Form II, the carbonyl stretching vibration band is relatively much stronger, indicating it results from a relatively higher polarizability as a consequence of the strong hydrogen bonding.

Nujol mull mid-infrared spectra comprised part of a detailed multi-technique study of two modifications of a novel anti-viral agent, Lamivudine [58], one of which, Form I, had an asymmetric cell containing five molecules, while the other, Form II, had only one molecule in the crystallographic asymmetric unit. Although different, the magnitude of the differences observed between the two solid-state mid-infrared spectra did not seemingly adequately reflect the signifi-

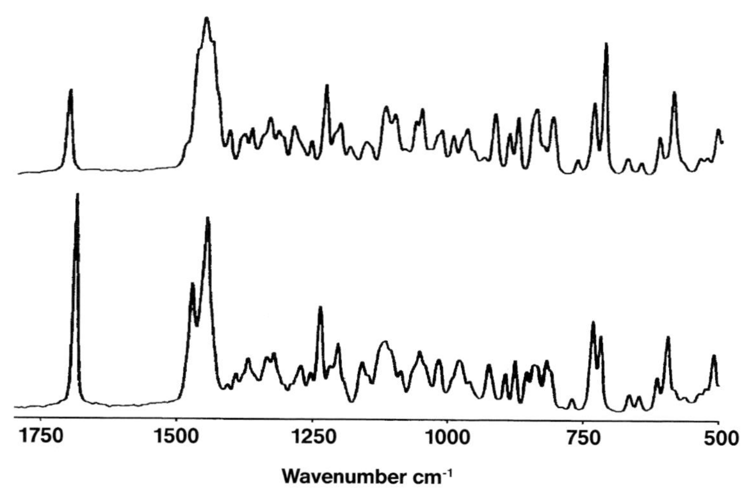

Fig. 5.16 FT-Raman spectra of Minaxolone: Form I (upper) and Form II (lower). (Reproduced from [57], Copyright ©1998, with permission from Elsevier).

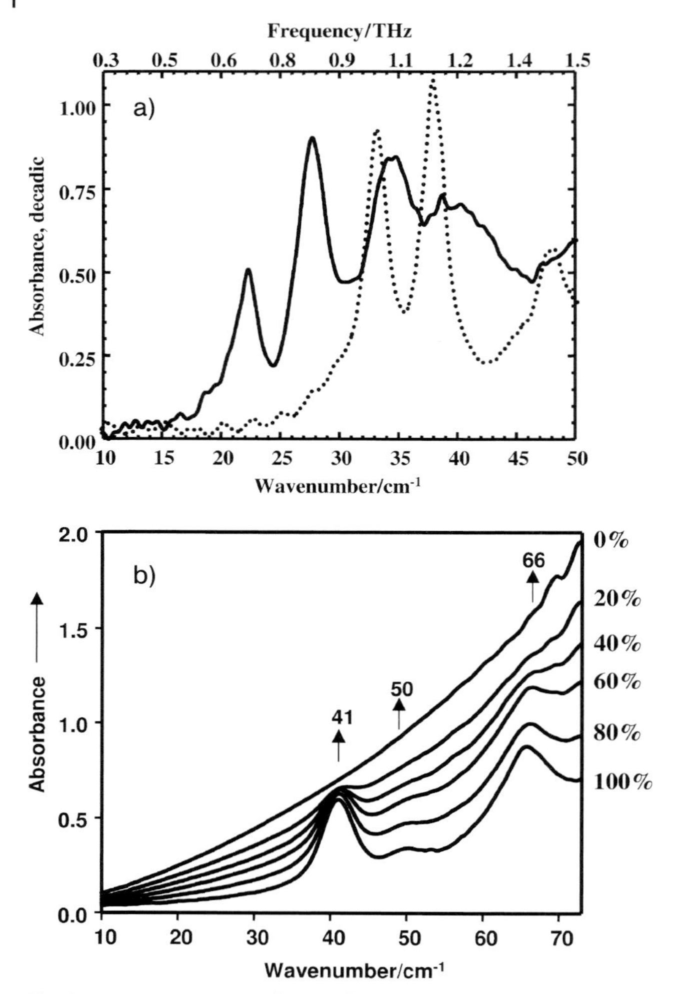

Fig. 5.17 (a) THz spectra of two polymorphs of sulfathiazole. Continuous line, polymorph II; dotted line, polymorph V. (Copyright ©2004, John Wiley & Sons Ltd. and IM Publications, reproduced with permission from [7]). (b) Series of THz spectra showing increasing indomethacin crystallinity. (Reproduced by kind permission of Teraview Limited, Cambridge, UK).

cant fundamental differences in symmetry between the two forms. However, Form I was a hydrate, containing one water molecule for every five molecules of Lamivudine – as revealed by a sharp peak at 3545 cm^{-1}.

As mentioned in the Introduction, recently there has been significant interest in the potential of THz spectroscopy, because of its sensitivity to lattice modes, for characterizing and assaying polymorphic forms of drug substances [7, 59, 60]. Figure 5.17 (a) compares the THz absorption spectra of two polymorphs

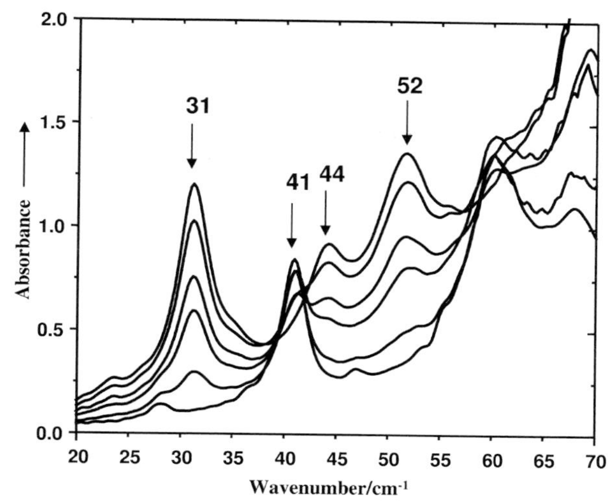

Fig. 5.18 THz spectra of mixtures of carbamazepine Form I and Form III polymorphs. As the concentration (%) of Form I increases from 0 to 20 to 40 to 60 to 80 to 100 the intensity of the bands at 31, 44 and 52 cm^{-1} increases, while that of the band at 41 cm^{-1} decreases. (Reproduced by kind permission of Teraview Limited, Cambridge, UK).

of sulfathiazole. The spectra are very distinct and distinguish readily the two forms. Figure 5.17 (b) shows the sensitivity of THz spectra to increasing indomethacin crystallinity; bands at 41, 50 and 66 cm^{-1} can be seen clearly to increase in intensity with increasing sample crystallinity.

Figure 5.18 illustrates a quantitative assessment of the technique, showing varying concentration mixtures of carbamazepine Form I in Form III. For a calibration developed over the range 0–10%, the limit of detection, LOD, of Form I was 0.44%, with a lower limit of quantification, LOQ, determined of 1.34% [61]. Distinct THz spectra of different crystalline polymorphic forms of CBZ, indomethacin and enalapril maleate have been reported, whereas amorphous indomethacin showed no absorption bands in this region, which is in line with the amorphous sample having no intermolecular modes of long-range order [59]. THz spectroscopy has also been reported for studying and distinguishing the crystalline structure of Forms I and II of the drug ranitidine hydrochloride, in both pure form and also in a range of marketed pharmaceutical products [60].

FT-Raman, dispersive Raman microscopy and FTIR microscopy have been used in a study of the polymorphism and devitrification of nifedipine [62]. A controlled humidity cell was employed in the study. The glassy nifedipine was found to convert into a metastable β-polymorph, which then transformed into the stable α-form, with relative humidity (RH) in the range 20 to 40% having little effect on the onset or rate of crystallization; however, an increase of RH to 60 or 80% led to both a reduced time for onset of crystallization and a significant increase in crystallization rate of the glassy drug.

5.4.2 Hydration/Drying

While vibrational spectra, in particular mid-infrared, display the effects of hydration, one must be careful in such studies, since experimental parameters, such as heating or purging, may alter the state of hydration being investigated. Figure 5.19 shows two examples of this, one mid-infrared and one Raman.

Figure 5.19(a) compares two mid-infrared spectra recorded from a sample of erythromycin [24]. The lower transmission spectrum, recorded as a conventional KBr disc, is characteristic of erythromycin dihydrate; the upper spectrum is a microtransmission spectrum. The differences between the spectra are a consequence of thermal dehydration; the microscope stage was purged with a downward flow of purge nitrogen, to minimize intrusion of atmospheric water and CO_2 vapor; the spectrum recorded was consistent with that of thermally dehydrated erythromycin. The lactone carbonyl-stretching band has shifted from 1714 to 1732 cm^{-1}, probably from a change in hydrogen bonding strength, while the ketone carbonyl band at 1704 cm^{-1} becomes more distinct and seemingly increases in relative intensity. Figures 5.19 (b) and (c) shows FT-Raman spectra recorded from theophylline monohydrate [4]. Theophylline monohydrate contains relatively labile water molecules that can be removed from the crystal lattice by heating. With the static sample (Fig. 5.19 b) the spectral changes recorded with time suggest that dehydration and conversion into the anhydrous form is occurring. However, this is caused by heating of the sample by the laser beam, which can be alleviated by examining a rotating sample housed in a spinning sample cell (Fig. 5.19 c). Providing such experimentally induced transformations are recognized and understood, and proper precautions are taken against their occurrence, then the effects of solid-state hydration/dehydration can be fully explored by either spectroscopic technique.

Figure 5.20 displays a series of FT-Raman spectra that show the transformation of anhydrous theophylline towards the monohydrate form on storage at 75% relative humidity over 12 days [4]. Such studies have importance in, for example, highlighting potential effects of long-term storage of a drug substance.

Another study using vibrational spectroscopy to elucidate a drying process [25] concerned a particular cephalosporin that is pharmacologically active in the pentahydrate form, which, however, showed decreased activity under certain drying conditions. A diffuse reflection accessory fitted with an environmental chamber was used to recreate in the laboratory the manufacturing plant conditions. Inadequate control of the drying step led to dehydration of the pentahydrate form to an unstable anhydrous form of a cephalosporin, which, in the presence of trace moisture levels, rehydrated rapidly to form a monohydrate, which has poor potency. Mid-infrared spectroscopy was used, since the spectra of the three cephalosporin forms of interest have very distinct spectra. For instance, in the pentahydrate, the carbonyl stretching vibration gives rise to a band at 1760 cm^{-1}, while in the monohydrate this vibration occurs at 1805 cm^{-1}, and is indicative of highly strained β-lactam ring [25].

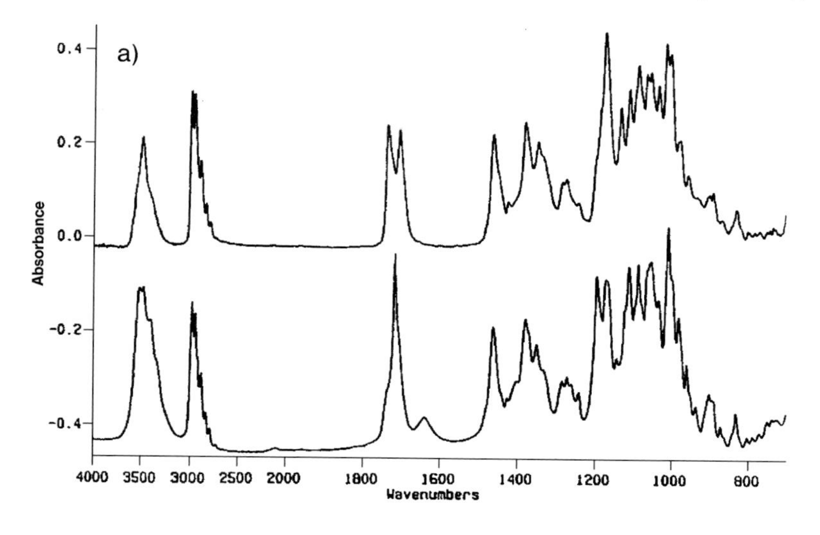

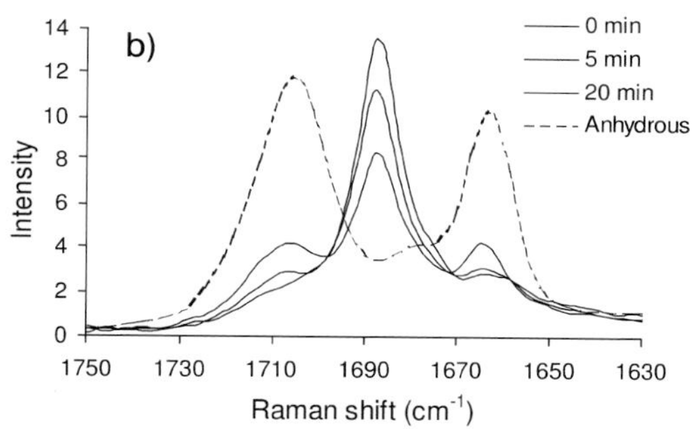

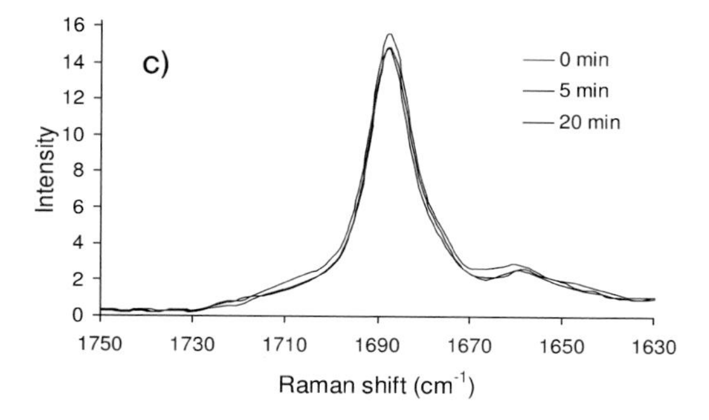

Fig. 5.19 (a) Mid-infrared spectra of erythromycin spectra: top, transmission microscopy; bottom, KBr disc. (b, c) FT-Raman spectra of theophylline monohydrate; (b) static sample, (c) sample in rotating sample cell. [(a) Copyright ©1995, reproduced from [24] with permission of Marcel Dekker, Inc., (b, c) Copyright ©2001, reproduced from [4] with permission of Russell Publishing].

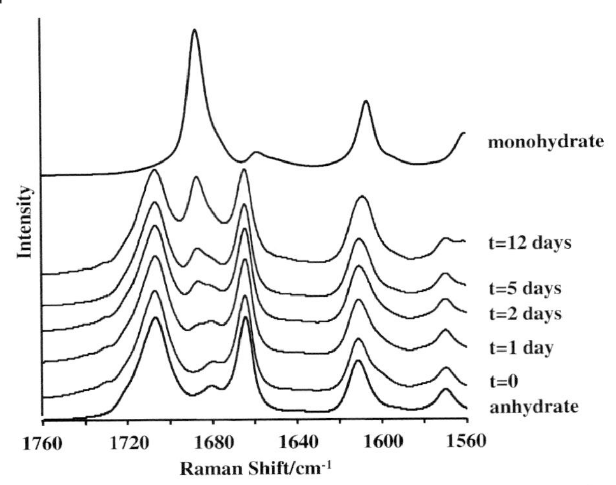

Fig. 5.20 FT-Raman spectra of a theophylline tablet, showing the effect of storage at 75% RH on the solid-state form of the drug. Conversion into the monohydrate form can be seen from the increase in intensity of the band at ~1685 cm^{-1}. (Copyright ©2001, reproduced from [4] with permission of Russell Publishing).

Both in-line and off-line monitoring using near-infrared spectroscopy has been reported for the study of dehydration and polymorphic form of theophylline during fluid bed drying of granules [64]. Another near-infrared spectroscopy study examined the role of excipients in hydrate formation from anhydrous theophylline in wet masses [65]. Silicified microcrystalline cellulose competed with the theophylline by absorbing water, but did not inhibit the hydrate formation, although, under the conditions used, the excipient retarded the hydrate formation of theophylline monohydrate. The studies indicated that α-lactose monohydrate might accelerate the process. Peaks at 1478 and 1972 nm attributed to OH modes were used in a near-infrared spectroscopy coupled with dynamic vapor absorption to study dehydration and rehydration transitions between solid-state forms of theophylline [66].

5.4.3 Quantitative Analysis and Process Monitoring

It is sometimes important within the pharmaceutical industry to be able to detect and quantify a polymorphic "impurity" in a drug substance, since small differences in form can significantly affect solubility and activity. With appropriate attention to reproducible sampling, both infrared and Raman spectroscopy may be used for quantification. The levels of detection and precision will depend very much on the sampling technique and the spectral characteristics of the component(s).

Bugay et al. [67] evaluated quantitative assays, by both diffuse reflection mid-infrared and X-ray diffraction, for determining the dihydrate content in the monohydrate form of cefepime·2HCl, a cephalosporin. For the mid-infrared assay covering the range 1–8% (wt/wt), a minimum quantifiable level (MQL) of 1.0% and a LOD of 0.3% (wt/wt) dihydrate were reported, although the assay was only calibrated and valid for samples within a particle size range of 125–590 μm. Particle size distribution was controlled by passing samples of the two materials through vibrating sieves, retaining material within the required particle size range, and then making homogeneous standard mixtures by using an acetone slurry. After filtering and vacuum drying to remove residual acetone, these were then examined as neat powders (without any alkali-halide diluent) in a diffuse reflection accessory sample cup. The dihydrate exhibits two distinct –OH stretching mode bands, at 3574 and 3432 cm^{-1}; the monohydrate has a single sharp absorption band at 3529 cm^{-1}. The mid-infrared assay was based on a measurement of the intensity of the dihydrate band at 3574 cm^{-1}.

In another reported example of a quantitative assay evaluation, a simple FTIR method was developed for estimating a polymorph concentration in a pharmaceutical intermediate, and then studying samples from a polymorphic transformation under crystallization slurry conditions [68]. A linear calibration curve was constructed from binary mixtures of the polymorphs over the range ca. 3–100% (wt/wt) for Form II in Form I. To be compatible with wet cake and slurry samples, the standard mixtures were examined as carefully prepared mineral oil mulls. The mid-infrared method was based on an absorbance intensity ratio measurement of the ratio of the band at 1109 cm^{-1} to that at 1058 cm^{-1}; the former being a characteristic of Form II, while the latter band, common to both polymorphs, served as an internal standard reference peak. Four replicate measurements were made for each standard mixture; an estimated LOQ of ca. 2.7% (wt/wt) of Form II in Form I was quoted.

Head and Rydzak [69] used multivariate chemometric analysis models developed from both diamond-element ATR mid-infrared and FT-Raman spectroscopic data to determine small amounts of a minor form in a polymorphic mixture. Acceptable standard errors of estimate of ca. 2.9% were deduced for both calibration and validation for the mid-infrared partial least squares (PLS) model, with ca. 3.4% for calibration and ca. 3.8% for validation for the Raman PLS model. Discriminant analysis models were also developed, from which it was possible to classify mixture samples as being either primarily one form or the other or essentially a mixture of the two polymorphs. A PLS model has also been developed from mid-infrared diffuse reflection data for determining ephedrine hydrochloride in mixtures of ephedrine hydrochloride and pseudoephedrine hydrochloride [70]. Standard errors of prediction of 0.74 and 0.11 wt% were derived for the concentration of ephedrine hydrochloride over the concentration ranges 0–50 and 0–5 wt%, respectively. Critical to the quality of the results was careful experimental procedure, which included preparation of homogeneous calibration mixtures, control of the sample surface, and the number of calibration standards. Langkilde et al. [71] demonstrated a simple linear calibration over the range 1.8–15.4% (wt/wt) from FT-Raman data of one polymorph in mixtures of two polymorphs of an API. Each polymorph had a distinct characteristic carbonyl band, and the calibration was based on the relative intensity of these, which occurred at 1716 and 1724 cm^{-1}. Samples for analysis were placed in 5 mm-diameter NMR tubes, which were rotated continuously during recording of the FT-Raman data. Near-infrared reflectance data have been used in the development of a PLS model for determining the total and crystalline miokamycin in a pharmaceutical preparation [72]; the principle active was amorphous miokamycin at a nominal level of 64.5%. Prior to NIR analysis, the samples were ground. Absolute errors of $<1\%$ were demonstrated for both total miokamycin over the range ca. 56–72 wt% and its crystalline polymorph over the approximate range 0–9 wt%. A standard error of prediction of 2% and a detection limit of 3–5% have been determined by principal component regression (PCR) from mid-infrared spectra recorded from prepared binary mixtures for both Form VIII and Form XII in Form XI of delavirdine mesylate [73]. The mixture samples were examined as mulls using FTIR spectroscopy.

Controlling process-induced transformations during pharmaceutical manufacturing is clearly extremely important. In a recent study, Wikström et al. [74] demonstrated the viability of *in situ* monitoring in real time of hydrate formation in a high-shear wet granulation using a dispersive Raman spectrometer fitted with a stainless steel immersion probe. The sapphire window at the end of the probe was placed just above the impeller in the mixing bowl. Spectra were collected every 15–30 s. The system was calibrated using mixtures of anhydrous theophylline and theophylline monohydrate; the process monitored was the solvent-mediated phase transformation during wet granulation of the anhydrate into the monohydrate. The region monitored was ca. 1500–1800 cm^{-1}, since in this region there is no significant interference from excipients or water. Discrete bands due to anhydrate were observed at 1664 and 1707 cm^{-1}, while the monohydrate had a characteristic band at 1686 cm^{-1}.

The potential for fiber-optic near-infrared spectroscopy to monitor and analyze a solid-state API during an industrial crystallization process has been demonstrated [75]. NIR spectra were sensitive to the polymorphic composition of the product, and could be used to optimize the cooling crystallization and filtering processes.

5.4.4 Tablets

The unambiguous characterization of a drug substance as a solid dosage form in a pharmaceutical product, such as an intact tablet or a capsule, is usually much more difficult than analyzing the bulk active ingredient, particularly if it is only present as a low dosage (weight percentage) [63]. The advantage of requiring no sample preparation and relative ease of non-contact, potentially non-destructive sampling by Raman spectroscopy makes it a good candidate for such measurements. However, one has to be aware of the possibility of sub-sampling, i.e., one must ensure that enough material is sampled so that measurements are reproducible and representative [63]. The volumes of material sampled in

Raman measurements can be small; the beam diameter at the laser focus (sampling point) in FT-Raman instruments varies from 100 to 1000 μm, and can be as small as 1 to 2 μm in a dispersive Raman microscope. Also, one must be careful of the sensitivity of samples to laser power, particularly with the highly focused beams in a microscope set-up, and ensure that thermally-induced transformations or degradation does not occur. Both sub-sampling and thermal transformations can be lessened or avoided by using rotating sample mounts, whereby the sample at the excitation beam focus is continually being refreshed.

Examples of the use of Raman spectroscopy for identifying a drug substance in a product inside commercial packaging have been demonstrated; they include sodium diclofenac tablets inside a transparent bubble pack [38] and Pepto-Bismol™ enclosed in a blister pack [76].

Taylor and Langkilde [63] have explored the use of FT-Raman spectroscopy for investigating the solid-state drug form in various commercial formulations and ascertaining how representative the spectra are. The concentration of the active substances present in these formulations ranged from 0.7 to 70% (wt/wt). They have demonstrated some of the problems that may be encountered when examining static samples and the importance of using sample rotation to improve repeatability and decreasing sample heating. They also recommend it as good practice to compare spectra from a sample recorded at different laser powers and at different scanning times if there is suspicion or potential that the sample may change during analysis. The FT-Raman spectra shown in Fig. 5.21 were recorded from a stationary sample of Unilair theophylline monohydrate capsule [63]. Clearly, with increasing time the laser power of 800 mW induces a change

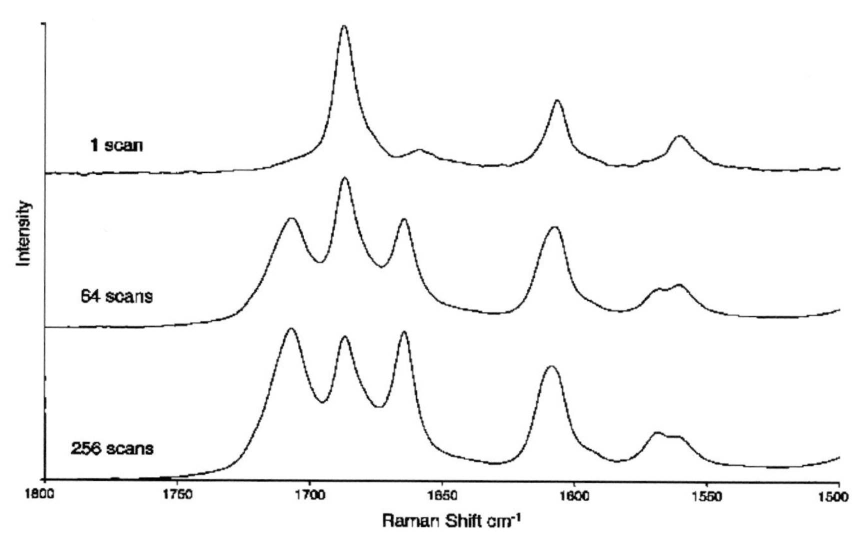

Fig. 5.21 FT-Raman spectra (800 mW, 1064 nm) of a stationary Unilair 200-mg capsule. (Copyright ©2000, John Wiley & Sons Ltd., reproduced with permission from [63]).

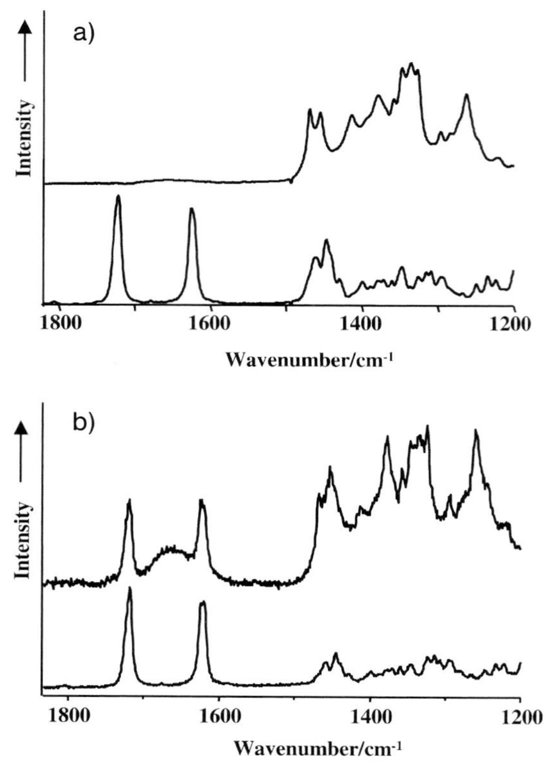

Fig. 5.22 (a) Top: FT-Raman spectrum obtained from a rotating sample of a 0.25-mg digoxin tablet; bottom: reference spectrum of digoxin. No API peaks were detected. (b) Top: spectrum recorded using a Raman microscope from a localized area on a 0.25-mg digoxin tablet; bottom: reference spectrum of digoxin. The API bands are clearly observed. (Copyright ©2001, reproduced from [4] with permission of Russell Publishing).

in the sample. When the sample was rotated and examined with the same laser power the spectrum did not change with time.

Although caution with sub-sampling was mentioned above, in some circumstances highly localized sampling can be advantageous [4]. This may be particularly so when the concentration of a drug substance is low. Figure 5.22 shows an example for a 0.25-mg digoxin tablet [the drug level is 0.25% (wt/wt)] [4]. The FT-Raman spectrum of Fig. 5.22 (a) was recorded using a rotating sample in which, because of the low ratio of drug to excipient, the drug cannot be detected. However, mapping the tablet with a Raman microscope and probing for the presence of the drug by monitoring the intensity of bands specific for the drug yielded the spectrum shown in Fig. 5.22 (b). In the localized small volume sampled the drug concentration relative to that of the excipient is apparently higher than elsewhere or the average over a larger area.

In the digoxin example above, the bands in the Raman spectrum attributable to any excipient present clearly do not preclude or unduly interfere with observation of the presence of the drug substance. Since, carbonyl functional groups and aromatic systems are common in drug substances but few excipients, e.g., α-lactose monohydrate and cellulose contain these, it is often possible to identify bands characteristic of an API in the region 1500 to 1880 cm^{-1} and in the aromatic C–H stretch region 3000 to 3500 cm^{-1} [63]. Carbonyl-containing excipients that will likely interfere include providone, cross providone, stearic acid and sodium stearyl fumarate; aromatic dyes will also likely cause overlap of bands. The extent of interference will depend on both the relative amounts of materials present and their relative Raman scattering cross-sections. Iron oxide, which is often used in tablet coatings or sometimes incorporated into a tablet core as a colorant, can be troublesome with Raman spectroscopy examinations, since it is fluorescent when excited with near-infrared excitation [63].

Two examples of the use of Raman spectroscopy to distinguish polymorphs in tableted formulations are shown in Figs. 5.23 and 5.24 [63]. The spectrum shown in Fig. 5.23 was recorded from a low dosage 2.5 mg strength prednisolone tablet (ca. 1.5% (wt/wt) drug substance in tablet total mass), and while the spectrum below 1500 cm^{-1} is dominated by bands characteristic of the excipients present, bands characteristic of the prednisolone are easily observed in the region 1500 to 1800 cm^{-1}. Moreover, the spectrum may be identified as that of Form II prednisolone (characteristic peaks are marked with an asterisk). In a

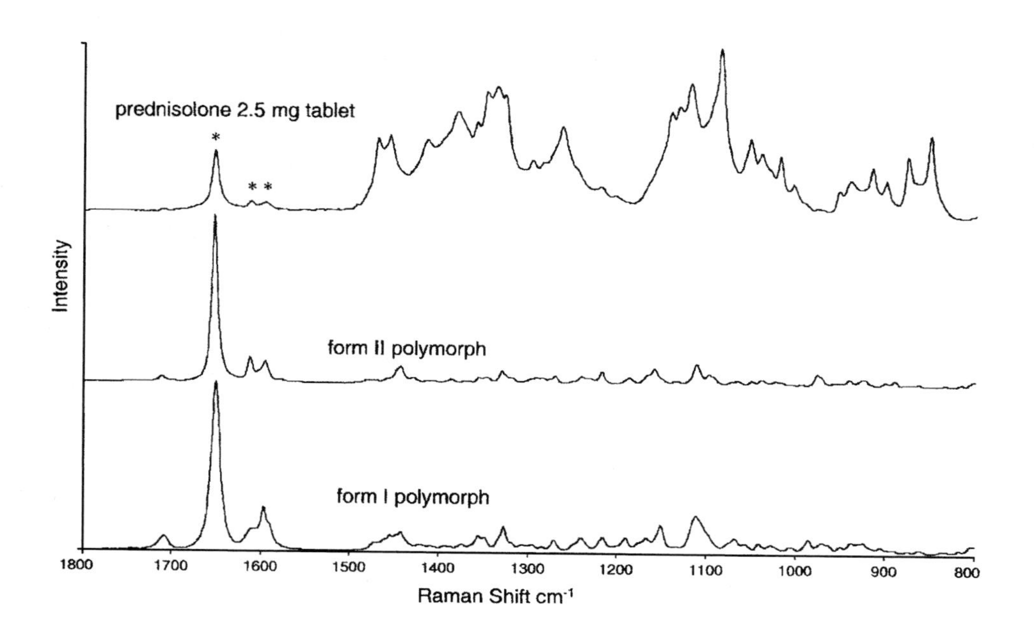

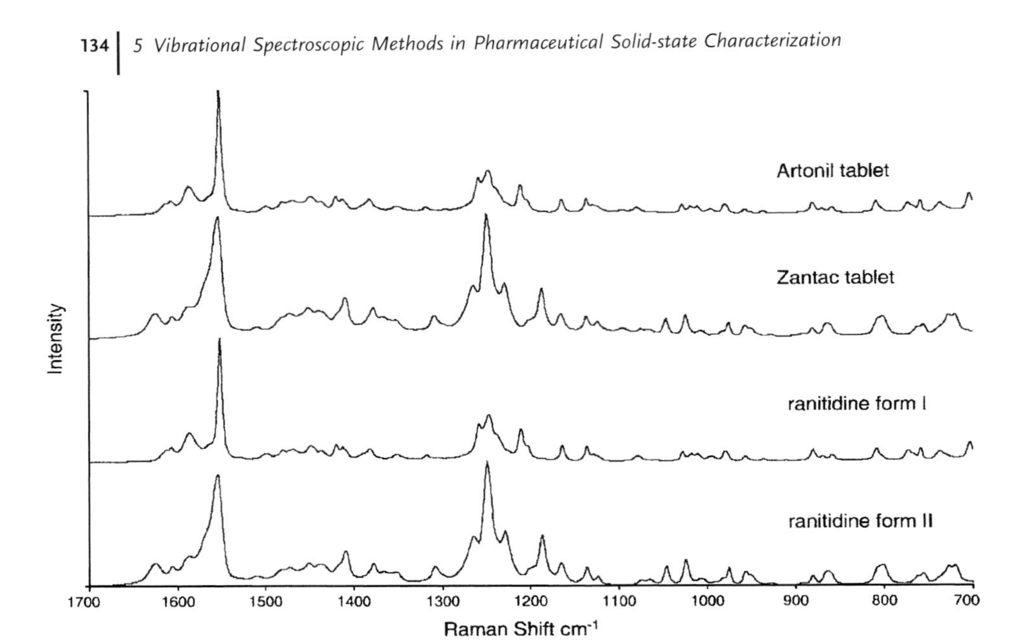

Fig. 5.23 Raman spectra of a 2.5-mg prednisolone tablet, and Form I and Form II polymorphs of prednisolone. (Copyright ©2000, John Wiley & Sons Ltd., reproduced with permission from [63]).

similar study [63], on a 2.5-mg Coumadin (warfarin sodium) tablet, it was also straightforward to confirm the drug substance as being present as the crystalline clathrate, rather than in the amorphous form. In a third study of a low dosage tablet, 5-mg enalapril maleate, by the same authors [63], however, the Raman spectrum in the region 1500 to 1800 cm^{-1} of the drug substance observed in the tablet was very different from that recorded from the pure drug powder, implying that the drug was present in a different form in the tablet. In Fig. 5.24 Raman spectroscopy can clearly readily distinguish the polymorphic form of ranitidine hydrochloride between tablets from two different manufacturers. The authors [63] also demonstrated how the form of theophylline in two commercial tablets, a Unilair 200-mg tablet and a Theo-Dur 450-mg tablet, could be recognized readily from Raman spectra in the region 1500 to 1800 cm^{-1}; the Unilair tablet was shown to contain the monohydrate form; the Theo-Dur tablet was shown to contain theophylline as the anhydrate.

5.5
Closing Remarks

While there are many other well established and important applications of infrared and Raman spectroscopy to characterizing and analyzing solid-state drug substances and other materials of interest within the pharmaceutical industry (such as process measurements and understanding, raw material identification, contaminant analysis, excipient and packaging material analysis, blend uniformity and dryer monitoring, synthesis product characterizations, and Raman microscopy as a tool for analyzing multiple-well plates and for high-throughput crystal form screening [77, 78]), we have in this chapter tried to contain our application examples and discussions within the confines of direct solid-state characterization of polymorphism, since this is overall intent of this book. For this reason, we have also excluded discussion about the chiroptical spectroscopy techniques of vibrational circular dichroism (VCD) [79] and Raman optical activity (ROA) [80]. Instrumentation for both these techniques is commercially available for determining parameters such as enantiomeric purity and configuration.

From the limited number of published application examples selected to illustrate their importance, near-infrared, mid-infrared and Raman, and in the future THz, spectroscopies all clearly have key and sometimes unique roles to play in characterizing, fingerprinting and analyzing quantitatively drug substances in the context of their solid-state form, provided that there is a full appreciation of any likely modifying effects of sample preparation and presentation. In addition, as with any study of solid-state API form, it is unwise, and rarely enough, to rely solely on evidence from one analytical technique. Vibrational spectroscopy techniques are merely part of a collection of techniques, which includes XPRD, solid-state NMR and thermal analysis, that is appropriate to providing information on the solid-state form of drug substances.

Acknowledgments

The authors are indebted to Lynne Taylor (Purdue University), Philip Taday (Teraview Ltd.) and Bob Lancaster for their considerable help in initially providing material or references for several of the applications examples used. We also acknowledge Roger Davey and his research group at the University of Manchester for helpful discussions.

References

1 USP 28 NF 23 *The United States Pharmacopoeia*, <197> Spectrophotometric Identification Tests, pp. 2295, United States Pharmacopoeial Convention, Inc., Rockville (**2004**).

2 D.H.W. Dickson, J.E. Page, D. Rogers, *J. Chem. Soc., Part I*, 443–447 (**1955**).

3 J.H. Giles, G. Shackman, M.B. Denton, *Am. Pharmaceut. Rev.*, 44–51, Winter **1999**.

4 L.S. Taylor, *Am. Pharmaceut. Rev.*, 60–67, Winter **2001**.

5 A.L. Enculescu, J.R. Steiginga, *Am. Pharmaceut. Rev.*, 81–88, Spring **2002**.

6 P.R. Griffiths, Far-infrared spectroscopy, pp. 229–239, in *Handbook of Vibrational Spectroscopy*, Vol. I, eds. J.M. Chalmers, P.R. Griffiths, John Wiley & Sons Ltd., Chichester (**2002**).

7 P.F. Taday, D.A. Newnham, *Spectrosc. Eur.*, 16(5), 22–25 (**2004**).

8 E.M. Ciurczak, J.K. Drennen, *Pharmaceutical Applications of Near-Infrared Spectroscopy*, Marcel Dekker, New York (**2002**).

9 P.R. Griffiths, Introduction to vibrational spectroscopy, pp. 33–439, in *Handbook of Vibrational Spectroscopy*, Vol. I, eds. J.M. Chalmers, P.R. Griffiths, John Wiley & Sons Ltd., Chichester (**2002**).

10 Infrared and Raman Spectroscopy, ed. B. Schrader, VCH, Weinheim (**1995**).

11 J.M. Hollas, *Modern Spectroscopy*, John Wiley & Sons Ltd, Chichester (**1997**).

12 D. Steele, Infrared spectroscopy: theory, pp. 44–70, in *Handbook of Vibrational Spectroscopy*, Vol. I, eds. J.M. Chalmers, P.R. Griffiths, John Wiley & Sons Ltd., Chichester (**2002**).

13 D.A. Long, *The Raman Effect*, John Wiley & Sons Ltd., Chichester (**2002**).

14 G. Keresztury, Raman spectroscopy: theory, pp. 71–87, in *Handbook of Vibrational Spectroscopy*, Vol. I, eds. J.M. Chalmers, P.R. Griffiths, John Wiley & Sons Ltd., Chichester (**2002**).

15 G. Socrates, *Infrared and Raman Characteristic Group Frequencies: Tables and Charts*, 3rd edition, John Wiley & Sons, Chichester (**2001**).

16 B. Smith, *Infrared Spectral Interpretation*, CRC Press, Boca Raton, FL (**1999**).

17 I.A. Degen, *Tables of Characteristic Group Frequencies for the Interpretation of Infrared and Raman Spectra*, Acolyte Publications, Harrow (**1997**).

18 N.P.G. Roeges, *Guide to the Complete Interpretation of Infrared Spectra of Organic Structures*, John Wiley & Sons Ltd., Chichester (**1994**).

19 G.C. Pimentel, A.L. McClellan, *The Hydrogen Bond*, W.H. Freeman & Co., San Francisco (**1960**).

20 R.L. McCreery, *Raman Spectroscopy for Chemical Analysis*, John Wiley & Sons, Inc., New York (**2000**).

21 Analytical Applications of Raman Spectroscopy, ed. M.J. Pelletier, Blackwell Science Ltd., Oxford (**1999**).

22 P.R. Griffiths, J.A. de Haseth, *Fourier Transform Infrared Spectrometry*, John Wiley & Sons, Inc., New York (**1986**).

23 E. Smith, G. Dent, *Modern Raman Spectroscopy*, John Wiley & Sons Ltd., Chichester (**2005**).

24 D.S. Aldrich, M.A. Smith, Pharmaceutical applications of infrared microspectroscopy, Chapter 9, pp. 323–375, in *Practical Guide to Infrared Microspectroscopy*, ed. H.J. Humecki, Marcel Dekker, Inc., New York (**1995**).

25 D. Clark, The analysis of pharmaceutical substances and formulated products by vibrational spectroscopy, pp. 3574–3589, in *Handbook of Vibrational Spectroscopy*, Vol. 5, eds. J.M. Chalmers, P.R. Griffiths, John Wiley & Sons Ltd., Chichester (**2002**).

26 F. Clarke, S. Hammond, *Eur. Pharmaceut. Rev.*, 1, 41–50 (**2003**).

27 E.N. Lewis, J.E. Carroll, F. Clarke, *NIR News*, 12(3), 16–18 (**2001**).

28 J. Breitenbach, W. Schrof, J. Neumann, *Pharm. Res.*, 16(7), 1109–1113 (**1999**).

29 C. Rustichelli, G. Gamberini, V. Ferioli, M.C. Gamberini, R. Ficarra, S. Tommasini, *J. Pharm. Biomed. Anal.*, 23, 41–54 (**2000**).

30 C.L. Anderton, Vibrational spectroscopy in pharmaceutical analysis, Chapter 6,

pp. 203–239, in *Pharmaceutical Analysis*, eds. D.C. Lee, M. Webb, Blackwell Publishing Ltd., Oxford (**2003**).

31 U.J. Griesser, M.E. Auer, A. Burger, *Microchem. J.*, 65, 283–292 (**2000**).

32 M. Szelagiewicz, C. Marcolli, S. Cianferani, A.P. Hard, A. Vit, A. Burkhard, M. von Raumer, U.C. Hofmeier, A. Zilian, E. Francotte, R. Schenker, *J. Therm. Anal. Calorim.*, 57, 23–43 (**1999**).

33 J.-O. Henck, E. Finner, A. Burger, *J. Pharm. Sci.*, 89(9), 1151–1159 (**2002**).

34 S. Williams, P. Tampkins, H. Jervis, *Raman Spectroscopy of Hydration State Changes of Active Pharmaceutical Ingredients*, VGI Application Note 507, Surface Measurements Systems Ltd., Alperton, UK.

35 K.P.J. Williams, I.P. Hayward, P. Tampkins, L.K. Pickard, Raman technology for today's spectroscopists – 2004 technology primer, *Spectroscopy*, pp. 38–47 June **2004** Supplement.

36 F. Adar, G. leBourdon, J. Reffner, A. Whitley, *Spectroscopy* 18(2), 34–40 (**2003**).

37 F. Adar, C. Naudin, A. Whitley, Raman technology for today's spectroscopists – 2004 technology primer, *Spectroscopy*, pp. 22–29 June **2004** Supplement.

38 P.J. Hendra, Fourier transform Raman spectroscopy, Chapter 3, pp. 73–108, in *Modern Techniques in Raman Spectroscopy*, ed. J.J. Laserna, John Wiley & Sons Ltd., Chichester (**1996**).

39 J.M. Chalmers, Infrared spectroscopy: sample preparation, pp. 402–415, in *Encyclopedia of Analytical Science*, 2nd edition, Vol. 4, Elsevier, Oxford (**2005**).

40 J.M. Chalmers, G. Dent, *Industrial Analysis with Vibrational Spectroscopy*, The Royal Society of Chemistry, Cambridge (**1997**).

41 R.W. Hannah, Standard sampling techniques for infrared spectroscopy, pp. 933–959, in *Handbook of Vibrational Spectroscopy*, Vol. 2, eds. J.M. Chalmers, P.R. Griffiths, John Wiley & Sons Ltd., Chichester (**2002**).

42 S.R. Byrn, *Solid-state Chemistry of Drugs*, Academic Press, New York (**1982**).

43 T.L. Threlfall, *Analyst*, 120, 2435–2460 (**1995**).

44 J.M. Chalmers, Mid-infrared spectroscopy: anomalies, artifacts and common errors, pp. 2326–2347, in *Handbook of Vibrational Spectroscopy*, Vol. 3, eds. J.M. Chalmers, P.R. Griffiths, John Wiley & Sons Ltd., Chichester (**2002**).

45 J.M. Chalmers, Mid-infrared spectroscopy of the condensed phase, pp. 128–140, in *Handbook of Vibrational Spectroscopy*, Vol. 1, eds. J.M. Chalmers, P.R. Griffiths, John Wiley & Sons Ltd., Chichester (**2002**).

46 T.L. Threlfall, Polymorphs, solvates and hydrates, pp. 3557–3573, in *Handbook of Vibrational Spectroscopy*, Vol. 5, eds. J.M. Chalmers, P.R. Griffiths, John Wiley & Sons Ltd., Chichester (**2002**).

47 W.C. McCrone, Polymorphism, Chapter 8, pp. 725–767, in *Physics and Chemistry of the Organic Solid State*, Vol. II, eds. D. Fox, M.M. Labes, A. Weissberger, Interscience, New York (**1965**).

48 Chirality in Industry: the Commercial Manufacture and Applications of Optically Active Compounds, eds. G. Sheldrake, A.N. Collins, J. Crosby, John Wiley & Sons Ltd., Chichester (**1992**).

49 J. Jacques, A. Collet, S.H. Wilen, *Enantiomers, Racemates and Resolutions*, John Wiley & Sons, Inc., New York (**1981**).

50 E.L. Eliel, S.H. Wilen, M.P. Doyle, *Basic Organic Stereochemistry*, John Wiley & Sons, Inc., New York (**2001**).

51 M. Kuhnert-Brandstatter, R. Ulmer, *Mikrochim. Acta*, 927–935 (**1974**).

52 S.C. Stinson, *Chem. Eng. News*, 76, 83–104 (**1998**).

53 G. Schwartzman, I. Asher, V. Folen, W. Brannon, J. Taylor, *J. Pharm. Sci.*, 67(3), 399–400 (**1978**).

54 A.L. Grzesiak, M. Lang, K. Kim, A.J. Matzger, *J. Pharm. Sci.*, 92(11), 2260–2271 (**2003**).

55 L.E. O'Brien, P. Timmins, A.C. Williams, P. York, *J. Pharm. Biomed. Anal.*, 36, 335–340 (**2004**).

56 R.K. Gilpin, W. Zhou, *Vib. Spectrosc.*, 37, 53–59 (**2005**).

57 R.K. Harris, A.M. Kenwright, R.A. Fletton, R.W. Lancaster, *Spectrochim. Acta A*, 54, 1837–1847 (**1998**).

58 R.K. Harris, R.R. Yeung, R.B. Lamont, R.W. Lancaster, S.M. Lynn, S.E. Staniforth, *J. Chem. Soc., Perkin Trans. 2*, 2653–2659 (**1997**).

59 C.J. Strachan, T. Rades, D.A. Newnham, K.C. Gordon, M. Pepper, P.F. Taday, *Chem. Phys. Lett.*, 390, 20–24 (**2004**).

60 P.F. Taday, I.V. Bradley, D.D. Arnone, M. Pepper, *J. Pharm. Sci.*, 92(4), 831–838 (**2003**).

61 Teraview Limited, Cambridge, UK.

62 K.L.A. Chan, O.S. Fleming, S.G. Kazarian, D. Vassou, G.D. Chryssikos, V. Gionis, *J. Raman Spectrosc.*, 35(5), 353–359 (**2004**).

63 L.S. Taylor, F.W. Langkilde, *J. Pharm. Sci.*, 89(10), 1342–1353 (**2000**).

64 S. Airaksinen, M. Karjalainen, E. Räsänen, J. Rantanen, J. Yliruusi, *Int. J. Pharm.*, 276, 129–141 (**2004**).

65 A. Jørgensen, S. Airaksinen, M. Karjalainen, P. Luukkonen, J. Rantaanen, J. Yliruusi, *Eur. J. Pharm. Sci.*, 23, 99–104 (**2004**).

66 K.L. Vora, G. Buckton, D. Clapham, *Eur. J. Pharm. Sci.*, 22, 97–105 (**2004**).

67 D.E. Bugay, A.W. Newman, W.P. Findlay, *J. Pharm. Biomed. Anal.*, 15, 49–61 (**1996**).

68 P.J. Skrdla, V. Antonucci, L.S. Crocker, R.M. Wenslow, L. Wright, G. Zhou, *J. Pharm. Biomed. Anal.*, 25, 731–739 (**2001**).

69 T. Head, J. Rydzak, *Am. Pharmaceut. Rev.*, 6(1), 78–84 (Spring **2003**).

70 Y.K. Dijiba, A. Zhang, T. Niemczyk, *Int. J. Pharm.*, 289, 39–49 (**2005**).

71 F.W. Langkilde, J. Sjöblom, L. Tekenbergs-Hjelte, J. Mark, *J. Pharm. Sci.*, 15, 687–696 (**1997**).

72 M. Blanco, A. Villar, *J. Pharm. Sci.*, 92(4), 823–830 (**2003**).

73 R.W. Sarver, P. A. Meulman, D.K. Bowerman, J.L. Havens, *Int. J. Pharm.*, 167, 105–120 (**1998**).

74 H. Wikström, P.J. Marsac, L.S. Taylor, *J. Pharm. Sci.*, 94, 1, 209–219 (**2005**).

75 G. Févotte, J. Calas, F. Puel, C. Hoff, *Int. J. Pharm.*, 273, 159–169 (**2004**).

76 C.J. Frank, Review of pharmaceutical applications of Raman spectroscopy, Chapter 6, pp. 224–275, in *Analytical Applications of Raman Spectroscopy*, ed. M.J. Pelletier, Blackwell Science Limited, Oxford (**1999**).

77 R. Hilfiker, J. Berghausen, F. Blatter, A. Burkhard, S.M. De Paul, B. Freiermuth, A. Geoffroy, U. Hofmeier, C. Marcolli, B. Siebenhaar, M. Szelagiewicz, A. Vit, M. von Raumer, *J. Therm. Anal. Calorim.*, 73, 429–440 (**2003**).

78 R.A. Storey, R. Docherty, P.D. Higginson, *Am. Pharmaceut. Rev.*, 100–105, Spring **2003**.

79 L.A. Nafie, R.K. Dukor, T.B. Freedman, Vibrational circular dichroism, pp. 731–744, in *Handbook of Vibrational Spectroscopy*, Vol. 1, eds. J.M. Chalmers, P.R. Griffiths, John Wiley & Sons Ltd., Chichester (**2002**).

80 W. Hug, Raman optical activity spectroscopy, pp. 745–758, in *Handbook of Vibrational Spectroscopy*, Vol. 1, eds. J.M. Chalmers, P.R. Griffiths, John Wiley & Sons Ltd., Chichester (**2002**).

6
Crystallography for Polymorphs

Philippe Ochsenbein and Kurt J. Schenk

6.1
Introduction

Crystalline matter consists of a regular 3D packing of its constituent atoms or molecules, whereas amorphous solids show only atomic organization at short distance. A crystal may be considered as the quintessence of a supra-molecular assembly whose synthons are the molecules [22]. Properties of condensed matter such as solubility, dissolution rate or thermal evolution are a function of the subtlest molecular interactions during crystallization. Polymorphism may be defined as the existence of distinct molecular packings of the very same molecule [17, 40]. This phenomenon, originally recognized from crystal morphologies of well-shaped inorganic crystals alone, long before a structure determination was possible [44], has become of great importance in the pharmaceutical industry, and also for regulatory reasons. Polymorphism may arise if several energetically comparable molecular associations are known and it is frequently observed in practice, when processes are transferred from laboratory to industrial scale. It is the consequence of different packings of rigid molecules or of the condensation of distinct conformations of the same molecular entity. This phenomenon must be controlled to obtain a reliable industrial procedure. Possible polymorphic phases, concomitant or not, must be looked for, identified and their respective thermodynamic stability evaluated [10, 23]. X-ray crystallography, thanks to the specific fingerprints of microcrystalline powders, is the preferred method for identifying polymorphic phases. Moreover, while the usual analytical tools, including DSC, IR, thermogravimetry, solubility and density measurements, can detect polymorphism at best, crystallography, by reaching atomic resolution, often furnishes an explanation for its origin. Indeed, X-ray crystallographic studies yield the most detailed possible characterization of condensed matter. Determination of the lattice constants reveals the periodicity of the crystal, while the assignment of the phases to the structure factors gives details of the structure at the atomic level. We shall illustrate, by way of several representative examples, the power of crystallography for understanding those steric and electronic phe-

Polymorphism: in the Pharmaceutical Industry. Edited by Rolf Hilfiker

ISBN: 3-527-31146-7

nomena as well as the characteristic packings of certain crystalline solids that may have given rise to polymorphism. The crucial role of the solvent also has to be highlighted since molecular crystals are generally grown from solution. The relative stabilities of polymorphs may also be – once their structures have been determined – estimated by molecular modeling and compared with the results of experimental techniques such as solubility measurements or DSC. We conclude by describing a structure determination from powder data. These modern techniques are well adapted to the study of polymorphs for which no single crystals are available.

6.2 Solving Difficult Crystal Structures with Parallel Experiments

We begin by emphasizing that growth, detection, isolation, delicate manipulation and mounting of a crystalline sample is far from the easiest part of a crystallographer's job. Quite often, with organic material, crystals grow as platelets or very thin needles that are always difficult to handle. Furthermore, ill-grown, defect-laden organic crystals (inhomogeneity, twinning, quasi- or non-periodicity, microdeformation or texture) often give rise to greatly entangled diffraction patterns that consist of broad, weak and overlapped features. Diffraction patterns obtained from these crystals may be difficult to interpret, and if it were not for Jens Richter's pioneering and unbeatable RECIPE program [57], our knowledge of twins, incommensurate systems, composite structures or other pathological pieces of matter might be less advanced than it is. However, with the advent of synchrotron radiation or beam intensifying devices (such as capillary optics or other collimators featuring total reflection coatings, microsources or similar), very small specimens are nowadays accessible for study (a few μm for inorganic samples and a few tens of μm for organic ones).

N-Decylammonium bromide (C10Br) (Fig. 6.3 a, below) is a moderately complex example that illustrates the analysis of a difficult sample using several experimental X-ray diffraction techniques. Knowing that most hydrochlorides and hydrobromides are twinned, we took a powder diagram at the Swiss-Norwegian Beam Line (SNBL) of the European Synchrotron Radiation Facility (ESRF). This looked quite perfect (Fig. 6.1) and three classical indexing programs (DicVol [15], Ito [66] and Taupin [63]) spontaneously suggested the unit cell I (Table 6.1) and space group $P2_1$. (The lattice constants have been obtained by whole pattern fitting with the help of the XND [13] program.) This cell provided indices for all but a few very faint lines – we did not know whether they were impurity lines or stemmed from C10Br. Notably, in this context, the black peaks in Fig. 6.1 would have never been detected even by an excellent laboratory X-ray source since they are both weak and too close to the main reflections.

Classical direct methods (SIRPow97 [2]) applied to the synchrotron data failed to give a model for C10Br. This failure was not unexpected because of the well-known pseudo-peaks of n-alkyl chains. We therefore decided to check the cell of

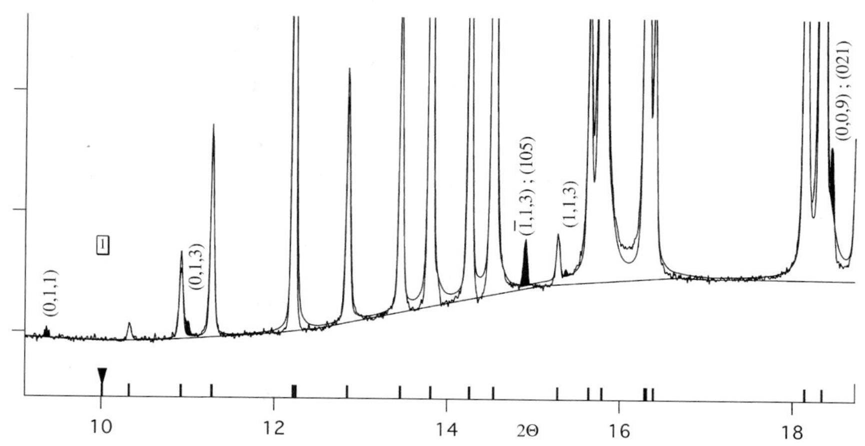

Fig. 6.1 Synchrotron (λ=1.09818 Å) powder diagram of C10Br. Filled features remain unexplained by cells I/Ib. Intensity ratio of black to non-black peaks is about 10^6. Indices above the filled black maxima correspond to a cell II/IIb.

Table 6.1 Unit cells of C10Br. I/II are from a Si-calibrated synchrotron powder diagram (λ=1.09816 Å), Ib/IIb from at winned specimen on a Stoe IPDS (MoKα).

Cell	*a* (Å)	*b* (Å)	*c* (Å)	β (°)
I	6.1087(3)	6.8904(2)	15.4899(9)	92.721(3)
Ib	6.125(1)	6.891(1)	15.544(3)	92.87(3)
II	6.1075(4)	6.8899(5)	30.972(2)	92.714(3)
IIb	6.129(1)	6.893(1)	31.121(6)	92.54(3)

a twinned sample by means of precession photographs (Fig. 6.2). Both ($h0l$) and ($hk0$) layers did not contradict the cell given above, but for technical reasons (short axis and Cu radiation) we were unable to obtain neat upper layers and their interpretation did not question the cell. Finally, we put a twinned sample on a Stoe IPDS and exposed 200 oscillation images. A naïve cell search again yielded cell Ib and so we undertook a structure determination in this cell. The SHELX97 program [62] found a structure and it could be refined, using quite a few restraints to $R_1 \approx 0.12$.

However, the resulting structure (Fig. 6.3 a) was clearly suspicious as the alternation between normal and such large anisotropic displacement parameters (ADP) indicates the presence of either dynamical or static disorder. In either case, the refined chain may be interpreted as the superposition of two chains (Fig. 6.3 b).

At this point, we used the RECIPE program for a thorough scrutiny of the diffraction pattern (Fig. 6.4). Two twin individua, some amorphous features (ring) and spurious peaks may clearly be identified. The two cells shown (in

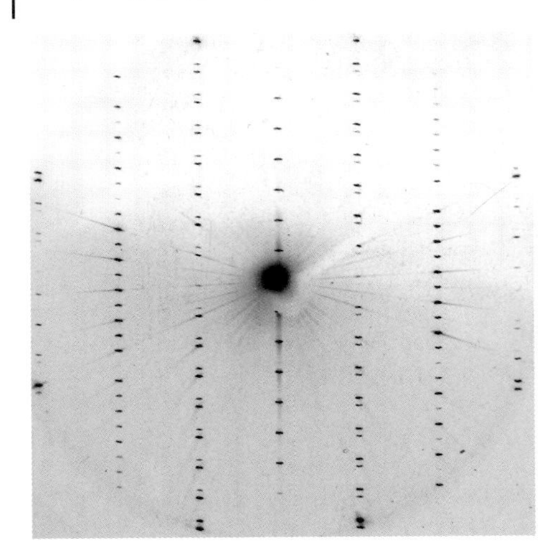

Fig. 6.2 (*h0l*) layer precession photograph of C10Br obtained with Ni-filtered CuKα radiation.

dark and light gray) correspond to the two twin individua (cell Ib, Table 6.1). At this stage it was not clear whether these twins were related by a σ_x mirror plane or a C_{2z} binary rotation. The utility of this kind of program is finally confirmed by comparing the reciprocal space projection with a precession photograph (Fig. 6.2). Indeed, careful inspection of Fig. 6.4 reveals that cells I/Ib do not account for all reflections in the (1*kl*) and (3*kl*) layers and that the reflections from the two individua are superimposed for (0*kl*) and (2*kl*). After the discovery of these discrepancies, everything was carefully re-assessed and the true cell turned out to be II/IIb, which explains the powder diagram very well (Fig. 6.1). The reason for the prior ill-assessment was the rather weak difference reflections that would have been missed without a finer analysis of reciprocal space (Fig. 6.5). This faintness of the difference reflection thwarted, incidentally, a routine structure solution from either powders or "single crystal" data. The solution had to be found by reconstructing the structure from its projection (Fig. 6.3).

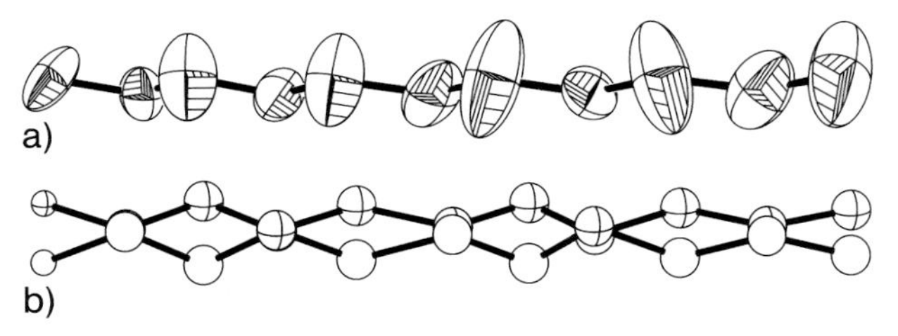

Fig. 6.3 (a) C10Br chain resulting from refinement of $(h,k,l=2m)$ reflections. (b) Model explaining the unusual ADPs.

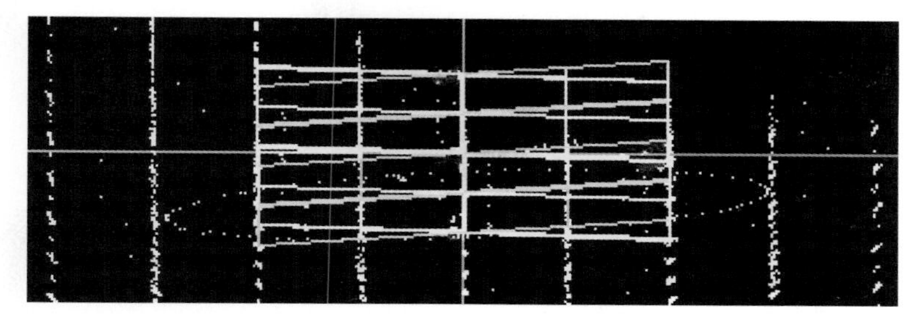

Fig. 6.4 ⟨a*,c*⟩ reciprocal space projection of C10Br obtained with the help of the RECIPE program. Data collection was carried out using MoKα radiation.

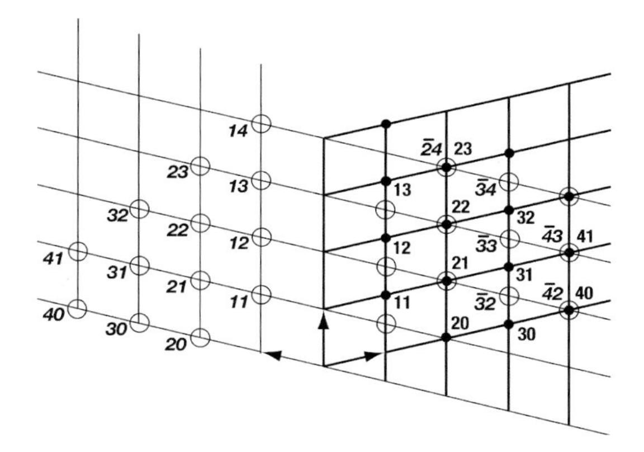

Fig. 6.5 Indexation of the reflections of both twin individua. Filled circles represent the reference twin and open circles the twin created by a σ_x or a C_{2z} operation. Several examples of pairs of overlapping reflections with the respective indices are given.

The expression below shows how one can find the indices of twin (2) overlapping with a reflection of the reference twin (1):

$$\begin{pmatrix} \bar{1} & 0 \\ \frac{1}{2} & 1 \end{pmatrix} \begin{pmatrix} 2 \\ 3 \end{pmatrix}_1 = \begin{pmatrix} \bar{2} \\ 4 \end{pmatrix}_2$$

The 2×2 matrix corresponds to the transposed inverse of the matrix expressing the axes of twin (2) as a function of the axes of the reference individuum. The intensities of both twin individua are then refined using a twin volume fraction.

In summary, this example shows that (a) it is very useful to have several X-ray wavelengths available and (b) it is helpful, even crucial, to be able to use different techniques.

6.3
Atropisomers and Desmotropes

Polymorphism generally arises for good structural reasons. The first of these is the existence, for a given compound, of molecular conformations of comparable energies [7]. Indeed, pharmaceutically important molecules are often flexible and possess several potential recognition-sites (such as amide, amine, alcohol or ketone groups) that provide various modes of molecular association in the solid state. Conformational polymorphism is often encountered in molecules consisting of alkyl chains of four or more atoms linking two chemical functionalities. The crystal structures of such molecules can easily be disordered. Notwithstanding that, even for very flexible molecules, molecular geometries of two polymorphs can be rather close. Moderately flexible molecules – these being more numerous – often show a type of polymorphism based on similar molecular geometries. This polymorphism is the manifestation of distinct modes of association between chemical entities and the faint conformational differences (e.g., the orientation of an aromatic ring) are just consequences of the packing. Nevertheless it can happen that some preferred molecular geometries distinguish themselves and may, for this reason, be frozen-in in the crystal by intermolecular interactions. We have described the example of the piperidine in a hydrazine group, (R_2N)–NH–C(=O), which exists in two orientations, corresponding to torsion angles of 0° and 180° about the H–N–N–(lone pair) group [47]. These two conformations give rise to polymorphism. One of the structures consists of dimers linked by anti-parallel association of cyclic NH $\cdots$ N hydrogen bonds, and the other is a kind of polymer in which the molecules are linked by NH $\cdots$ O=C amide hydrogen bonds. The CCDC (Cambridge Crystallographic Data Centre) codes VUHSEW [20] and FIFNOX [32] represent further examples of the two orientations of the hydrazine moiety, albeit for different molecules. Polymorphism may also be due to the coexistence of rotamers or atropisomers (conformational stereoisomer of a reference compound that converts into the reference compound only slowly on the NMR or laboratory time scale) in solution. Crystallography is very useful in these cases since mixtures of atropisomers often modify the expected chemical shifts and/or give rise to additional NMR signals that are difficult to assign. Notably, crystal structures of atropisomers are very rare: a survey in the 1st October 2004 release ($\sim$300 000 entries) of the CCDC [1] yielded four hits for the qualifier *atropisomer* and ten hits for the qualifier *rotamer*. Known solid-state structures of a couple [35, 67] or more [24, 25] atropisomers are even rarer. In the example below, a pair of atropisomers crystallize concomitantly [8], yielding two distinct acicular habiti that can be distinguished in orthoscopy (one of the crystals being slightly colored; see Chapter 7). The molecule clearly adopts different conformations in the two structures that are due to the respective orientations of the styrene, methyl ester and anilide groups around the central σ-bond (Fig. 6.6). Both structures display an intramolecular CH $\cdots$ O hydrogen bond. This kind of weak interaction (the energy is roughly half that of a classical hydrogen bond) has been intensively studied for the past decade [21].

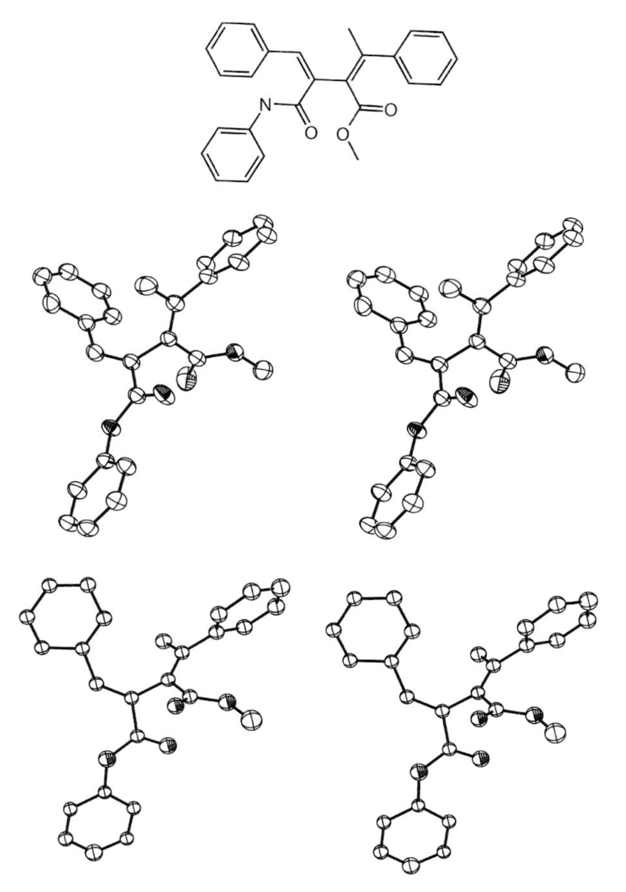

Fig. 6.6 Diagram of the molecule giving rise to the two atropisomers, which are shown as stereoviews.

Atropisomerism does not always arise because of steric hindrance, but can also be caused by non-covalent intramolecular bonds. An example is a pseudo-peptide, in which a hydroxyl group connects via a hydrogen bond to an oxygen atom of either an amide or a sulfonamide. The resulting, very different conformations may be separated by chromatography and eventually lead to two crystal structures. Interconversion between the two atropisomers reaches equilibrium in DMF after more than a week.

In many molecules certain arrangements of atoms are energetically almost equivalent and, therefore, interconvert quite easily. One of the best-known examples is the keto–enol tautomery ($H_3C–CH=O \leftrightarrow H_2C=CH–OH$). If two desmotropes (two crystals of a compound in two different tautomers) can be crystallized separately, then the crystallization process has effectively separated two entities of the same compound that coexist in solution. Strictly speaking, desmo-

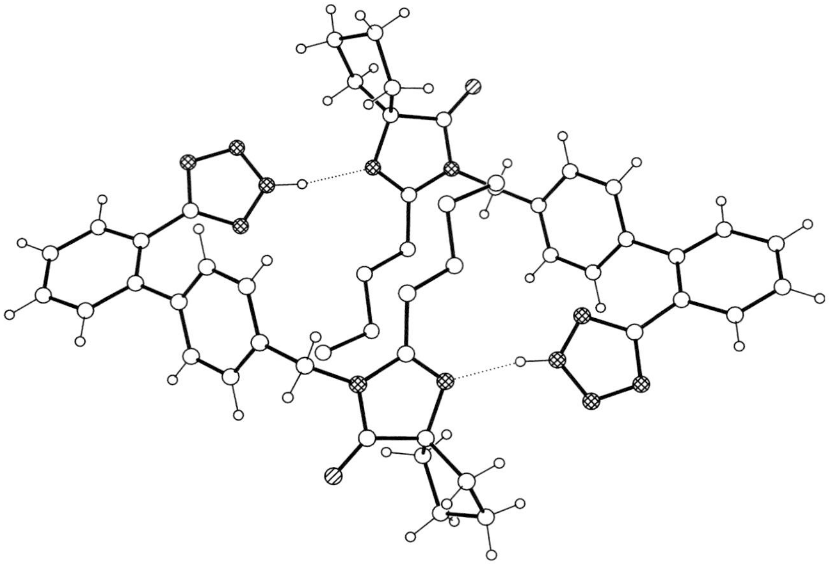

Fig. 6.7 Dimer of molecules building-up parapolymorph II of Irbesartan. Hydrogen bonds (dotted lines) are between the imidazole and a nitrogen on the tetrazole ring.

tropes are not polymorphs since their primary structures are different. But, these isomers are almost identical and similarities with other polymorphs are so striking that we consider them as polymorphs nevertheless – parapolymorphs (παρά = almost, beside) actually. There are few examples of desmotropy in the literature; a case within a diazole ring is described in [30].

Here we present the example of the solid-state tautomerism of Irbesartan [5]. Originally demonstrated by solution [29] and solid state [6] NMR, it occurs within a tetrazole ring on which the hydrogen is located either adjacent or opposite to the carbon atom. In the solid state [12] the desmotropes build up very different structures indeed, and it is truly surprising what a single proton can cause to the structural arrangement. Despite being rather different in terms of symmetry (form II is triclinic, whereas modification I is trigonal), both structures (Figs. 6.7 and 6.8) contain hydrogen bonds of comparable strength, namely a N–H···N between tetrazole and imidazole and a C–H···O between the imidazole and the biphenyl moiety (Table 6.2). The rest seems to be van der Waals interactions. Clearly, it is the different position of the tetrazole proton that induces the molecules to bend and pack differently in the two desmotropes.

Parapolymorph II (Fig. 6.7) consists of [011] chains of molecules linked by the two hydrogen bonds cited above, whereas in parapolymorph I (Fig. 6.8) the hydrogen bonds help build up helix-like [001] chains.

Another case of desmotropy occurs in a 2-(carbonylamino)pyridine moiety (Fig. 6.9). Analysis of bond distances in the two crystal structures clearly reveals

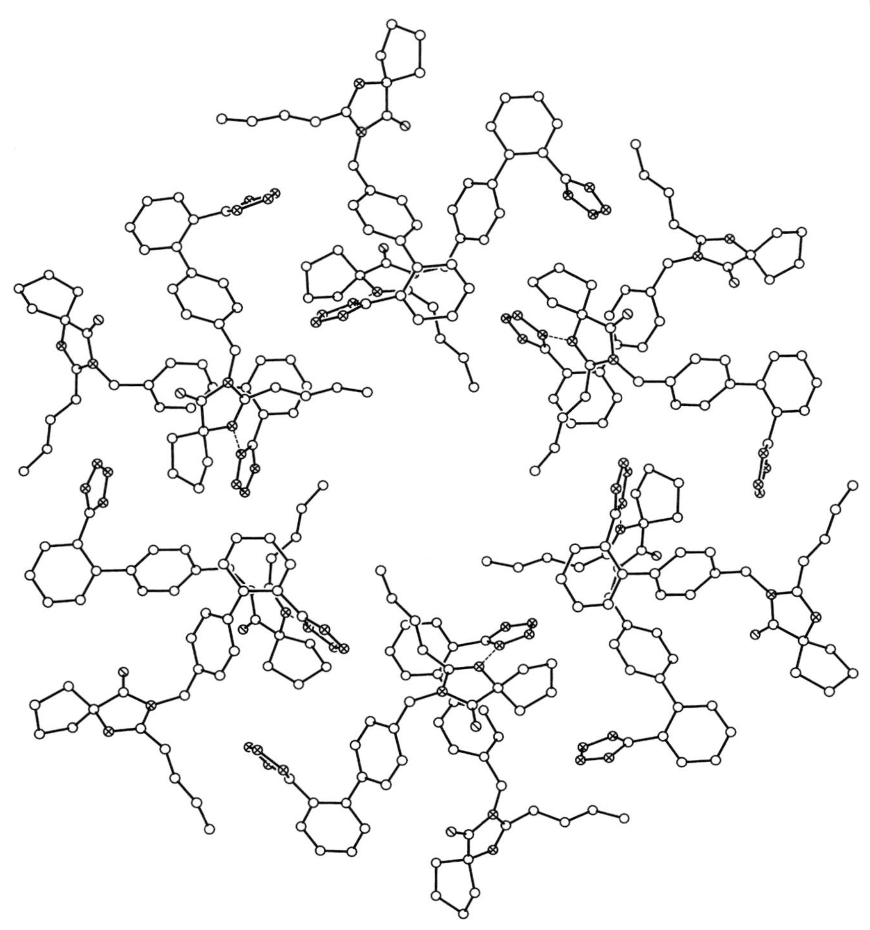

Fig. 6.8 View, along the *c*-axis, of the trigonal structure of parapolymorph I of Irbesartan. The imidazole also connects to the tetrazole ring by hydrogen bonds (dotted lines), but to an adjacent nitrogen.

Table 6.2 Hydrogen bonds participating in the example of desmotropy

Parapolymorph	Donor	H	Acceptor	D–H	H···A (Å)	D–H···A (°)	D–A (Å)
II	N_4	H_{4c}	N_2	0.90	1.90	167.14	2.783
II	C_{20}	H_{20a}	O_1	0.96	2.44	148.835	3.296
I	N_{24}	H_{24a}	N_3	0.90	1.85	163.9	2.726
I	C_{22}	H_{22a}	O_1	0.96	2.61	170.8	3.563

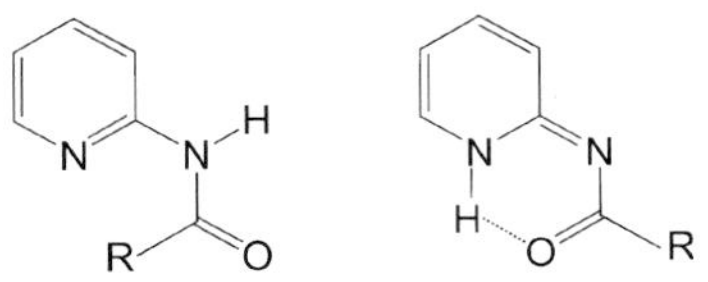

Fig. 6.9 Partial diagram of a molecule of a 2-(carbonylamino)pyridine moiety exhibiting solid-state tautomery.

two distinct sequences. Figure 6.9 depicts their Kekulé structures [–N=C–NH–C(=O)– and –NH–C=N–C(=O)]. In one of the desmotropes, an intramolecular hydrogen bond is observed between the pyridine nitrogen and the carbonyl, which leads to a rather different molecular orientation. Optimization of the complete isolated molecules containing this fragment by DFT (BLYP) with a double numerical basis set with polarization indicates that this intramolecular bond is probably the cause of the amazing stabilization ($\sim$70 kcal mol^{-1}) of this tautomer. The energy difference becomes only slightly smaller if one considers dimers since the less stable tautomer (a) associates itself in an anti-parallel manner, whereas the more stable one only establishes a few hydrophobic interactions. This is another example in which lattices energies are sufficient to stabilize an otherwise unstable molecular conformation [9]. Twinning can achieve similar effects [26, 27].

6.4 Salts

Many pharmaceutically important molecules are developed as salts. Popular salts are those of fumaric, sulfuric and hydrochloric acids. The sulfates we have studied formed two distinct schemes, namely polymeric and catemeric. In the polymeric case each acid individually coordinates one base and builds up infinite ribbons in the structure, whereas in the catomeric case one sulfate coordinates two bases and the second acid does not bind to any base. Fumaric acid molecules often associate with each other, although other schemes are also possible. For an equimolar salt: base ratio, two schemes seem to be favored. We will demonstrate this for a rather rigid molecule, of which only the quaternary nitrogen is shown in the following figures. In the first, fumaric acid lies on crystallographic inversion centers (Fig. 6.10). This special position means that one of the fumaric acid molecules is twice coordinated to nitrogen, whereas the carboxylic functions of the second acid are not involved. The arrangement of fumaric acids can be described as an undulated polymer (herringbone patterns). In the second scheme (Fig. 6.11), in which the fumaric acid molecules are crystallographically independent, a linear polymer is formed in which each acid individually co-ordinates a base. A third modification has also been isolated, in which one fumaric acid builds up molecular layers featuring isolated molecules (Fig. 6.12).

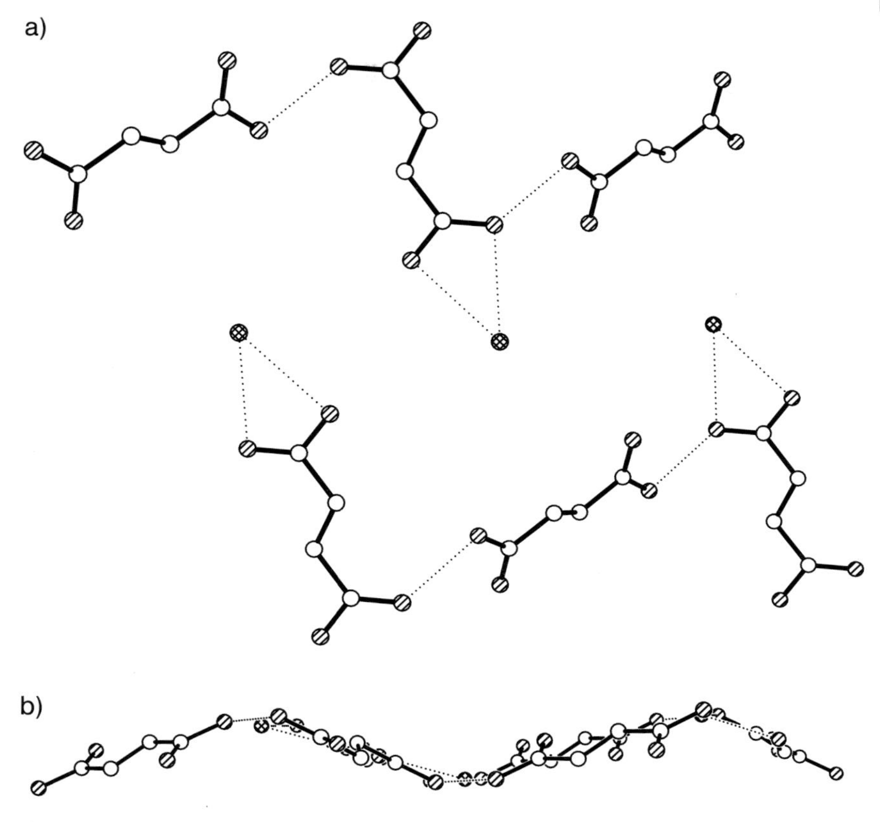

Fig. 6.10 Undulated layer of fumaric acid molecules lying on centers of inversion: (a) top view and (b) side view. Only the quaternary nitrogens (cross-hatched) of the base are shown.

6.5 Influence of Solvents

The role of the solvent is crucial in crystallization, since it may favor one phase over another [14]. Several structures may also be formed concomitantly during crystallization. This happens quite frequently for common solvents miscible with water, such as alcohols or acetone. Understanding such conglomerates is often difficult, since the real stoichiometries cannot be determined with certainty. Analytical methods such as thermogravimetry or calorimetry probe the whole of the sample and yield, therefore, an average composition only. Observation with a binocular followed by manual isolation of the various morphologies is often useful in such cases, although two crystals displaying different habiti may be built up according to the same structure. In the example below we describe three structures of a hydrochloride, where crystallization in ethanol can

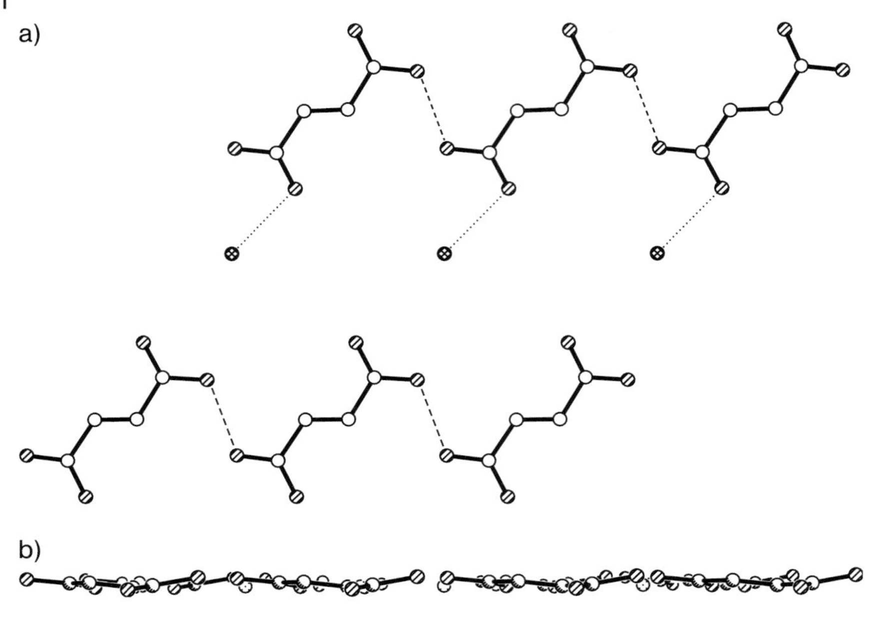

Fig. 6.11 Planar layer of crystallographically independent fumaric acid molecules: (a) top view and (b) side view. Only the quaternary nitrogens (cross-hatched) of the base are shown.

lead to three different species in a concomitant manner. The solvent-free structure is depicted in Fig. 6.13. A disordered ethanol molecule is present in the second modification (Fig. 6.14). The third modification (Fig. 6.15) is a dihydrate in which the chloride ion lies in an unexpected position. Chlorine does not directly coordinate the quaternary nitrogen; the salt bridge is established via a water molecule. We have characterized, by crystallography, another case of such a phenomenon that is responsible for a polymorphism. This dimorphism is observed for a monohydrated hydrochloride compound in which the nitrogen of a protonated piperidine is coordinated by either chlorine or a water-oxygen in the two crystal structures, respectively. Water incorporation into a hydrochloride salt bridge has been encountered for a few structures (see, e.g., CCDC code ESUCEA [37]). Energy calculations by force field methods suggest a greater stability when the salt bridge is established classically.

The stability of the solvated phases is, in general, unpredictable, but the existence of these very unstable phases that often lose their solvent molecules within seconds can nowadays be ascertained thanks to the latest generation of novel X-ray detectors. Indeed, suspensions (i.e., powders and mother liquor) can be sampled directly from the reaction vessel and be prepared as flat specimens. Subsequent data collection can be achieved within a few seconds before complete evaporation of the liquid. If channels or pores accepting solvent molecules are present, the host structures might survive after desolvation, i.e., they retain

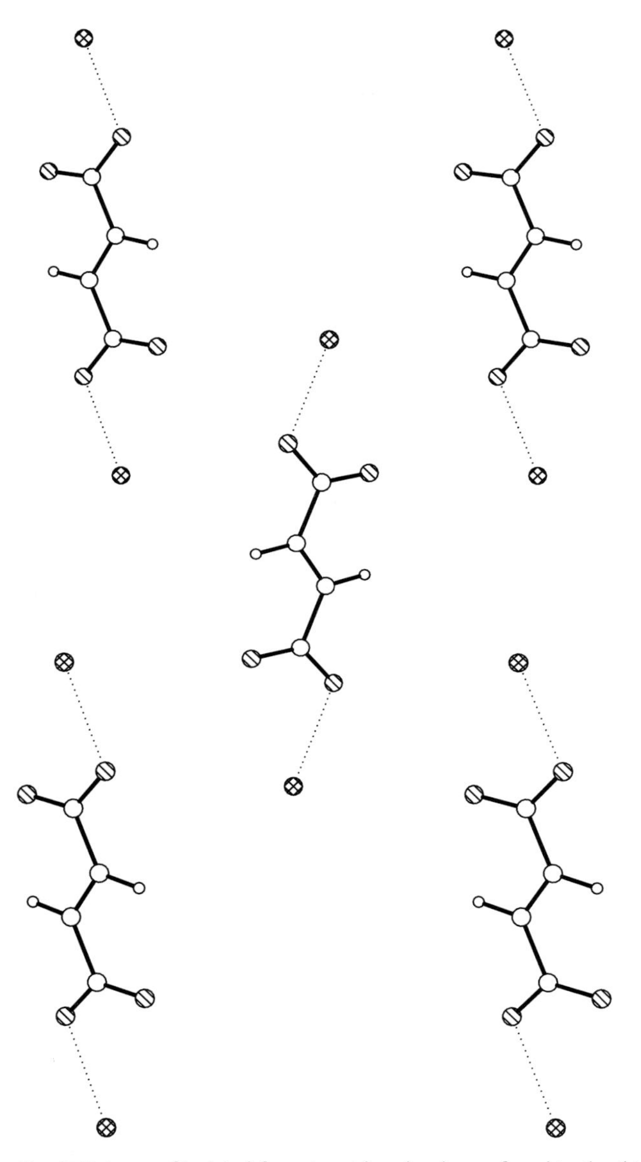

Fig. 6.12 Layer of isolated fumaric acid molecules as found in the third modification. Only the quaternary nitrogens (cross-hatched) of the base are shown.

their crystal lattice. Analyzing such phases is often problematic since results depend on the moment the sample is taken. Such high energy structures, called isomorphic desolvates [59], always transform into other structures after an undetermined time.

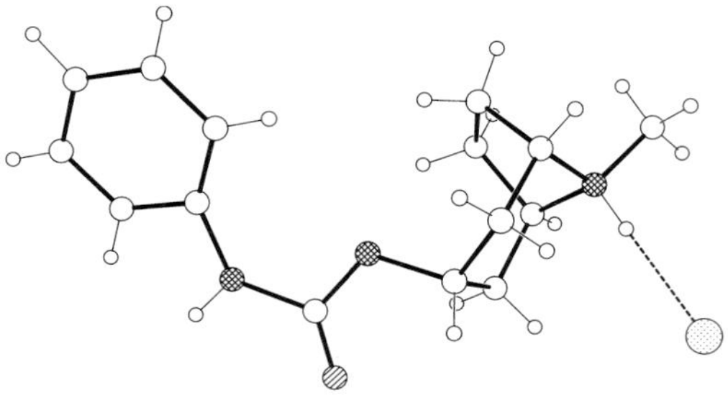

Fig. 6.13 Solvent-free form of the hydrochloride. Chlorine: dotted, oxygen: hatched, nitrogen: cross-hatched.

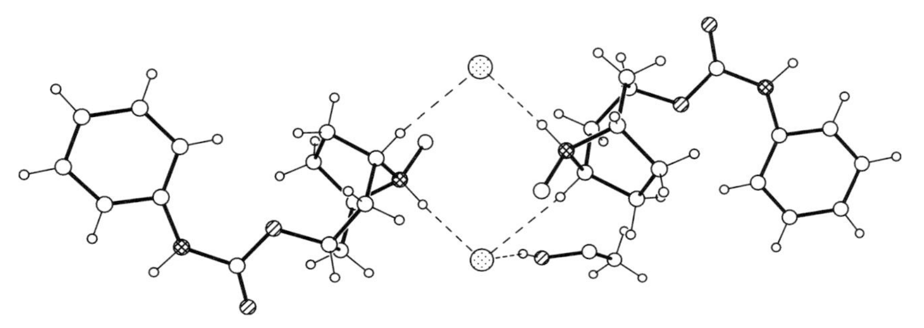

Fig. 6.14 Ethanolate modification of the hydrochloride. Chlorine: dotted, oxygen: hatched, nitrogen: cross-hatched.

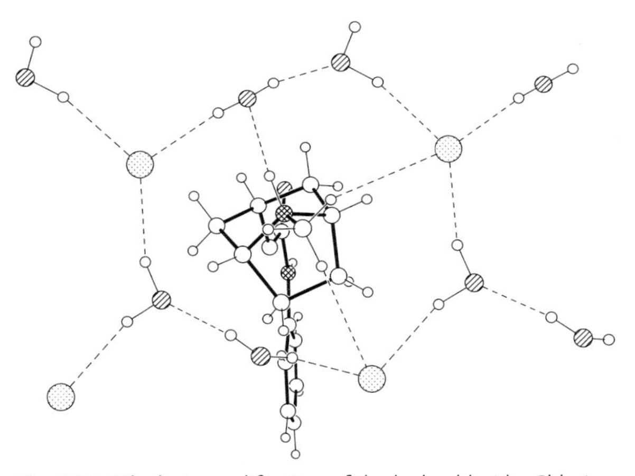

Fig. 6.15 Dihydrate modification of the hydrochloride. Chlorine and water build an unusual circular pattern, the former connecting to the methyl protons, the latter to the quaternary nitrogen proton. The water hydrogens have been located in difference maps. Chlorine: dotted, oxygen: hatched, nitrogen: cross-hatched.

6.6 Isolation of a Furtive Species

In this case, a well-known molecule formed a furtive species. The example is not truly about polymorphs, yet it convincingly illustrates how complex solvation can be and how crystallography in the widest sense (i.e., diffraction and orthoscopy teamed up with an enlightened human eye) can be used to understand it. The soothing effects of salicylic acid were known in a mystical-intuitive manner to our farthest ancestors [58]. By 2004, many powerful drugs, against a broad spectrum of illnesses, based on acetylsalicylic acid (ASA), had been elaborated, but the structure of one of the most intriguing ones, namely the complex Ca(II)[(ASA)$_2$(urea)] (CC), has only been elucidated very recently [48]. CC corresponds to *catena*-bis(μ_2,η^2-acetylsalicylato)-μ_2-urea-calcium (Fig. 6.16). The most remarkable feature of its structure is the largest Ca–O(urea) distances [2.503(6) and 2.519(6) Å] reported to date. Thus, the Ca^{2+}–O bonds acquire an almost van der Waals like character – an essential point in understanding the solvation behavior of CC.

CC has been claimed to be amazingly soluble in water (25% [64], 231 mg mL^{-1} of solution at 37 °C [53]) contrary to other ASA complexes, e.g., Cu_2ASA_4 [65]. Indeed, a very large amount of CC can be dissolved in water, until grains remain on the surface of the mother liquor. Then, however, careful scrutiny by means of a binocular reveals a fascinating story (Fig. 6.17). For, after some time, even these undissolved grains begin to disappear and, successively, two new habiti nucleate in the heart of the mother liquor, first thickish joists and about 30 s later sturdy laths. The latter turned out to be ASA, and were the only ones remaining in the mother liquor for several days, before they also degraded to salicylic acid. After about 90 s of coexistence, the joists began to disappear, but, by acting swiftly under

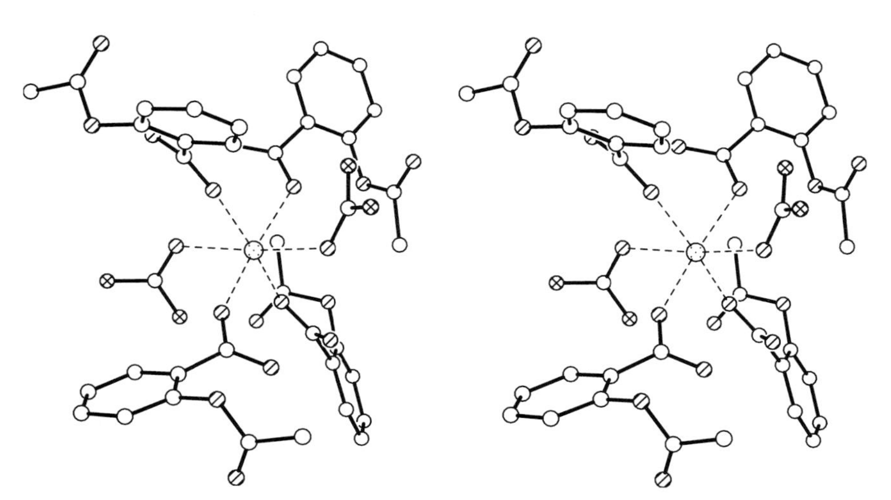

Fig. 6.16 Stereoview of Ca(II)(ASA)$_2$(urea) (CC); the Ca octahedra (approx. D_2 symmetry) share corners and form columns.

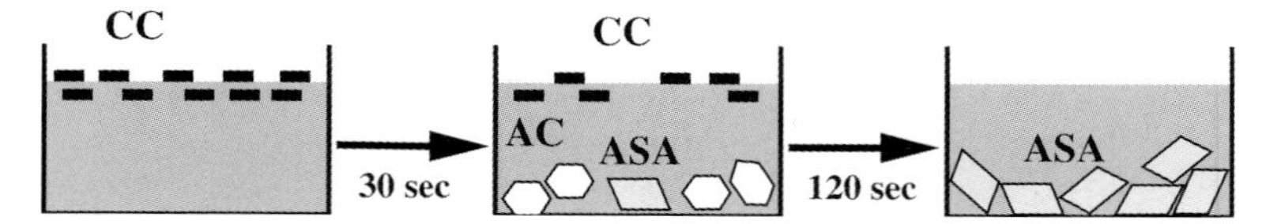

Fig. 6.17 Schematic representation of the degradation of Carbasalatum Calcicum (CC).

a binocular, crystals of this furtive form could be isolated and, being stable enough in air, analyzed by X-ray diffraction, revealing that it is *catena*-bis(μ_2,η^2-acetylsalicylato-aquo)-calcium dihydrate (AC, Fig. 6.18 [42]). Crystallography has thus established the following sequence of events: In a supersaturated aqueous solution, coördinated urea in CC becomes rapidly unstable and is replaced by water molecules. This instability might be mirrored in the particularly long Ca–O(urea) distances. However, the thus-formed AC does not survive for long either and the acetylsalicylato groups are quickly exchanged by water molecules, the resulting dissolved species being $[Ca(H_2O)_6] \cdot ASA_2$. Therefore, the high solubility of CC is only a kinetic effect.

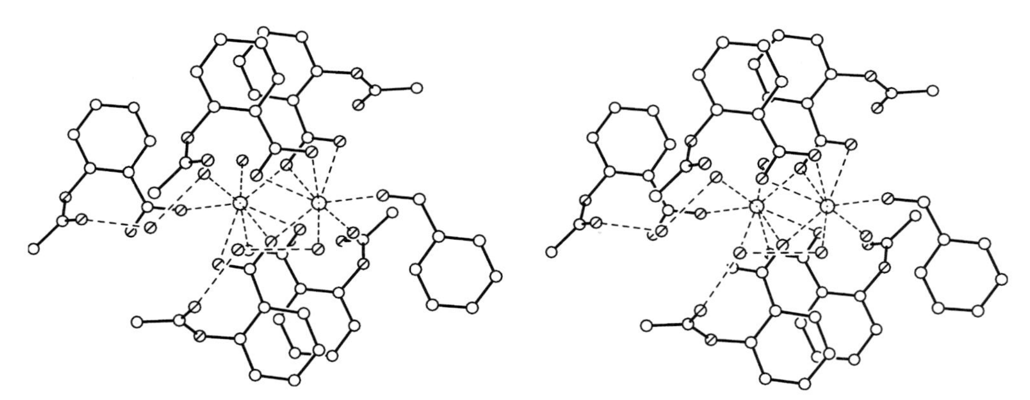

Fig. 6.18 Stereoview of Ca(II)(ASA)$_2$(aquo)$_2$ dihydrate. The Ca dodecahedra (approx. C_{2v} symmetry) share edges and form columns.

6.7
Mizolastine Polymorphs

The Mizolastine molecule builds up several polymorphs and pseudopolymorphs. It is a case study of a polymorphism that could only be understood by means of single-crystal structure determinations [50]. The hydrated phase arising from a crystallization in a mixture of alcohol and water shows normal behavior by DSC, i.e., a single melting point above 150 °C whatever the heating rate (1, 5 or 10 °C min^{-1}). The water content, established by Karl Fischer analysis, suggests that Mizolastine could form clathrates (i.e., non-stoichiometric

Table 6.3 Unit cells of Mizolastine crystallized from ethanol/water. Cells I and II are those of the two polymorphs. Cell III corresponds to the transform of cell I, and cell IV is that of the bilayer.

Cell	*a* (Å)	*b* (Å)	*c* (Å)	α (°)	β (°)	γ (°)	
I	9.4556	11.874	20.0575	88.655	82.657	81.996	$P\bar{1}$
II	11.8715	17.7034	21.0147		92.505		$P\frac{2_1}{n}$
III	11.874	9.456	21.053	70.89	92.30	98.00	$P\bar{1}$
IV	11.8728	8.84	21.034		92.403		$P\bar{1}$

hydrates) in this mixture of solvents. The morphology of the sample is ill-developed; it consists of entangled aggregates of inhomogeneously sized and irregularly shaped crystals. This problematic powder gives rise to an X-ray powder diagram characterized by weak and broad reflections. Unsurprisingly, all attempts at indexing the diagram failed. After careful observation by means of a binocular it was possible to isolate two types of habiti whose structures could be determined. The hydrated Mizolastine phase is therefore not unique, the concomitant polymorphs belong to different crystal systems (monoclinic and triclinic, Table 6.3) and their calculated powder diagrams are clearly different.

Only knowledge of the crystal structures can explain why this material crystallizes as a mixture of polymorphs. The moderately flexible molecule is made up of four ring systems (benzimidazole, piperidine, pyrimidone and fluorophenyl) that are separated by no more than one atom. Despite several chemical recognition sites, the same molecular association is shown to be present in all four crystal structures so far determined, namely anti-parallel dimers made up of pyrimidinones connected by cyclic N–H $\cdots$ N hydrogen bonds (Fig. 6.19). Both monohydrated modifications are layer structures in which water molecules coordinate two nitrogen atoms that belong to a benzimidazole and a pyrimidone of two neighboring molecules (Fig. 6.20). These two modifications are order–disorder (OD) structures [19].

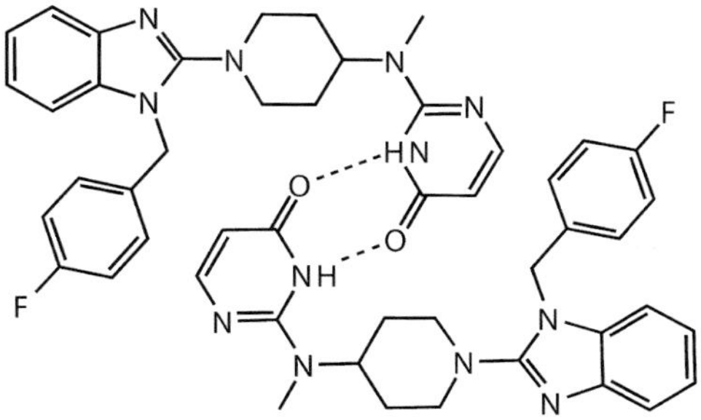

Fig. 6.19 Chemical formula of a folded dimer of Mizolastine in the solid state.

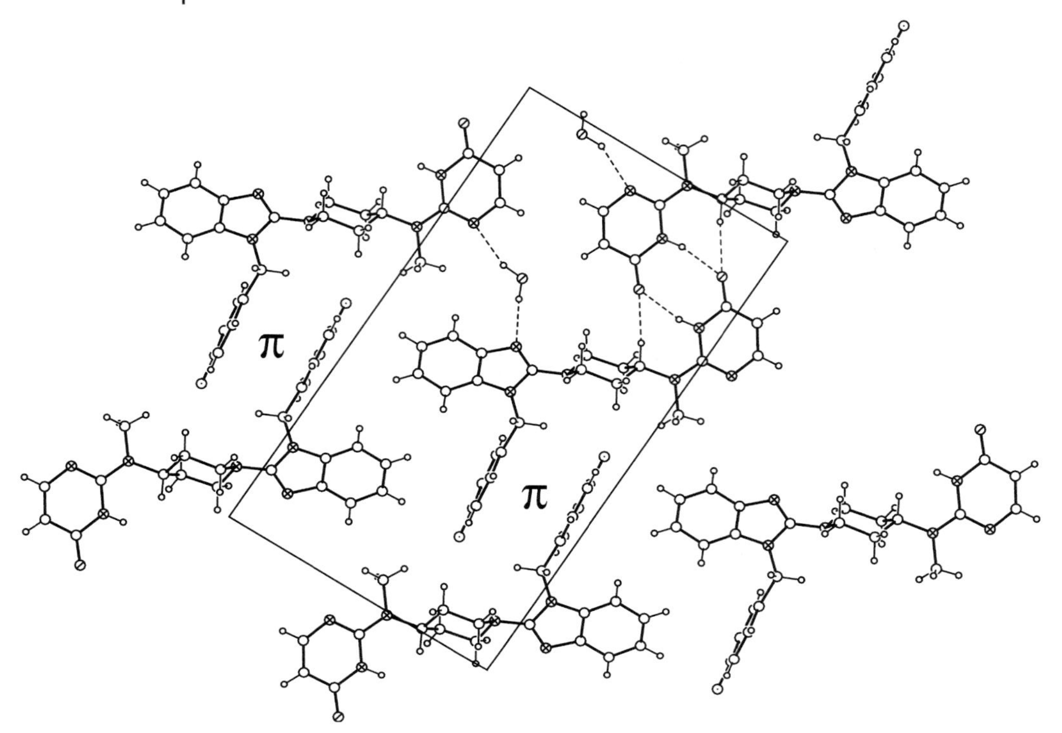

Fig. 6.20 Mizolastine dimer building up the layer in both monohydrated modifications. Hydrogen bonds (dotted) and π–π (denoted π) interactions are indicated.

To clearly see this, one needs to transform cell I by means of the matrix:

$$\begin{pmatrix} 0 & 1 & 0 \\ \bar{1} & 0 & 0 \\ \bar{1} & 0 & 1 \end{pmatrix}$$

This transformation leads to cell III (Table 6.3) and the layer group (cell IV) is seen to be $P\bar{1}$. The layer is formed by hydrogen-bonded cyclic dimers of folded molecules (Fig. 6.20). The supramolecular synthon corresponds to one bilayer of molecules. The major difference between the structures lies in their layer stacking.

The two polymorphs represent so-called MDO (Maximal Degree of Order) structures [19] of this OD structure that can be characterized by means of the OD groupoid shown below and the lattice constants of cell IV.

$$P\ \bar{1}\ (1)$$

$$\{1\ (2_2)\ 1\}$$

Indeed, polymorph I is the MDO structure with the cell $\mathbf{a}_3=\mathbf{a}_{IV}$, $\mathbf{b}_3=-0.1\mathbf{a}_{IV}+\mathbf{b}_{IV}+0.15\ \mathbf{c}_{IV}$, $\mathbf{c}_3=\mathbf{c}_{IV}$, ($\alpha_3=70.24°$, $\gamma_3=98.00°$) and polymorph II is the MDO structure with the cell $\mathbf{a}_2=\mathbf{a}_{IV}$, $\mathbf{b}_2=2\mathbf{b}_{IV}$, $\mathbf{c}_2=\mathbf{c}_{IV}$, ($\beta_2=\beta_{IV}$). In polymorph I it is the $-0.1\mathbf{a}_{IV}+\mathbf{b}_{IV}+0.15\mathbf{c}_{IV}$) translation that has a continuation in all *n*-tuples of layers and in polymorph II the 2_2 partial operation (PO) [19].

When Mizolastine is crystallized from pure ethanol, it displays another interesting feature, which follows Ostwald's famous Law of Stages [46]. A homogeneous oil is formed that remains stable for several weeks at room temperature. Under these conditions, nucleation of a metastable modification can then be initiated by mechanical means (e.g., by slightly stimulating the oil with a metallic needle). This modification had to be handled below 10 °C and a preliminary X-ray experiment at 140 K revealed an unusually large lattice constant of 65 Å. The orthorhombic broadsword-like crystals were, however, not stable in the drop, but subject to an ongoing cycle of nucleation and dissolution. After several weeks, the drop would vanish and leave crystals of another non-solvated phase that can be obtained from several solvents but with very different crystal morphologies, and probably corresponds to the thermodynamically stable one. Since it displays many modifications, the Mizolastine molecule has been investigated by molecular modeling, by semi-empirical (AM1, [3]) and force field (SYBYL, [60], Dreiding, [43]) methods. Conformational analyses by dynamic and torsional grid searches have been performed prior to crystal packing simulations [31, 39]. Lattice energy calculations were carried out following the method described in Refs. [4, 61]. The results suggest that anti-parallel folded dimers are definitively more stable energetically (by more than 3 kcal mol^{-1}) than an extended dimer based on a 180° rotation of the $-N(CH_3)-$ pyrimidinones single bond.

6.8 Solid Solutions

The story of solid solutions began in the Bronze Age, when artisans learnt to influence mixtures of copper and tin in a controlled way to obtain CuSn alloys suited for a particular application. A more scientific approach was developed in the 1930s with the advent of X-ray powder diffraction. Substitutional (α-brass), interstitial (austenite steel) and defect lattice (β-NiAl) solid solutions played a key rôle in the advance in almost every domain of human technology. The study of their complicated phase diagrams, transformations and kinetics also led to many theoretical concepts [11]. The phenomenon, as the example below demonstrates, is, however, also encountered with organic molecules, provided the participating species show significant structural similarities. This is the case, for example, with regioisomers or stereoisomers, compounds that one typically wishes to separate for industrial purposes. Note that structurally similar molecules can be both flexible – we present such an example – or quasi-rigid, as with methyprylon [49], CCDC codes BEPHUY and BEPHUZ. The rather fasci-

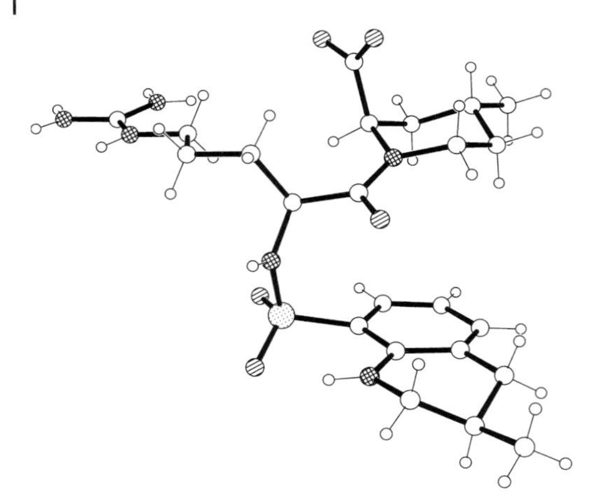

Fig. 6.21 Molecular structure of argatroban in a crystal grown from the pure (*R*)-diastereomer. The methyl group of the tetrahydroquinoline lies in an equatorial position. Oxygen: hatched, nitrogen: cross-hatched, sulfur: dotted.

nating peculiarity of a solid solution lies in the internal constraints created while accommodating the two species on a common site in the crystal [18]. This is reflected by the fact that, when crystallized as isolated species, the two components generally do not share a common crystal structure. The pseudopeptide *argatroban* (a synthetic thrombin inhibitor) is an example of such a situation [45]. Argatroban is made up of an L-arginine chain substituted by methyl pipecolic acid and by methyl tetrahydroquinoline. This rather bulky molecule has four chiral centers, three of which are determined. The last chiral center is located on the quinoline moiety where it connects the methyl group. This fourth chiral atom is responsible for the existence of the two diastereomers, named (*R*)-argatroban (Fig. 6.21) and (*S*)-argatroban (Fig. 6.22).

As established by HPLC, attempts at increasing the yield of the desired isomer during crystallization of argatroban in alcohol lead only to a 2:1 mixture of

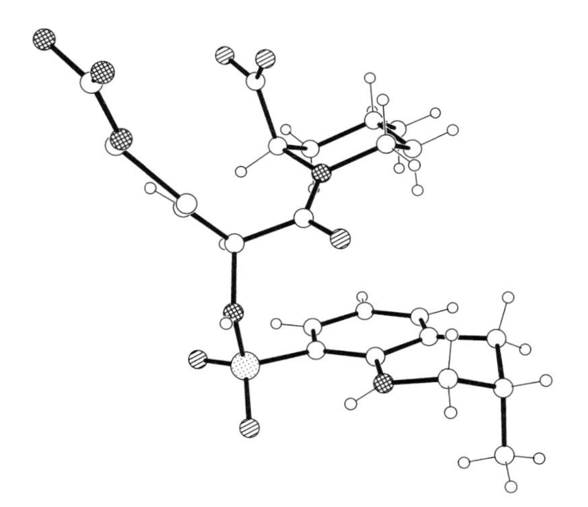

Fig. 6.22 Molecular structure of (*S*)-argatroban in the crystal. The methyl group of the tetrahydroquinoline lies in an axial position. Oxygen: hatched, nitrogen: cross-hatched, sulfur: dotted.

Table 6.4 Structural data for stereoisomers of argatroban.

	a (Å)	*b* (Å)	*c* (Å)	*α* (°)	*β* (°)	*γ* (°)	
SS	6.4303(9)	11.0297(14)	19.382(3)	90	93.822(2)	90	$P2_1$
R	6.4278(9)	11.0390(15)	19.399(3)	90	94.291(2)	90	$P2_1$
S	9.251(2)	15.452(3)	21.365(4)	90	90	90	$P2_12_12_1$

(*R*)-argatroban (major component) and (*S*)-argatroban (minor component). The crystal structure of this mixed crystal (argatroban), as well as those of the isolated diastereomers, is available. It is remarkable that (*R*)-argatroban and argatroban crystallize in very similar structures (Table 6.4) with respect to metric (monoclinic) and space group ($P2_1$) but that (*S*)-argatroban builds up a rather different crystal structure. Both molecular structures of the pure isomers exhibit the same main feature in the solid state, namely intramolecular non-covalent interactions corresponding to a hydrophobic collapse. These interactions link the piperidine and quinoline moieties. Apart from the flexible alkyl chain, their main discrepancy is the position of the methyl substituting the quinoline: it lies in an axial position in (*S*)-argatroban and it is located equatorially in (*R*)-argatroban. (*S*)-Argatroban coordinates several organic solvent molecules and crystalline water that build up into a complicated 3D hydrogen bonding network in its orthorhombic crystal structure, whereas the structure of (*R*)-argatroban is only monohydrated and stabilized by many rather strong N–H · · · O hydrogen bonds. The structure of argatroban itself corresponds to a solid solution, which, crystallographically, is disordered. Indeed, the fourth chiral atom shows two positions in the solid state: one of absolute configuration (*R*) below the quinoline mean plane, and the other of absolute configuration (*S*), above said plane (Fig. 6.23).

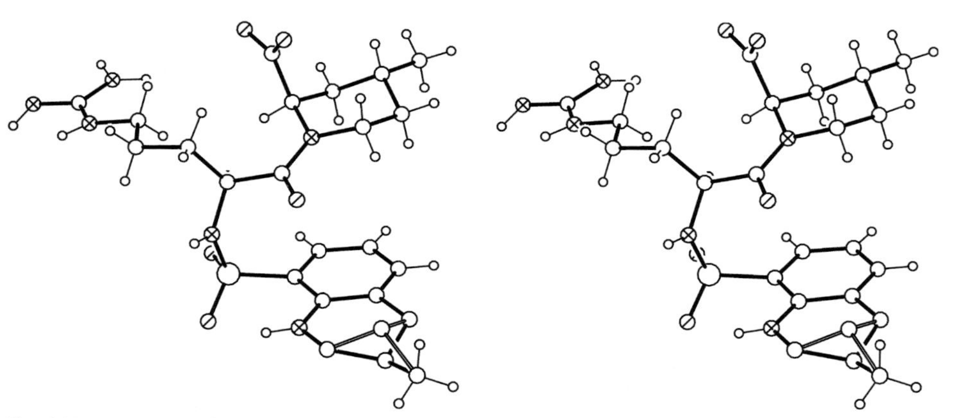

Fig. 6.23 Stereoview of the structure of a solid solution of argatroban. The non-resolved chiral centre is twice disordered. A population ratio of 2:1 was refined at 150 K.

It is therefore necessary to invert chirality (i.e., break a chemical bond) to change the "S" position into an "R" position, which demonstrates that argatroban effectively corresponds to a solid solution and not a mere kind of disorder. Hence, for the (*S*)-argatroban molecule, the methyl group is no longer axial. Instead, the "recessive" species accommodates its molecular conformation to fit the crystal structure of the "dominant" molecule [(*R*)-argatroban]. This 2:1 (*R*)/(*S*) solid solution adopts the (*R*) structure whereas the (*S*) species forms a unique structure. Since the (*S*) isomer is considerably distorted in the solid solution, owing to its incorporation in an un-natural environment, its structure must probably be considered of higher potential energy. This hypothesis is further confirmed by X ray powder diffractometry, which shows that the structure of the solid solution is preserved after crystallization of a 1:1 (*R*)/(*S*) mixture. Crystallographic study of the argatroban solid solution convincingly explains why there is no way of industrially separating the two argatroban species.

6.9 Structures from Powder Data

Growing single crystals suitable for the structure determination of pharmaceutically relevant molecules is often difficult, usually owing to (a) low molecular symmetry of the compound, (b) strong intermolecular hydrogen bonds favoring crystal growth in a preferred direction, resulting in needle-like crystal morphologies, and (c) low solubility of the sample (which is often a salt) in organic solvents. When studying polymorphism, this issue becomes even more crucial since it is generally not possible to grow sufficiently large single crystals for each polymorphic modification. When single crystals are not available, things become, unsurprisingly, much more involved. Sometimes it is preferable though to deal with a good powder (high scattering power, good crystallinity, isotropic on all levels) rather than with one or several entangled paracrystals. In general, one must invest as many weeks in a structure analysis from powders as one invests hours for a normal single crystal. Fortunately, the past decade has witnessed spectacular progress on the experimental as well as on the theoretical-computational front of powder diffractometry. There are two major bottlenecks associated with structure elucidation from powders: (a) indexation of the maxima and (b) finding a structural model. There are powerful tools available for overcoming these difficulties, but the success rate is still low, judging from the number of structures stored in the CCDC that have been solved from powders. Resolution is clearly the key to reliable indexation. Superior resolution can be achieved with either an experiment on a synchrotron or a rather long wavelength (e.g., Cu or Cr). Synchrotron radiation has considerably enhanced structural information that can be derived from powder diffraction data [33]. By providing very narrow and accurately positioned diffraction intensities, synchrotron radiation can resolve low symmetry powder patterns containing severely overlapped peaks. Moreover, the high brilliance of the beam is adapted to the low scattering power characteristic of organic molecules [38]. Half a dozen different mathematical approaches are available for indexing a powder pattern, the most powerful ones taking into account a few impurity lines. Notably, one can only trust a unit cell after the successful refinement of a meaningful structure. But, prior to this, one has to find a structural model and this is hampered, often enough, by a cut-off in the scattering angle, which is too low. Classical direct methods rely on a certain minimum resolution and fail quite regularly if this is not available. Access to low temperature data can help circumvent this problem. With the advent of so-called reverse Monte Carlo (MC) (or direct-space) techniques, this deficiency has lost some of its negative consequences. The basic principle of direct-space approaches is the simulation of powder diffraction patterns while varying the positions, orientations and internal degrees of freedom of molecular fragments within the unit cell, until an optimal match with the experimental pattern is achieved [34]. For flexible molecules, the limiting factor in direct-space approaches is the large number of torsion angles to be considered, which results in prohibitively long computing times. Application of this method to disordered structures is also limited because in such a case additional atomic positions have to be considered and it is generally not possible to define the initial molecular model. A flexible hydrochloride of 29 non-hydrogen atoms was our first case of a crystal structure successfully solved from high-resolution powder diffraction data, in a joint project with MSI Inc., for the set-up using the *Powder Solve* software [28]. The synchrotron diagram of this compound (Fig. 6.24), recorded on beam line BM1 at the ESRF, was readily indexed in a triclinic unit cell; the final agreement factor for the pattern reached $R_{wp}=0.08$. Several million trial structures were generated before the program could solve the crystal structure. Molecular flexibility was modeled by allowing selected torsional degrees of freedom (DOF) within the fragments to vary. In addition to three rotations and six translations (3+3 from the counter ion), seven torsions in a nine-atom long chain connecting a hetero five-membered ring to a quinazoline bicycle were defined. The calculations were quite discriminating; they only produced two structures with R_{wp} factors below 30% that turned out to be equivalent by crystallographic translation.

Owing to hydrophobic interactions, the structure essentially consists of dimers of parallel molecules arranged in a head-to-tail fashion, with strong lateral hydrogen bonds (Fig. 6.25) giving an overall "scorpion-like" molecular conformation. This structure has been solved, as described in 1999, allowing 16 DOF. In 2005 we are able to deal with more complicated molecules, consisting of up to 40 DOF. Despite such success, however, powder experiments provide limited information about finer structural features, especially the centrosymmetricity of the structures. But this question is of particular importance in the example above, since the molecule contains a chiral center. To gain complementary information, we undertook a Second Harmonic Generation [41] (SHG) study of several samples. An observed SHG signal indicates a non-centrosymmetric point group (except for the point groups 432, 422 and 622, but, luckily, few organic molecules crystallize in them). By irradiating our samples with a Ti/Sapphire laser

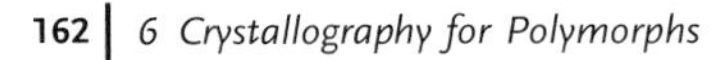

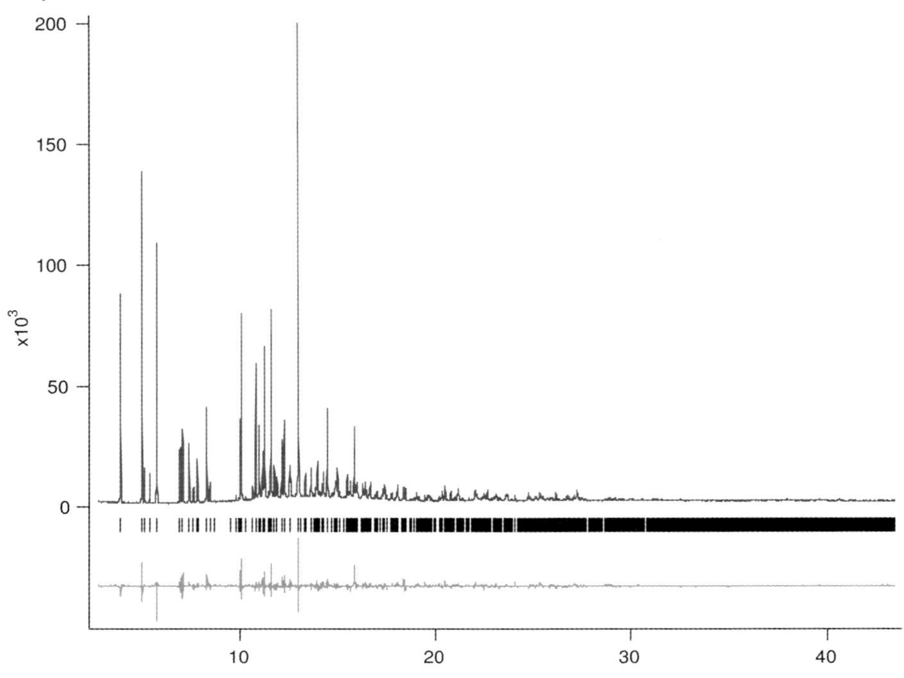

Fig. 6.24 Observed, calculated and difference reflection profiles resulting from Rietveld refinement. Data were collected at a $\lambda = 0.80193(1)$ Å. Final R were $R_p = 0.067$ and $R_{wp} = 0.091$.

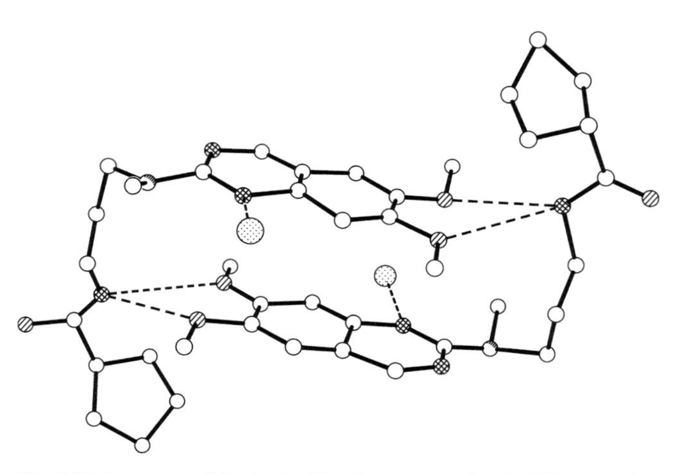

Fig. 6.25 Structure of the hydrochloride compound solved from synchrotron powder data. The chlorine coordinates a nitrogen atom of the fused cycle via an ionic bond at a distance of 3.1 Å. Oxygen: hatched; nitrogen: cross-hatched; chlorine: dotted.

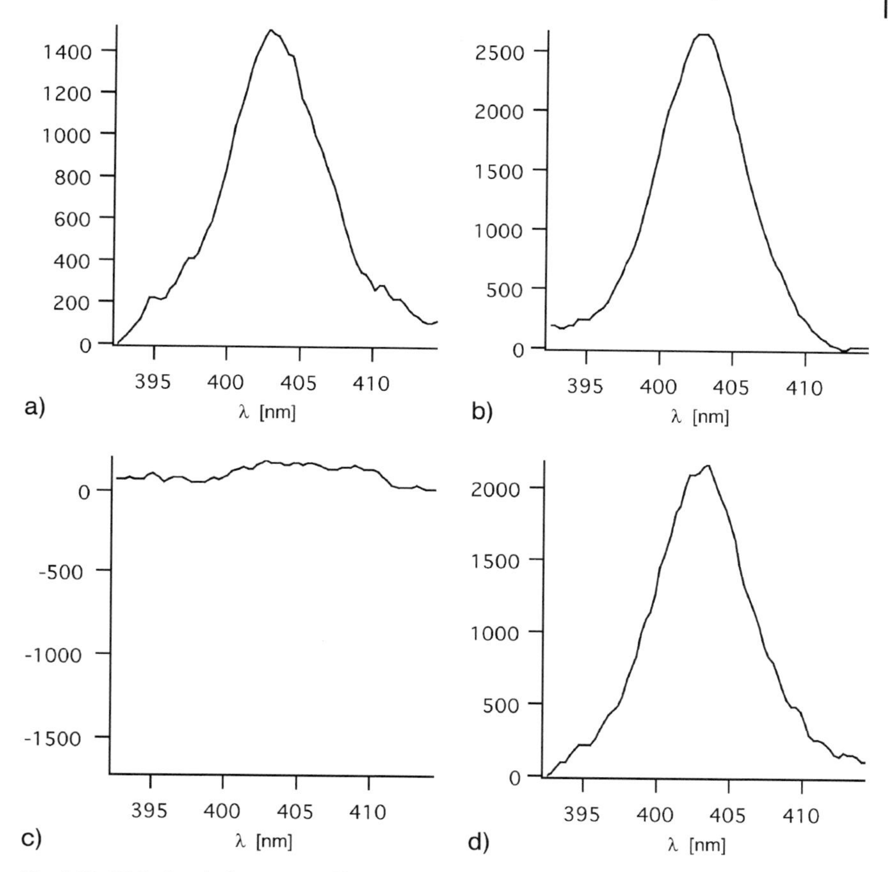

Fig. 6.26 SHG signals for (a) (*S*)-Alfuzosine, (b) (*R*)-Alfuzosine, (c) Alfuzosine racemate, (d) a thoroughly ground 1:1 conglomerate of (*S*)- and (*R*)-Alfuzosine.

(λ=800 nm), we observed in the backscattering region a surprisingly strong signal in the UV for the pure (*S*) and (*R*) enantiomers (for which crystalline powders were also available) and for a mechanical 1:1 (*S*)/(*R*) conglomerate (obtained by grinding), but no signal for the sample that was used for the structure described above (Fig. 6.26). From this we may conclude that (a) there are no spurious effects from surfaces or interfaces, (b) the (*R*) and (*S*) samples are true enantiomers and (c) that the compound studied effectively corresponds to a racemate.

High-resolution powder diffractometry has also enabled us to determine the structures of certain carbohydrates. These are highly flexible molecules owing to the existence of glycoside bonds linking the monosaccharide units. To reduce the number of DOF in the MC procedure, the relative orientations of two trisaccharide units were kept invariant according to the geometries established from solution NMR [51]. Solid-state NMR has also successfully been applied to this problem [56]. Even with the precision of high-resolution synchrotron data combined with the continuously increasing power of modern hard- and software, determining crystal structures by direct-space approaches remains a huge piece of work. With the area detectors available today, such research should, in our opinion, only be undertaken after all attempts to obtain structural information from even a minute single crystal have failed.

6.10
"Behind Every Structure There is a Crystal"

The number of structures deposited with the Cambridge Crystallographic Data Centre or the Protein Data Base keeps growing. This boom testifies that crystallography is and remains an indispensable science at the interface of major disciplines such as physics, chemistry and biology. Every crystallographer taking interest in a structure, be it that of a small molecule or a protein, is well aware that (s)he may have to investigate several crystals before one will yield the details of its essence, and that this task will require intuition, experience, skill and patience. Often, an initial victory may later be spoilt by the appearance of a polymorph that has escaped detection and subsequently complicates matters.

It is not uncommon for polymorphs to share one or several lattice constants, occasionally almost quantitatively. This is certainly the case for OD structures, since they contain a clearly identifiable common structural element. But the similarity is sometimes – as in our Mizolastine example – obscured by an unlucky choice of unit cells. Polymorphs that are topologically very different may have no lattice constant in common. This is the case for the recently discovered polymorph of acetic acid [52], in which the synthons build up trigonal channels, whereas in the well-known *vinaigre* [36] they form endless chains.

We may say that, in general, different polymorphic states are not related by 3D group–subgroup relations, not even with a (hypothetical), higher symmetric mother phase. But we have presented above some cases in which part of the molecular disposition is preserved or at least cognate. Therefore, we believe in the viability of a phenomenological approach of "polymorphic transitions" based on the identification of Landau type order parameters partially leaving invariant the crystal edifice, and reconstructive parameters pertinent to the rearranging of the lattice [16].

"Behind every structure there is a crystal." Despite the hesitant beginnings of theories of nucleation [14], systematic application of this paradigm in the inverse sense is far from being achieved. Programs conceived for predicting crystal packing, despite being able to explain – but rarely predict – certain possible associations of two chemical entities, are in no situation to claim the contrary. In this spirit, let us close by restating that the study of polymorphism has given new life to a simple, regretfully neglected, tool, namely careful orthoscopic observation of crystals [55], which may lead to detection, isolation and characterization of polymorphs and even of furtive species.

References

1 F. A. Allen, O. Kennard, G. D. Watson, in *Structure Correlation* (Eds. H.-B. Bürgi, J. D. Dunitz), VCH, Weinheim, **1994**, pp. 71–110.

2 A. Altomare, M. C. Burla, G. L. Cascarano, C. Giacovazzo, A. Guagliardi, A. G. G. Moliterni, G. Polidori, EXPO, *J. Appl. Cryst.*, **1999**, *32*, 339–340.

3 M. J. S. Dewar, E. G. Zoebisch, E. F. Healy, J. J. P. Stewart, *J. Am. Chem. Soc.*, **1985**, *107*, 3902–3909.

4 D. Buttar, M. H. Charlton, R. Docherty, J. Starbuck, *J. Chem. Soc., Perkin Trans. II*, **1998**, 763–772.

5 M. Bauer, in *Cristallisation et Polymorphisme*, Techniques de l'Ingénieur, Paris **2004**, *AF3 640*, 1–12.

6 M. Bauer, R. Harris, R. Rao, D. Apperley, C. A. Rodger, *J. Chem. Soc., Perkin Trans. II*, **1998**, 475–481.

7 J. Bernstein, *Polymorphism in Molecular Crystals*, IUCR Monographs on Crystallography, Oxford University Press, New York, **2002**, p. 151.

8 J. Bernstein, R. J. Davey, J. O. Henck, *Angew. Chem. Int. Ed.*, **1999**, *38*, 3440–3461.

9 J. Bernstein, A. T. Hagler, *J. Am. Chem. Soc.*, **1978**, *100*, 673–681. J. Bernstein, in *Organic Solid State Chemistry*, Vol. 32: *Studies in Organic Chemistry* (Ed. G. R. Desiraju), Elsevier, Amsterdam, **1987**, p. 471.

10 N. Blagden, R. J. Davey, *Cryst. Growth Des.*, **2003**, *3(6)*, 873–885.

11 C. P. Brock, W. B. Schweizer, J. D. Dunitz, *J. Am. Chem. Soc.*, **1991**, *113*, 9811–9820.

12 Z. Böcskei, K. Simon, R. Rao, A. Caron, C. A. Rodger, M. Bauer, *Acta Crystallogr., Sect. C*, **1998**, *54*, 808–810.

13 J.-F. Bérar, *IUCr Satellite Meeting on "Powder Diffractometry"*, Toulouse, **1990**.

14 N. Blagden, R. J. Davey, H. F. Lieberman, L. Williams, R. Payne, R. Roberts, R. Rowe, R. Docherty, *J. Chem. Soc., Faraday Trans.*, **1998**, 1035–1044.

15 A. Boultif, D. Louër, *J. Appl. Cryst.* **1991**, *24*, 987–993.

16 M. Bonin, personal communication.

17 W. C. McCrone, in *Physics and Chemistry of the Organic Solid State* (Eds. D. Fox, M. M. Labes, A. Weissberger), Interscience, New York, **1965**, Vol. II, p. 725.

18 J. Jacques, A. Collet, S. H. Willen, in *Enantiomers, Racemates and Resolutions*, Wiley, New York, **1994**, p. 104.

19 K. Dornberger-Schiff, *Lehrgang über OD-Strukturen*, Akademie Verlag, Berlin, **1966**.

20 P. G. Dunbar, G. J. Durant, W. P. Hoss, W. S. Messer, Jr., A. A. Pinkerton, S. Periyasamy, S. Ghodsi-Hovespian, B. Ellerbrock, B. A. Kroa, *Med. Chem. Res.*, **1991**, *1*, 119.

21 G. R. Desiraju, *Acc. Chem. Res.*, **1991**, *24*, 290–296.

22 G. R. Desiraju, *Angew. Chem. Int. Ed. Engl.*, **1995**, *34*, 2311–2327.

23 R. G. Davey, N. Bladgen, G. D. Potts, R. Docherty, *J. Am. Chem. Soc.*, **1997**, *110*, 1767–1772.

24 C. Drexler, M. W. Hosseini, J.-M. Planeix, G. Stupka, A. De Cian, J. Fischer, *J. Chem. Soc., Chem. Commun.*, **1998**, 689–690.

25 The crystal structure of the 3α-β atropisomer described in [24] has been determined (C. Drexler, P. Ochsenbein, M. Bonin, K. J. Schenk, unpublished results).

26 R. J. Davey, S. J. Maginn, S. J. Andrews, A. M. Buckley, D. Cottier, P. Dempsay, J. E. Rout, D. R. Stanley, A. Taylor, *Nature*, **1993**, *366*, 248–250.

27 R. J. Davey, S. J. Maginn, S. J. Andrews, S. N. Black, A. M. Buckley, D. Cottier, Ph. Dempsey, R. Plowman, J. E. Rout, D. R. Stanley, A. Taylor, *J. Chem. Soc., Faraday Trans.*, **1994**, *90*, 1003–1009.

28 G. E. Engel, S. Wilke, O. König, K. D. M. Harris, F. J. J. Leusen, *J. Appl. Cryst.*, **1999**, *32*, 1169–1179.

29 M. El-Hajji, **2005**, personal communication.

30 C. Foces-Foces, A. L. Llarmas-Saiz, R. M. Claramunt, C. Lopez, J. Elguero, *J. Chem. Soc., Chem. Commun.*, **1994**, 1143–1145.

31 R.J. Gdanitz, in *Theoretical Aspects of Computer Modeling of the Molecular Solid State* (Ed. A. Gavezzotti), Wiley, Chichester, **1997**, pp. 185–199.
32 R. Gilardi, *Acta Crystallogr., Sect. C*, **1987**, *43(5)*, 1002–1003.
33 J.B. Hastings, W. Thomlinson, D.E. Cox., *J. Appl. Crystallogr.*, **1984**, *17*, 85–95.
34 K.D.M. Harris, M. Tremayne, B.M. Kariuki, *Angew. Chem. Int. Ed.*, **2001**, *40*, 1626–1651.
35 K. Harano, M. Yasuda, Y. Ida, T. Komori, T. Taguchi, *Cryst. Struct. Commun.*, **1981**, *10(1)*, 165–171.
36 R.E. Jones, D.H. Templeton, *Acta Crystallogr.*, **1958**, *11*, 484–487.
37 S. Kodato, J.T.M. Linders, X.-H. Gu, K. Yamada, J.L. Flippen-Anderson, J.R. Deschamps, A.E. Jacobson, K.C. Rice, *Org. Biomol. Chem.*, **2004**, *2(3)*, 330–336.
38 K.D. Knudsen, P. Pattison, A.N. Fitch, R.J. Cernik, *Angew. Chem.*, **1998**, *37(17)*, 2340–2343.
39 H.R. Karfunkel, R.J. Gdanitz, *J. Comp. Chem.*, **1992**, *13*, 1171–1183.
40 A.I. Kitaigorodskii, *Molecular Crystals and Molecules*, Academic Press, New York, **1973**, p. 71.
41 S.K. Kurtz, T.T. Perry, *J. Appl. Phys.*, **1968**, *39(8)*, 3798–3813.
42 W.H. Lawrence, *US Patent*, 2,003,374, **1935**.
43 S.L. Mayo, B.D. Olafson, W.A. Goddard III, *J. Phys. Chem.*, **1990**, *94*, 8897–8909.
44 E. Mitscherlich, *Ann. Chim. Phys.*, **1822**, *19*, 350–419.
45 T. Matsuzaki, Y.T. Osano, *Anal. Sci.*, **1989**, *5*, 123–125.
46 W. Ostwald, *Z. Phys. Chem.*, **1897**, *22*, 289–330; W. Ostwald, *Grundriß der Allgemeinen Chemie*, Leipzig, **1899**.
47 P. Ochsenbein, *Colloque "Polymorphisme et Morphologie"*, CPE Lyon, **2002**, 16–18 December.
48 P. Ochsenbein, M. Bonin, O. Masson, D. Loyaux, G. Chapuis, K.J. Schenk, *Angew. Chem. Int. Ed.*, **2004**, *43*, 2694–2697.
49 W.E. Oberhansli, *Helv. Chim. Acta*, **1982**, *65(3)*, 924–933.
50 P. Ochsenbein, M. Bonin, K.J. Schenk, D. Loyaux, *Bull. Czk. Slov. Cryst. Assoc.* **1998**, *5(B)*, 289.
51 P. Ochsenbein, Ph. Sizun, M. El-Hajji, Acta Crystallogr. Sect. A, **2005**, *C279*.
52 P. Ochsenbein, M. Bonin, K.J. Schenk, to be submitted.
53 E.L. Parrott, *J. Pharm. Sci.*, **1962**, *51(9)*, 897–900.
54 R.R. Pfeiffer, K.S. Yang, M. A. Tucker, *J. Pharm. Sci.*, **1970**, *59(12)*, 1809–1812.
55 L. Pasteur, *Ann. Chim. Phys. (Paris)*, **1853**, *38*, 437.
56 M. Rajeswaran, T.N. Blanton, N. Zumbulyadis, D.J. Giesen, C. Conesa-Moratilla, S.T. Misture, P.W. Stephens, A. Huq, *J. Am. Chem. Soc.*, **2002**, *124(48)*, 14 450–14 459.
57 J. Richter, RECIPE – Reciprocal Space Viewer, *IPDS 2.89*, **1997**, Stoe & Cie GmbH, Darmstadt.
58 B. Roueché, *The Medical Detectives*, Washington Square Press, New York, **1982**, pp. 64–84.
59 G.A. Stephenson, E.G. Groleau, R.L. Kleemann, W. Xu, D.R. Rigsbee, *J. Pharm. Sci.*, **1998**, *87(5)*, 536–542.
60 *SYBYL Molecular Modeling Software*, Tripos, Inc., 1699 Hanley Road, St. Louis, MO.
61 J. Starbuck, R. Docherty, M.H. Charlton, D. Buttar, *J. Chem. Soc., Perkin Trans. II*, **1999**, 677–691.
62 G.M. Sheldrick, *SHELX97*, Program for the solution and refinement of crystal structures, Universität Göttingen, **1997**.
63 D. Taupin, *J. Appl. Cryst.*, **1968**, *1*, 178–181.
64 P. Voisin, *Brevet français*, 8,814,911, **1990**.
65 B. Viossat, J.-Cl. Daran, G. Savouret, G. Morgant, F.T. Greenaway, N.-H. Dung, V.A. Pham-Tran, J.R.J. Sorenson, *J. Inorg. Biochem.*, **2003**, *96*, 375–385.
66 J.W. Visser, *J. Appl. Cryst.*, **1969**, *2*, 89–95.
67 B. Zimmer, V. Bulach, C. Drexler, S. Erhardt, M.W. Hosseini, A. De Cian, *New J. Chem.*, **2002**, *26(1)*, 43–57.

7
Light Microscopy

Gary Nichols

7.1
Introduction

The value of using light microscopy as a routine technique for materials characterization in the pharmaceutical and fine chemical industries has been largely ignored in preference for more sophisticated instrumental methods. Indeed, it has been noted [1] that

> "the optical microscope in the hands of most organic crystal chemists and chemical crystallographers, at least, has been relegated from a principal research tool to merely an aide in choosing and mounting crystals for diffraction experiments".

There is no doubt that a light microscope is a very useful tool for specimen preparation, but it can also be used for so much more when studying the solid-state properties of materials.

Several more recent publications [2–5] have rightly included light microscopy in overviews of the range of analytical techniques used to characterize the solid form properties of pharmaceuticals and fine chemicals. These reviews concentrate primarily on the use of thermomicroscopy to study polymorphism together with the descriptions of some fundamental optical properties of compounds. These are, of course, very important applications, but the light microscope can be used to gather considerably more valuable information from samples. This chapter aims to redress the balance and demonstrate that light microscopy, especially polarized light microscopy, has a major role to play in the characterization of a wide range of materials.

Applications for polarized light microscopy are not restricted just to understanding the solid-state properties of research compounds; it is also an indispensable analytical technique for troubleshooting and problem solving during the development of organic and inorganic chemical compounds from discovery, through development and into full-scale production.

Polymorphism: in the Pharmaceutical Industry. Edited by Rolf Hilfiker

ISBN: 3-527-31146-7

When using multi-techniques to investigate the solid-state properties of compounds, polarized light microscopy adds great value when used to help understand and explain phenomena that have been discovered using other techniques. By simply looking at samples at different magnifications, invaluable information about them can be gained; information that cannot be gained using non-microscopical techniques! Polarized light microscopy, which in itself is a powerful stand-alone technique, should be considered as a primary tool to support other solid-state characterization techniques, such as X-ray diffraction, thermal analysis, solid-state NMR, and vibrational spectroscopy. Throughout this chapter, emphasis is placed upon the practical application of polarized light microscopy and how observations can be used to complement other analytical techniques.

7.2
Why Use a Light Microscope to Study Solid-state Properties?

The purpose of a microscopical examination is to magnify the image of a specimen to resolve its fine detail and to reveal information that is invisible to the naked eye. When polarizing filters are added to a light microscope it becomes an analytical tool with which to determine the optical properties of crystals and to perform chemical microscopy experiments. Chemical microscopists occupy a unique position in a solid-state characterization laboratory because they can directly observe the physicochemical properties of minute amounts of sample both rapidly and inexpensively and then relate these observations to the crystal structure and crystal properties.

Different solid forms have different optical properties, such as refractive index, color, extinction angle, and optical dispersion that are readily determined using optical crystallographic methods. The optical properties of a crystal are controlled by its crystal structure and chemistry and so they can provide valuable analytical data to support structural data derived using other techniques. Other phenomena that are readily observed and determined using polarized light microscopy include particle size, particle shape, presence of inclusions, twinning, crystal growth and dissolution, mechanical behavior and mesomorphism. Thermomicroscopy is used to observe thermally induced events (such as melting point, polymorphic conversions, desolvation, and sublimation) as small amounts of sample are heated or cooled.

In addition to the light microscope, other types of microscope are used to probe and analyze the physical and chemical properties of samples. These include electron microscopes with X-ray elemental microanalyzers, atomic force microscopes, FT-infrared and Raman microscopes, near-infrared microscopes, acoustic microscopes, and X-ray microscopes.

Henry Sorby pioneered the use of polarized light microscopy to examine crystals during the mid 19th century to study minerals in rock thin-sections [6]. Since then, many of the optical crystallographic methods developed and used by mineralogists have been adapted for the microscopical investigation of a wide range of substances in the organic and inorganic chemical industries. Therefore, it is no coincidence that some of the examples described in this chapter have geological origins.

Much analytical information can be obtained from small amounts of sample using microscopical methods. To help understand larger scale processes (such as crystallization, dissolution, drying and milling), simple micro-scale experiments performed on the stage of a light microscope can be highly informative. Therefore, the microscopist needs to be creative in developing novel approaches to elucidate and trouble-shoot many scale-up and production issues.

7.3 Polarizing Light Microscope

A polarizing light microscope is a compound light microscope that has been modified by placing a polarizing filter (the polarizer) below the specimen and a second polarizing filter (the analyzer) above the specimen (Fig. 7.1). The vibration direction of the polarizer is usually aligned so that its plane of polarization is west to east (i.e., left to right when viewed through the microscope). The vibration direction of the analyzer is aligned 90° to that of the polarizer.

The polarizer is normally left in the light path at all times and is located between the illuminator and the specimen (typically it is incorporated into the condenser)

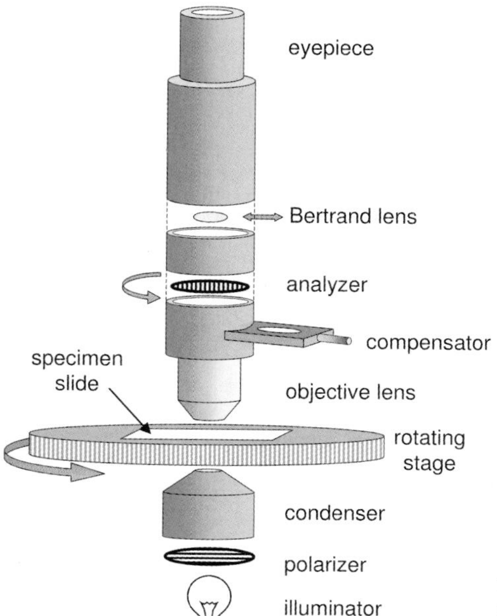

Fig. 7.1 Schematic diagram of the basic components of a transmitted polarizing light microscope.

for when specimens are to be viewed using plane polarized light. The analyzer is designed so that it can be inserted into the light path to allow specimens to be viewed between crossed polarizers. Some microscopes will allow the analyzer to be rotated relative to the polarizer so that specimens can be viewed between uncrossed polarizers, or for characterizing optically active specimens, or for using specialized compensator plates. To achieve a uniformly illuminated field of view, the microscope should be adjusted (if possible) for Köhler illumination [7].

To enable a comprehensive examination of the optical properties of crystals using polarized light, the microscope should also be equipped with a rotating specimen stage so that individual crystals can be aligned with the vibration directions of the polarizer and analyzer. At least one eyepiece should have cross-hairs that are aligned to indicate the vibration directions of the polarizer and analyzer. The rotating stage should have a vernier scale graduated in degrees of rotation around its periphery so that angular measurements, such as the angles between crystal faces and the angles at which crystals go into extinction when viewed between crossed polarizers, can be made by reference to the cross-hairs in the eyepiece. A compensator (such as sensitive tint plate) that can be inserted into the light path below the analyzer is a useful accessory to determine other characteristic optical properties of crystals.

When selecting and specifying a polarizing light microscope to investigate the solid form properties of crystals, the basic requirements are:

- upright stand for transmitted light;
- trinocular head (for a pair of eyepieces and a camera);
- high intensity illuminator with an adjustable field diaphragm;
- circular rotatable stage with a vernier scale graduated in degrees;
- focusable condenser with an aperture diaphragm;
- polarizer;
- analyzer (rotatable and removable);
- low, medium and high power strain-free objectives;
- rotating nosepiece with objective centration screws;
- eyepiece with a focusable cross-hair;
- eyepiece with a focusable calibrated measuring scale (reticle);
- Bertrand lens (retractable);
- compensator plate (1st order red plate);
- camera.

7.4 Photomicrography

Photomicrography is the recording of an image of a small object formed by a microscope; it should not be confused (as is so often the case) with microphotography, which is the recording of the image of a large object that requires magnification to view it [8]. During examination of a sample, a photographic record of what it looks like is extremely useful for future reference and for illustrations in reports. The camera is mounted either onto the microscope using a purpose built trinocular head or by using an adapter to connect it to one of the viewing eyepieces.

Until the recent availability of high performance and relatively low cost digital cameras, photomicrography (especially color photomicrography) using conventional photographic media, required a reasonably high level of technical skill and experience to achieve satisfactory and reproducible results. For the faithful reproduction of the color of a specimen, great care was needed to control the illumination to ensure that it matched the color sensitivity of the film. Even with a high intensity light source, the images of some specimens (such as those viewed between crossed polarizers) are frequently of low brightness, so a long exposure time was needed; this in itself could cause poor color reproduction.

Digital cameras have literally revolutionized photomicrography and many microscopists use them routinely. Exposure control is easy, color control is easy, results are almost instantaneous and image processing software can be used to produce a perfect picture every time. Electronic archiving of photomicrographs is also easy and very convenient. For those microscopists with a collection of treasured 35 mm transparency slides, scanners can be used to produce exceptionally high quality digital copies.

Although digital imaging has greatly simplified photomicrography, the best results will only be accomplished with a well-prepared specimen, a correctly aligned microscope, and the understanding and application of the basic skills of photomicrography [9]. No matter how good the microscope and camera are, it is the skill and proficiency of the microscopist that ensures good quality results!

A scale bar should be included on photomicrographs to indicate a known length represented in the image for each magnification. It is not adequate just to quote a magnification because if the original photomicrograph is reproduced at a different size, then the magnification value will be incorrect. A scale bar will be correct, regardless of subsequent magnification changes. The length represented by the scale bar is determined using an accurately known measuring scale (a stage micrometer or stage graticule) that is placed on the specimen stage. Digital cameras designed specifically for photomicrography will usually have a function to automatically superimpose a scale bar onto a photomicrograph.

7.5 Specimen Preparation

As with any analytical technique, correct specimen preparation for light microscopy is critical to ensure that the results of examinations are meaningful and of value. Samples produced to investigate the solid-state properties of a compound may consist of a few single crystals or may be powders consisting of many particles. Ideally, the sample to be examined should be representative of the bulk and the specimen preparation method must not modify the sample in any way.

Powders are relatively easy to prepare for microscopical examination and require little more than being dispersed into a liquid on a glass microscope slide. However, to get the best possible preparation of a sample, the ultimate purpose of the study should be understood from the outset. For example, to conduct a particle size analysis, good contrast between the powder particles and the background is essential (if you cannot see the particles, how can you measure them?).

Contrast between a sample and its surroundings can be most easily achieved by mounting the sample in a liquid that has a refractive index very different to that of the particles (Section 7.7). Many organic solids have refractive indices in the range 1.50 to 1.70, so to ensure that they are visible when prepared for microscopy a liquid that has a refractive index significantly different to the specimen is used. Silicone fluid is a good mounting liquid because it has a refractive index of about 1.40, is relatively cheap and is practically inert as very few compounds will dissolve in it. Other liquids that are possible mounting media include Nujol (mineral oil), water, or immersion oil. Remember, whatever liquid is chosen, it must *not* be a solvent for the particles being examined!

7.5.1 Permanent and Temporary Mounts

Microscope slide preparations are either permanent or temporary. Permanent mounts are retained for future reference and temporary mounts are those be examined within a few minutes of being prepared and will normally be discarded.

7.5.1.1 Permanent Mounts

These are prepared using mounting media with known optical properties that become hard, optically clear, chemically stable and colorless at room temperature. There are many types of media available for preparing permanent mounts and these will cure either by solvent loss, heating (thermosetting), cooling (thermoplastic), or by exposure to ultraviolet light. Thermosetting and solvent-loss media are unsuitable for mounting some organic crystals because they may dissolve them or cause chemical or thermal degradation. Thermoplastic media (such as Cargille Meltmount™ [10]) that become fluid at elevated temperature may also react with some organic compounds. Ultraviolet curing media (such as Norland™ Optical Adhesive [11]) harden rapidly at room temperature upon exposure to UV light and are often more suitable for producing permanent mounts of organic particles. However, whichever medium is selected, a test preparation should be made to confirm that there is no adverse reaction with the sample.

7.5.1.2 Temporary Mounts

These are usually prepared using a non-volatile and non-hardening liquid (e.g., water or silicone oil). The specimen could also be a slurry, consisting of crystals dispersed in the mother liquor, and so would not require another mounting medium. To retain a permanent record of what the specimen looks like, photomicrographs at different magnifications should be taken.

7.5.2 Preparation of Temporary Mounts

Most likely, for examining the solid-state properties of samples, most preparations for light microscopy will use temporary mounts. One method of preparing a suitable powder is as follows:

1. Add one or more drops of the chosen mounting liquid onto a clean microscope slide (note that powders containing large particles will need more liquid to fill the space between the slide and coverglass).
2. Use a micro-spatula to add a small amount of the powder (enough to cover a dot about 2 mm in diameter) into the liquid and disperse it by stirring with a circular and zigzag motion without crushing particles. Allow the dispersion to stand for a few seconds to allow any air bubbles to burst.
3. Carefully place a coverglass onto the dispersion and avoid trapping air bubbles by laying the coverglass at a slight angle (if necessary, use a needle to slowly lower the coverglass onto the preparation).
4. The space between the slide and coverglass needs to be completely filled with the sample–liquid dispersion. If there is too little liquid, the preparation will continue to move until the space has been filled; another drop of liquid can be added to the edge of the coverglass and it will be drawn under by capillary action. Excess liquid is undesirable because it can spill onto the microscope stage and/or contaminate the objective lens. To remove excess liquid, use a twist of soft tissue paper or a small triangle of filter paper to wick away the liquid from the edge of the coverglass.

7.5.3 Examination of Tablets

The use of polarized light microscopy is not restricted to the examination of powders. A novel method has been developed to examine cross-sections through pharmaceutical tablets [12]. This method has been adapted from the long-established technique used by geologists to prepare thin-sections of rocks for petrographic examination. Tablets are gently hand ground in a dry state and are mounted onto microscope slides using UV curing adhesive [11] for final polishing. These cross-sections reveal the spatial distribution of the components in the tablets so that any post-compression changes (e.g., polymorph conversions or changes into the hydration state of components) can be investigated.

7.6 Observations Using Polarized Light Microscopy

Some of the most informative observations about the optical properties of crystals are made using polarized light microscopy. Many books that describe the application of optical crystallography to study crystals have been published, such as [13–16]. It is not within the scope of this chapter to describe the many specialized techniques used by optical crystallographers, so these texts should be consulted for more comprehensive information (in particular, the book by Elizabeth Wood [13] is an excellent, inexpensive introduction). Crystals of every substance, regardless of whether they are pharmaceuticals, minerals, high explosives, agrochemicals, pigments, or ceramics, will have optical properties that can be used to characterize and identify them.

A polarizing light microscope can be configured to examine specimens under different conditions without the need for specialized accessories. These conditions are easily attained and include: non-polarized light, plane polarized light, crossed polarizers, uncrossed polarizers, and circularly polarized light. Each of these is described later, with examples of specific applications.

7.6.1 Polarized Light

Light waves that vibrate in all directions perpendicular to the direction of propagation are non-polarized (Fig. 7.2). Ordinary, *non-polarized light* has limited value in the optical characterization of crystals, except for general observations, such as the measurement of particle size and the determination of crystal shape. When non-polarized light is passed through a polarizing filter, only the light that vibrates parallel to its privileged vibration direction is transmitted as *plane polarized light*. When plane polarized light is directed towards the analyzer

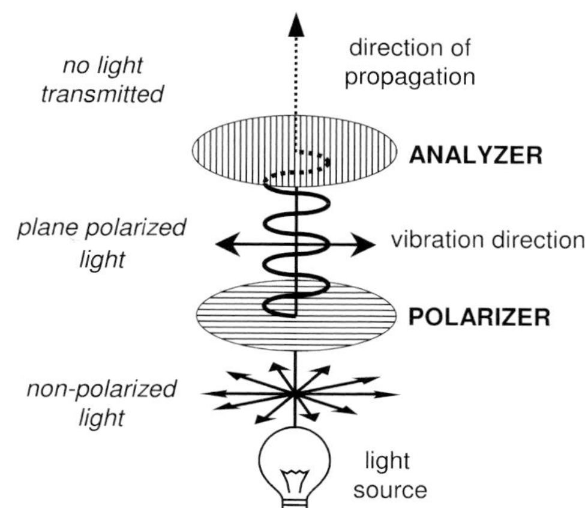

Fig. 7.2 Schematic diagram showing how a polarizer and analyzer affect non-polarized light.

(which has its vibration direction aligned 90° to that of the polarizer), it is extinguished, no light is transmitted and an empty field of view is black; this is the condition known as *crossed polarizers*.

7.6.2
Crystal Studies with Plane Polarized Light

Plane polarized light has great diagnostic value to the microscopist because it has a precisely known vibration direction. This enables the optical properties of single crystals to be determined at different orientations by rotating the specimen stage. Two specific applications using plane polarized light are:

- Measurement of the principal refractive indices of anisotropic crystals by orientating crystals relative to the vibration direction of the polarizer (Section 7.7.1). If non-polarized light is used to examine crystals, the refractive index measured is the average value for each orientation observed. For an isotropic substance, plane polarized light is not essential for refractive index determination because its refractive index is independent of particle orientation.
- Determination of the absorption colors along specific vibration directions in a colored anisotropic crystal. Most colored compounds will show a different color, or a change in the intensity of the color, as they are rotated in plane polarized light (this effect is known as dichroism or pleochroism and is described in Section 7.14). In non-polarized light, an average color would be seen for all orientations.

Cubic crystals, like liquids and non-strained glass, are isotropic and light travels through them at the same speed (for a specific wavelength and temperature) in all directions. Consequently, cubic crystals have a single refractive index (n). All other crystals have either two or three principal refractive indices because they are anisotropic. When polarized light enters an anisotropic crystal, it is resolved into two mutually perpendicular vibration directions that correspond to slow and fast directions. As these two waves pass through the crystal, the slow wave becomes retarded relative to the fast wave.

A convenient way of representing the refractive index for different vibration directions is to construct a three-dimensional diagram, known as an *indicatrix*. The refractive index is drawn with a radius having a length that is proportional to its value. For an isotropic substance, the indicatrix is a sphere (Fig. 7.3 a) because the refractive index is the same in all directions and the only variable is its size, which changes with an increase or decrease in refractive index.

The indicatrix for anisotropic crystals is ellipsoidal. Tetragonal, trigonal, and hexagonal crystals have two mutually perpendicular principal refractive indices, ordinary (omega, n_ω) and extraordinary (epsilon, n_ε), either of which can be the fast or slow direction. The ordinary ray vibrates parallel to two identical principal crystallographic axes and so it behaves isotropically. The ray velocity surface for omega is a sphere and it has a circular section (C). The extraordinary ray

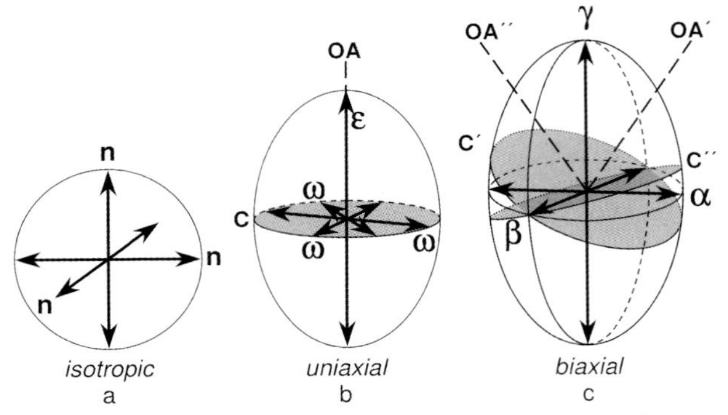

Fig. 7.3 Ray velocity surface diagrams for (a) the indicatrix for an isotropic substance, (b) the uniaxial indicatrix, and (c) the biaxial indicatrix. The isotropic circular sections are shown (C).

vibrates at right angles to the circular section. The indicatrix resulting from the combination of the ordinary and extraordinary rays is an ellipsoid of revolution (Fig. 7.3 b). The unique axis that is perpendicular to the isotropic circular section is the optic axis (OA) and it is parallel to the principal symmetry crystallographic axis. Crystals of this type are characterized by having one optic axis and are classed as *uniaxial*. Viewing a uniaxial crystal down the optic axis will enable its characteristic interference figure to be inspected (Section 7.6.3.3).

Orthorhombic, monoclinic and triclinic crystals have three mutually perpendicular principal refractive indices: alpha (n_α, or nX), beta (n_β, or nY), and gamma (n_γ, or nZ), where α has the lowest value (fast direction) and γ has the highest value (slow direction). The indicatrix is a triaxial ellipsoid (Fig. 7.3 c) constructed with the three vibration directions for alpha, beta and gamma, each having different axial lengths. There are two inclined intersecting circular sections (C′ and C″) within the triaxial ellipsoid and these are isotropic with diameters corresponding to the beta vibration direction. Each circular section has a corresponding optic axis (OA′ and OA″) perpendicular to it. The optic axes lie in a plane, called the optic axial plane, which is occupied by the alpha and gamma refractive indices. Crystals of this type, which are characterized by having two optic axes, are classed as *biaxial*.

The angle between the two optic axes, the optic axial angle (called 2V), is related to the three principal refractive indices and is a constant for a specific compound; 2V can be estimated by observing a centered biaxial interference figure (Section 7.6.3.4), but this method is not sufficiently accurate. A crystal rotation device, on which a single crystal is rotated about the axis corresponding to the n_β vibration direction, can be used to measure 2V directly and accurately [15]. Alternatively, 2V can be calculated if the three principal refractive indices are known [15]. If just two of the three principal refractive indices and 2V are known, then the third index value can be calculated [15]. The line that bisects the optic axial angle is called the bisectrix (either acute or obtuse). Viewing a biaxial crystal down the acute bisectrix will enable its characteristic interference figure to be inspected (Section 7.6.3.4).

The uniaxial and biaxial indicatrices contain a great deal of optical crystallographic data and they are an invaluable way to summarize pictorially the complex optical properties of crystals. The specialized books mentioned at the beginning of this section give more detailed accounts of how to utilize fully the indicatrix.

For uniaxial and biaxial crystals, there is a relationship between their crystallographic axes and their principal vibration directions. The *c*-axis for tetragonal, trigonal and hexagonal crystals always coincides with the epsilon vibration direction. For orthorhombic crystals, any crystallographic axis (*a*, *b*, or *c*) can coincide with any vibration direction (n_α, n_β, or n_γ). Due to the lower symmetry of monoclinic crystals, only the *b*-axis will coincide with a vibration direction (n_α, n_β, or n_γ). For triclinic crystals, which have the lowest symmetry, no vibration direction need coincide with any crystallographic axis.

7.6.3
Crystal Studies with Crossed Polarizers

The condition known as *crossed polarizers* is when the analyzer (with its vibration direction aligned perpendicular to that of the polarizer) is inserted into the microscope above the specimen (Fig. 7.1). With no anisotropic specimen on the specimen stage, the field of view will be black. When an anisotropic substance is placed between the crossed polarizers, it will show bright interference colors if it is not at an extinction position and is not being viewed down an optic axis. An isotropic specimen (such as a cubic crystal or a glass particle) examined between crossed polarizers will not be seen because it will be black against the black background. The remainder of this section describes optical phenomena that can be observed using crossed polarizers.

However, by rotating one of the polarizing filters (usually the analyzer) by a few degrees (say, 10–15°) relative the other, the condition of *uncrossed polarizers* is achieved and the field of view is gray, rather than black.

When viewed with uncrossed polarizers, both isotropic and anisotropic particles are visible simultaneously. For anisotropic crystals, interference colors and extinction positions will be seen, but they are not identical to those seen between fully crossed polarizers. Therefore, uncrossed polarizers should be used to inspect mixtures but not to measure optical properties.

7.6.3.1 Interference Colors

Interference colors (or polarization colors) seen in crystals viewed between crossed polarizers are the result of the constructive and destructive interference of white light as one wave is retarded relative to the other (Section 7.6.2) after they have traveled through a doubly refracting crystal and are then recombined in the analyzer. The colors, also known as Newton's colors, are shown pictorially on the Michel-Levy chart of interference colors [17]. The Michel-Levy chart

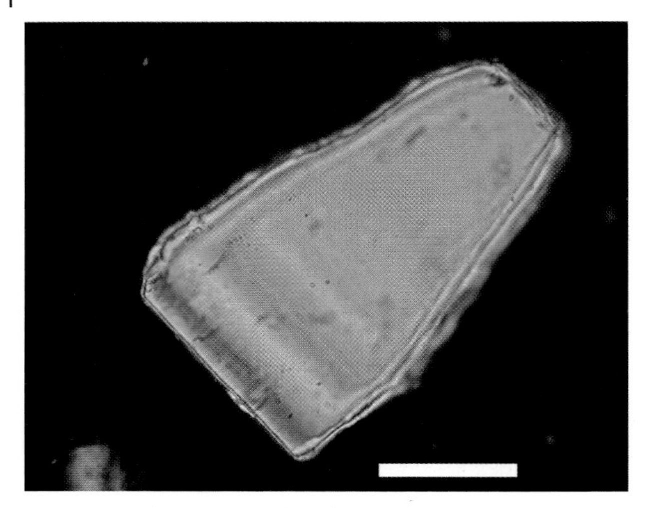

Fig. 7.4 A wedge-shaped crystal of lactose monohydrate showing interference color bands. Scale bar = 50 μm.

relates birefringence, crystal thickness and retardation, and if two of these are known then the third can be read from the chart.

The sequence of colors is divided into orders (i.e., 1st, 2nd, 3rd, etc.) and these are often seen at the edges of crystals with beveled edges. The low order colors are seen at the thin edges and the interference colors increase towards the higher orders as the crystal gets thicker. The lactose monohydrate crystal shown in Fig. 7.4 has a pale 5th order green at its centre, with the 2nd order red band along the lower left diagonal edge. The colored bands correspond to changes in thickness and are analogous to contours on a map, such that closely-spaced colored bands occur along steeply inclined surfaces.

Crystals having more than one refractive index are doubly refracting and are said to be birefringent. Birefringence is the numerical difference between the highest and lowest refractive indices. For a specific substance having a fixed chemical composition, the birefringence is constant. However, the interference colors observed will vary depending upon the orientation of the crystal and its thickness. Consequently, a range of interference colors may be seen for randomly orientated particles of a compound (Fig. 7.5). For this reason, interference colors alone cannot and should not be used to distinguish between different compounds or polymorphs.

Crystals will usually display normal interference colors (i.e., colors that follow the sequence of colors of Newton's scale [15]) and as such are not particularly diagnostic in their own right. However, some colored crystals will show abnormal, or anomalous, interference colors that do not match Newton's scale [15]. In addition, crystals that display optical dispersion or have dispersed extinction (Section 7.6.3.2) could be of diagnostic value.

Fig. 7.5 Different interference colors shown by a drug compound having crystal fragments of different sizes and different orientations viewed with uncrossed polarizers (scale bar = 250 μm).

Many pharmaceutical companies have adopted the microscopical detection of birefringence, as indicated by the presence of interference colors, as a rapid screening tool to locate automatically crystals grown in multi-well plates on high-throughput crystallization salt and polymorph screening systems [18]. Image analysis is used to locate potential new solid forms and those wells that contain anisotropic crystals are subsequently analyzed using Raman spectroscopy and/or powder X-ray diffraction. Some high-throughput solid-form screening systems also incorporate the facility to heat the multi-well plates so that the loss of birefringence, as crystals melt or undergo a phase transition, is detected by image analysis [19].

7.6.3.2 **Extinction**

When birefringent crystals are viewed between crossed polarizers and are rotated on the specimen stage, they become black every 90° due to extinction. This happens when the vibration directions in the crystal are aligned with the vibration directions of the polarizer and the analyzer. Optically active compounds (such as sucrose crystals in some orientations) are an exception because they rotate the plane of polarization and so the polarizers need to be slightly uncrossed to achieve complete extinction.

The angle at which a crystal goes into extinction is referenced against the cross-hairs in the microscope eyepiece, which must be aligned exactly parallel to the vibration directions of the analyzer and polarizer. To check that the cross-hairs are aligned correctly, an elongated crystal having straight extinction is used as a test specimen (e.g., orthorhombic crystals of ammonium sulfate grown from a water drop on a microscope slide) and is rotated to its extinction position. The analyzer is then removed so that the crystal can be seen in plane polarized light relative to the cross-hairs. If necessary, the eyepiece is turned until one of the cross-hairs is exactly parallel with a long edge of the crystal.

When the extinction position is determined in relation to the shape of a crystal, it can be used as an indication of its crystal system. The extinction shown by crystals being viewed between crossed polarizers can be diagnostic and is described as straight, inclined or symmetrical [20]. If an elongated crystal goes into extinction when one of its long sides is parallel to a cross-hair, it shows straight extinction (Fig. 7.6 a). If the extinction position is at an angle to a cross-hair, it shows inclined, or oblique, extinction (Fig. 7.6 b). The reported extinction angle should not exceed 45°. If the extinction position occurs when the angle between two adjoining crystal faces is bisected by a cross-hair, it is called symmetrical extinction (Fig. 7.6 c). Undulose extinction can sometimes be seen in strained or curved crystals and is recognized as a dark region that sweeps across a crystal or particle as it is rotated between crossed polarizers (Fig. 7.6 d).

Crystals that show straight extinction only could be orthorhombic or one of the uniaxial crystals (conoscopic examination of the interference figure will confirm which it is; see Section 7.6.3.3). However, crystals that show both straight extinction (when viewed down the *a*- or *c*-axes) and inclined extinction (when viewed down the *b*-axis) in different orientations are likely to be monoclinic, whereas those crystals showing inclined extinction for all orientations are likely to be triclinic. Uniaxial and biaxial crystals could both show symmetrical extinction, so this is not diagnostic. Some orthorhombic crystals may be rhomb- or diamond-shaped plates, similar to that shown in Fig. 7.6 (c), and may appear to have inclined extinction if the long edges go into extinction at an angle to a cross-hair.

Some crystals show incomplete extinction and will not go completely black at any point during rotation of the specimen stage. Instead, they may show a change of polarization color (e.g., yellow to blue); this is known as dispersed or anomalous extinction [15]. It occurs when the extinction position is different for different wavelengths (colors) of light. Indeed, for triclinic crystals, it is possible that no orientation will have complete extinction [16]. Consequently, when a

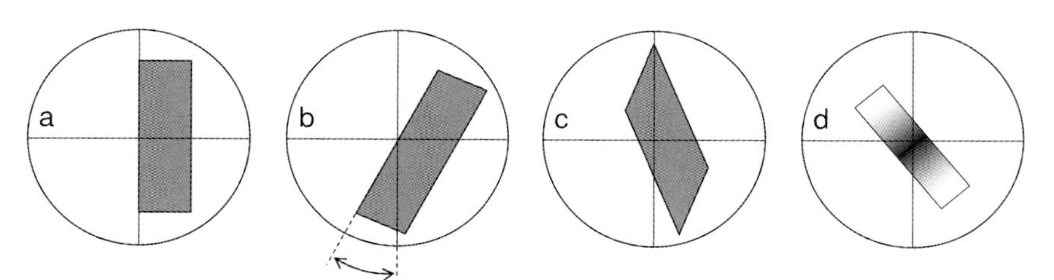

Fig. 7.6 Diagram showing (a) straight extinction, (b) inclined extinction (with the extinction angle indicated), (c) symmetrical extinction and (d) undulose extinction.

crystal is at the extinction position for, say, blue light, it will not be at extinction for the other colors and so will appear yellow (because white minus blue is yellow).

The stable monoclinic polymorph of paracetamol (Form I), when viewed down the *b*-axis (parallel to n_α), shows dispersed extinction [21]. The metastable orthorhombic polymorph of paracetamol (Form II) is readily distinguished from Form I because it does not show dispersed extinction. During development of a laboratory-scale process to prepare pure samples of Form II, it was vital to be able to identify rapidly any contamination by crystals of Form I [21]. By searching for crystals showing the very distinctive dispersed extinction, it was possible to locate single crystals of Form I in microscope slide preparations containing hundreds of crystals. No other analytical technique would have this level of specificity and sensitivity.

If a particle shows bright interference colors, but has no extinction position, then it may be polycrystalline and, at a higher magnification, it will be seen to have many crystallites at different extinction positions at any one time. Alternatively, it may be a twinned crystal (Section 7.13) that is being viewed perpendicularly to the twin plane so that as one part of the twin is at extinction the other part is not at extinction.

When selecting suitable crystals for single-crystal XRD, the crystallographer will look for crystals that show complete extinction as an indication of a good quality, strain-free specimen. Crystals not showing complete extinction are not necessarily of poor quality, but could be rejected just because they show dispersed extinction, or are being viewed down an optic axis. To confirm that dispersed extinction is the cause and the crystal is not strained, it should be examined orthoscopically using monochromatic light, rather than white light, to confirm that it does actually have an extinction position.

7.6.3.3 Interference Figures

Most polarized light microscopical observations are conducted with the microscope configured as an orthoscope [14]. Orthoscopic viewing is the normal viewing mode and has a near-parallel beam of light to illuminate the specimen. However, additional and highly informative optical information can be acquired from anisotropic crystals between crossed polarizers using conoscopic viewing. When configured as a conoscope [14], a convergent beam of light illuminates the specimen and light travels through it in more than one direction at one time. This produces a pattern known as an interference figure that shows the optical properties of a crystal and reveals if it is uniaxial or biaxial. An isotropic crystal or glassy particle will not show an interference figure.

Interference figures are viewed between crossed polarizers on suitably orientated crystals at high magnification using an auxiliary lens, known as a Bertrand lens (Fig. 7.1). If the microscope is not fitted with a Bertrand lens, the interference figure can be viewed by removing an eyepiece and looking directly into the empty eyepiece tube.

To inspect a centered interference figure, it is necessary to search a prepared slide (orthoscopically) for crystals that are viewed directly down an optic axis or down the acute bisectrix (Section 7.6.2). Such crystals (or fragments) show a low order interference color and do not go into complete extinction as they are rotated through 360° (i.e., they remain bright). An interference figure (viewed using white light) has two distinctive features: black brushes (or isogyres) and colored interference bands of equal retardation (or isochromes) that are centered around "eyes" (or melatopes) from which an optic axis emerges [15]. The number of colored bands increases with birefringence and/or specimen thickness. The color seen at the centre of an interference figure is the color that the crystal has when viewed orthoscopically. Note that most crystals in a prepared slide will show an off-centered interference figure and all that is seen is part of an isogyre as it sweeps across the field of view. Detailed analytical interpretation of interference figures is beyond the scope of this chapter, so the reader is referred to specialized texts, e.g. [14–16].

Interference figures can be used to distinguish between polymorphs, as exemplified by calcium carbonate. Calcite is the trigonal form and its centered uniaxial interference figure has concentric colored rings, with the optic axis emerging from the centre of a black cross (Fig. 7.7 a). Aragonite is the metastable ortho-

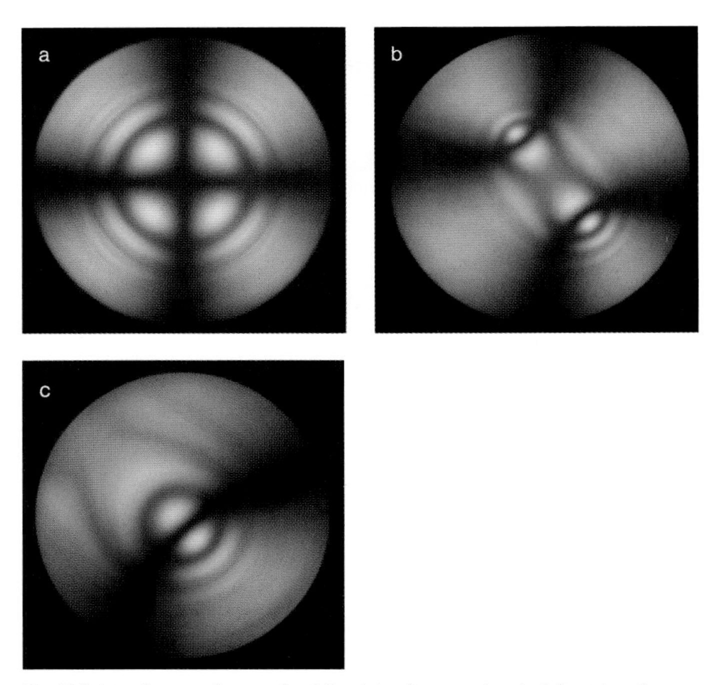

Fig. 7.7 Interference figures for (a) calcite (centered uniaxial optic axis figure), (b) aragonite (centered biaxial acute bisectrix figure), and (c) sucrose (biaxial optic axis figure).

rhombic form and its centered biaxial interference figure has two curved isogyres (Fig. 7.7 b), when it is rotated 45° from its extinction position. The two optic axes (emerging from the centers of the isogyres) are separated by a distance related to the optic axial angle, 2V (this is about 18.5° for aragonite). Crystals having a greater 2V would have a greater separation of the isogyres. As a biaxial crystal is rotated into its extinction position, the isogyres will combine to form a cross that can sometimes look similar to a centered uniaxial interference figure.

If a biaxial crystal is orientated so that it is viewed directly down just one of its optic axes, the interference figure will be a single isogyre surrounded by isochromes, as shown by sucrose (Fig. 7.7 c).

7.6.3.4 Compensator Plates

To aid the determination of the slow and fast vibration directions of crystals, especially for elongated crystals, an optical accessory known as a compensator plate can be used when viewing them between crossed polarizers. Many different types of compensator plate are available (such as a quartz wedge and a mica plate), but the most frequently used is the gypsum plate, also known as the sensitive tint plate or first-order red plate [16]. This plate is made from a clear crystal of gypsum that gives a retardation corresponding to the first order red on the Michel-Levy chart [17]. The red (more strictly a magenta color) is superimposed on the field of view and is sensitive to very slight changes in the vibration directions in crystals that show low order interference colors when examined between crossed polarizers.

The compensator plate is inserted in the microscope above the specimen, but below the analyzer (Fig. 7.1). In addition to assisting the determination of fast and slow vibration directions, compensators can also be used to establish the optic sign of crystals when their interference figures are being inspected. Full details about the many uses of compensator plates are given in the specialized optical crystallography books mentioned previously [14–16]. One non-crystallographic application for the gypsum plate is to provide an aesthetically pleasing magenta background to images viewed between crossed polarizers (e.g., see Figs. 7.17 and 7.24).

7.6.3.5 Use of Circularly Polarized Light

When viewing randomly orientated anisotropic crystals between crossed polarizers, some will inevitably be at extinction. If an image analyzer is employed to detect birefringent crystals using crossed polarizers, e.g., on a multi-well plate (Section 7.6.3.1), then those at extinction will not be imaged. To see those crystals, they need to be rotated out of extinction, but this action will cause some of the other crystals to go into extinction. Uncrossed polarizers (see Section 7.6.3) could be used to see all of the crystals, but not all will show interference colors at their maximum intensity. To overcome this, circularly polarized light should be used [22]. Figure 7.8 compares three photomicrographs of the same field of

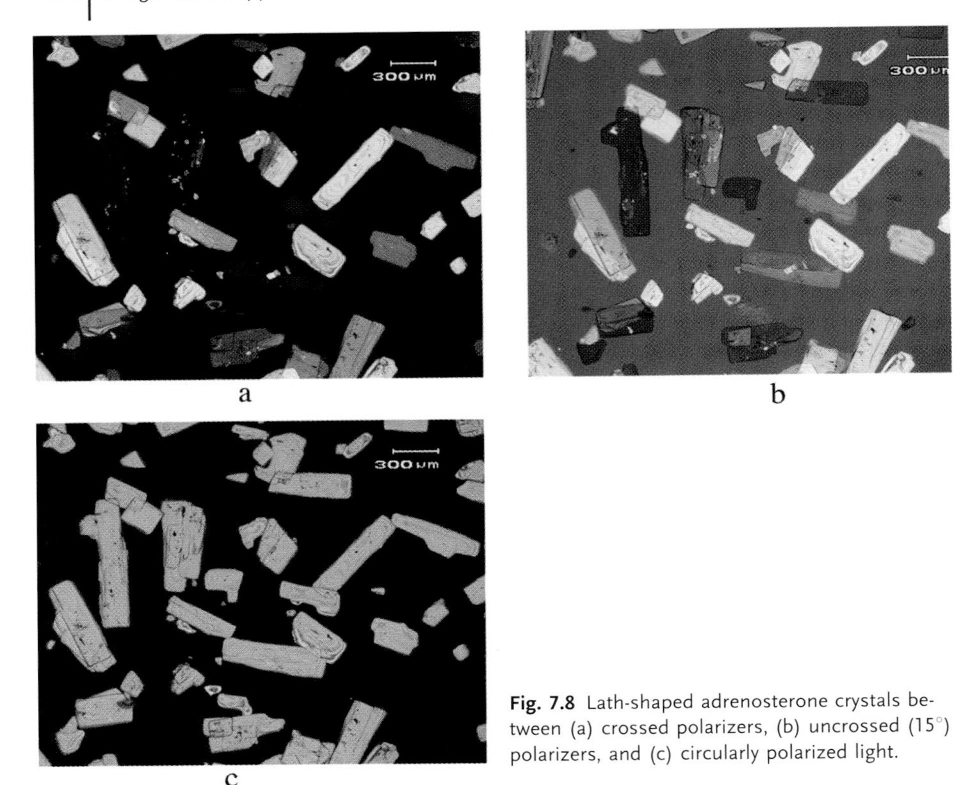

Fig. 7.8 Lath-shaped adrenosterone crystals between (a) crossed polarizers, (b) uncrossed (15°) polarizers, and (c) circularly polarized light.

view containing anisotropic crystals (adrenosterone) seen between (a) crossed polarizers, (b) with slightly uncrossed polarizers and (c) with circularly polarized light. With circularly polarized light, notably, all of the crystals, regardless of orientation, are visible, none is at extinction, and the correct interference colors are seen. To achieve circularly polarized light, two crossed quarter wave plates (orientated at 90° to each other, but at 45° to the polarizers) are inserted into the light path between the polarizer and analyzer, with one below the specimen and one above it.

7.6.4 Crystallinity

Polarized light microscopy is frequently used for the detection of crystallinity in samples. Indeed, the United States Pharmacopoeia (USP) test for crystallinity [23] describes a crystalline substance as one that shows interference colors and extinguishes every 90° of rotation. For most samples examined, the USP test is adequate. However, there are exceptions and some crystalline materials would fail if this was the only test used. Crystals belonging to the cubic system, regardless of orientation, will always be at extinction. Uniaxial crystals orientated so that they are viewed directly down an optic axis (e.g., the basal section of an hexagonal crystal) will not show interference colors (Section 7.6.2). A biaxial crystal viewed down, or close to, an optic axis may show a dull interference color without any extinction positions. Some twinned crystals will also not show complete extinction (Section 7.6.3.2).

However, non-crystalline and crystalline substances do have distinctive features that can be used to distinguish between them. Non-crystalline particles will often consist of irregular-shaped, angular fragments with sharp edges showing no cleavage, perhaps with conchoidal (shell shaped) fracture surfaces. Being optically isotropic, they will not display interference colors between crossed polarizers. Crystals and crystalline particles will often have well-developed crystal faces (or at least traces of crystal faces), straight edges and perhaps distinct crystal cleavage. When examined between crossed polarizers, isotropic crystals will not display interference colors, whereas anisotropic crystals will. Non-crystalline materials having many internal cracks (perhaps due to desolvation) may show interference colors because the polarized light is depolarized by reflection or scattering along the cracks.

To the naked eye, a powder consisting of small particles that sparkle in the light could be described as crystalline. However, when examined between crossed polarizers, the particles may be found to consist of glassy fragments.

Ideally, powder X-ray diffraction should be used to support polarized light microscopy to distinguish unequivocally between crystalline and non-crystalline samples [24]. A crystalline material will have a diffraction pattern with sharp, well-defined peaks, whereas a poorly crystalline material will have broad, poorly defined peaks, and a non-crystalline material will have a broad, diffuse, low intensity peak (or halo).

The drug particles shown in Fig. 7.9 are hexagonal prisms and are crystalline according to the USP test because they display interference colors and have ex-

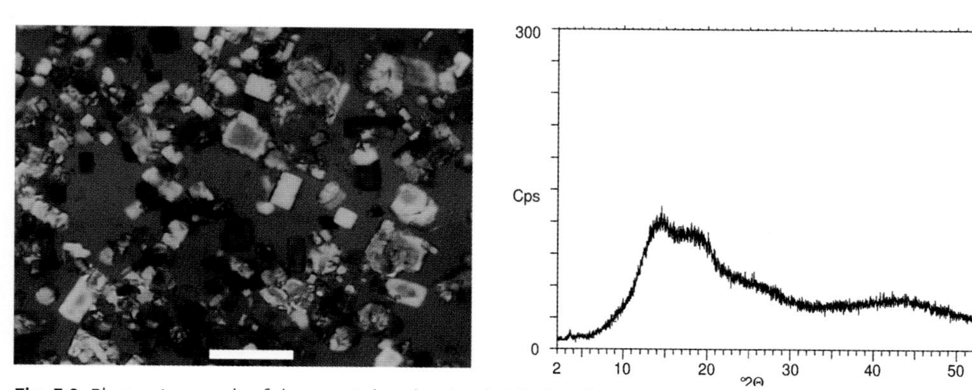

Fig. 7.9 Photomicrograph of drug particles showing bright interference colors and extinction positions between slightly uncrossed polarizers (scale bar=200 μm) and the corresponding powder X-ray diffraction pattern.

tinction positions every 90° of rotation. However, the powder X-ray diffraction pattern shows that they are highly disordered and practically amorphous. The disorder results from the prolonged drying of a toluene solvate at elevated temperature and they contain about 0.4% residual toluene.

7.7 Refractive Index

The refractive index of a specimen has a profound affect upon its visibility and is a property that can be used to characterize and identify it. Therefore, to the microscopist, refractive index is perhaps the most important optical property of a transparent substance. The visibility of a specimen can be improved by changing the refractive index of its mounting medium so that both are very different (Section 7.7.1). Therefore, it is sometimes necessary to know the refractive index of a specimen so that the optimum imaging conditions are selected.

Refractive index is the ratio of the speed of light in a vacuum to the speed of light in the substance and it will always be greater than 1. For any substance, its refractive index is a physical constant that varies with both wavelength and temperature. Accurate measurement of refractive index is usually made using standardized conditions with monochromatic light (589.3 nm) and corrected to 25 °C. An increase in refractive index has a corresponding decrease in the speed at which light travels through a substance (Section 7.6.2).

Substances that have more than one principal refractive index are doubly refractive, or birefringent. Birefringence is the numerical difference between the highest and lowest values for the refractive indices of an anisotropic crystal. For example, calcite ($CaCO_3$; trigonal) has two principal refractive indices (n_ω 1.658 and n_ε 1.486), and so it has a birefringence of 0.172. An isotropic substance, such as sodium chloride (NaCl; cubic), has one refractive index (n 1.544) and is, therefore, non-birefringent.

Refractive indices and other optical properties have been measured and published for many inorganic and organic substances and are exceedingly useful for reference purposes when identifying and characterizing unknown samples [25, 26]. Considering that numerous pharmaceutical compounds have had their solid-state properties characterized, only a few have had their optical properties published, e.g., [27].

The refractive index of a substance is controlled by its atomic or molecular structure [15] and so, by determining the refractive indices of a crystal, insight into its crystal structure is possible. Different polymorphs have different crystal structures and will, therefore, have different refractive indices, as exemplified by Forms I and II of paracetamol [21]. For crystals having a high birefringence, it may be possible to predict how the molecules are packed within the crystal, even if the crystal structure has not been determined [28]. This knowledge becomes more valuable when the predicted molecular packing is correlated with the crystal shape (e.g., needles or plates) observed microscopically. Planar molecules tend to stack parallel to each other to give needles, and elongated molecules tend to pack in rows to give sheet-like structures (e.g., plates and flakes) [28].

During the last 80 years, attempts have been made to calculate the refractive indices of inorganic crystals and, more recently, organic and molecular crystals from their experimentally derived crystal structures [29, 30]. These calculations are made to establish a more precise relationship between the molecular structure of a crystal and its optical properties. These complex calculations take into account the crystal environment and its effect on the molecular polarizability that is suitably distributed over the molecule. So far, these calculations are reasonable but imprecise, being accurate to only one or two decimal places; this does not match the greater accuracy possible when measuring the refractive index of crystals using the immersion technique (Section 7.7.1).

7.7.1 Measuring Refractive Indices

The principal refractive indices of microscopic crystals and particles can be measured by successively immersing them in liquids of known refractive index, using the immersion technique [15]. When the refractive index of the particle is very different to that of the liquid, the particle will be visible because it has contrast, or relief. The greater the refractive index difference, the greater is the contrast. However, as the difference decreases, the particle becomes less visible and when the refractive index of the liquid matches that of the crystal, the particle will become invisible. Therefore, because the refractive index of the liquid is known, then the solid will have the same refractive index.

Most commercially available refractive index liquids [10] are organic and there is a possibility that an organic crystal will dissolve before its refractive index can be measured. If the compound dissolves slowly, then it should be possible to obtain a refractive index that approximates the true value. However, if the compound dissolves rapidly, then it will be necessary to saturate the liquid with the test compound and determine the actual value for the liquid using a refractometer before re-measuring the index of the crystal.

Several methods can be used to determine the match point at which the refractive index of a solid matches that of the immersion liquid. These include the Becke test, phase contrast microscopy, dispersion staining, oblique illumination, and differential interference contrast microscopy. Of these methods, the Becke test is the easiest to apply and is sensitive to differences in values of refractive index of about 0.001. The Becke test (Section 7.7.2) is used to determine if the immersion liquid has a refractive index that is higher or lower than that of the specimen studied. If no polarizer (i.e., non-polarized light) is used, the refractive index that is measured for an anisotropic crystal is an average value.

The single refractive index of a glass or a cubic crystal is most easily determined because its value is independent of orientation and so does not have to be aligned to the polarizer. However, for anisotropic crystals, the measurement of refractive index is more complex and time consuming because the two (uniaxial) or three (biaxial) principal values are determined by carefully aligning them with respect to the polarizer [15].

7.7.2
The Becke Test

The Becke test [16] utilizes the bright halo of light (called the Becke line) that is seen near the interface between a transparent particle and the surrounding mounting medium. The Becke line will move across the interface as the specimen is focused up and down. The easiest way to remember how to apply the Becke test is *high to low.* When the objective lens is lowered from a high position to a low position (relative to the specimen), the Becke line moves from the higher refractive index to lower refractive index, and vice versa. At the match point, the Becke line becomes indistinct and the specimen contrast is at a minimum. Figure 7.10 shows the application of the Becke test for an isotropic particle mounted in a liquid having a greater refractive index. For maximum sensitivity, the illumination should be monochromatic (e.g., Na_D; $\lambda = 589.3$ nm), a medium to high power objective should be used and the aperture diaphragm should be almost closed.

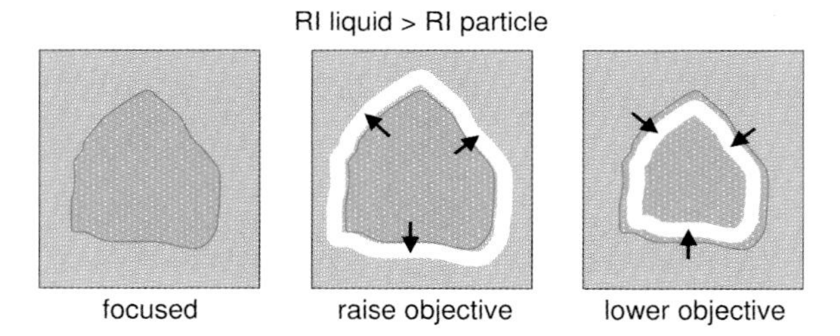

Fig. 7.10 Diagram showing the Becke line test (where the refractive index of the mounting medium is greater than the particle).

7.7.3
Dispersion Staining

Dispersion staining utilizes the difference in dispersion of the refractive index of a particle and that of the mounting liquid [22] and optically stains particles with a colored border (a colored Becke line). A special objective lens, called a dispersion staining objective [31], is used to observe particles that have been immersed in refractive index liquids having high optical dispersion [10]. It is a highly sensitive, simple and elegant technique used to determine the refractive index of transparent particles crystals and to analyze mixed particles. Dispersion staining has been used successfully as a quality control tool to rapidly detect trace amounts of contaminants in antibiotic compounds [32].

7.8
Particle Size

Light microscopy is a well-established technique for the measurement of particle sizes from about 1 μm up to about 400 μm across. For the accurate measurement of particles in finely divided powders, especially those with particles less than about 5 μm across, the use of a scanning electron microscope should be considered.

When a measuring scale is fitted in one of the microscope eyepieces, particles can be measured as they are being viewed (Fig. 7.11). The measuring scale must be calibrated using a calibrated stage micrometer scale [17]. For spherical particles (which are rarely encountered in pharmaceutical powders), the diameters are sufficient to describe their particle size. For non-spherical particles, various statistical diameters (such as Feret's diameter, Martin's diameter and projected area diameter) can be determined by light microscopy [17].

When comparing the sizes of particles in powder samples, it may not always be necessary or appropriate to perform a comprehensive particle size distribution analysis. A simple estimation of the particle size distribution for a powder can be performed rapidly by light microscopy. The particles are dispersed into a mounting fluid (Section 7.5.2) that provides adequate contrast on a microscope slide and are viewed at different magnifications. Alternatively, a slurry of the powder, suspended in the mounting medium, can be prepared in a vial and an aliquot portion taken and dropped onto a microscope slide. The sizes of the smallest and the largest particle seen are noted and the average size is estimated. Although this method is not precise, with experience it can be fairly reproducible and is a rapid way to identify significant particle size differences between samples.

Manual microscopy methods to determine a particle size distribution are slow, tedious and (for most powders) inaccurate because so few particles are measured. For this reason, a quicker, more efficient method is to use an image analyzer. The prepared slide is scanned using a computer-controlled, motorized specimen stage and images of the particles are captured by a camera and ana-

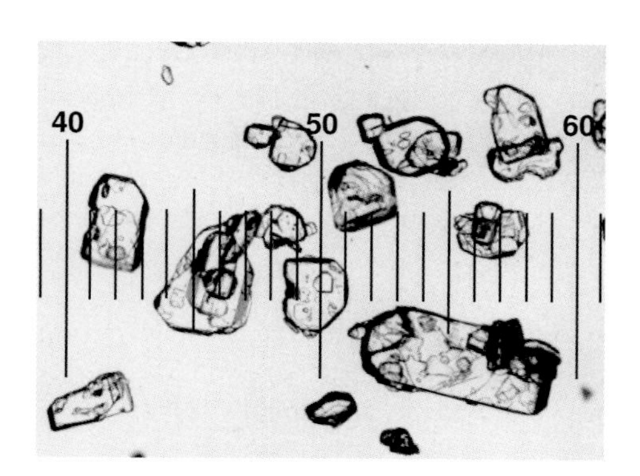

Fig. 7.11 Measuring the sizes of platy crystals using a calibrated eyepiece measuring scale (each small division represents 10 μm).

lyzed automatically by a computer. However, the results obtained are only as good as the sample preparation and the individual particles should not touch or overlap their neighbors. Ideally, the particles must not flocculate, they should adopt a random distribution of orientations and many thousands of particles may need to be measured to give a statistically valid result. These criteria are often very difficult to achieve and are subject to operator variability.

Consequently, most particle size analyses now performed in the pharmaceutical industry use non-microscopical techniques, principally laser diffraction. However, light microscopy is still a valuable method that is used to support particle size analyses, especially to assess the size range of the particles in a powder prior to laser diffraction analysis and to investigate and troubleshoot unexpected or aberrant results.

To overcome the problems associated with dispersing powders in a random and homogeneous manner in fluids on microscope slide preparations, automated systems have been developed that use dry dispersion for both static and dynamic measurements. The Malvern PharmaVision [33] deposits a monolayer of particles onto a glass plate and uses automated microscopy to scan the plate and image analysis to measure particle size. An alternative method is the Sympatec QicPic [34] that measures particle size as a moving powder aerosol passes through a rapidly pulsed light beam. Both of these systems also acquire particle shape data from the two-dimensional images.

7.9 Particle Shape

The external shape, or habit, of a crystal is determined by its internal structural symmetry. Crystallographers have developed a nomenclature of specific terms to describe the many different forms and habits that exist for each of the seven crystal systems. The term "habit" has a specific meaning and is used to describe the outward appearance of a crystal that consists of a group of symmetry related faces, known as forms [35]. For routine microscopical examination of pharmaceutical and fine chemical compounds, it is often not practical or even necessary to identify these specific crystal forms and habits (such as an orthorhombic prism, a dihexagonal bipyramid or a tetragonal sphenoid). This is especially true for crystals that look different just because they are distorted due to the uneven development of certain crystal faces during growth. There is value, however, in describing the shape of a crystal in much simpler terms, which is more easily determined and understood, and is ideal for monitoring batch-to-batch variations. The shape of a crystal can influence its pharmaceutical performance, such as suspension syringeability, tableting behavior, or dissolution properties [36].

Six basic crystal shapes have been recognized [22, 37] and crystals having these shapes can occur in any of the seven crystal systems. The US Pharmacopoeia (monograph 776) has also adopted these shape descriptions [23] and they are described below and shown diagrammatically in Fig. 7.12.

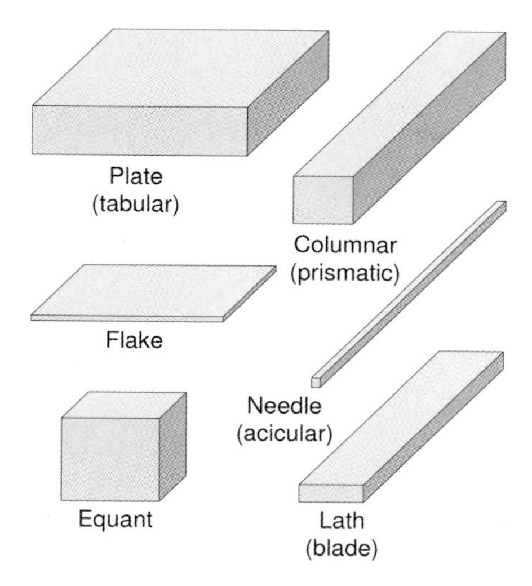

Fig. 7.12 Diagram showing the six basic crystal shapes.

- *Equant* crystals are equi-dimensional (such as cubes or spheres).
- *Plates* are flat, tabular crystals and have similar breadth and width; thicker than flakes.
- *Flakes* are thin, flat crystals that have similar breadth and width; thinner than plates.
- *Laths* are elongated, thin and blade-like crystals (may be ribbon-like if flexible).
- *Needles* are acicular, thin and highly elongated crystals having similar width and breadth.
- *Columns* are elongated, prismatic crystals, with greater width and thickness than needles.

When characterizing crystal shapes, descriptions such as cubic, hexagonal or rhombic should be used with great caution. These are very specific crystallographic terms and their use implies knowledge of the crystal system to which a crystal belongs. The descriptions, cuboid, blocky, hexagon-shaped and rhomb-shaped would be more appropriate and just as informative.

Some crystals will have a distinctive, sometimes characteristic geometrical shape, such as square or diamond; this is especially true for platy and flaky crystals where the straight edges are defined by very narrow faces. A crystal could therefore be described as square, rectangular or diamond-shaped plate, a pointed lath, a triangular flake, and so on. Thus, when identifying crystal shapes, such terms will make the descriptions more complete and meaningful.

Crystals in a powder sample could have a range of shapes, depending upon the growth rate of the different crystal faces, e.g., a powder having abundant hexagon-shaped plates may also contain some triangular- or diamond-shaped plates. Powders that have been milled might not show any evidence of the original shapes of the crystals, and the particles are likely to be irregularly-shaped and angular fragments.

Crystals are three-dimensional (3D) objects and light microscopy allows them to be observed directly so that their shapes can be determined simply by looking at them. For suitably orientated crystals, it is possible to measure some interfacial angles using the graduated rotating microscope stage. However, this task is made easier and more accurate if a single crystal is first bonded to the end of a needle on a rotation accessory, known as a spindle stage [38].

Ideally, crystals in a microscope slide preparation will present all possible orientations for inspection so that their lengths, breadths and thicknesses can be measured. This situation is only true for spheres; thus, for most preparations, the crystals will tend to rest in a stable position and will adopt a preferred orientation so that they lay on their largest faces. Consequently, they present a two-dimensional view to the observer. Alternatively, if the crystals are mounted in a fairly viscous liquid (such as silicone oil, or even treacle) they can be moved and rolled (as shown in Fig. 7.13) by gently moving the coverglass with the tip of a needle probe so that their thickness can be seen and measured using an eyepiece measuring scale.

Alternatively, the thickness of a crystal can be measured by focusing through it using the fine focus control. Most microscopes will have a graduated fine focus control and this needs to be calibrated (they will usually be 1 or 2 μm per division) so that vertical measurements can be made [39]. The vertical movement can be calibrated using an object having a precisely known thickness (perhaps a coverglass that has been accurately measured using a micrometer) and noting how many fine focus divisions are required to focus from the upper to the lower surfaces (allowing for the apparent thickness due to the refractive index of the glass).

During the development and production scale-up of a compound, the crystal shape or the length to breadth aspect ratio can vary between batches because of changes in the solvent, the saturation of the crystallizing solution, the cooling rate, or the stirring speed. A difference in the crystal shape for a compound may be recognized by light microscopy and this could be indicative of a differ-

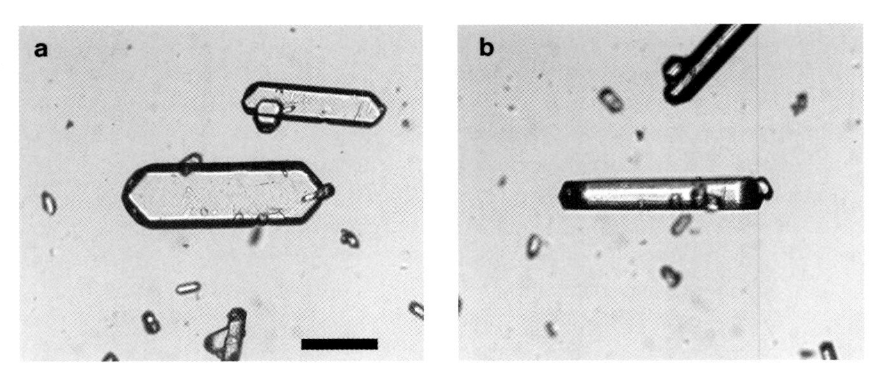

Fig. 7.13 Lath-shaped crystal (a) rolled onto its side (b) to measure its thickness. Scale bar = 250 μm.

ent polymorphic form. In this case, another analytical technique (such as X-ray diffraction, Raman spectroscopy, or solid-state NMR) should be used to confirm that the crystal structure is actually different to that expected.

Specimens analyzed by powder X-ray diffraction, especially for quantitative analysis, should consist of randomly orientated particles so that the powder pattern has no preferred orientation effects that give rise to unexpected changes in peak intensities [24]. If preferred orientation is suspected, the powder should be examined microscopically to establish if it has particles with significantly different shapes to those in a reference sample.

When examining crystals manually by microscopy, their 3D shapes can be characterized simply and quickly. The use of an image analyzer speeds up this process and removes any operator subjectivity [40] to derive shape descriptors, such as circularity and length–breadth aspect ratio. Image analysis has been used to characterize and classify the 3D shapes of sucrose crystals [41] and for the in-process monitoring of the particle shape of L-glutamic acid during batch crystallization [42]. However, image analyzers are not yet as sophisticated as the microscopist's eye–brain combination, which can rapidly recognize the shapes of crystals and classify them according to the six basic shapes described in this section (Fig. 7.12).

In addition to the shape descriptions given above, the terms listed below can be used to further describe the microscopical appearance of particles to provide a more comprehensive characterization [43–45].

- *Angular*: having sharp corners and edges (e.g., on fragmented crystals).
- *Rounded*: having indistinct corners and edges.
- *Irregular*: shapeless, lacking any recognizable symmetry.
- *Granular*: irregularly shaped, but equi-dimensional.
- *Euhedral*: crystalline with well-formed faces.
- *Anhedral*: crystalline, but without faces.
- *Polycrystalline*: composed of many tightly interlocking crystals.
- *Microcrystalline*: a polycrystalline material that requires the use of a microscope to see the small interlocking crystals.
- *Cryptocrystalline*: a polycrystalline material having crystallites that cannot be resolved using a light microscope; but does give an X-ray diffraction pattern.
- *Transparent*: clear, but may have a few cracks or inclusions.
- *Translucent*: mostly cloudy due to fine cracks or being polycrystalline.
- *Opaque*: no transmission of light due to scatter or could have a very high refractive index.
- *Colored*: has a characteristic color.
- *Lamellar*: consisting of stacked plates (Fig. 7.14).
- *Foliated*: consisting of stacked leaf-like flakes (Fig. 7.14).
- *Spherulitic*: a sphere-shaped mass with needles radiating from the centre. Fragments are usually fan-shaped with radial crystals (Fig. 7.14).

A powder will often consist of an intimate mixture of primary particles and assemblages of particles. Such assemblages are referred to as aggregates or ag-

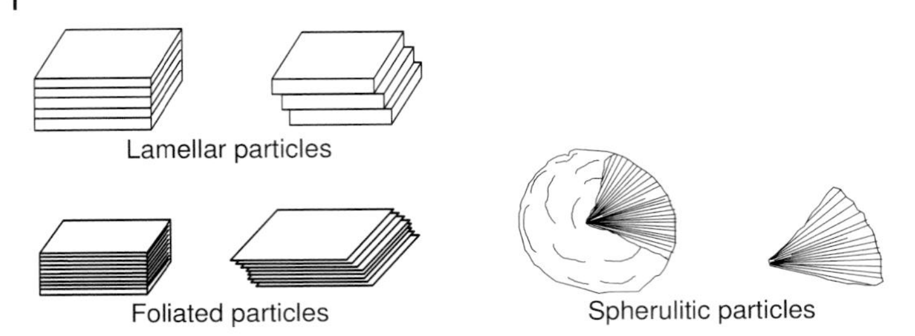

Fig. 7.14 Sketches showing examples of lamellar, foliated and spherulitic particles.

glomerates. However, because these terms are frequently interchanged, it has been recommended that when describing powder particles the term aggregate should be discontinued and agglomerate used exclusively [46]. Where necessary, a distinction between soft and hard agglomerates is possible using a simple microscopical test to determine ease of friability.

7.10 Comparing Powder Samples

When examining powder samples by light microscopy, it is beneficial to take photomicrographs of each so that they can be compared with other batches to establish if there are any differences. These could be changes in particle shape or particle size that might affect the performance of the powder on a manufacturing plant. Comparison of batches using photomicrographs is made easier if they are all recorded at the same magnification.

When reporting the findings of a microscopical examination, a photomicrograph should be supported with a brief description of the sample, including

- description of crystal shape;
- size range and average size of particles;
- description of how crystals associate with others;
- any observations about crystallinity (are the particles birefringent?);
- photomicrograph (with scale bar and magnification);
- record of mounting medium used (e.g., silicone oil);
- the microscope used.

An example of a microscopical examination is that performed on a sample of aspartic acid. The powder (shown in Fig. 7.15) consists of a mixture of colorless laths, prisms, needles, and plate-shaped fragments. The laths, prisms and needles range in length from about 50 to about 600 μm (typically about 150 μm). These particles have length to breadth ratios ranging from about 2:1 to about 15:1. The platy fragments are up to about 450 μm across. When viewed between crossed polarizers, the particles show bright interference colors, which

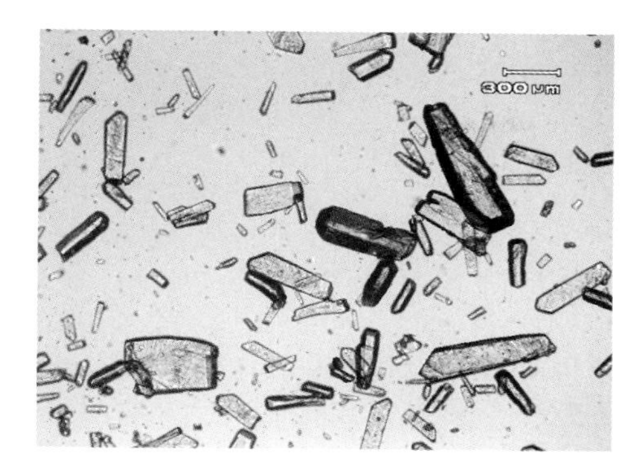

Fig. 7.15 Aspartic acid powder.

extinguis every 90° of rotation, indicating that the particles are crystalline. The powder was mounted in silicone oil and examined using a Nikon Optiphot POL microscope (serial number: 123456).

7.11 Thermomicroscopy

Thermomicroscopy (also known as hot stage microscopy and fusion microscopy) allows the direct observation of small samples as they are heated or cooled in a micro-furnace, or hot stage, that is placed on the specimen stage of a polarizing light microscope (Fig. 7.16). It is, without doubt, the most frequently used microscopical technique for the characterization of the solid-state properties of organic compounds, especially the polymorphism of pharmaceutical compounds. Many books and papers have been published on this very important technique, of which books by McCrone [47] and Kuhnert-Brandstätter [48] are highly recommended reading for their wealth of practical guidance.

As an analytical technique, thermomicroscopy provides a unique insight into thermally induced events, such as melting, crystallization, sublimation, desolvation, polymorph phase transitions, and thermotropic mesomorphism. In addi-

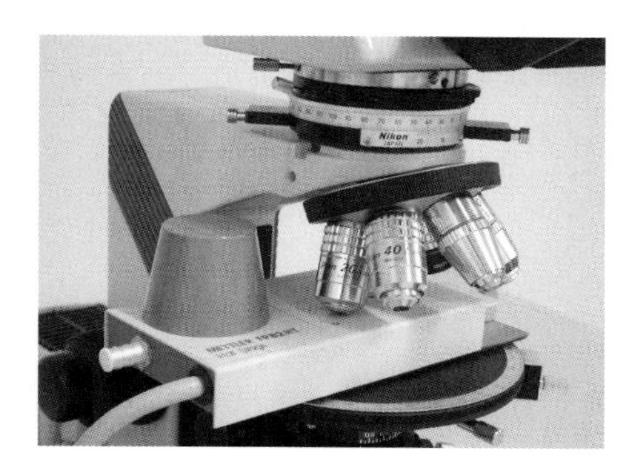

Fig. 7.16 Microscope heating stage (Mettler) mounted on the rotating stage of a polarizing light microscope.

tion, it complements other thermal analysis techniques, such as differential scanning calorimetry, differential thermal analysis and thermogravimetric analysis, by providing a rapid and non-ambiguous method to interpret thermograms.

Some novel experiments can be performed using a hot stage to elucidate the physicochemical properties of compounds. For example, thermomicroscopy is ideally suited to the study and characterization of mixtures of enantiomers and polymorphs of the racemates using the Kofler contact method [48–50]. Some organic compounds, such as vitamin K, DDT and TNT, have crystal lattices that expand anisotropically when heated [47]. Unusual recrystallization behavior in fusion preparations of these compounds is due to grain growth (analogous to the annealing of metals) and could be misinterpreted as a solid–solid polymorphic conversion.

7.12
The Microscope as a Micro-scale Laboratory

Not only is the polarizing light microscope an excellent tool for examining and characterizing crystals, but it can also be used as a micro-scale laboratory to observe the growth of crystals from solution, from the melt, or by sublimation from vapor. In addition, solvent-mediated polymorphic conversions can be observed [47]. Indeed, the conditions to grow crystals of the metastable orthorhombic polymorph of paracetamol (Form II) from solution were established using microexperiments performed on the specimen stage of a light microscope [51].

When good quality crystals of organic compounds are to be grown for optical characterization or for single-crystal X-ray diffraction studies, high-boiling solvents such as benzyl alcohol and nitrobenzene, or a mixed fusion with thymol, have been used with great success [47]. These crystallizations are easily performed on a microscope slide and suitable crystals are then selected and isolated whilst being observed with the microscope.

Microscopes have been adapted to be integral components in some laboratory-scale crystallizers. For example, a novel microcrystallization apparatus uses a microscope to monitor crystal shape variations induced by changing agitation rate, growth rate and the addition of shape modifiers [52]. An inverted microscope has been used to investigate the polymorphs of the explosive HMX (cyclotetramethylene tetranitramine) to enable the construction of the stability diagram at different temperatures in a saturated solution [53].

An excellent example of a polymorphic compound that exhibits a solution phase conversion and a solid–solid phase conversion is potassium nitrate. In one simple experiment, which is readily observed by polarized light microscopy, Ostwald's Rule of Stages is demonstrated and concomitant polymorphs are observed [54]. When a slightly under-saturated drop of an aqueous solution of potassium nitrate is placed on a microscope slide (without a coverglass) and is allowed to slowly evaporate at room temperature, rhomb-shaped crystals of the metastable trigonal form (β) will appear first at the edge of the drop (Fig. 7.17-1). After a short time,

Fig. 7.17 Growth of two polymorphs (β and α) of potassium nitrate at the edge of a drop of aqueous solution that show a solution phase (1 and 2) and solid–solid phase (3) polymorphic conversion. Crossed polarizers with a sensitive tint plate. Scale bar = 250 μm.

a second phase, the room temperature stable orthorhombic form (α) will appear as broad, highly birefringent plates (Fig. 7.17-1). As the α-form grows, a solution-phase polymorph conversion causes the β-form to dissolve adjacent to the advancing crystal front (Fig. 7.17-2). When the advancing α-form touches the residual crystals of the β-form, an instantaneous solid–solid conversion occurs (Fig. 7.17-3). Eventually, the entire crop of crystals will convert into the α-form.

7.13 Twinning

When viewed using plane polarized light, some crystals will appear to be single entities, but upon more detailed inspection using crossed polarizers, it becomes apparent that they are in fact two crystals that have grown together and share a common plane, the twin plane. Polarized light microscopy is ideally suited for the examination of twinned crystals because of the changes to the optical properties when compared with non-twinned crystals. The phenomenon of twinning occurs in crystals belonging to all crystal systems and is common in many metals, minerals, inorganic and organic crystals [15].

Twinning is not a random crystallization event, but occurs in specific, crystallographically controlled orientations, and several types are known [15], including common contact twins (e.g., Fig. 7.18), interpenetration twins (e.g., Fig. 7.19), and multiple twins. Crystals that have grown under stressful, uncontrolled conditions, in undisturbed environments, or in the presence of impurities may develop as twins [55, 56]. Twinning is most frequently encountered during crystal-

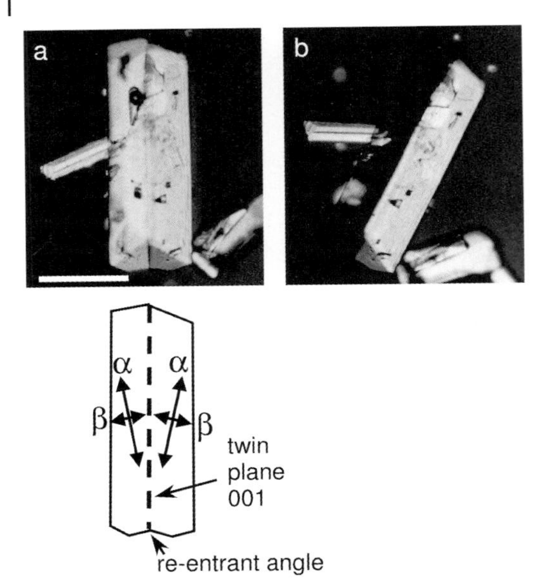

Fig. 7.18 Photomicrographs showing contact twinning in a crystal of a drug compound viewed between crossed polarizers with both parts at equal brightness (a) and when rotated clockwise through about 30° (b) so that the left half is at extinction. Scale bar = 100 μm. The schematic diagram shows the orientation of the principal refractive indices for each half of the twinned crystal shown in (a) with the twin plane and the re-entrant angle indicated.

lization from solution or fusion, but it can also be induced by mechanical stress, as in L-lysine monohydrochloride dihydrate [57]. The occurrence of twinning in molecular crystals is relatively common and has been reported for many compounds, e.g., stearic acid [58] and saccharin [59]. Agglomerated crystals, however (Section 7.9), will generally have randomly orientated contacts with their neighbors at edges and faces.

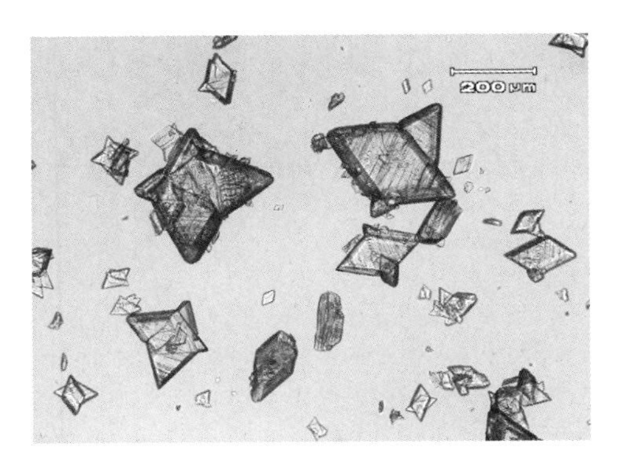

Fig. 7.19 Diamond-shaped platy crystals of a developmental drug compound with star-like interpenetration twins.

Twinning manifests itself in several ways and is recognized when a suitably orientated crystal (i.e., with the twin plane normal to the microscope stage) is viewed between crossed polarizers. The extinction position for each crystal in a twin may be different (perhaps by several degrees of stage rotation) and there is frequently a re-entrant angle (a V-shaped groove or notch coincident with the twin plane, as shown in Fig. 7.18). For isotropic substances, or for those twins in which the individual crystals go into extinction at the same time, polarized light observations have limited value. In these cases, the occurrence of twinning will have to be confirmed by crystal shape, the presence of a re-entrant angle, or perhaps X-ray crystallography.

Twinning should not be considered just an oddity or freak of nature, but as an indicator of a change in a crystallization process. Twinning adds either a plane or axis of symmetry [28] and so a twinned crystal has a higher symmetry than its non-twinned counterpart. Consequently, twinned crystals may tend towards a more equant shape and may be larger than the non-twinned crystals; therefore, exploitation of twinning may have commercial benefits as a way to engineer a specific crystal shape or to control particle size. Twinning is sometimes caused by impurities that cause localized defects and the impurities become trapped along the twin plane. A change in the level of an impurity can also induce twinning, as in adipic acid [60]. Due to the crystalline disorder along the twin plane, twinned crystals may be more soluble than the non-twinned crystals. Consequently, using polarized light microscopy, twinning confirmed in pharmaceutical substances and fine chemicals could have distinct advantages when developing products.

Crystals grown for single-crystal X-ray structure determination must be free from disorder and twinning. Although software solutions are available for crystallographers to unravel the complex data from twinned crystals, their use is time consuming, challenging and demands a high level of skill to correctly interpret the results [61]. Therefore, polarized light microscopy should be considered as a technique to screen crystals for the occurrence of twinning so that unsuitable crystals are rejected prior to crystallographic analysis.

7.14 Color and Pleochroism

Many organic and inorganic substances are colored, and the colors can be used to distinguish between different polymorphic forms. For example, mercuric iodide has a red tetragonal phase that is stable at room temperature, an orange tetragonal phase that is metastable at room temperature, and a yellow orthorhombic phase that is stable above 127 °C [62]. The transition of the red and yellow phases is reversible (with hysteresis) upon cooling and is readily observed by light microscopy and the naked eye. Another example is the polymorphic compound 5-methyl-2-[(2-nitrophenyl)amino]-3-thiophenecarbonitrile, which occurs in seven solid forms with red, orange and yellow colors [63]. This phenom-

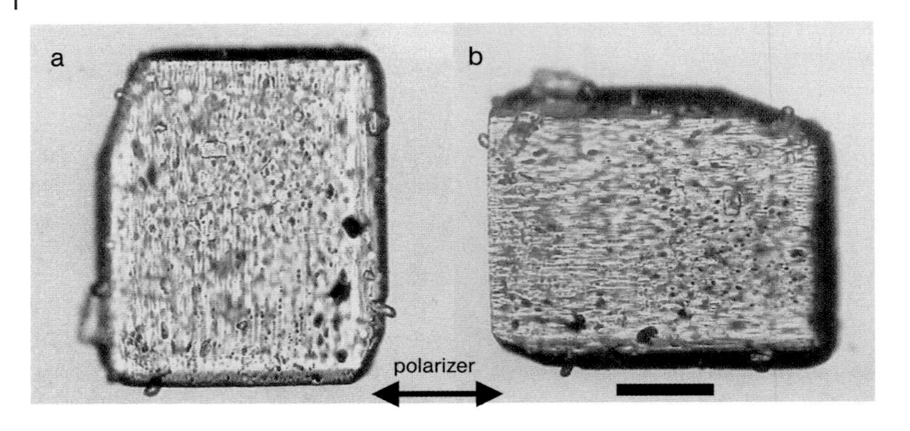

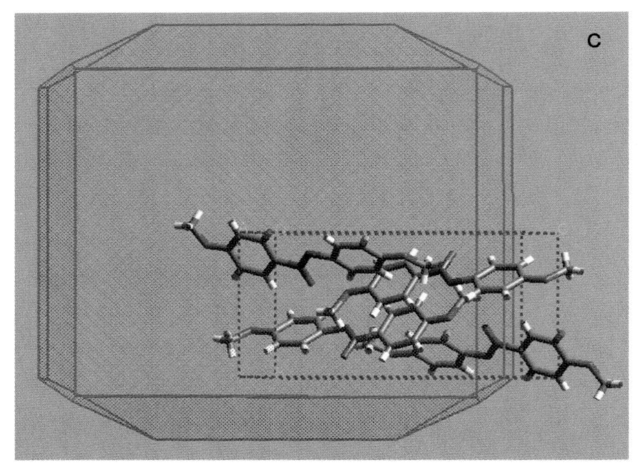

Fig. 7.20 Photomicrographs of a single platy crystal of *p*-azoxyanisole viewed with plane polarized light showing (a) colorless to (b) yellow pleochroism when rotated through 90° (scale bar = 50 µm). The crystal packing diagram (with four molecules shown) and morphology model (c) show that the yellow color is observed when the azo-groups are aligned with the vibration direction of the polarizer.

enon has been described as "color polymorphism" and the cause of the color differences has been attributed to conformational changes for each of the polymorphs.

When viewed using plane polarized white light, transparent colored crystals will frequently show different colors or a change in color intensity during rotation of the microscope stage [15]. This phenomenon is due to the selective absorption of light for specific vibration directions in colored crystals. It is known as pleochroism when the three principal vibration directions in pleochroic biaxial crystals show different colors and dichroism when just two principal vibration directions, in either biaxial or uniaxial crystals, show different colors. Generally, the term pleochroism is used to describe crystals that display different colors.

Colorless crystals and isotropic crystals cannot be pleochroic. Some colored crystals may appear to be non-pleochroic if the absorption is similar in different orientations. In addition, if a highly birefringent, colorless crystal is immersed in a liquid that almost matches one of its principal refractive indices and it is rotated in plane polarized light, it may appear to be pleochroic because of a dramatic change in contrast. The color shown by some crystals may be caused by colored impurities and these can often be seen microscopically as discrete inclusions within crystals or as small areas on the crystal surfaces.

Organic crystals that display pleochroism have chromophores [15] and the absorption of light is greatest when it is vibrating along the direction of the chromophoric groups, such as the azo (–N=N–) group (e.g., *p*-azoxyanisole, shown in Fig. 7.20) or a polyene chain $(–CH=CH–)_n$. Therefore, in the absence of supporting single-crystal X-ray data, the observation of pleochroism in a crystal can give valuable clues about the crystal packing and alignment of the molecules in its structure. For example, the strong yellow color for *p*-azoxyanisole occurs only when the crystal is orientated as shown in Fig. 7.20 (b), indicating that the azo-groups are aligned in one direction (parallel to the polarizer); this is confirmed by inspection of the morphology model and crystal packing diagram [64].

7.15 Fluid Inclusions

Transmitted light microscopy is ideal for examining inclusions that occur within organic and inorganic crystals. Inclusions are impurities (liquid, gas or solid) that become trapped inside crystals as they grow (usually too rapidly) from a solution or a melt. Mineralogists study the compositions of fluid inclusions in minerals as a way to understand the thermal and chemical history of minerals that trap the inclusions as they form rocks and mineral deposits from either a solution or a magma [65].

It can sometimes be difficult to prevent solvent-filled inclusions, but slow and controlled crystal growth can help to minimize them. Solvent-filled inclusions could potentially be the cause of post-crystallization changes during storage. Liquid leaking from inclusions in crystals could induce a solvent-mediated polymorphic conversion or may cause agglomeration or caking due to partial recrystallization. Mullin [56] gives an excellent review of inclusions that occur in both organic and inorganic crystals. As solvent-filled inclusions can have deleterious effects upon the post-crystallization behavior of batches of crystals, it is beneficial if they are examined microscopically to see if they are present. If a powder has a slight discoloration, the color may be due to small inclusions filled with colored impurities or residues of a colored mother liquor.

Solvent-filled cavities in crystals produced under laboratory conditions or in large-scale crystallizers can be the cause of batch failure if measured residual

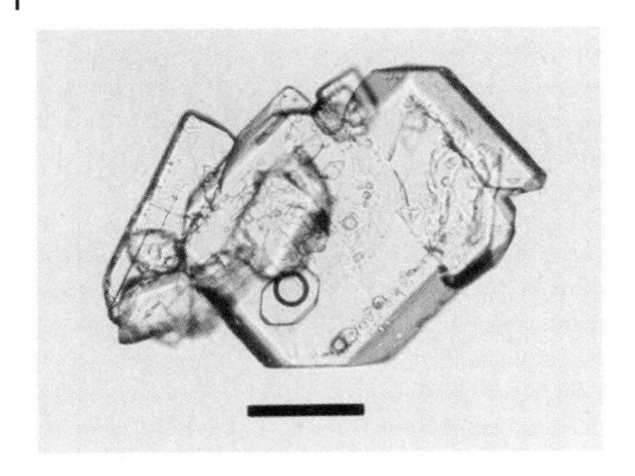

Fig. 7.21 Agglomerated platy crystals of a drug compound showing a large fluid filled negative crystal inclusion with a vapor bubble. Scale bar =100 μm.

solvent levels are too high. A quick microscopical examination may reveal that the solvent is located in inclusions rather than being structurally bound in the crystal lattice as a solvate or as a hydrate. When examined by light microscopy, the solvent may be seen in cavities inside the crystals. Very often, a solvent-filled cavity will have a small vapor bubble that forms when the trapped solvent cools and contracts; this observation confirms the presence of a liquid.

To examine crystals for the presence of solvent-filled inclusions, they should be immersed in a liquid (such as silicone oil) and viewed using plane polarized light or uncrossed polarizers. If crossed polarizers were used, the inclusions would not be seen because the liquid is optically isotropic and would not produce interference colors. Sometimes it may be difficult to distinguish an inclusion from another defect inside the crystal (such as a crack) or a rough area on the surface of the crystal. By focusing slowly through a crystal, inclusions will be seen, perhaps dispersed randomly or along specific lines or planes that are related to the crystal structure. To reduce or eliminate the disturbing effects of surface roughness on a crystal that may obscure the solvent-filled inclusions, it should be immersed in a mounting medium that has the same refractive index as the crystal (Section 7.7). This makes the crystal become almost invisible and the inclusions are easily seen because of high image contrast.

Inclusions are not always large; some can be just one or two micrometers across and will require high magnification to make them visible. If small vapor bubbles are present, they will often be seen to vibrate rapidly inside the solvent-filled inclusions because of Brownian movement.

Sometimes, the inclusions may occur as "negative crystals", which are cavities with facets that correspond to the external crystal faces of the parent crystal (Fig. 7.21). Negative crystals usually form when the temperature of the parent crystal has been raised (e.g., during drying) sufficiently for the trapped solvent to dissolve the crystal from the inside [56].

7.16 Mechanical Properties of Crystals

During the development of a pharmaceutical compound, it is very desirable to understand as early as possible how it will respond during milling and compaction. In early development, only small amounts of powder will be available for investigation, so predictive methods are applied by measuring mechanical properties (such as hardness, elasticity, plasticity, and brittleness) on single crystals using a nanoindenter [66].

Some very simple qualitative experiments can also be performed on the stage of a light microscope to indicate if a powder will undergo brittle fracture, plastic deformation, or is soft or hard. To conduct these experiments, a small amount (say, 50 to 100 μg) of powder is placed in a pile on a clean microscope slide and a second slide is gently placed on top. The powder is then sheared between the two slides whilst sliding and rotating one slide relative to the other and applying light pressure with the fingers. The resulting sheared powder is then examined microscopically to see how it has deformed. Powders that have undergone brittle deformation will appear as finely divided dust particles, whereas powders that deform plastically will be smeared or may have elongated, sticky, plastic rods that have rolled between the two slides.

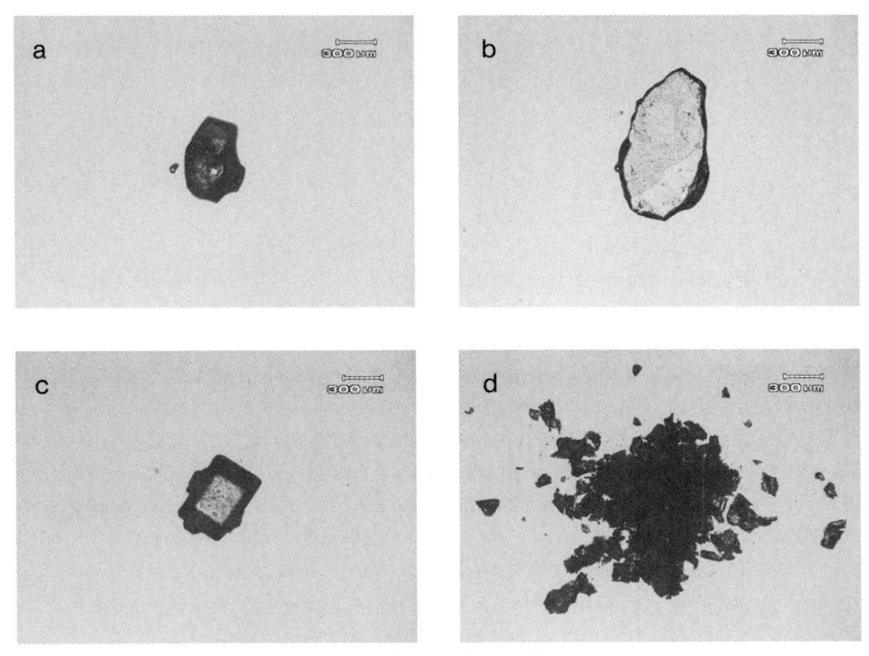

Fig. 7.22 Photomicrographs showing a soft crystal (camphor) before (a) and after (b) being squashed, and a hard crystal (sucrose) before (c) and after (d) being crushed.

To establish if a crystal is soft or hard, it is placed on a microscope slide under a coverglass and pressed using a needle probe. Soft crystals will deform and spread out, whilst hard ones will shatter into many fragments (Fig. 7.22).

Different polymorphs or salts of a compound may have quite different mechanical properties and these may be critical when selecting the solid form to progress into development. For example, the commercial form of paracetamol (monoclinic, Form I) undergoes brittle fracture, whilst the metastable form (orthorhombic, Form II) deforms plastically [51]. Unlike Form I, Form II can be made into tablets by direct compression without the need for binders.

7.17 Pseudomorphs

The shape of a crystal may be altered if it undergoes a chemical or physical change (such as desolvation or a phase transformation). This results from a change in its internal structure and can cause the destruction of the crystal due to a dramatic change in volume or it may change shape, as shown by the thermally induced transformation of Form II into Form I of terephthalic acid [67]. However, the alteration can be non-destructive and, although the crystal has changed internally, it may retain its original external shape; this is called a pseudomorph (=*false form*) (not to be confused with pseudopolymorph!). Such a change, which may be due to desolvation or a solid-state polymorph conversion, can often be detected by polarized light microscopy.

Many solvated or hydrated crystals will lose their solvent of crystallization when heated in air. The crystals may become polycrystalline, due to a change in density or a rearrangement of the crystal structure, and will be cloudy or opaque when viewed using transmitted light. This is an example of *alteration pseudomorphosis* [68] and is demonstrated by heating oxalic acid dihydrate above 60 °C so that it dehydrates to give opaque crystals (Fig. 7.23) that are white in reflected light because they scatter the light.

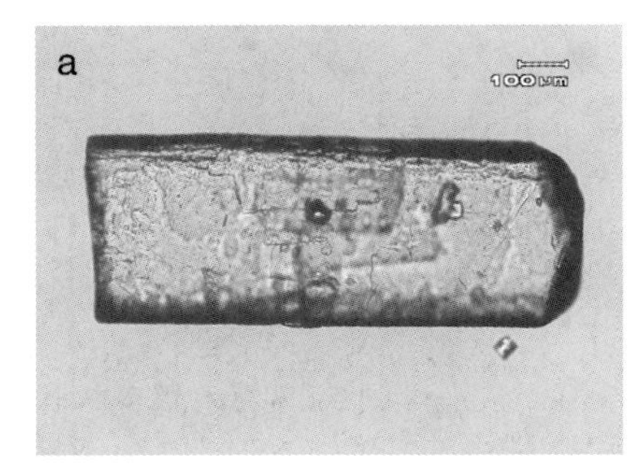

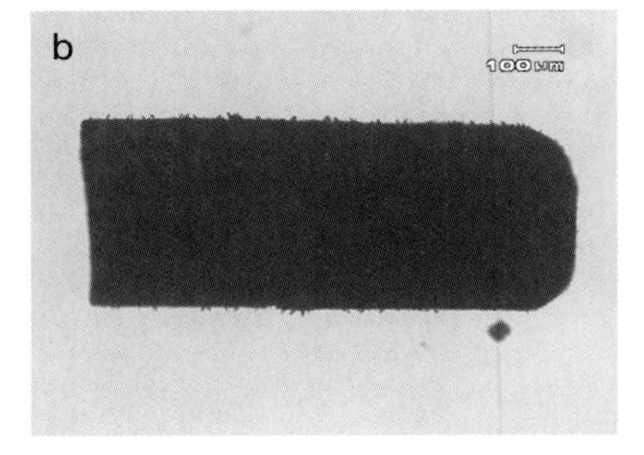

Fig. 7.23 Oxalic acid dihydrate crystal (a) before dehydration at 60 °C to give an opaque pseudomorph of anhydrous oxalic acid (b).

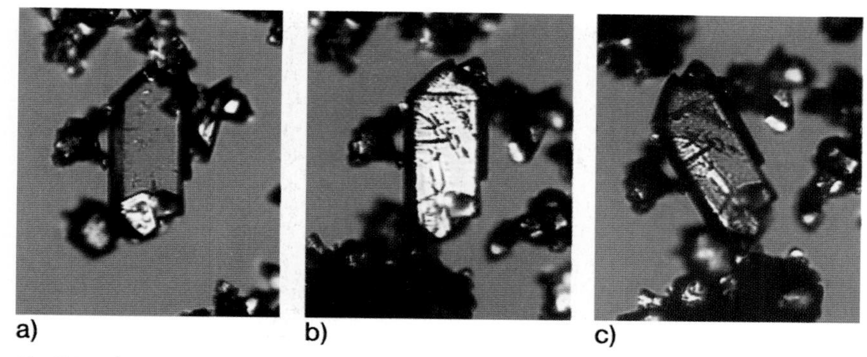

Fig. 7.24 Photomicrographs showing a non-cloudy pseudomorph of paracetamol I after thermally induced conversion from Form II. (a) Clear crystal of orthorhombic paracetamol Form II showing straight extinction at 25 °C. (b) After conversion into monoclinic Form I and no longer at extinction. (c) When rotated by about 30° the crystal goes into extinction and is seen to be non-cloudy, but is slightly cracked.

A change that produces a pseudomorph without the loss of solvent, such as a polymorphic transition, is called *paramorphosis* [68]. A pseudomorph may be cloudy and polycrystalline, such as that produced by the thermally induced dehydration of inosine dihydrate [69], or it may be a non-cloudy, single crystal, such as that resulting from the solid-state conversion of paracetamol II into I (Fig. 7.24).

For some compounds, a metastable form may be converted into a stable form so rapidly that it cannot be observed directly. In this instance, the only evidence of its existence is the presence of a polycrystalline pseudomorph that is detected by polarized light microscopy, such as shown by TNT [70].

7.18 Mesomorphism

Polarized light microscopy is an important analytical technique for the investigation and characterization of liquid crystals, or mesophases [71]. Compounds that form mesophases due to a change in temperature (e.g., *p*-azoxyanisole) are thermotropic [72] and those that form mesophases by the addition of a solvent, such as water, to an amphiphilic compound (e.g., sodium lauryl sulfate) are lyotropic [73]. Some mesomorphic compounds exhibit both thermotropic and lyotropic behavior, e.g., fenoprofen sodium [74].

To study the thermotropic and lyotropic properties of compounds by polarized light microscopy, thin films are prepared between a microscope slide and cover glass [71]. Thermotropic compounds, when heated and cooled using a microscope heating stage, are characterized by their distinctive, birefringent textures. Lyotropic compounds can be examined in several ways, but a quick method is to place a small amount on a microscope slide, gently compress it under a coverglass and add just enough solvent to surround it. This method will allow a

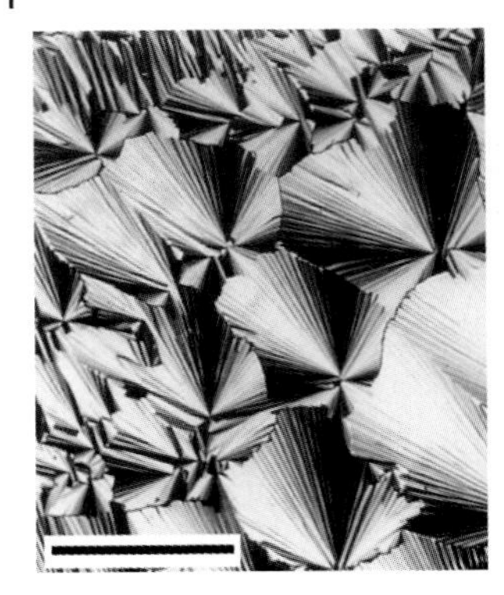

Fig. 7.25 Lyotropic mesophase of sodium lauryl sulfate in water, showing (left) a well-developed fan-like texture (hexagonal phase) and (right) the same field of view after moving the coverglass. Scale bar=200 μm.

solvent concentration gradient to develop so that different phases can be observed. At the edge of the coverglass, where evaporation is greatest, another concentration gradient will develop. Confirmation that a compound has entered the liquid crystalline state is achieved by touching and moving the coverglass with a needle probe, whereupon the birefringent sample will shear-flow under the influence of pressure (Fig. 7.25).

During screening of pharmaceutical compounds for polymorphs and solvates, the potential for mesophase formation should be considered and actively sought. Polarized light microscopy and thermomicroscopy provide a rapid means to test compounds and to identify mesophases that would otherwise be missed using an automated screening robot. Hartshorne [71] has described some simply applied microscopical techniques to investigate liquid crystals.

Mesophases are novel materials and could be patented and exploited commercially as products or as mechanisms for enhanced drug delivery, e.g., cromolyn sodium [75]. As with solid crystalline substances, liquid crystalline compounds can also exist in different polymorphic forms and these have characteristic microscopic textures that are readily recognized by experienced researchers. To confirm the structures and to classify each mesophase found, polarized light microscopy observations are usually supported by X-ray diffraction analysis.

7.19
Identification of Contaminants and Foreign Matter

A polarizing light microscope is an incredibly powerful tool for carrying out micro-scale chemical tests to identify trace amounts of inorganic and organic substances [76] and for the identification of particulate contamination [77]. During the microscopical investigation of samples to characterize their solid-state properties, foreign particles and contaminants, such as fibers, inorganic substances, plastic, metal particles, and glass chips, could be found. Some of these particles may be process related (e.g., fibers from filters, diatoms from filter aids, catalyst residues, carbon from clarifiers, inclusions, or inorganic by-products from a chemical reaction). An experienced microscopist would be able to recognize and identify these types of particle, thereby enabling their sources to be located and further occurrences eliminated. However, as an aid to identification, reference to previously prepared microscope slides of known specimens or to a library of photomicrographs, such as *The Particle Atlas* [22], is highly recommended.

The ability of the microscopist to recognize and identify a wide range of particulate contaminants is a valuable skill and an asset to his or her employer; an ability that may be critical if a complete characterization of samples is to be achieved. For example, consider a compound being synthesized that has a step using hydrochloric acid as reactant and this is subsequently followed by the addition of sodium hydroxide to adjust the pH. If the purity of the product is unexpectedly low, then a quick microscopical examination (using uncrossed polarizers) may reveal the presence of a few small, isotropic crystals of a second substance, sodium chloride, which is an unwanted by-product. Powder X-ray diffraction could also be used to identify the sodium chloride, but it would have to be present at a relatively high level, whereas infrared spectroscopy would be unsuccessful because NaCl is not detectable using this technique.

Conclusion

Look at your samples!

References

1 J. Bernstein, J.-O. Henck, *Cryst. Eng.* **1998**, 1(2), 119–128.

2 S. R. Byrn, R. R. Pfeiffer, J. G. Stowell, *Solid-State Chemistry of Drugs*, 2nd edition, SSCI, West Lafayette, IN, **1999**.

3 J. Bernstein, *Polymorphism in Molecular Crystals*, Oxford University Press, Oxford, **2002**.

4 T. L. Threlfall, *Analyst* **1995**, 120(10), 2435–2460.

5 L. Yu, S. M. Reutzel, G. A. Stephenson, *Pharm. Sci. Technol. Today* **1998**, 1(3), 118–127.

6 *The Hutchinson Dictionary of Scientists*, Helicon Publishing Ltd., Oxford, **1997**.

7 S. Bradbury, *An Introduction to the Optical Microscope*, Oxford University Press and Royal Microscopical Society, Oxford, **1984**.

8 *RMS Dictionary of Light Microscopy*, Oxford University Press and Royal Microscopical Society, Oxford, **1989**.

9 J. G. Delly, *Photography through the Microscope*, 9th edition, Eastman Kodak Company, Rochester, NY, **1988**.

10 Cargille Laboratories, 55 Commerce Road, Cedar Grove, NJ 07009, USA.

11 Norland Products Inc., 2540 Route 130, Suite 100, PO Box 637, Cranbury, NJ 08512, USA.

12 J. Smoliga, Characterization of pharmaceutical drug tablets by polarized light microscopy and X-ray powder diffraction, **2001**, assa international, *Second International Workshop on the Physical Characterization of Pharmaceutical Solids* (IWPCPS-2), Lancaster, Pennsylvania, USA, and personal communication.

13 E. A. Wood, *Crystals and Light – An Introduction to Optical Crystallography*, 2nd edition, Dover Publications, New York, **1977**.

14 F. D. Bloss, *An Introduction to the Methods of Optical Crystallography*, Holt, Rinehart and Winston, **1961**.

15 N. H. Hartshorne, A. Stuart, *Crystals and the Polarizing Microscope*, 4th edition, Edward Arnold, London, **1970**.

16 E. E. Wahlstrom, *Optical Crystallography*, 3rd edition, John Wiley and Sons, New York and London, **1960**.

17 W. C. McCrone, L. B. McCrone, J. G. Delly, *Polarized Light Microscopy*, Ann Arbor Science Publishers, Ann Arbor, **1978**.

18 S. L. Morissette, Ö. Almarsson, M. L. Peterson, J. L. Remenar, M. J. Read, A. V. Lemmo, S. Ellis, M. J. Cima, C. R. Gardner, *Adv. Drug Delivery Rev.* **2004**, 56, 275–300.

19 J. McCabe, The use of increased throughput technologies to investigate solid state properties of pharmaceutical compounds, **2004**, IQPC Polymorphism and Crystallisation Conference, London.

20 E. M. Chamot, C. W. Mason, *Handbook of Chemical Microscopy*, Volume I, 1st edition, John Wiley and Sons, New York, **1931**.

21 G. Nichols, *Microscope* **1998**, 46(3), 117–122.

22 W. C. McCrone, J. G. Delly, J. A. Brown, S. J. Palenik, I. M. Stewart, *The Particle Atlas*, Volumes I–VI, Ann Arbor Science Publishers, Ann Arbor, **1972–1980**; also published as a CD-ROM available from McCrone Research Institute, Chicago, **1993**.

23 USP 27 2004, *The United States Pharmacopeia*, Monograph 776 (Optical Microscopy). United States Pharmacopeial Convention, Inc., 12601 Twinbrook Parkway, Rockville, MD 20852.

24 R. Jenkins, R. L. Snyder, *Introduction to X-ray Powder Diffractometry*, John Wiley and Sons, New York, **1996**.

25 A. N. Winchell, H. Winchell, *The Microscopical Characters of Artificial Inorganic Solid Substances: Optical Properties of Artificial Minerals*, 3rd edition, Republished by McCrone Research Institute, Chicago, **1989**.

26 A. N. Winchell, *The Optical Properties of Organic Compounds*, 2nd edition, Republished by McCrone Research Institute, Chicago, **1987**.

27 D. D. Jordan, *J. Pharm. Sci.* **1993**, 82(12), 1269–1271.

28 C. W. Bunn, *Chemical Crystallography*, 2nd edition, Oxford University Press, Oxford, **1961**.

29 W. L. Bragg, *Proc. Roy. Soc. A* **1924**, 105, 370–386.

30 R. W. Munn, M. Andrzejak, P. Petelenz, A. Degli Espoti, C. Taliani, *Chem. Phys. Lett.* **2001**, 336, 357–363.

31 S.-C. Su, *Microscope* **1998**, 46(3), 123–146.

32 J. A. Diamond, *Microscope* **1974**, 22(3), 209–212.

33 Malvern Instruments Ltd., Enigma Business Park, Grovewood Road, Malvern, Worcestershire WR14 1XZ, UK.

34 Sympatec GmbH, System-Partikel-Technik, Am Pulverhaus 1, D-38678 Clausthal-Zellerfeld, Germany.

35 F. C. Phillips, *An Introduction to Crystallography*, 4th edition, Longman Group Ltd., **1971**.

36 J. K. Haleblian, *J. Pharm. Sci.* **1975**, 64(8), 1269–1288.

37 N. H. Hartshorne, A. Stuart, *Practical Optical Crystallography*, 2nd edition, Edward Arnold Ltd., London, **1969**.

38 M. E. Gunter, R. Weaver, B. R. Bandli, F. D. Bloss, S. H. Evans, *Microscope* **2004**, 52(1), 23–39.

39 S. Bradbury, *Basic Measurement Techniques for Light Microscopy*, Oxford University Press and Royal Microscopical Society, Oxford, **1991**.

40 M. N. Pons, H. Vivier, V. Delcour, J.-R. Authelin, L. Paillères-Hubert, *Powder Technol.* **2002**, 128, 276–286.

41 N. Faria, M. N. Pons, S. Feyo de Azevedo, F. A. Rocha, H. Vivier, *Powder Technol.* **2003**, 133, 54–67.

42 J. Calderon De Anda, X. Z. Wang, K. J. Roberts, *Chem. Eng. Sci.* **2005**, 60, 1053–1065.

43 D. S. Aldrich, M. A. Smith, Pharmaceutical applications of infrared microspectroscopy, in *Practical Guide to Infrared Microspectroscopy*, ed. H. J. Humecki, Marcel Dekker, New York, **1995**, Chapter 9, 323–375.

44 H.H. Read, *Rutley's Elements of Mineralogy*, 26th edition, Thomas Murby and Co., London, **1970**.

45 C.S. Hurlbut, Jr., *Dana's Manual of Mineralogy*, 18th edition, John Wiley and Sons, New York, **1971**.

46 G. Nichols, S. Byard, M.J. Bloxham, J. Botterill, N.J. Dawson, A. Dennis, V. Diart, N.C. North, J.D. Sherwood, *J. Pharm. Sci.* **2002**, 91(10), 2103–2109.

47 W.C. McCrone, *Fusion Methods in Chemical Microscopy*, Interscience Publishers, Inc., New York, **1957**.

48 M. Kuhnert-Brandstätter, *Thermomicroscopy in the Analysis of Pharmaceuticals*, Pergamon Press, Oxford, **1971**.

49 J. Jacques, A. Collet, S.H. Wilen, *Enantiomers, Racemates and Resolution*, Wiley and Sons, New York, **1981**.

50 A. Burger, J.M. Rollinger, P. Brüggeller, *J. Pharm. Sci.* **1997**, 86(6), 674–679.

51 G. Nichols, C.S. Frampton, *J. Pharm. Sci.* **1998**, 87(6), 684–693.

52 M. Bayard, *Microscope* **2003**, 51(4), 189–200.

53 A.S. Teetsov, W.C. McCrone, *Microsc. Cryst. Front* **1965**, 15(1), 13–29.

54 J. Bernstein, R.J. Davey, J.-O. Henck, *Angew. Chem. Int. Ed.* **1999**, 38, 3441–3461.

55 P.H. Egli, L.R. Johnson, Ionic salts, in *The Art and Science of Growing Crystals*, ed. J.J. Gilman, John Wiley and Sons, New York, **1963**, Chapter 11, 194–213.

56 J.W. Mullin, *Crystallization*, 3rd edition, Butterworth-Heinemann, Oxford, **1993**.

57 R. Bandyopadhyay, D.J.W. Grant, *Pharm. Res.* **2002**, 19(4), 491–496.

58 R.S.N. Swamy, I.V.K. Bhagavan Raju, *Cryst. Res. Tech.* **1987**, 22(5), K94–K98.

59 H.F. Lieberman, L. Williams, R.J. Davey, R.G. Pritchard, *J. Am. Chem. Soc.* **1998**, 120, 686–691.

60 L. Williams-Seton, R.J. Davey, H.F. Lieberman, R.G. Pritchard, *J. Pharm. Sci.* **2000**, 89(3), 346–354.

61 J.L. Flippen-Anderson, J.R. Deschamps, R.D. Gilardi, C. George, *Cryst. Eng.* **2001**, 4, 131–139.

62 M. Hostettler, H. Birkedal, D. Schwarzenbach, *Helv. Chim. Acta*, **2003**, 86, 1410–1422.

63 L. Yu, *J. Phys. Chem. A*, **2002**, 106, 544–550.

64 Crystal structure for p-azoxyanisole (PAZOXN03) downloaded from Cambridge Structural Database, v5.24, Cambridge Crystallographic Data Centre, Cambridge, UK, and modeled using C2 Morphology, Cerius^2 v4.6, Accelrys, Inc., San Diego, California.

65 E. Roedder, *Microscope* **1992**, 40(1), 59–79.

66 L.J. Taylor, D.G. Papadopoulos, P.J. Dunn, A.C. Bentham, N.J. Dawson, J.C. Mitchell, M.J. Snowden, *Org. Process Res. Dev.* **2004**, 8, 674–679.

67 R.J. Davey, S.J. Maginn, S.J. Andrews, S.N. Black, A.M. Buckley, D. Cottier, P. Dempsey, R. Plowman, J.E. Rout, D.R. Stanley, A. Taylor, *J. Chem. Soc., Faraday Trans.* **1994**, 90(7), 1003–1009.

68 E.S. Dana, *A Textbook of Mineralogy*, 4th edition (revised by W.E. Ford). John Wiley and Sons, New York, **1922**.

69 A.L. Gillon, R.J. Davey, R. Storey, N. Feeder, G. Nichols, G. Dent, D.C. Apperley, *J. Phys. Chem. B* **2005**, 109, 5341–5347.

70 J. Haleblian, W. McCrone, *J. Pharm. Sci.* **1969**, 58(8), 911–929.

71 N.H. Hartshorne, *The Microscopy of Liquid Crystals*, Microscope Publications, Chicago, **1974**.

72 D. Coates, G.W. Gray, *Microscope* **1976**, 24(2), 117–150.

73 F.B. Rosevear, *J. Am. Oil Chem. Soc.* **1954**, 31, 628–639.

74 T. Rades, C.C. Mueller-Goymann, *Eur. J. Pharm. Biopharm.* **1998**, 46, 51–59.

75 J.S.G. Cox, G.D. Woodard, W.C. McCrone, *J. Pharm. Sci.* **1971**, 60(10), 1458–1465.

76 H.F. Schaeffer, *Microscopy for Chemists*, Dover Publications, New York, **1966**.

77 W.C. McCrone, *Am. Lab.* **2000**, 32(8), 17–23.

8
The Importance of Solvates

Ulrich J. Griesser

8.1
Introduction

In many stages of industrial processing, substances are exposed to solvents or solvent vapors. The most prominent solvent-based procedures in the chemical and pharmaceutical industry are precipitation, crystallization or recrystallization from a suitable solvent or solvent mixture to separate or purify the desired substance. In most of these cases the aim is to harvest a single-component crystalline solid, preferably in the desired solid form and poor in impurities. Other solvent-based processes are wet granulation, spray-drying, lyophilization, coacervation etc. However, the "solvent of crystallization" often becomes "entrapped" in the solid to some extent, which is almost invariably regarded as a nuisance and may cause trouble in industrial crystallization. When an organic solvent cannot be completely removed from the product by a suitable drying procedure we talk about "residual solvents", which are generally considered to be impurities and regulated by authorities such as the ICH guideline Q3C on residual solvents in pharmaceuticals [1].

A solvent may be associated with a crystalline solid in different ways (Fig. 8.1). The binding of solvent molecules to the surface by weak interactions (hydrogen bonding, van der Waals, dipole–dipole), i.e., by physisorption, is reversible and variable. In addition, the affinity to individual crystal faces is different and, therefore, the amount of surface adsorbed solvent or water in crystalline materials depends on their morphology besides many other parameters. The solvent (mother liquor) may also become simply physically entrapped in a growing crystal, which is called liquid inclusion [2]. Since these pockets are filled with a saturated solution of the mother liquor, other kinds of impurities also remain associated with the crystal in this way. In both cases, however, the amount of solvent in crystalline solids of small molecules is usually low (<0.5%). Larger amounts of solvent may adsorb in localized disordered regions or defects that may arise by any kind of mechanical stress (e.g., grinding, granulation). Disorder phenomena and amorphicity are regarded as the main rea-

Polymorphism: in the Pharmaceutical Industry. Edited by Rolf Hilfiker

ISBN: 3-527-31146-7

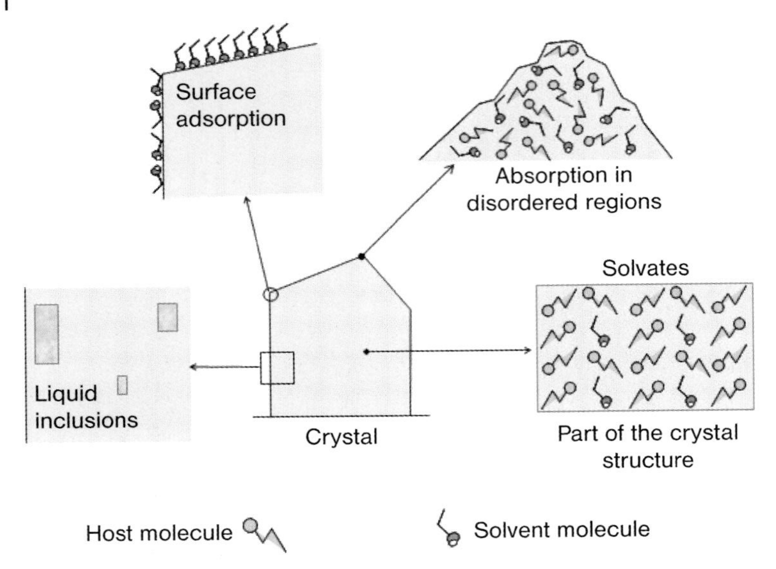

Fig. 8.1 Simple illustration of different principles of solvent associations with crystalline solids.

sons for enhanced residual solvent contents. High amounts of solvents can more or less "dissolve" in these highly energetic metastable areas. Unlike in the liquid-like amorphous state, (small) molecular entities in a crystalline solid mostly pack closely to maximize intermolecular interactions and provide no extra space for a solvent uptake. Yet we know that compounds can crystallize and pack "together" with a solvent, forming a new solid phase, a "solvate", where the solvent molecule is part of the crystal structure. The metaphors host (organic molecule) and guest (solvent) are frequently used in this context, symbolizing the structural picture of the usually much smaller solvent molecules that occupy relatively small spaces in the structural assembly of the larger organic partner molecule. Water represents the smallest molecule in solvates. Owing to this feature and its extraordinary ability to form hydrogen bonds, water is unrivaled as solvate former. In contrast to the other principles of solvent association, solvate formation is inevitably linked to a significant change in the material properties, which may entail various problems and may become a major concern in the development of industrial products.

The present chapter briefly highlights the features of crystalline solvates. Solvates with water (hydrates) and organic solvents are discussed jointly to convey the principles of the phenomenon as a whole. Due to the experience of the author in this area, emphasis is placed on small organic drug molecules and the implication of solvates in pharmaceutical development.

8.2
Terminology and Classification of Solvates

8.2.1
General Terms and Definitions

The terminology of crystalline solids that contain a solvent as a second component is not uniform and very confusing. This becomes particularly evident from the recent dispute and probably ongoing expositions on the terms "pseudopolymorphism" and "co-crystal" [3–7]. Various terms such as solvate, molecular compound, molecular complex, packing complex, inclusion compound, solid solution, channel compound, clathrate and lattice compound have been and are used for a specific phenomenon or to address these multi-component crystals in general. Some of the terms (e.g., hydrate, molecular compound, solid solution) were coined long before scientists were able to gain structural information about crystalline solids. Since the middle of last century, knowledge about crystal structures and molecular interactions has increased almost exponentially due to the rapid development of X-ray diffraction and spectroscopic techniques. This process resulted in several structural based terms such as inclusion compound, clathrate or channel compound. Unfortunately, different and sometimes contradictory preferences for certain terms have now been established in the scientific disciplines concerned (e.g., mineralogy, inorganic, organic, or supramolecular chemistry).

In this chapter the term "*solvate*" is used as a general term for crystalline solvent adducts because it is definitely the most widely accepted descriptor for these materials. Since adducts with water represent most solvent adducts, the term "*hydrate*" is used to mark this subclass of solvates more specifically. A "crystalline solvate" may be defined in the broadest sense as a solid where a solvent is coordinated in or accommodated by the crystal structure. However, on closer examination we have to recognize that this term also shows some flaws considering the current definitions of its origin, namely "solvation". The term "solvation" is commonly used to describe the process by which solvent molecules surround and interact with each dissolved molecule or ion [8]. The formation and stability of the solvent shell results from intermolecular forces, which are particularly strong in solutions of electrolytes, due to the strong electrostatic interactions of ions with dipolar solvent molecules, especially water. For aqueous solutions the term hydration is used. Now, any aggregate that consists of a solute ion or molecule with one or more solvent molecules is termed solvate [8], which is definitely not consistent with the situation in most crystalline solvates of organic compounds where solvent molecules participate in hydrogen-bonding networks or act as space fillers with no strong interactions between the solvent and the "host" molecules. Crystalline hydrates with high coordination (hydration, solvation) numbers, such as $AlCl_3 \cdot 6\,H_2O$ or $NiSO_4 \cdot 6\,H_2O$ etc., rather resemble the state of a solute in a solution. We may of course ask the question: What is the solute (according to the definitions) in a solid where the solvent acts more as a glue in forming the three-dimensional network of an organic molecule? The IUPAC definition [9] of "solvation" is broader and includes specific polymers:

> "Any stabilizing interaction of a solute (or solute moiety) and the solvent or a similar interaction of solvent with groups of an insoluble material (i.e., the ionic groups of an ion-exchange resin). Such interactions generally involve electrostatic forces and van der Waals forces, as well as chemically more specific effects such as hydrogen bond formation."

The term "solvate" is not mentioned in this collection of terms used in organic physical chemistry, but if we just use the phrase "stabilizing interaction of a solvent and a partner molecule" it also describes well what happens during the formation of crystalline solvates.

The present status of terms, without extended discussion on their possible origin, can be summarized as follows. *Pseudopolymorphism* is a frequently used general term for all kinds of crystal forms of a substance where one or more solvents are present in the crystal lattice in a stoichiometric or a non-stoichiometric amount, whereas *polymorphism* [10–12] is the general term for different crystal forms of exactly the same chemical composition. Pseudopolymorphism is a historically grown term [11, 13, 14] that, unluckily, has been misused to describe exclusively solvates that retain their crystal structure upon desolvation (isomorphic desolvates). Notably, the term "solvatomorphism" has emerged recently [15] as a substitute for pseudopolymorphism. This purely artificial term lacks any fundamental explanation and discussion in the scientific community so far and it is advisable to bypass the use of such terms unless clear conventions exist.

The fact that solvent and solute crystallize together ("co-") and that the solvent becomes an essential part of the structure sufficiently justifies the use of the term "*co-crystal*" for all kinds of solvates and not only for "molecular complexes" (stoichiometric solvates, see below) to be at one with Dunitz [5]. Almarsson and Zaworotko [16] have stated that

> "The primary difference (between a solvate and a co-crystal) is the physical state of the isolated pure components: if one component is a liquid at room temperature, the crystals are referred to as solvates; if both components are solids at room temperature, the products are referred to as co-crystals."

Whether the melting point is an acceptable criterion for such a differentiation is questionable. For instance, 10-undecylenic acid (mp 24 °C) can either be liquid or solid at room temperature. Is a solid adduct of a compound with this substance a co-crystal or a solvate? Anyway, the term co-crystal has become a heavily used term for targeted multi-component crystals [17], especially in the context of crystal engineering and is an elegant term to clinch all kinds of arrangements and phenomena of multi-component crystals, including those for which debatable or inconsistent terms were/are used or for borderline cases. Its preferential use for adducts of two components, of which neither one is used as a solvent in the classical sense, is doubtlessly convenient and similar to the use of solvate as a more specific term for a "solvate containing an organic solvent and not water" when mentioned along with hydrate.

Clearly, a concise terminology (and classification) of "crystalline solvent adducts" can only be established by a multidisciplinary approach that considers all kinds of information, including phenomenology, thermodynamics and the structural features.

8.2.2 Types of Solvates

Applied science often requires pragmatic solutions and therefore it is common, especially in the context of pharmaceutical solids, to subdivide solvates into two main classes: stoichiometric and non-stoichiometric solvates (hydrates) [12, 18]. This classification is definitely not perfect but it is relatively easy to apply and also implies practical consequences. The scheme in Fig. 8.2 illustrates the relation of these two groups of solvates to the main classes of binary (multinary) systems. The circles of the two classes overlap to stress the existence of possible cases that do not allow a clear classification.

8.2.2.1 Stoichiometric Solvates

Stoichiometric solvates are regarded as molecular compounds. The solvate is an individual phase and the binary phase diagram shows a eutectic and/or a peritectic with the parent components (compound and solvent). The term implies a fixed, although not necessarily integral, ratio of solvent to compound. The solvent in stoichiometric hydrates is usually an integral part of the crystal structure

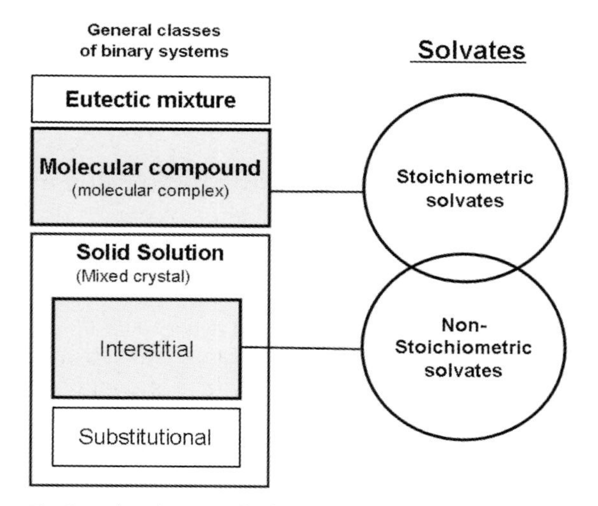

Fig. 8.2 Classification of solvates in relation to the classical types of binary systems.

and is essential for the maintenance of the molecular network. Desolvation of stoichiometric solvates always leads to a different crystal structure or results in a disordered or amorphous state.

Morris [19] has shown that hydrates with one mole-equivalent water (monohydrates) are most frequently encountered among organic drug substances [set of about 6000 substances of the Cambridge Structural Database (CSD)] and that the frequency of higher coordination numbers (di-, tri-, ... hydrate) decreases exponentially. The number of hemihydrates and trihydrates is about equal but only a sixth of the total number of monohydrates.

8.2.2.2 Non-stoichiometric Solvates

Non-stoichiometric solvates are a type of inclusion compound. In the pharmaceutical business, and particularly for patent issues, these solvates often cause problems and puzzles. Therefore, this class requires a more detailed description. Non-stoichiometric solvates may be regarded as interstitial solid solutions or interstitial co-crystals. The crystal structure only constitutes in the presence of the solvent, which is usually located in certain structural voids (often channels) and acts more or less as a space filler of these voids. Particularly large and awkwardly shaped molecules that cannot pack closely form such structures. The most important feature of this class of solvates is that the structure is retained, while the solvent content can take on all values between possibly zero and a multiple of the molar compound ratio. The amount of solvent in the structure depends on the partial pressure of the solvent in the environment of the solid and on the temperature. Exactly this situation is expressed by the term non-stoichiometric, but not the fact that a solvate may crystallize in uncommon yet well-defined solvent/compound ratios such as the cyclohexanone solvate of oxyphenbutazone (0.25 mol mol^{-1}) [20] and caffeine 0.8 hydrate [21, 22]. Stoichiometric hydrates with ratios of, for example, 1/4, 4/5, 3/4 are rare but not impossible. However, in most cases odd numbers of incorporated solvent molecules are a sign of a non-stoichiometric solvate. Good examples for such solvates are a recently reported series of dipeptide structures, which represent intriguing channel structures [23]. These channels (nanotubes) can host not only water but also molecules of different solvents. Single-crystal X-ray diffraction analysis at low temperatures often enables determination of the exact position of the disordered solvent molecules, showing odd occupancies from which ratios of, for example, 0.88, 1.26, 2.47 were derived [23]. Obviously, such numbers merely result from the "snapshot" of a continuous series of possible solvent/compound ratios. It is common practice to round odd numbered solvent ratios determined by any analytical method in a solvate to integer ratios. This may lead to wrong denotations and causes much confusion. For example, caffeine 0.8 hydrate is still quoted as a monohydrate in many textbooks and even in pharmacopoeias. Many non-stoichiometric solvates are also wrongly labeled with integer numbers (e.g., hemi-, mono-, di-hydrate), pretending a stoichiometric solvate. Chloroquine phosphate is one of many examples of non-stoichiometric hydrates that are specified as dihydrate. In the Cambridge Structural Database we can find two structures, one labeled as "dihydrate" another as "monohydrate" [24, 25], that are clearly isostructural ($P2_1/c$), except for small differences in the lattice parameters due to the different "states of hydration". Figure 8.3 shows the calculated powder X-ray diffraction patterns (PXRD) and the moisture sorption isotherm of the compound. The diffraction pattern exhibits a characteristic but unequal shift of the reflections of the "dihydrate" to smaller angles (larger interplanar distances) than for the "monohydrate", indicating the expansion of the structure at higher hydration levels. This very typical situation can be found in many pharmaceutical solvates/hydrates.

Another term in the context of non-stoichiometric solvates is "*isomorphic desolvate*" (or "desolvated solvate") [26–29], which describes the fact that non-stoichiometric solvates can desolvate to a one-component phase without changing the main structural characteristics. An isomorphic desolvate can only be obtained via the isostructural solvate and largely retains the crystalline lattice of the solvate, frequently with a very high degree of disorder. Furthermore, it is usually metastable and easily takes up the original solvent or even another solvent. Their behavior resembles that of zeolites. Since a desolvated solvate represents a

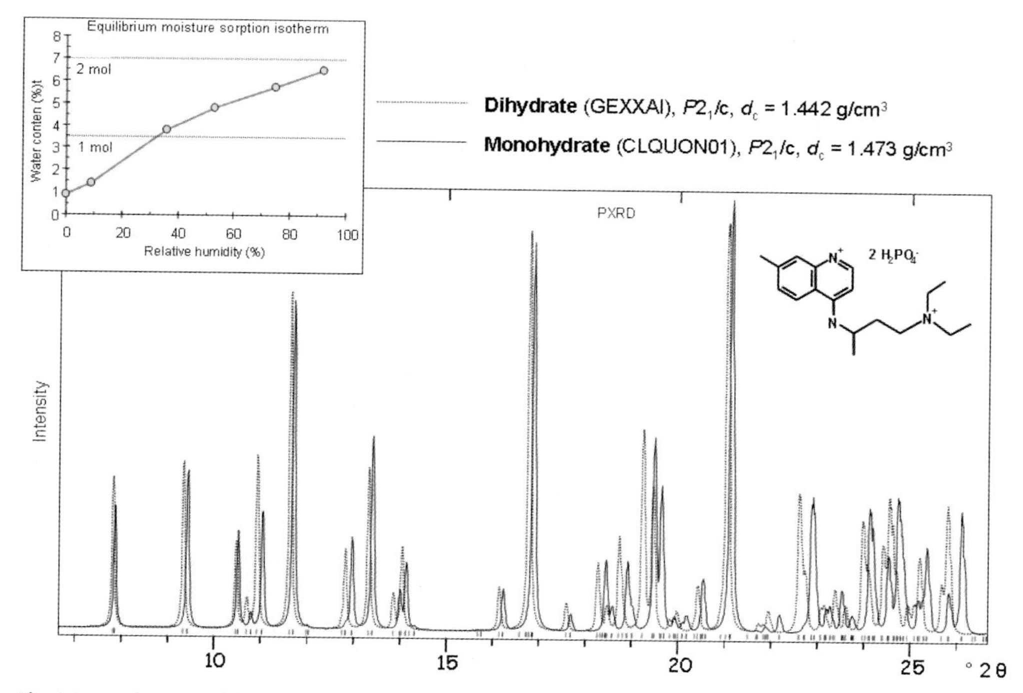

Fig. 8.3 Powder X-ray diffraction patterns calculated from the single-crystal structure data of two structures of chloroquine phosphate labeled as mono- and dihydrate. Inset on upper left: moisture sorption isotherm of the substance at 25 °C.

one-component phase (which is usually stable only at low partial pressure of an incorporated solvent or low humidity), it may considered as an individual polymorphic form. To draw an extreme comparison with this case: whether a house is occupied or empty it is still the same house. A dry sponge is also not regarded as a different object than a wet sponge. Therefore, a desolvated solvate and its isostructural solvate cannot be regarded as different forms.

Similar but more difficult is the situation for different solvates of a substance that crystallizes in essentially identical structures but can contain different solvents. The solvates are isostructural (i.e., isomorphous), which means that the individual *"isostructural solvates"* (syn: isomorphic solvates) crystallize in the same space group with only small distortions of the unit cell dimensions and the same type of molecular network of the host molecule. Due to the lack of systematic studies on solvates, only a fairly a limited number of publications (e.g., [30–34]) highlight this phenomenon, which is probably more common than one might assume.

Reasonably, supramolecular chemists and crystallographers use different and often more specific structural terms to describe solvates than those who apply non-crystallographic techniques to characterize solvates. An example of such a specific structural term is "clathrate", which is frequently used for solvates where the solvent molecules are entrapped within voids of the structural network of the host molecules without significantly interacting with them. Clathrate means "cage" and is applicable only to a structure where the solvent is entrapped in three-dimensional closed voids (cages). However, the term is often used as a general descriptor for various kinds of "inclusion compounds" and one can even find the term "channel clathrate" in the scientific literature, which is clearly an oxymoron.

8.2.3 Classification Models of Hydrates

Various classification concepts of hydrates have been discussed in the past. The association of water molecules in organic hydrates was reviewed and classified by Jeffrey [35] and Clark [36]. Morris [19, 37] applied a more convenient structural classification to hydrates. All these classifications focus more or less exclusively on the location and self-association of the water molecules in the crystal lattice (isolated, channels, planes, ion-coordinated) and thus require knowledge of the crystal structure. An ideal classification of solvates should also consider their stability (e.g., behavior under different conditions) and their thermodynamic features. A non-structural based classification of hydrates, mainly regarding the moisture sorption/desorption behavior of compounds, was given by Gal [38] in 1968, and Kuhnert-Brandstätter and Grimm [39] grouped different types of organic hydrates according to their thermal behavior. The so-called "Rouen 96 model" proposed by Petit and Coquerel [40] considers the mechanism and structural changes upon dehydration of organic hydrates. This intriguing model was the first attempt to go beyond a "static" picture of the arrangement of water in a crystal structure by considering the reorganization principles and changes when the solvent leaves the structure. In a complementary approach Galwey [41] systematically explored and reviewed solid-state dehydration reactions of various inorganic hydrates and classified them according to the water evolution type (WET) and structural criteria. All these contributions highlight the diversity and complexity of solvates and the need for more research to both understand and, one day, predict the nature of these solid species. An important contribution towards this goal comes from supramolecular chemistry and systematic surveys (e.g., [42, 43]) of the vastly rising number of crystal structures and the understanding of molecular interactions and recognition principles [44, 45]. However, only the combination of structural principles and reliable thermochemical data, phase diagrams, vapor–solid interaction behavior, kinetics etc. will allow us to get there.

8.3 Statistical Aspects and Frequency of Solvates

Desiraju et al. [46, 47] as well as Görbitz and Hersleth [48] have analyzed the solvent-forming propensities of organic and organometallic compounds from CSD entries [49]. These studies showed clearly that water is at the top of the solvate-forming solvents (>11 000 hits) followed by dichloromethane (>2800 hits), methanol, benzene, acetonitrile, toluene and tetrahydrofuran (about 1000 to 1400 hits respectively). The total number of different solvents in the different adducts was 300. Organometallic structures show a much higher tendency (about twice) to form solvates. When only organic structures are considered, methanol, benzene, dichloromethane, ethanol and acetone are top of the ranking list. Because some solvents are traditionally used more than others, a usage-correction was applied [45], which revealed that the likelihood of forming a solvate is highest for dimethylformamide (DMF), dimethyl sulfoxide (DMSO) and dioxane. These solvents can form multiple (strong or weak) hydrogen bonds with an organic solute molecule and so the extrusion of the solvent in the aggregation and nucleation process becomes disadvantageous from an enthalpic point of view and the solvent remains associated with the solute in the crystallization process. However, only a minority of the organic CSD compounds (~15%) form solvates, which may explain the importance of the entropic gain in eliminating solvent molecules to the bulk solution during nucleation.

Görbitz and Hersleth [48] found that 8.1% of the organic compounds in the CSD represent hydrates, while 7% form solvates with organic solvents (organometallic compounds: 10.5% hydrates and 17.3% solvates). From a time-dependent analysis of the CSD entries they found that not only the relative number of solvates in general but also those that contain more than one solvent (heterosolvates) has increased dramatically in recent years. This increase correlates with the growing size and complexity of newer molecules. Interestingly, the ratio of hydrates decreased from 18% to 7% in the last 50 years. One conclusion of this study was that the use of solvent mixtures may improve the chances for obtaining crystals of organic compounds with high molecular volumes.

Over almost two decades we carefully collected data on the solid-state properties of a few thousand pharmaceutically relevant organic compounds, with special focus on those drug substances listed in the Pharmacopoeia European (PhEur). The 1997 edition of PhEur contained 559 well-defined organic drug compounds [50]. For more than 55% of them either polymorphs or solvates were known. In a newer evaluation of a larger set of data (PhEur edition 4.02, 808 solid organic compounds [51]) this fraction increased only slightly to 57%. As shown in Fig. 8.4, 29% of the compounds are known to form hydrates, 10% other solvates and 36% may occur in different polymorphic forms (i.e., different non-solvated forms). These statistics represent the present status of knowledge and the number of compounds that may appear in different crystal forms can only increase in the future when more experimental work is performed on these compounds. The relatively low occurrence of solvates with organic solvents coincides with the results of the CSD searches mentioned above, but the occurrence of hydrates among drug compounds is much higher. This can probably be attributed to the fact that drugs must have hydrophilic properties, at least to some extent, to dissolve in the body fluids and to be effective. Secondly, about 45% of the surveyed compounds are used as salts, preferentially as hydrochlorides (39% of all salts). Figure 8.5 illustrates that salts exhibit an expected higher occurrence of hydrates than non-salts, whereas the

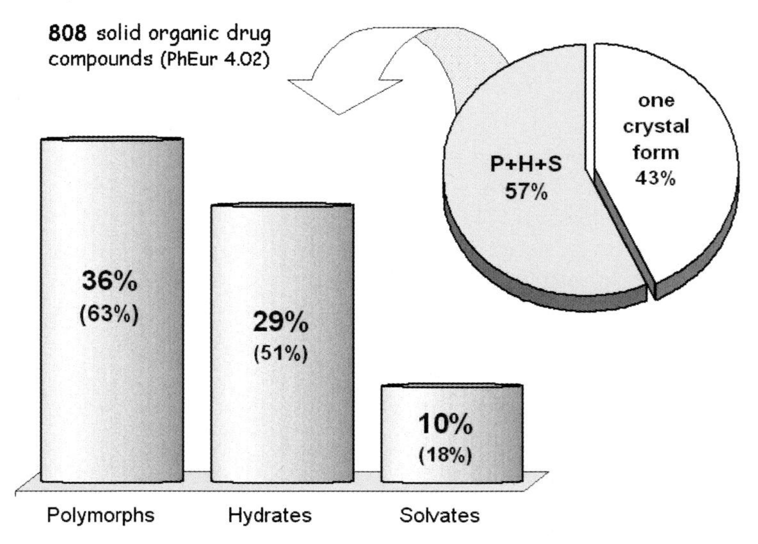

Fig. 8.4 Frequency of the occurrence of polymorphs (P), hydrates (H) and solvates containing organic solvents (S) among drug substances of the Pharmacopoeia Europaea. Upper percentages in the bars refer to the total number of compounds (808), those in parenthesis refer to the 57% (pie diagram) of substances that can exist in more than one solid form. The latter ratio exceeds 100% because a substance can exist in different solid-state species.

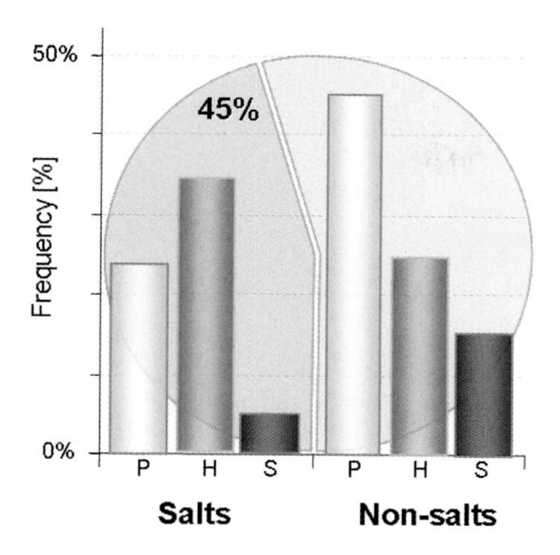

Fig. 8.5 Frequency of polymorphs (P), hydrates (H) and solvates with organic solvents (S) among substances of the Pharmacopoeia Europaea divided into salts and neutral compounds.

latter tend to form more solvates as well as polymorphs. This result clearly shows the propensity of water to interact and coordinate with ions and the lower ability of less polar solvents to bind strongly to the solute in the presence of counter-ions. The importance of hydrates in drug substances is also highlighted by the fact that almost 50% of the substances known to form adducts with water are specified (pharmacopeial monograph) and used in a crystalline hydrate form. In 15 cases the PhEur specifies anhydrate and hydrate in extra monographs. On the contrary, for only 18% of the substances known to form polymorphs is it mentioned that "the substance may show polymorphism". Only three compounds are official in another solvate form: cefatrizin propylene-glycol solvate, doxycyclin hydrate (doxycyclin hydrochloride hemiethanolate hemihydrate) and warfarin sodium (which for some reason is called a "clathrate" in the pharmacopoeia but is a simple 2-propanol-solvate according to recent studies [91]). This analysis also confirms that solvate formation increases with molecular size. Furthermore, the likelihood of hydrate formation decreases with the lipophilicity of the drug molecule, whereas solvate formation with organic solvates increases. Further trends and interrelations between molecular properties and solvate/hydrate formation could be ascertained [51].

In rare cases, incredibly high numbers of solvates can be observed. The unsurpassed candidate is sulfathiazole with over one hundred solvates described [52]. For most of them the crystal structure could be determined, showing two types of structures, one where the solvent molecules are essential parts of the hydrogen-bonded framework structure and a second type in which the solvent basically fills the cavities (channels or layers) in the structural assembly of the sulfathiazole molecules. These types align well with the simple classification of stoichiometric and non-stoichiometric hydrates discussed above. Concerning isostructurality the author stated that the sulfathiazole solvate structures show unusual structural diversity. A class of substances prone to solvate formation is corticosteroids. For instance, fluocortolon [53] has been reported to form 19 solvates (out of 31 tested) and one of its esters (fluocinolon pivalate [54]) forms at least 12. The formation of multiple solvates is not rare; however, it requires more attention and systematic work.

8.4 Generation and Characterization of Solvates

In the past, solvates were mostly obtained accidentally during crystallization from solvents by one of the various crystallization techniques (e.g., [55]) that are applied to purify or isolate a substance or, for instance, to grow single crystals for a crystal structure determination. Since it is imperative to establish the crystal forms of an active pharmaceutical ingredient (API) to satisfy the regulatory authorities [56, 57], solvates of drug compounds are now preferentially discovered in systematic polymorph screenings. The idea of such screens is to maximize the number of possible crystallization variables, which means that numerous crystallization experiments have to be performed properly to find all possible forms. Automated crystallization systems and strategies have been developed to speed up this process, allowing thousands of crystallization experiments in a short time [58–61]. Apart from the limited number of pharmaceutically acceptable solvents [1], hazardous and toxic solvents are also used because their specific features might increase the probability of finding new polymorphs or solvates.

Solvent crystallizations may yield highly unstable solvates that immediately desolvate on harvesting. Such unstable solvates survive only at low temperatures or when kept in the mother liquor and are often overlooked. Therefore, the statement that an individual polymorph was obtained by crystallization from a certain solvent does not necessarily mean that this polymorphic form nucleates directly in this solvent. The polymorph may have formed during the isolation procedure of an unstable solvate by desolvation. Attention to such "transient" solvates is important in order to understand formation principles of polymorphic forms.

It is unlikely that a stable hydrate will be overlooked by any of the screening programs, which always require careful and gentle drying procedures in the final stage to prevent desolvation. Since the presence of water can hardly be avoided during experimental work, unless specific efforts are undertaken, very stable hydrates may mask the existence of other forms and crystallize even with traces of water in a solvent or they form by the uptake of atmospheric moisture. Less stable or very unstable hydrates can sometimes be discovered only by chance or under rather extreme conditions. The recently discovered dihydrate of paracetamol, which crystallized at a pressure of 1.1 GPa from water, highlights these extremes [62]. Freeze-drying may also be a successful route to potential hydrate forms – as demonstrated by Yu et al. [63], who found a hemihydrate of the important pharmaceutical excipient mannitol, which was not detected by earlier solid-state screenings or freeze-drying experiments. That it is also possible to produce a solvate with this method was demonstrated by Otsuka et al. [64], who obtained a dioxane-solvate of phenobarbital by lyophilization.

Often, solvates can simply and quickly be generated by suspending the substance in the solvent or by wetting and grinding. Vapor sorption is a common method to generate hydrates. Exposure of a specific polymorph to water vapor may even result in a metastable hydrate form [10, 65] and it is also worth trying to transform solvates with organic solvents in moist air, with the aim of finding new hydrates or polymorphs.

The most common methods for characterizing solid-state forms, including solvates, are discussed in other chapters of this book and have been reviewed extensively (e.g., [13, 66, 67]). Compared to true polymorphs, the successful characterization of solvated crystal forms requires additional methods and strategies and doubtlessly needs more effort and time. The associated solvent is in most cases highly volatile and desolvation may already occur under mild conditions. Therefore, a comprehensive characterization of solvates often requires measurements under non-ambient conditions (dry atmosphere, certain partial pressure conditions, low and high temperature). Also, combinations of several methods can be very advantageous in the characterization of solvates, such as coupling of TGA or DSC with IR or mass spectrometry. These methods become particularly important for mixed solvates.

Moisture sorption experiments are an imperative part in any solid-state screening program and are a key technique in the characterization of hydrates (Chapter 9). Combination with powder X-ray diffraction allows the unequivocal recognition of stoichiometric and non-stoichiometric hydrates.

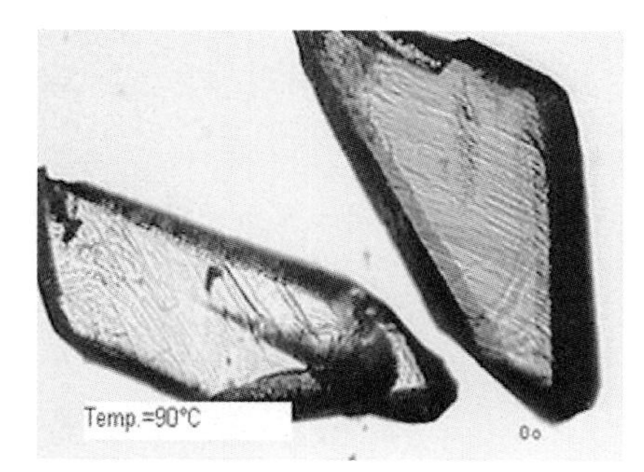

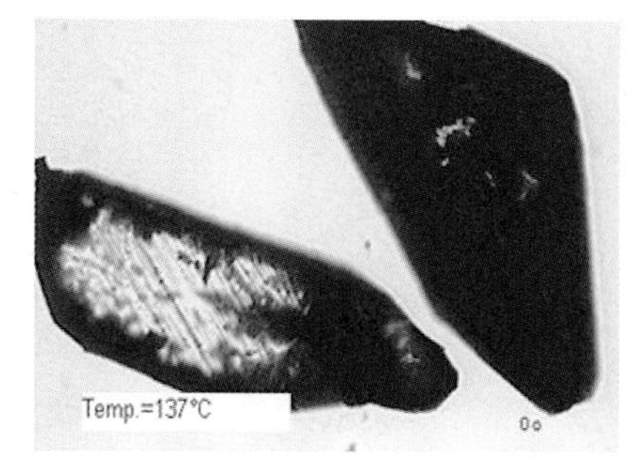

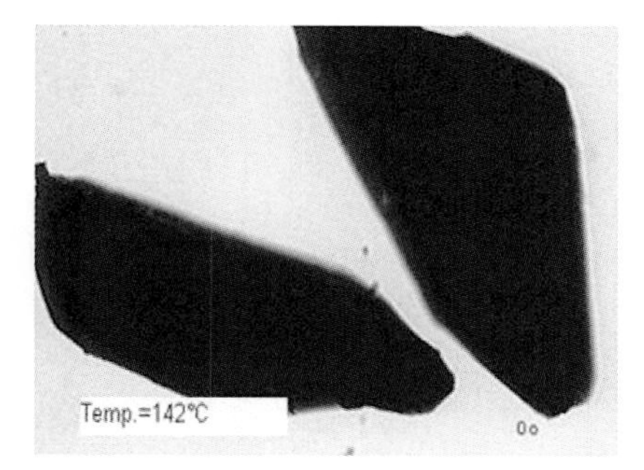

Fig. 8.6 Microphotographs of the desolvation process of a stoichiometric hydrate upon heating, showing a classical "pseudomorphosis".

Hot-stage polarized light microscopy is also helpful in the characterization of solvates. In many cases it even allows a simple, quick distinction between stoichiometric and non-stoichiometric solvates. As mentioned above, desolvation of a stoichiometric solvate is always associated with a significant structural change. Mostly, the solvate breaks up ("recrystallizes") into numerous small crystallites of the unsolvated crystal form during this process, whereas the original shape of the crystal is maintained. This can easily be recognized microscopically by the loss of transparency due to light scattering on the surfaces of the numerous crystallites. An example of this process called "pseudomorphosis" (NB not pseudopolymorphism!) is shown in Fig. 8.6 and further examples are given in [10]. Non-stoichiometric solvates, though, mostly desolvate without, or with only minor, observable changes. The crystals remain transparent on desolvation but fissures and cracks may appear. Therefore, it may be possible to determine crystal structures of the solvated and unsolvated state on one single crystal, as demonstrated recently [29].

8.5 Stability and Solubility of Solvates

The phase stability of solvates/hydrates, i.e., their potential to desolvate or to form, may differ extremely and is one of their most critical properties regarding handling and processing. Their stability depends on the temperature and the partial pressure of the solvent in the vicinity of the solid. In a solution of its own solvent, only temperature (and total pressure) must be considered. As the degree of solvatization increases with decreasing temperature so the solvates stabilize as the temperature decreases. Thus crystallization experiments at lower temperatures favor the formation of solvates. Outside the solvent the conditions are quite different for hydrates and solvates with organic molecules. The stabilizing solvent partial pressure becomes practically zero for solvates with organic solvents but not for hydrates due to atmospheric moisture.

There is no definition of what specifies a stable, moderately stable or unstable solvate. We may refer to the condition where the solvent partial pressure is at its maximum, i.e., their stability in the incorporated solvent (solubility experiments) or to an atmospheric condition at zero solvent partial pressure (thermogravimetry), both as a function of temperature. In contrast to organic solvents, the adjustment of a condition between these extremes is rather simple for water vapor and is routinely applied to characterize hydrates (moisture sorption experiments). Therefore, the stability of stoichiometric hydrates is often represented by partial pressure/composition phase diagrams at isothermal conditions, also called equilibrium moisture isotherms (Chapter 9). The relative humidity where a stoichiometric hydrate becomes unstable and starts to dehydrate (critical relative humidity) is an easily accessible criterion for the stability of a stoichiometric hydrate. However, the problem is the kinetics of desolvation, which usually do not allow us to reach the equilibrium in a reasonable time and depend on many variables such as size of the crystals, crystal defects, the dynamics in the atmosphere, desolvation mechanism and more. Therefore, critical relative humidities determined by vapor sorption experiments may vary strongly, as has been demonstrated with caffeine hydrate [22]. Nevertheless, the information we can derive from partial pressure/composition phase diagrams allows us to derive valuable stability information about stoichiometric hydrates. The additional determination of a composition/temperature diagram as mentioned earlier [68] would characterize the stability range of a hydrate almost comprehensively and characterize the temperature above which a solvate cannot exit at atmospheric pressure and maximal partial pressure of the solvent. An exemplary and comprehensive study that conveys this was performed by De Kruif et al. [69] on citric acid. The determination of a composition/temperature phase diagram with differential scanning calorimetry (DSC) was demonstrated by Suzuki et al. [70]. Thermogravimetric analysis is an imperative part in the characterization and stability determination of solvates. The analysis gives the thermal stability of a solvate at dry atmospheric conditions but it may be also performed at elevated humidities.

Solubility studies at different temperatures in the solvent gives the stability range of solvate and unsolvated form and allow the determination of the transition temperature between these phases from a van't Hoff plot. This approach was demonstrated on various compounds forming hydrates or solvates with organic solvents by Shefter and Higuchi [71] in a classic paper and has been frequently applied since then. It became a rule that a solvate is always the most stable and therefore the least soluble form in its own solvent. Figure 8.7 demonstrates a typical situation, showing the order of solubility as stable hydrate → metastable hydrate → anhydrous form → solvate with organic solvent. From the intersection of the solubility lines of the anhydrous form and the hydrates the transition temperature can be established. In its own solvent the solvate becomes the most stable and least soluble form. However, in reality it is problematic to achieve equilibrium solubility values for metastable phases because they often transform faster into the solvate than they dissolve [65].

Van Tonder et al. [72] studied the solubility and intrinsic dissolution rates of several solvates of niclosamide in aqueous media. Both values were lower than that of the anhydrous form. This can be explained by the fast transformation of the solvates into the least-soluble hydrate. Conversely, the pentanol and toluene solvates of glibenclamide have higher solubility and dissolution rates than two non-solvated polymorphs [73]. This demonstrates the importance and the problems one has to expect when stable hydrates are involved.

Another approach to determine the stability of solvates is the use of solvent mixtures. Zhu and Grant have examined the influence of the water activity parameter a_w in organic solvent/water mixtures on the stability of the hydrates of theophylline and ampicillin [74]. A solvate is stabilized with increasing activity of the solvent of crystallization. Water activities, adjusted by mixing water with a water-miscible organic solvent, correlate with the relative humidity in the vapor. The advantage of this method is that it is often possible to achieve the equi-

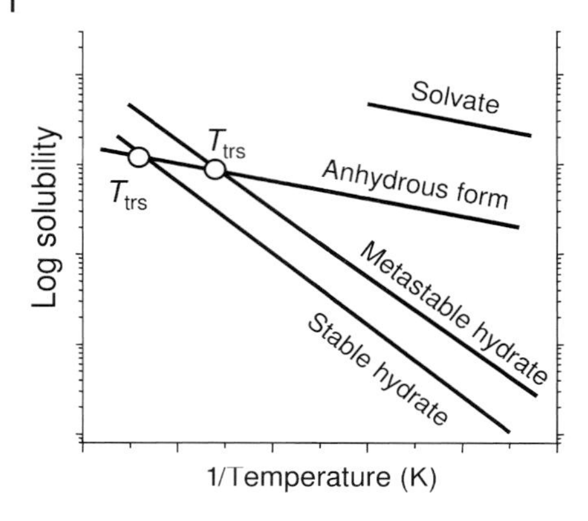

Fig. 8.7 Schematic representation of the solubilities of different solvates and an anhydrous form (succinylsulfathiazole) in an aqueous solution, plotted according to van't Hoff (Shefter and Higuchi [71]). T_{trs}: transition temperatures between the anhydrous form and the hydrates.

librium in the liquid medium but usually not in water vapor for kinetic reasons. The activity parameter represents the principal thermodynamic factor that determines the stability of a solvate.

Solvates also often cause problems in solvent-mediated transformation experiments (also called slurry conversion experiments). Such experiments are applied to find or to verify the stable anhydrous form or the thermodynamic transition point of enantiotropically related polymorphs. Solvents forming a solvate with the substance under investigation are not suitable for such experiments because this solvate will very likely form. For polymorphic compounds that also form hydrates with strongly bound water this experimental approach becomes even less applicable, because the hydrate may form with the traces of water that are present in almost every solvent.

The stabilizing and protecting role of water in crystalline hydrates is well known for a series of β-lactam antibiotics. Cefixime trihydrate [75] is a very intriguing example. The partially dehydrated hydrate becomes disordered, which enables a hydrolytic reaction. Although the anhydrous form is chemically stable it cannot be used because it readily picks up water and rehydrates on storage. To stabilize the drug compound, the hydration state must be maintained during drug formulation, which requires a careful control of temperature, humidity and time. Similar reports exist of ampicillin trihydrate [76]. Vitamin B_{12} (cyanocobalamine) is more stable to light and temperature in the hydrated form [77]. Byrn et al. [78] demonstrated on a series of non-stoichiometric solvates of prednisolone 21-*tert*-butylacetate that the presence of the solvent in voids (tunnels) protects the structure. Removing the solvent enables the penetration of oxygen into the voids and subsequent solid-state oxidation. Interestingly, water showed no protecting effect.

8.6 Processing of Solvates

Knowing the solubility difference between a solvate (almost exclusively a hydrate) and another solid-state candidate, the temperature and moisture dependent physical and chemical stability as well as the kinetics of these processes (which would deserve more attention than given here) allows us to estimate whether a solvate is a useful candidate for development. The solubility issue may become attractive in the case of poorly soluble compounds. Hydrate stability is critical for both classes of hydrates, but more for stoichiometric hydrates than non-stoichiometric ones. For a highly soluble, stable and bioavailable molecule, the risk of affecting the bioavailability by different crystal form is relatively low in general.

The next step is to consider the possible process that allows maintaining the features of the selected material (see also Chapter 13). Many steps in industrial processing may cause dehydration and rehydration cycles that may lead to a different polymorph, disordered and amorphous phases or another solvate. Theoretical approaches to processing-induced transformations (PITs) during pharmaceutical processing have been introduced by Morris et al. [68]. The examples given herein cover several hydrate problems and highlight the most problematic processes. In a later review Zhang et al. discussed the more practically oriented aspects of this topic [79]. Only a very few issues will be addressed here.

Literature on the compaction and tableting behavior of hydrates versus anhydrous forms is very rare. Burger et al. [80] compared the tableting behavior of modification I and a stable hemihydrate of sulfametrol. The hydrate showed better consolidation properties and yielded tablets with a higher tensile strength due to a better plastic deformation than the anhydrous form. Although the hemihydrate shows an about 20% lower solubility and thus also a slower dissolution rate, the liberation of the drug substance is about equal in tablets with a low excipient content (10%). This was explained by the faster disintegration time of the hemihydrate tablet. However, at higher excipient-to-drug ratios the liberation of the hemihydrate is worse than that of the anhydrous form. A study of the compaction behavior of thiamine-hydrochloride hydrates [81] also revealed that the less stable "monohydrate", which shows a non-stoichiometric behavior, possesses better bulk, consolidation and tablet properties than the stable hemihydrate due to a higher plasticity. Sun and Grant [82] showed that the monohydrate of 4-hydroxybenzoic acid shows a superior plastic deformation compared with the anhydrous form. The water molecules in the hydrate structure are arranged in layers, which represent slip planes and provide greater plasticity. Nevertheless, it seems that the differences of crystal forms in the compaction behavior can be balanced by excipients unless the active drug is not highly dosed.

Non-stoichiometric hydrates usually entail other problems than stoichiometric hydrates. The often weakly-bound water may become available ("free water") and interact with other components in a drug (e.g., excipients). Changes in mass of several percent may also result in erroneous analytical results. Since no real phase change is connected with the variation in water/solvent content, the physical properties can be expected to be affected only moderately.

When a compound forms different solvates including a stable hydrate, the hydrate will press home an advantage in the widely unavoidable aqueous environment. At sufficient water activity, water can then strongly interact with the solvate, which is metastable in this environment, and induce a transformation by becoming a part of the structure. Such moisture-induced transformations proceed at a high rate, particularly in contact with an aqueous solution. Obviously, such a solvate can never be a suitable candidate for processing. Moisture may also catalyze the transformation of a solvate to be desolvated into an unsolvated form.

Pentamidine isothionate exemplifies that the nucleation of a solvate may be the critical process that determines whether one or another form results from a crystallization process [83]. In this case a trihydrate may form during freeze-drying, which dehydrates to a metastable form in the secondary drying stage. At lower concentrations and fast freezing rates the nucleation of the hydrate is less favored and the stable form with a low degree of order results.

8.7 Relevance, Problems and Potential Benefits

The observable trend [84] that new drug entities are becoming larger and therefore less soluble, as well as less absorbable and bioavailable, requires extra efforts and strategies in the pharmaceutical industry to obtain and market a (preferentially orally) active drug. The major solubility-improving strategy is doubtlessly the selection of a suitable salt. From the statistical data (see above) it becomes clear that both a larger molecular size as well as the change to a salt enhances the propensity of solvate (particularly hydrate) formation dramatically. New drug entities are, therefore, expected to show a higher ratio of solvates (hydrates). Considering that solvates that contain an organic solvate will play a subsidiary role compared to hydrates, we may overstate this as a "hydrate problem" in the pharmaceutical industry, reflecting the multilayered nature of hydrates and the extra effort needed to handle them.

Indinavir sulfate is an excellent example that stresses the problems we can meet when we administer salts of poorly soluble bases as well soluble salt [85]. The protease inhibitor is formulated as the sulfate salt in capsules for oral administration and is well bioavailable. The solubility in water at a pH <3.5 exceeds 100 mg mL^{-1} but drops to about 0.02 mg mL^{-1} at pH 7.0. This means that the solubility in the blood is 5000-fold lower than in the stomach. The concentration in urine right after the glomerular filtration in the kidney is close to saturation but becomes supersaturated during the following reabsorption of water in the loop of Henle. Finally, the base crystallizes as the worst soluble form, i.e., as the monohydrate in leaflet-like crystals and forms kidney stones.

Solvates of organic solvents that are regarded as unsafe are definitely not suitable candidates for the development of a final dosage form. But they can have technical features that can be used in manufacturing raw drug materials with desired or improved properties. Solvates can, for instance, be ideal precursors for the manufacturing of an individual polymorphic form or sometimes they can even be the only way. Zantosteron [86] and pilocain hydrochloride [87] are two representative examples.

Another potential technical feature is connected with the formation of small crystallites of often very homogeneous particle size distribution upon desolvation (pseudomorphosis, Fig. 8.6) of stoichiometric solvates. Particles of the original solvate crystals easily break apart to yield a very fine, highly crystalline and homogeneous powder. However, desolvation reactions are versatile and one may also get metastable or fully/partly disordered forms. The applicability of this particle size reduction method [88] as an alternative to often problematic milling processes depends on the structural features of the individual solvate.

Like other high-energy forms such as amorphous forms and metastable polymorphs, a solvate (with a non-toxic organic solvent) may also exhibit a significantly higher solubility than the stable form or an existing hydrate in aqueous media. This offers a viable strategy of enhancing the dissolution rate and bioavailability of poorly soluble drug compounds classified in class II or IV of the Biopharmaceutics Classification System (BCS) [89]. Possible problems of this strategy have been addressed above.

Finally, solvate formation, like polymorphism, opens perspectives for supramolecular chemistry and crystal engineering. Crystal structures of different solvates enable the study of molecular conformations, recurring patterns in hydrophobic aggregation and inclusion as well as molecular recognition phenomena in general [48, 90].

8.8 Patents

Because the crystalline forms of a substance cannot yet be predicted, and because of their potential industrial and technical benefits, solid-state forms can be protected as intellectual property by a patent (Chapter 14). A representative list of patented crystal forms of drug substances can be found in Byrn's classic book on solid-state chemistry [10]. Several past [11] and ongoing patent litigations concern solvates, particularly hydrates. In fact, more information about solid-state forms of newer drug compounds can now be found in patents than in peer-reviewed journals. Sometimes, amazingly high numbers ($\gg$20) of crystal forms are claimed for one compound. Such high numbers are only possible if the compound can form many different solvates, because the number of unsolvated forms (polymorphs) very rarely exceeds five. One striking problem is that solvates or hydrates are frequently claimed to be individual stoichiometric solvates with a defined solvent content, which on closer inspection are isostructural and represent only one non-stoichiometric solvate phase. One can of course artificially create (and claim?) an infinite number of forms from a non-stoichiometric solvate when the criteria for a discrete crystalline form is reduced to a measured solvent content in a sample. The generally enhanced complexity of the solid-state properties of solvates compared to polymorphs, connected with insufficient analytical data and the confusion in terminology, provides a lot of space for controversial issues and discussions. It is therefore extremely important to specify crystal forms and in particular solvates more accurately, based on several appropriate analytical techniques.

8.9 Conclusions

Solvate formation is a common phenomenon among drug substances and represents a challenging area for industry and basic research. Industry is faced with the problems of the complex behavior of these multicomponent crystals entities, although there are also potential benefits to exploit. The growing interest in supramolecular assemblies and principles of molecular recognition has attracted more attention to the relevance of solvent–molecule interactions in crystals. Except for water, this has often been ignored in the past in favor of the molecular structure.

This chapter is far from comprehensive and there are certainly many solvate/hydrate systems that do not fit into the picture drawn here. However, the area requires more systematic work and the interplay of various scientific disciplines in order to draw the ultimate picture of this intriguing and important phenomenon.

References

1 ICH Q3C Impurities: *Guideline for Residual Solvents*, International Conference on Harmonization of Technical Requirements for Registration of Pharmaceuticals for Human Use, Geneva, Switzerland, July **1997**.

2 J.W. Mullin, *Crystallization*, 3rd edn. Butterworth-Heinemann, Oxford, **1993**, pp. 260–263.

3 G.R. Desiraju, *Cryst. Eng. Commun.* **2003**, *5*, 466–467.

4 K.R. Seddon, *Cryst. Growth Des.* **2004**, *4*, 1087.

5 J.D. Dunitz, *Cryst. Eng. Commun.* **2003**, *5*, 506.

6 G.R. Desiraju, *Cryst. Growth Des.* **2004**, *4*, 1089–1090.

7 J. Bernstein, *Cryst. Growth Des.* **2005**, *5*, 1661–1662.

8 C. Reichardt, *Solvents and Solvent Effects in Organic Chemistry*, 3rd edn., Wiley-VCH, Weinheim, **2002**.

9 P. Müller, *Pure & Appl. Chem.* **1994**, *66*, 1077–1184.

10 S.R. Byrn, R.R. Pfeiffer, J.G. Stowell, *Solid-State Chemistry of Drugs*, SSCI, West Lafayette, IN, **1999**.

11 J. Bernstein, *Polymorphism in Molecular Crystals*, Oxford University Press, Oxford, **2002**.

12 J. Haleblian, W.C. McCrone, *J. Pharm. Sci.* **1969**, *58*, 911–929.

13 T. Threlfall, *Analyst* **1995**, *120*, 2435–2460.

14 W.C. McCrone, *Physics and Chemistry of the Organic Solid State*, D. Fox, M.M. Labes, A. Weissberger, Eds., Wiley-Interscience, New York, **1965**, Vol. 2, pp. 725–767.
15 H.G. Brittain, *Spectroscopy* **2000**, *15*, 34–39.
16 Ö. Almarsson, M.J. Zaworotko, *Chem. Commun.* **2004**, 1889–1896.
17 P. Vishweshwar, J.A. McMahon, M.L. Peterson, M.B. Hickey, T.R. Shattock, M.J. Zaworotko, *Chem. Commun.* **2005**, 4601–4603.
18 S.R. Vippagunta, H.G. Brittain, D.J.W. Grant, *Adv. Drug Deliv. Rev.* **2001**, *48*, 3–26.
19 K.R. Morris, in: *Polymorphism in Pharmaceutical Solids*, H.G. Brittain, Ed., Marcel Dekker, New York, **1999**, pp. 125–181.
20 M. Stoltz, A.P. Lötter, J.G. van der Watt, *J. Pharm. Sci.* **1988**, *77*, 1047–1049.
21 R. Gerdil, R.E. Marsh, *Acta Crystallogr., Sect. C* **1960**, *13*, 166–167.
22 U.J. Griesser, A. Burger, *Int. J. Pharm.* **1995**, *120*, 83–93.
23 C.H. Goerbitz, *Acta Crystallogr., Sect. C* **2004**, *60*, o810–o812; C.H. Goerbitz, *Acta Crystallogr., Sect. E* **2004**, *60*, o626–o628; C.H. Goerbitz, *Acta Crystallogr., Sect. E* **2004**, *60*, o647–o650. C.H. Goerbitz, *Chem. Eur. J.* **2001**, *7*, 5153–5159.
24 S. Furuseth, J. Karlsen, A. Mostad, C. Romming, R. Salmen, H.H. Tonnesen, *Acta Chem. Scand.* **1990**, *44*, 741–745.
25 J.M. Karle, I.L. Karle, *Acta Crystallogr., Sect. C* **1988**, *44*, 1605–1608.
26 R.R. Pfeiffer, K.S. Yang, M.A. Tucker, *J. Pharm. Sci.* **1970**, *59*, 1809–1814.
27 G.A. Stephenson, E.G. Groleau, R.L. Kleemann, W. Xu, D.R. Rigsbee, *J. Pharm. Sci.* **1998**, *87*, 536–542.
28 G.A. Stephenson, B.A. Diseroad, *Int. J. Pharm.* **2000**, *198*, 167–177.
29 R.L. Te, U.J. Griesser, K.R. Morris, S.R. Byrn, J.G. Stowell, *Cryst. Growth Des.* **2003**, *3*, 997–1004.
30 T. Hosokawa, S. Datta, A.R. Sheth, N.R. Brooks, V.G. Young, D.J.W. Grant, *Cryst. Growth Des.* **2004**, *4*, 1195–1201.
31 R.K.R. Jetti, R. Boese, P.K. Thallapally, G.R. Desiraju, *Cryst. Growth Des.* **2003**, *3(6)*, 1033–1040.
32 G.A. Stephenson, J.G. Stowell, P.H. Toma, D.E. Dorman, J.R. Greene, S.R. Byrn, *J. Am. Chem. Soc.* **1994**, *116*, 5766–5773.
33 M.R. Caira, G. Bettinetti, M. Sorrenti, *J. Pharm. Sci.* **2002**, *91*, 467–448.
34 S.A. Bourne, M.R. Caira, L.R. Nassimbeni, I. Shabalala, *J. Pharm. Sci.* **1994**, *83*, 887–892.
35 G.A. Jeffrey, *Acc. Chem. Res.* **1969**, *2*, 344–352; G.A. Jeffrey, *J. Inclusion. Phenom.* **1984**, *1*, 211–222.
36 J.R. Clark, *Rev. Pure Appl. Chem.* **1963**, *13*, 50–90.
37 K.R. Morris, N. Rodriguez-Hornedo, in *Encyclopedia of Pharmaceutical Technology*, J. Swarbrick, J.C. Boylan, Eds., Marcel Dekker, New York, **1993**, Vol. 7, pp. 393–441.
38 S. Gal, *Chimia* **1968**, *22*, 409–448.
39 M. Kuhnert-Brandstätter, H. Grimm, *Mikrochim. Acta* **1968**, 115–126; M. Kuhnert-Brandstätter, H. Grimm, *Mikrochim. Acta* **1968**, 127–139.
40 S. Petit, G. Coquerel, *Chem. Mater.* **1996**, *8*, 2247–2258.
41 A.K. Galwey, *Thermochim. Acta* **2000**, *355*, 181–238.
42 L. Infantes, S. Motherwell, *Cryst. Eng. Commun.* **2002**, *4*, 454–461.
43 A.L. Gillon, N. Feeder, R.J. Davey, R. Storey, *Crystal Growth Design* **2003**, *3*, 663–673.
44 G.R. Desiraju, T. Steiner, *The Weak Hydrogen Bond – In Structural Chemistry and Biology*, Oxford University Press, Oxford, **1999**.
45 G.R. Desiraju, C., V.K. Sharma, in: *The Crystal as Supramolecular Entity*, G.R. Desiraju, Ed., Wiley, Chichester, **1996**, pp. 31–61.
46 J.A.R.P. Sarma, G.R. Desiraju, in: *Crystal Engineering*, K.R. Seddon, M. Zaworotko, Eds., Kluwer, Norwell, MA, **1999**, pp. 325–356.
47 T. Nanagia, G.R. Desiraju, *Chem. Commun.* **1999**, 605–606.
48 C.H. Goerbitz, H.P. Hersleth, *Acta Crystallogr., Sect. B* **2000**, *56*, 526–534.
49 E.H. Allen, O. Kennard, *Chem. Des. Autom. News* **1993**, *8*, 31–37.
50 U.J. Griesser, A. Burger, Abstract book of the XVIII Congress and General

Assembly of the International Union of Crystallography, Glasgow, Scotland, **1999**, p. 400.

51 U. J. Griesser, D. Braun, in preparation.

52 A. L. Bingham, D. S. Hughes, M. B. Hursthouse, R. W. Lancaster, S. Tavener, T. L. Threlfall, *Chem. Commun.* **2001**, 603–604.

53 M. Kuhnert-Brandstätter, P. Gasser, *Microchem. J.* **1971**, *16*, 419–428.

54 M. Kuhnert-Brandstätter, P. Gasser, *Arch. Pharm.* **1971**, *304*, 926–932.

55 K. J. Guillory, in: H. G. Brittain, Ed., *Polymorphism in Pharmaceutical Solids*, Marcel Dekker, New York, **1999**, Vol. 95, pp. 183–226.

56 ICH Q6A Specifications: *Test Procedures and Acceptance Criteria for New Drug Substances and New Drug Products: Chemical Substances.* International Conference on Harmonisation of Technical Requirements for Registration of Pharmaceuticals for Human Use, Geneva, 1999. (www.ich.org)

57 S. Byrn, R. Pfeiffer, M. Ganey, C. Hoiberg, G. Poochikian, *Pharm. Res.* **1995**, *12*, 945–954.

58 R. A. Storey, R. Docherty, P. D. Higginson, *Am. Pharm. Rev.* **2003**, 100–105.

59 S. L. Morissette, O. Almarsson, M. L. Peterson, J. F. Remenara, M. J. Read, A. V. Lemmo, S. Ellis, M. J. Cima, C. R. Gardner, *Adv. Drug Deliv. Rev.* **2004**, *56*, 275–300.

60 R. Hilfiker, J. Berghausen, F. Blatter, A. Burkhard, S. M. De Paul, B. Freiermuth, A. Geoffroy, U. Hofmeier, C. Marcolli, B. Siebenhaar, M. Szelagiewicz, A. Vit, M. von Raumer, *J. Therm. Anal. Calorim.* **2003**, *73*, 429–440.

61 R. Hilfiker, J. Berghausen, F. Blatter, S. M. De Paul, M. Szelagiewicz, M. von Raumer, *Chim. Oggi* **2003**, *21*, 75–78.

62 F. P. A. Fabbiani, D. R. Allan, W. I. F. David, S. A. Moggach, S. Parsons, C. R. Pulham, *Cryst. Eng. Commun.* **2004**, *6*, 504–511.

63 L. Yu, N. Milton, E. G. Groleau, D. S. Mishra, R. E. Vansickle, *J. Pharm. Sci.* **1999**, *88*, 196–198.

64 M. Otsuka, M. Onoe, Y. Matsuda, *Pharm. Res.* **1993**, *10*, 577–582.

65 A. Burger, U. J. Griesser, *Sci. Pharm.* **1989**, *57*, 293–305; A. Burger, U. J. Griesser, *Eur. J. Pharm. Biopharm.* **1991**, *37*, 118–124.

66 H. G. Brittain, in: *Polymorphism in Pharmaceutical Solids*, H. G. Brittain Ed., Marcel Dekker, New York, **1999**, Vol. 95, pp. 227–278.

67 U. J. Griesser, J. G. Stowell, in: *Pharmaceutical Analysis*, D. C. Lee, M. Webb Eds., Blackwell Publishing, Oxford, **2003**, pp. 240–294.

68 K. R. Morris, U. J. Griesser, C. J. Eckhardt, J. G. Stowell, *Adv. Drug Deliv. Rev.* **2001**, *48*, 91–114.

69 C. G. De Kruif, J. C. Van Miltenburg, A. J. Sprenkels, G. Stevens, W. De Graaf, H. G. M. de Wit, *Thermochim. Acta* **1982**, *58*, 341–354.

70 E. Suzuki, K. Shirotani, Y. Tsuda, K. Sekiguchi, *Chem. Pharm. Bull.* **1985**, *33*, 5028–5035.

71 E. Shefter, T. Higuchi, *J. Pharm. Sci.* **1963**, *52*, 781–791.

72 E. C. van Tonder, M. D. Mahlatji, S. F. Malan, W. Liebenberg, M. R. Caira, M. Song, M. M. De Villiers, *AAPS PharmSciTech* **2004**, *5*, 1–10.

73 M. S. Suleiman, N. M. Najib, *Int. J. Pharm.* **1989**, *50*, 103–109.

74 H. J. Zhu, D. J. W. Grant, *Int. J. Pharm.* **1996**, *139*, 33–43; H. J. Zhu, C. M. Yuen, D. J. W. Grant, *Int. J. Pharm.* **1996**, *135*, 151–160.

75 S. Kitamura, S. Koda, A. Miyamae, T. Yasuda, Y. Morimoto, *Int. J. Pharm.* **1990**, *59*, 217–224.

76 E. Shefter, H.-L. Fung, O. Mok, *J. Pharm. Sci.* **1973**, *62*, 792–794.

77 J. Haleblian, *J. Pharm. Sci.* **1975**, *64*, 1269.

78 S. R. Byrn, P. A. Sutton, B. Tobias, J. Frye, P. Main, *J. Am. Chem. Soc.* **1988**, *110*, 1609–1614.

79 G. G. Z. Zhang, D. Laww, E. A. Schmitt, Y. Qiu, *Adv. Drug Deliv. Rev.* **2004**, *56*, 371–390.

80 A. Burger, S. Sturm, R. D. Bolitschek-Dialer, *Pharm. Ind.* **1988**, *50*, 1396–1405.

81 M. Hellemann, U. J. Griesser, 4th International Symposium on Solid Oral Dosage Forms, Malmö, Sweden, May 13–15, **2001**.

82 C. Sun, D.J.W. Grant, *Pharm. Res.* **2001**, *18*, 274–280.

83 S. Chongprasert, U.J. Griesser, A.T. Bottorff, N.A. Williams, S.R. Byrn, S.L. Nail, *J. Pharm. Sci.* **1998**, *87*, 1155–1160.

84 C.A. Lipinski, F. Lombarda, B.W. Dominy, P.J. Feeney, *Adv. Drug Deliv. Rev.* **2001**, *46*, 3–26.

85 R. Tessadri, U.J. Griesser, M. Prillinger, K. Wurst, *Mater. Sci. Forum* **2004**, *443/444*, 407–410.

86 W.L. Rocco, C. Morphet, S.M. Laughlin, *Int. J. Pharm.* **1995**, *122*, 17–25.

87 A.C. Schmidt, V. Niederwanger, U.J. Griesser, *J. Therm. Anal. Calorim.* **2004**, *77*, 639–652.

88 K. Sekiguchi, I. Horikoshi, I. Himuro, *Chem. Pharm. Bull.* **1968**, *16*, 2495–2502.

89 L.X. Yu, G.L. Amidon, J.E. Polli, H. Zhao, M. Mehta, D.P. Conner, V.P. Shah, L.J. Lesko, M.-L. Chen, V.H.L. Lee, A.S. Hussain, *Pharm. Res.*, **2002**, *1*, 921–925.

90 C.H. Goerbitz, H.P. Hersleth, *Acta Crystallogr., Sect. B* **2000**, *56*, 1094–1102.

91 A.R. Sheth, W.W. Brennessel, V.G. Young, Jr., F.X. Muller, D.J.W. Grant, *J. Pharm. Sci.* **2004**, *93*, 2669–2680.

9
Physical Characterization of Hygroscopicity in Pharmaceutical Solids

Susan M. Reutzel-Edens and Ann W. Newman

9.1
Introduction

The terms hygroscopic and hygroscopicity are widely used in the pharmaceutical literature to describe the moisture uptake of materials. Water vapor, an ever present component of the environment, can have profound, and often detrimental, effects on physicochemical processes of interest to the pharmaceutical and fine chemical industries, such as crystallization of lyophilized cakes, direct compaction, powder caking, coating and packaging material permeability, and solid-state stability. In terms of solid dosage forms and excipients, knowledge of moisture adsorption phenomena will give useful information for selecting excipients, such as disintegrating agents, direct compression carriers and binders, as well as humidity control during production and storage. The objectives of this chapter are to review basic concepts of water–solid interactions, to present approaches taken to characterize such interactions, on empirical, surface energy, and molecular structural levels, and to identify risks/solid-state issues encountered in the pharmaceutical industry that arise from these interactions.

9.1.1
Definition of Hygroscopicity

Finding a workable definition of hygroscopicity applicable to pharmaceutical systems is not straightforward. Definitions from common sources are given in Table 9.1 to demonstrate the variability of the term. A common theme is the sorption and retention of moisture. One definition requires that deliquescence not occur [1], while others do not mention a change in physical form. This aspect can be an important consideration during the development of a pharmaceutical material.

Polymorphism: in the Pharmaceutical Industry. Edited by Rolf Hilfiker

ISBN: 3-527-31146-7

Table 9.1 Definitions of *hygroscopic* from common sources.

Definition	Ref.
1. Readily taking up and retaining moisture. 2. Moisture taken up and retained under some conditions of humidity and temperature.	[2]
Readily absorbing, becoming coated with, and retaining moisture, but not enough to make a liquid.	[1]
A solid that can adsorb atmospheric moisture. There is both a kinetic and a thermodynamic component to this process. The kinetic component determines the rate of water uptake, while the thermodynamic component determines the energy of this process.	[3]
A substance that can remove moisture from the air.	[4]
Descriptive of a substance that has the property of adsorbing moisture from the air.	[5]

9.1.2
Classification of Hygroscopic Behavior

Attempts have been made to classify hygroscopic behaviors based on data obtained from adsorption isotherms, i.e., the curves obtained by plotting the weight change of a sample versus the relative humidity (RH) or water vapor pressure. The shape of the isotherm is determined by the specific conditions for adsorption onto a surface, such as pore size and heats of adsorption. The common isotherm classification types (I–V) are relevant to any adsorbate, including water, and are summarized in Fig. 9.1. It has been theorized that the isotherm of a "hygroscopic solid" should resemble a Type II isotherm where multilayer adsorption occurs [6]. Giles [7] proposes class H, L, C, and S curves for adsorption based on the slope of the initial portion of the isotherm, with L curves indicative of Langmuir isotherms.

Adsorption isotherms may be classified; however, there remains *no universally accepted definition for hygroscopicity*. This is because there are both thermodynamic driving forces and kinetic rate components to the term. Hygroscopicity can describe both the amount of moisture in a substance and the rate of moisture uptake when a sample is placed in a known RH. An excellent review by Umprayn and Mendes [8] discusses the hygroscopicity of pharmaceutical solids, as well as the thermodynamic and kinetic aspects of moisture sorption. Methods to measure and describe hygroscopicity in various organic and inorganic systems have been reviewed by Van Campen et al. [9]. Parameters used to evaluate the hygroscopicity of various systems include the critical RH (CRH) [8, 9], the hygroscopicity potential (HP) [10], the hygroscopicity coefficient [11], and heats of absorption [12]. Griffin [13] has introduced the terms "equilibrium hygroscopicity" and "dynamic hygroscopicity" to describe the amount and rate of uptake, respectively. The classification scheme for hygroscopicity proposed by Callahan [14, 15] (Table 9.2) is used for pharmaceutical excipients.

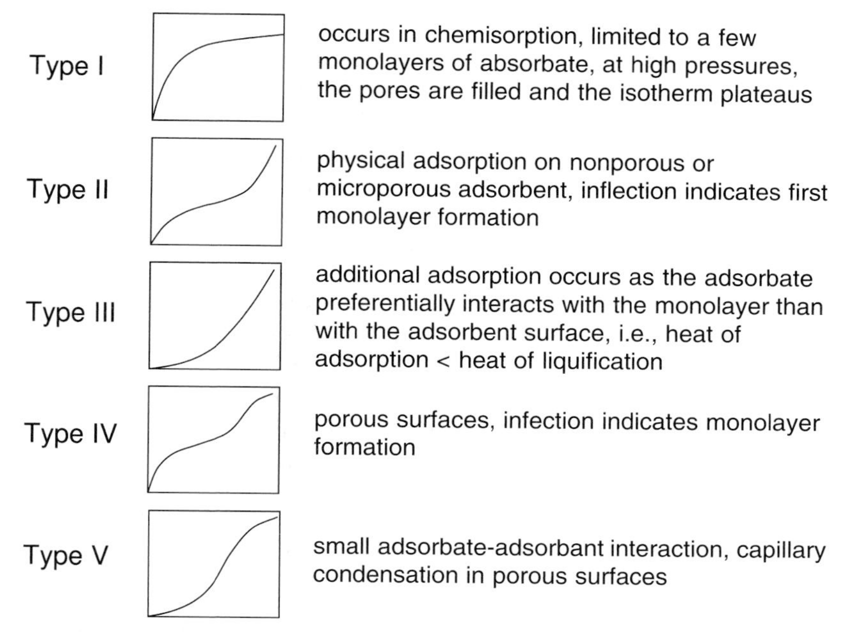

Fig. 9.1 Types of isotherms.

Table 9.2 Hygroscopicity classification scheme.

Class I	Non-hygroscopic	Essentially no moisture increase below 90% RH; less than 20% increase in moisture content above 90% RH in one week
Class II	Slightly hygroscopic	Essentially no moisture increase below 80% RH; less than 40% increase in moisture content above 80% RH in one week
Class III	Moderately hygroscopic	Moisture content does not increase >5% below 60% RH; less than 50% increase in moisture content above 80% RH in one week
Class IV	Very hygroscopic	Moisture content will increase as low as 40–50% RH; greater than 20% increase in moisture content above 90% RH in one week

Current approaches to classifying hygroscopic behaviors provide a limited view of water sorption in pharmaceutical solids. The absolute level of water uptake is an important consideration and certainly needs to be obtained. However, the rate of uptake as well as the RH and temperature also need to be considered [16]. How the water sorbs and the location of the water on a molecular level will be specific for a system and is useful information to have during development.

9.2 Water–Solid Interactions

When a solid is exposed to water vapor, the water molecules may attach to the surface of the solid via van der Waals, ion–dipole, or specific hydrogen-bonding interactions with the functional groups on the surface of the solid. Experimental studies [17] of adsorbed species on well-defined crystalline surfaces, using vibrational spectroscopy, NMR spectroscopy, and dielectric relaxation, have shown an especially strong tendency of water to form hydrogen bonds. Indeed, the ability of water to act as both a Lewis acid and Lewis base accounts for the wide range of interactions with pharmaceutical compounds. The Cambridge Structural Database has been used to closely examine the diverse solid-state environments of water molecules in crystalline hydrates of organic compounds. Desiraju [18] has suggested that the incorporation of water into crystal structures helps to balance the mismatch between the number of hydrogen bond donors and acceptors in the host molecules. Ferraris and Franchini-Angela [19], in analyzing the geometry and environments of water molecules in crystalline hydrate structures determined by neutron diffraction, attributed the "quasi-normal" distribution of H-bond lengths and angles to the "strain-absorbing" ability of water. Jeffrey and Maluszynska [20] reported that three-coordinate water molecules were more common than four-coordinate ones. Gillon et al. [21], in drawing the same conclusion, suggested that while water prefers to maximize its hydrogen-bonding interactions within a hydrate structure, severe steric interactions often preclude two H-bond donors and two acceptors from forming a four-coordinate, i.e., tetrahedral, geometry. The states of water associated with solids have been reviewed in detail by Zografi [16, 22, 23].

Water can interact with crystalline solids by sorption at the solid surface, incorporation into the lattice, deliquescence, and capillary condensation (in samples with micropores). This is shown schematically in Fig. 9.2. Deliquescence and capillary condensation will lead to condensed (bulk) water that may dissolve water-soluble compounds. Water sorption on surfaces can be in the form of individual molecules, clusters, monolayers, and multimolecular layers, which will eventually lead to condensed water [22]. In solid-state systems, water is not

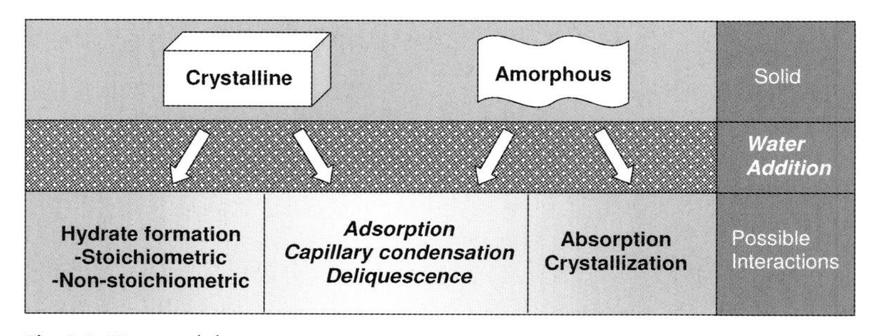

Fig. 9.2 Water–solid interactions.

static. Even when hydrogen bonding is present, water can move along the surface or within the crystal lattice over a period of time [22, 24].

In amorphous solids, various physical changes can occur during water sorption. Water can dissolve in the solid due to the disordered state of the system, where it can act as a plasticizer and significantly lower the glass transition (T_g) temperature of the material. A lower T_g can result in greater mobility and crystallization of the amorphous material. The effect of water (or other additives) can be described by the Gordon-Taylor equation [25]. Crystalline materials can change form, convert into crystalline hydrates, or deliquesce. Approaches to determine the changes that occur and examples of these changes are presented to illustrate the utility of understanding water uptake at a molecular level.

9.3 Characterizing Water–Solid Interactions

A thorough evaluation of hygroscopicity is routinely performed for drug substances and excipients during the early stages of pharmaceutical development. This entails determining both the equilibrium moisture content as a function of RH and the rate at which equilibrium is attained. To fully appreciate the impact that water–solid interactions can have on physical and chemical properties, the underlying mechanisms by which moisture is sorbed (or desorbed) must be understood. For this purpose, several methodologies have been used to characterize water–solid interactions at empirical, energetic, and molecular levels.

9.3.1 Moisture Sorption Analysis

Moisture sorption isotherms, which are plots of the equilibrium water content of a solid as a function of RH at constant temperature, are commonly used to assess the hygroscopicity of pharmaceutical solids. Sorption–desorption isotherms can be obtained gravimetrically by measuring the mass change of a sample with changes in RH. A pre-weighed sample is placed in a constant RH environment, e.g., a closed desiccator containing a saturated solution of an electrolyte, then periodically removed and weighed [26]. This process is repeated until the sample has reached equilibrium. The conventional desiccator method as described here has to a certain extent given way to automated techniques; however, it is still widely used to equilibrate samples when *in situ* analysis at specified humidities is not possible.

The development of dynamic vapor sorption instruments, which operate in a closed system at controlled temperature and either ambient or controlled pressure, has significantly automated moisture sorption analysis. In dynamic vapor sorption, a sample is placed on a microbalance and exposed to a continuous flow of air or N_2 of a predetermined RH [27, 28]. An isotherm is then calculated from the equilibrium moisture uptake at each partial pressure or RH. Because the mass must stabilize at each RH increment for an isotherm to be accurate

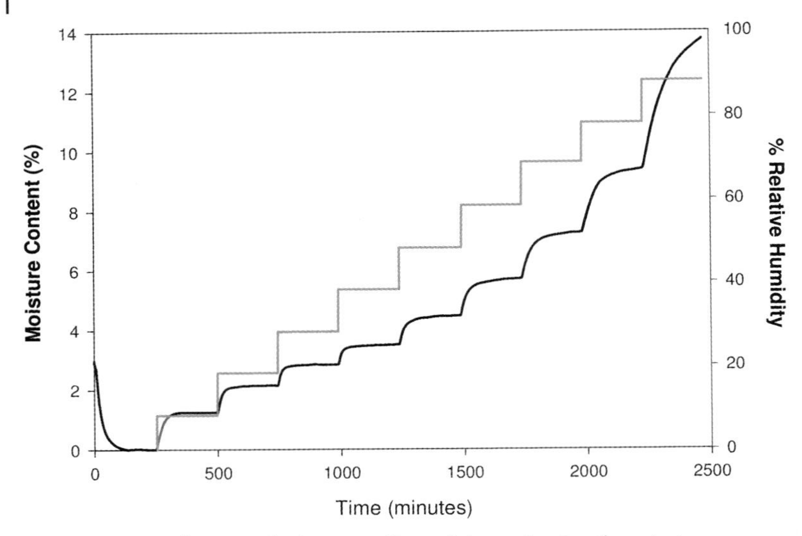

Fig. 9.3 Sorption kinetics of microcrystalline cellulose, showing the extent and rate of water uptake.

(Fig. 9.3), the duration of the experiment will depend in part on the nature of the moisture uptake. Whereas surface adsorption is typically relatively fast, bulk absorption via vapor diffusion is frequently quite slow.

Figure 9.4 shows several moisture sorption isotherms that are characteristic of many different types of pharmaceutical solids. Whereas substantial water uptake, particularly at high RHs, is commonly observed for amorphous (Fig. 9.4 a) and deliquescent materials (Fig. 9.4 b), highly crystalline solids (Fig. 9.4 c) can have very low (less than 0.1%) reversible affinities for moisture sorption. In cases of moderate water uptake, the equilibrium water content may experience stepwise jumps when stoichiometric hydrates form (Fig. 9.4 d–g) and/or gradual increases as water is continuously incorporated into non-stoichiometric or channel hydrates (Fig. 9.4 f and h). The overall percent weight change observed during hydrate formation will, of course, depend on the molecular weight of the compound. That is, a monohydrate formed by a low molecular weight compound will produce a larger percent weight increase than a monohydrate formed by a higher molecular weight compound. The percent weight change will also depend on the extent to which solvent (water) is lost from the sample during the initial equilibration at low RH. Because samples can partially or completely desolvate under these conditions, the water that is not lost during initial drying needs to be added to the sorbed water for the determination of absolute stoichiometry. Weight corrections for the initial water content may be accomplished using equilibrium moisture content (EMC) or relative to dry weight (RTDW) calculations [14, 29]. Quite often, the moisture sorption analysis of solvated materials is initiated at ambient RH, so as to avoid initial drying, which can change the crystal form or crystallinity of the sample.

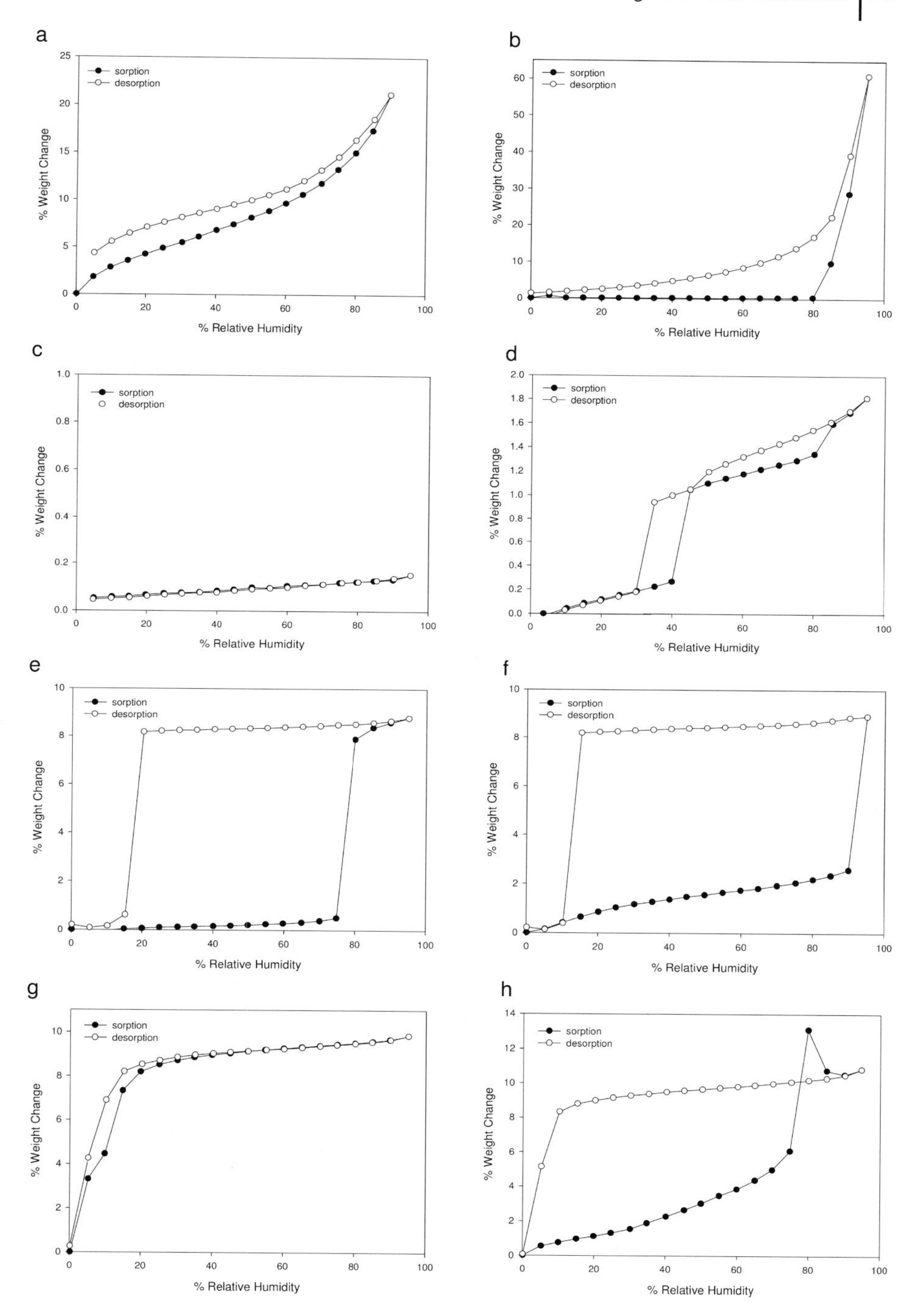

Fig. 9.4 Moisture sorption behaviors observed for pharmaceutical solids.

When the kinetics and/or path of hydration and dehydration are different, the sorption and desorption isotherms will not coincide and hysteresis is observed. In Fig. 9.4(h), for example, absorbed water causes the material to crystallize (producing a spontaneous weight loss) in a different form, so the adsorption and desorption isotherms reflect the kinetic stability of different hydrates. It is good practice to use orthogonal techniques, e.g., powder X-ray diffraction, to identify forms present at different regions of the moisture sorption isotherm to ensure that the sorption–desorption curves are interpreted correctly. Likewise, care should be exercised in interpreting moisture uptake curves on initial lots of materials that are of questionable crystallinity and/or purity, since the presence of phase impurities, crystalline defects and, especially, low levels of amorphous material can significantly alter the moisture sorption behavior.

Moisture sorption isotherms can provide insights into the potential for phase transformations and, to a certain extent, the thermodynamic RH stability relationships between different crystal forms at a given temperature. One potential pitfall in using a moisture balance alone to determine critical moisture contents for physical transitions is the time-dependent nature of the measurement. In cases where conversion of an anhydrate to a hydrate (or of a lower hydrate to a higher hydrate) is observed and found to be reversible (no hysteresis on desorption), the thermodynamic RH stability relationships will be readily apparent. If hysteresis is observed, however, then the water activity at which the relative thermodynamic stability of the forms reverses cannot be determined from the moisture sorption–desorption isotherms alone. In these cases, a slurry equilibration, which accelerates the slow transformation kinetics of crystals surrounded by vapor, can be used to determine the specific RH at which the hydrate (or higher hydrate) becomes thermodynamically more stable [30, 31].

When crystal forms differing in hydration state show extreme kinetic stability, i.e., no conversions are apparent in the moisture sorption isotherms, the (higher) hydrate should not be assumed to be the more stable form at high RHs. Although hydrated crystal forms are generally more stable in water than anhydrates [32, 33], cases of the opposite stability relationship have been reported. The anhydrous form of LY334370 HCl, for example, is more stable than the dihydrate crystal form in water at ambient temperature [34]. The moisture sorption isotherms, collected at ambient temperature, showed both of these crystal forms to be remarkably stable at all RHs (Fig. 9.4c and g). When suspended in water, however, the dihydrate immediately converted into the anhydrate, presumably by a solution-mediated process. In this case, the temperature at which the relative thermodynamic stability would be affected by water activity has most likely been exceeded. Mannitol hydrate is another example of a metastable hydrate that converts into anhydrous polymorphs upon exposure to moisture [35].

9.3.2
Surface Energy Approaches

Moisture sorption analysis is most often used to establish the amount of moisture sorbed or desorbed as a function of RH and, in many cases, the reversibility of the sorption process. More recently, attempts have been made to obtain thermodynamic information from the sorption isotherms. Sacchetti [36], for example, has shown that thermodynamic functions (μ, ΔH, ΔS, and ΔG) can be calculated from the low vapor pressure region of moisture sorption isotherms. Willson and Beezer [37] have also reported that thermodynamic parameters (K, ΔH_v, ΔS, and ΔG) can be calculated from equilibrium moisture sorption isotherms collected at several temperatures, provided ΔH_v is constant over the temperature range of interest. Both approaches, of course, are limited to systems that do not change during the moisture sorption experiment.

Isothermal microcalorimetry can be used to study the surface energetics of solid materials directly as a function of RH. At any selected humidity, the measured calorimetric response for water sorption will be related to the amount and energetics of water adsorption at various levels of surface coverage. Although isothermal microcalorimetry is a nonspecific technique, with some assumptions, the calorimetric signal can be deconvoluted into contributions due to water vapor sorption/desorption and other energetic events. Markova and Wadsö [38] showed that both differential heats of adsorption and heat capacities measured as a function of water activity nicely parallel the moisture sorption isotherms of crystalline morphine sulfate. Lechuga-Ballesteros et al. [39] used microcalorimetry to show a change in the mechanism of water–solid interactions for amorphous solids that could be related to the physical stability of the glassy phase. Water vapor sorption was reversible up to a threshold RH, at which point the water vapor saturates the binding sites within the amorphous phase. Above this RH, not only did the rates of water absorption–desorption change (hysteresis was observed), but the heat of water sorption approached the enthalpy of condensation as well. This change in absorption mechanism, wherein water becomes a solvent, was likely responsible for the irreversible physical change in the sample.

Inverse gas chromatography (IGC) is gaining in popularity in the pharmaceutical industry for the characterization of surface energy [40]. IGC studies of crystalline drugs have shown that both the dispersive and polar components of the surface free energy of adsorption of polar probes decrease as RH increases. By examining the influence of water on the retention behavior of different organic vapors, changes in the structural and chemical features of the surface can be evaluated. For example, Sunkersett et al. [41] performed IGC experiments with carrier molecules that could compete for the same sites as water on paracetamol and carbamazepine crystals to test the hypothesis that water shields the high-energy interaction sites from organic probes. Here, competitive sharing of the same chemical site by the water and the carrier molecule was demonstrated only in cases where the adsorption of the carrier was decreased at higher RH. Using molecular modeling techniques and attachment energy calculations to identify likely cleavage planes of the crystals, the specific interaction sites were identified. Ohta and Buckton [42] conducted a similar type of investigation on amorphous cefditoren pivoxil, which revealed that water preferentially adsorbs to the most hydrophilic sites at the lowest relative humidities. IGC vapor sorption, like the gravimetric and calorimetric approaches, has also been used to detect low amounts of amorphous material. Newell et al. [43] evaluated the effect of absorbed water on the surface energy of amorphous lactose and showed that the dispersive surface energy loss due to water sorption was reversible at low humidities only. At higher relative humidities, the sorption process was no longer reversible, indicating that the surface was reorganizing well before observable structural collapse.

9.3.3 Molecular Level Approaches

Studies of thermodynamic parameters and surface energies certainly improve the understanding of water–solid interactions; however, to truly relate the events observed in moisture sorption isotherms to structural changes, the mechanisms of water uptake and the nature of chemical interactions in the solid state need to be evaluated at a molecular level. Diverse approaches have been taken to examine water–solid interactions at the molecular level. Since the most common methods used to characterize water–solid interactions in amorphous pharmaceutical systems have been thoroughly reviewed [44], this section details analogous studies of crystalline solids.

9.3.3.1 Stoichiometric Hydrates

When the dominant mechanism of water uptake in a crystalline solid is absorption, several tools may be used to study water–solid interactions in the bulk crystal. X-ray crystallography is a particularly powerful technique for characterizing water–solid interactions in crystalline hydrates, since water molecules contribute to the coherence of the crystal structure. The use of this technique for establishing the precise location and site occupancy of water molecules that are regularly incorporated within a crystal lattice has been reviewed for different types of hydrates [29]. For stoichiometric hydration, water–solid interactions may be characterized for compositions representing each "step" along the moisture sorption curve. Structural information obtained by X-ray crystallography can, in some cases, be correlated to energy differences associated with different types of water in the crystal lattice. For example, Morris and Rodriguez-Hornedo [29] showed that ten waters of hydration residing in channels in the crystal structure of calteridol tetradecahydrate were lost more easily than the four water molecules associated with the calcium ions, thus explaining the stepwise weight losses observed by TGA.

Isotope exchange experiments may also be used to probe the structural and dynamic aspects of water–solid interactions in crystalline hydrates. Using raffinose pentahydrate and trehalose dihydrate as model compounds, Ahlqvist and Taylor [24] have demonstrated that Raman spectroscopy can be used to monitor the H/D exchange process. They found that while some of the water of hydration is dynamic and in constant exchange with the atmosphere, D_2O cannot access all parts of the crystal structures in these materials. To facilitate analysis of the incomplete exchange, the water contributions to the Raman ν_{OH} band were differentiated from those of the sugar hydroxyl groups by stopping the exchange experiments at various time points, removing the H_2O/D_2O by drying *in vacuo*, and measuring the extent of H/D exchange of the hydroxyl groups only. The number of exchangeable waters of hydration and hydroxyl groups were then correlated to the crystal structures. In both cases, the incomplete exchange was rationalized in terms of the location of specific waters in channels and the accessibility of hydroxyl groups to the D_2O vapor. Interestingly, both of the waters of hydration in trehalose dihydrate contribute to the structure and are energetically similar (both are lost from the crystal structure over the same temperature range), yet only one (W2) was found to exchange.

Because water absorption can have a profound influence on molecular dynamics in the solid state, NMR relaxometry, which is sensitive to molecular mobility, has been widely used to study water–solid interactions. Hydration affects solid-state ^1H and ^{13}C spin lattice relaxation times differently for different materials [45]. With β-estradiol hemihydrate, for example, Andrew and Kempka [46] found that, between 77 and 260 K (where water is presumed to be static on the NMR timescale), the dominant relaxation mechanism is C3 reorientation of a single methyl group. Above 260 K, however, water migration is sufficiently rapid to generate dipolar relaxation. While water absorption usually facilitates relaxation, Te et al. [47] showed that the water of hydration in thiamine hydrochloride (vitamin B1) hydrate actually *reduced* molecular mobility. In this study, the crystal structures of thiamine hydrochloride hydrate and its isomorphic desolvate were compared. Both hydrogen-bonding interactions to the water and less free volume (relative to the desolvate) likely limited the mobility of the thiamine in the hydrate crystal.

9.3.3.2 **Non-Stoichiometric/Channel Hydrates**

While characterizing water–solid interactions for stoichiometric hydrates is usually straightforward, establishing the structural changes induced by water sorption (or desorption) in non-stoichiometric hydrates can be challenging, due to the subtle changes to the crystal lattice and/or the continuous change of occupancies of partially populated water sites with changes in RH. In a compelling example of structural characterization of non-stoichiometric hydrates by X-ray crystallography, Steiner and Koellner [48] studied the crystal structures of β-cyclodextrin (β-CD) hydrate over an RH range, where fast, reversible diffusion (estimated at 1/1000 of the bulk water value) of water molecules occurs, despite the conspicuous absence of continuous diffusion channels in the crystal structure. The water molecules located in the β-CD cavity were affected by changes in the RH to a greater extent

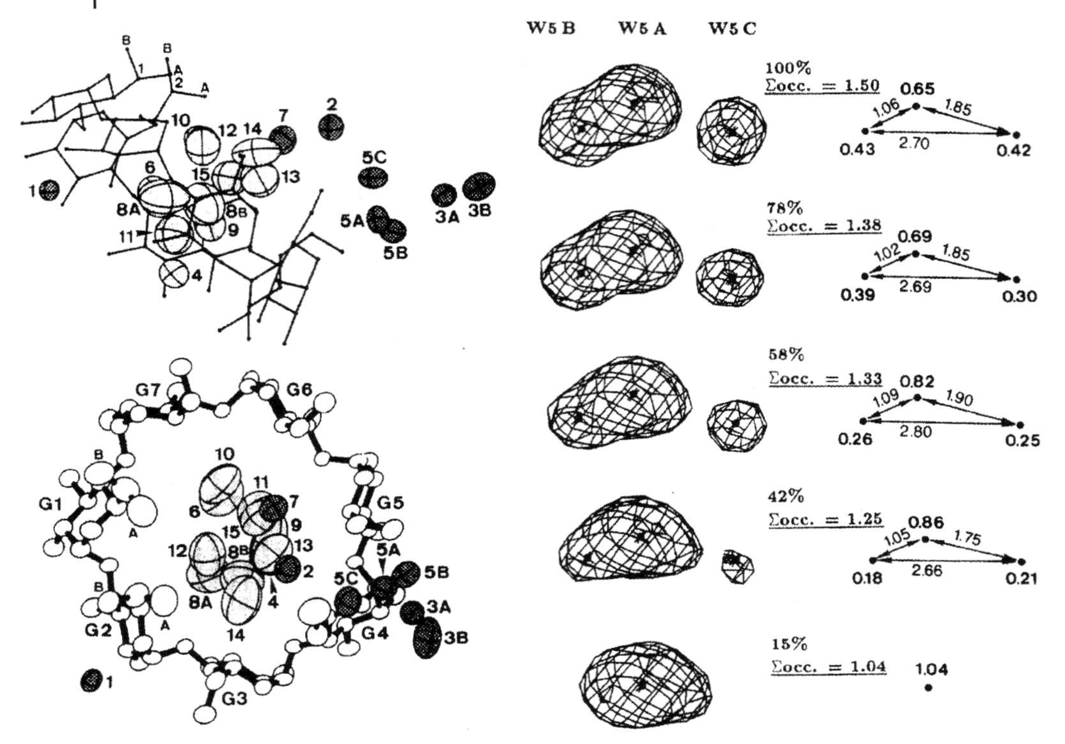

Fig. 9.5 Left: Asymmetric unit of β-CD hydrate at 58% RH, showing the location of the interstitial water relative to the β-CD cavity. Right: Electron density distributions at five RHs, reflecting the continuous change of site occupancy for the interstitial W5 waters. The population of sites W5B and W5C increases at the expense of site W5A as the RH is increased [48].

than those in the interstices; however, the continuous process of dehydration was amply revealed by careful examination of the well-refined interstitial waters (W5A/B/C) (Fig. 9.5). In this case, crystallography observed the superimposition of two unit cells, one with only the W5A site occupied and the other with the W5B and W5C sites equally populated. The cell volume smoothly diminished with decreasing RH, as the proportion of unit cells containing water at the W5A site gradually increased and those containing water at the W5B/C sites gradually decreased, resulting in a net loss of water. Because the crystal structure of β-CD lacks continuous channels, fast diffusion of water molecules was ascribed to transient positional fluctuations in the lattice.

Steiner et al. [49] later used a combination of H_2O/D_2O and $H_2{}^{16}O/H_2{}^{18}O$ exchange experiments to determine the mechanism of rapid water diffusion in β-CD dodecahydrate. The diffusion of D_2O and $H_2{}^{18}O$ into the β-CD crystal was followed by Raman spectroscopy and mass spectrometry, respectively. By observing complete H/D and $^{16}O/^{18}O$ exchange at comparable rates, the mechanism

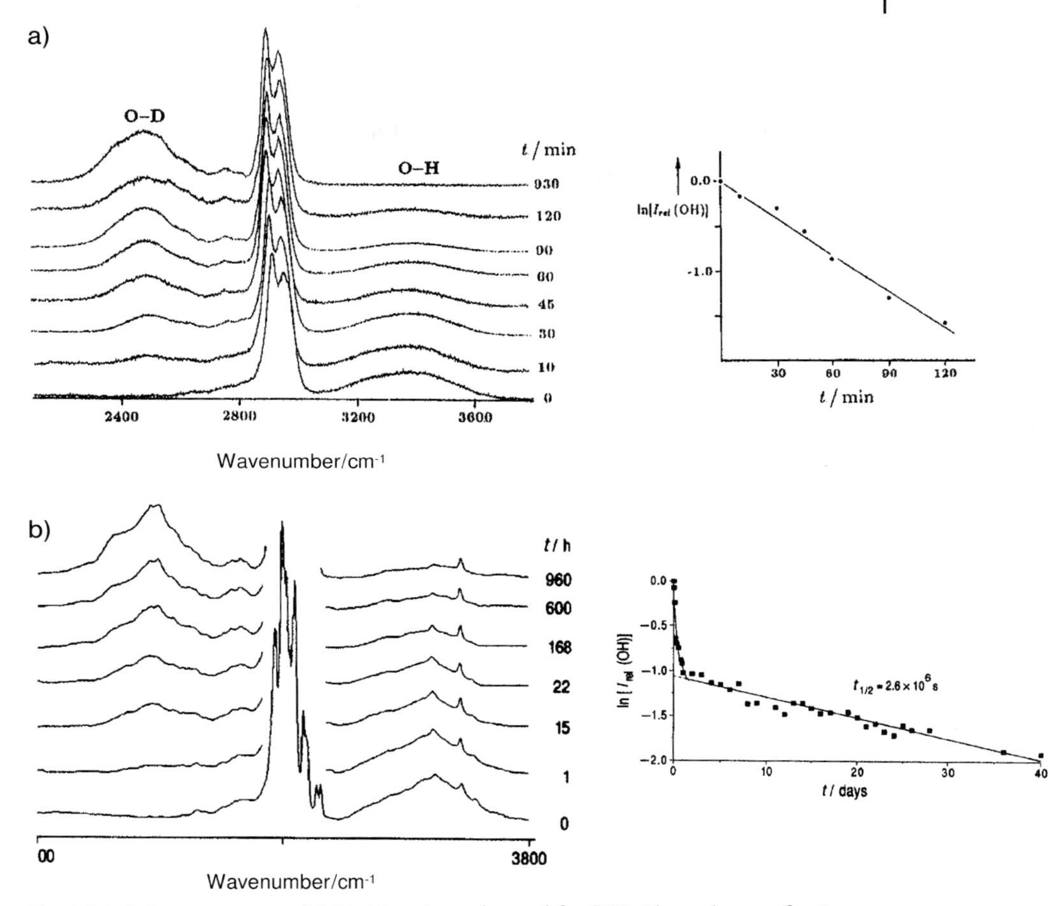

Fig. 9.6 Left: Raman spectra of β-CD (a) and α-CD (b) as a function of time of exposure to D_2O vapor. Right: Plots of integrated Raman νOH intensity, showing two distinct exchange rates for α-CD; only one is observed for β-CD. The exchange of water molecules in the α-CD crystal is predominantly in the beginning and H/D exchange is towards the end of the overall exchange process [49, 50].

was proposed to be a combination of diffusion of intact water molecules, initiating at the crystal surface and proceeding through the crystal lattice, and equally rapid exchange between water molecules and the β-CD hydroxyl groups. Rapid diffusion of intact water has also been observed for α-cyclodextrin hexahydrate (α-CD); however, the rate of H/D exchange was much slower than that of β-CD (Fig. 9.6) [50]. Evidence of structural rigidity (static hydrogen bond networks and fixed hydroxyl groups) in α-CD was obtained by Raman spectroscopy to explain the rate differences.

Following their investigation of H/D exchange in the stoichiometric hydrates of different sugars, Ahlqvist and Taylor [51] reported a similar study of the chan-

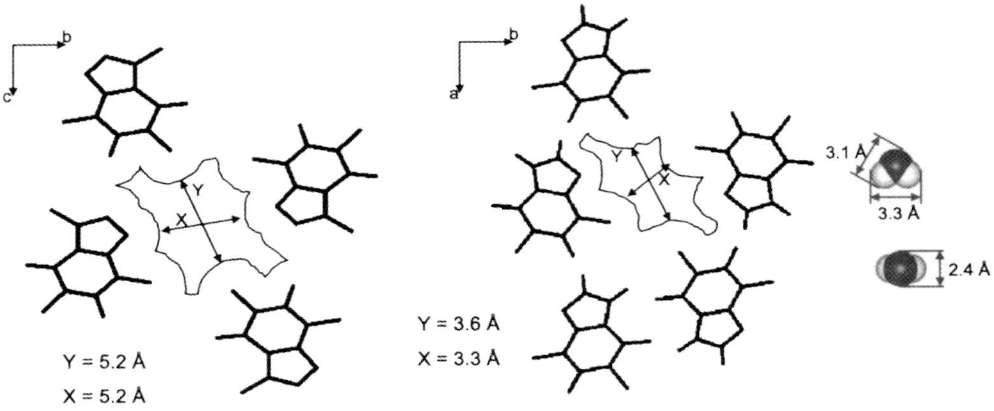

Fig. 9.7 Crystal structures of caffeine 4/5 hydrate (left) and theophylline monohydrate (right), along with contours defining the hydrate water channels and water molecules drawn as van der Waals representations [51].

nel hydrates formed by theophylline, caffeine, and sodium cromoglycate, wherein they attempted to correlate the mobility of water molecules inside lattice channels with structural parameters. Based on earlier work of Perrier and Byrn [52], it was suggested that the size of the water channels would be an important factor affecting H/D exchange rates of caffeine 4/5 hydrate and theophylline monohydrate. Indeed, the rate of H/D exchange in theophylline monohydrate was much slower than in caffeine 4/5-hydrate, presumably due to its smaller channel size (comparable to the size of a water molecule) (Fig. 9.7). The surprisingly slow kinetics of H/D exchange for sodium cromoglycate, considering the size of the channels in its crystal structure, were attributed to strong coordination of the water molecules to the sodium ions. Because the unit cell of sodium cromoglycate, when exposed to high RH, expands to accommodate more water [53], an equally plausible explanation for the slow kinetics may be that the hydrate structure, in adjusting to a specific water content to minimize the non-occupied volume, makes diffusion of D_2O molecules through the channel more difficult. In any case, the mobility of the water molecules in a channel hydrate clearly depends on the dimensions of the hydrate channel or, as observed for β-CD, the propensity for considerable positional fluctuations, which temporarily open paths for diffusion.

Applications of solid-state NMR spectroscopy to the structural characterization of water–solid interactions have also emerged. Because of the technical challenges encountered with solid-state ^{1}H and ^{17}O NMR spectroscopy, most structural studies of crystalline hydrates have employed ^{13}C CP/MAS NMR spectroscopy to indirectly observe the effects of water by their influence on the ^{13}C environments of the host molecules. One example, where this technique has been used in conjunction with X-ray crystallography to probe non-stoichiometric moisture sorption behavior, is LY297802 tartrate [54]. As shown in Fig. 9.4(f),

LY297802 tartrate experiences a continuous uptake of up to 0.5 molar equivalents of water from 0 to 90% RH. ^{13}C CP/MAS NMR spectra collected for samples of the non-stoichiometric hydrate dried at different conditions to alter the water content showed significant, yet systematic, peak shifting (Fig. 9.8). For the local chemical environments of the ^{13}C nuclei to change without altering the packing motif (no changes were observed by XPRD), water had to be incorporated into the structure, presumably by hydrogen-bonding interactions. Indeed, the water of crystallization was located in the crystal structure, bound by two relatively long OH···O hydrogen-bonding interactions to the crystallographically-inequivalent tartrate anions (Fig. 9.8).

The crystal structure of LY297802 tartrate also revealed that this hydrate possesses one common feature of non-stoichiometric hydrates: channels. In this case, the channels are somewhat corrugated and appear to be just large enough to permit water molecules to migrate to their hydrogen bonding sites (Fig. 9.8). Solid-state NMR spectroscopy, which in many respects complements X-ray crystallography [55], provided unique insights into both the progression of hydration in LY297802 tartrate and the mobility of the water in the corrugated channels. At sub-stoichiometric water compositions, only a fraction of the water sites can be occupied at any one time. Therefore, if the water is statically disordered, both hydrogen-bonded and non-hydrogen bonded tartrate carbonyl groups should be

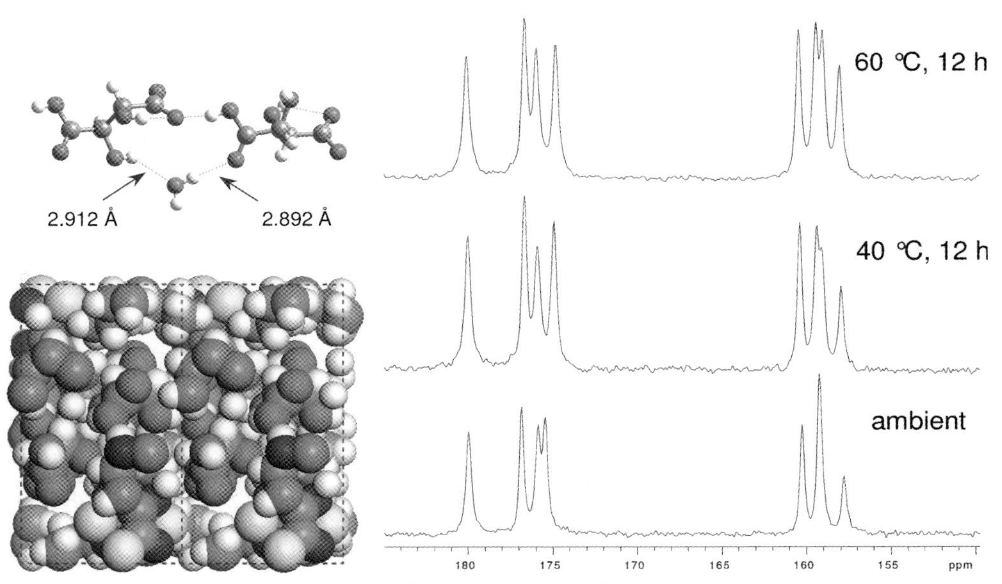

Fig. 9.8 Left: Hydrogen-bonding interactions binding water molecules in the channel hydrate of LY297802 tartrate, along with a crystal packing diagram. The water molecule is omitted in the crystal packing diagram for clarity. Right: Solid-state ^{13}C NMR spectra of the non-stoichiometric hydrate, showing systematic changes as the water content is altered.

observed, i.e., the ^{13}C resonances should be split. In fact, no peak splitting is observed, revealing that all particles are "hydrated" at the same time, and to the same extent, as water is absorbed into the polycrystalline solid. The observation of a single solid-state environment for each carbonyl ^{13}C at the various water compositions further reveals that the water molecules in the crystal are highly mobile, i.e., hopping rapidly (on the NMR timescale) between hydrogen bonding sites along the corrugated channels. The gradual shifting of the ^{13}C resonances simply reflects the changes in the average occupancy of water in the hydrogen bonding sites at different water compositions.

9.3.3.3 **Isomorphic Desolvates**

The examples presented in the previous section focused on investigations of water–solid interactions in non-stoichiometric hydrates, where the waters of hydration are reversibly sorbed into crystals with minimal changes to the crystal structure or crystallinity. A similar situation can arise when a stoichiometric hydrate is dehydrated at low RHs or high temperatures and an isomorphic desolvate is obtained (Fig. 9.4g). By definition, an isomorphic desolvate is formed when solvent is removed from a solvated material and the three-dimensional crystal structure of the parent solvate is retained. Due to the voids in the unit cell that are created by the loss of the solvent, an isomorphic desolvate is in a higher energy state than the parent solvate. Stephenson et al. [56], in studying the isomorphic desolvates formed by spirapril hydrochloride monohydrate and erythromycin A dihydrate, showed that the internal energy can be reduced by processes that increase packing efficiency, such as sorption of guest molecules or structural relaxation (possibly, collapse). The relevance of isomorphic desolvates of stoichiometric hydrates to the broader topic of water uptake in crystalline solids is that, if introduced to water vapor, these materials can be extremely hygroscopic.

Because water is not a component of isomorphic desolvates, the approaches taken to understand the structural basis of water uptake at a molecular level will necessarily be different for these materials. The techniques, which have been meaningfully applied to characterize the hygroscopicity of isomorphic desolvates, include powder and single crystal X-ray diffraction, isothermal microcalorimetry, and solid-state NMR spectroscopy, all of which were featured in the investigation of spirapril hydrochloride monohydrate and erythromycin A dihydrate [56]. Using X-ray powder diffraction, the spirapril hydrochloride monohydrate crystal structure was found to be minimally affected by the dehydration process. The unit cell volume of erythromycin A, however, was significantly reduced upon drying. These results were consistent with the greater structural relaxation and molecular "loosening" observed for erythromycin A by isothermal microcalorimetry and ^{13}C CP/MAS NMR spectroscopy, respectively. By comparing the crystal structures of the parent hydrates, the minimal disruption to the crystal lattice of spirapril hydrochloride was attributed to both the greater structural rigidity imparted by hydrogen-bonding interactions and the greater ease of molecular diffusion through its larger tunnels.

9.4
Significance of Water–Solid Interactions in Pharmaceutical Systems

Molecular level studies of water–solid interactions have yielded tremendous insight into the diverse effects that water absorption can have on both crystal structure and molecular dynamics. In many cases, the behaviors encountered in pharmaceutical systems are directly related to the specific crystalline forms of the API and/or excipients. Ahlneck and Zografi [23] have reviewed the mechanisms of water–solid interactions in amorphous and activated crystalline systems at the molecular level. They have suggested that the effects of water on the solid-state properties of otherwise crystalline drugs are directly linked to the extent to which the solid contains localized regions of higher energy, molecular disorder, and molecular mobility. Water will tend to concentrate in these regions, where it can plasticize the solid sufficiently to induce physical and/or chemical processes [57, 58]. Thus, the problems encountered with drug products can be due to changes in the drug substance, the excipients, or both, which may be caused by, but are not readily attributable to, water sorption. This section highlights several case studies, where physicochemical stability, dissolution, and physical-mechanical property issues encountered during drug development have been explained in terms of water–solid interactions (or lack thereof) on a molecular level.

9.4.1
Physicochemical Stability

Water absorption can enhance the molecular mobility of solids, leading to solid-state reactivity. Solid-state reactions of drug substances can include oxidation, cyclization, hydrolysis, and deamidation [59]. While higher energy sites are frequently the source of physicochemical instability, cases have been reported where the reactivity with, or mediated by, water directly depends on the crystalline state. For example, certain crystal forms of moisture-sensitive drugs feature reactive conformers or surfaces, which may be more susceptible to hydrolysis [60]. In other cases, crystal forms are inherently permeable, or may be rendered so, to reactive species. Desolvated channel solvates, for example, have been reported to be chemically unstable to oxidation [61]. Presumably, molecular oxygen diffuses into the tunnels vacated by the solvent, where it reacts with the host molecules.

While the physical characteristics of an API can present significant challenges in their own right, introducing excipients may substantially add to the complexity of developing a robust dosage form. Water can mediate drug–excipient interactions through the vapor phase or facilitate interactions at the drug–excipient interface by plasticizing both components [23]. By either mechanism, the effectiveness of the excipients and/or the physicochemical stability of the API can be compromised. The ability of superdisintegrants to enhance dissolution, for example, may be inhibited by other hygroscopic tablet components, which compete for the available water [62, 63]. Herman et al. [64] attributed the increased rate of hydrolysis of methylprednisolone sodium succinate in freeze-dried formulations prepared with mannitol to the crystallization of this excipient and the subsequent migration of water into the regions of amorphous drug substance within the freeze-dried cake. Jain et al. [65] reported a study that contradicts the notion that anhydrous lactose provides greater stability to drugs compared with lactose monohydrate. The argument is made that as the anhydrous lactose sorbs water, en route to converting into the monohydrate, a more intimate contact occurs between the drug and excipient, which compromises the stability of moisture-sensitive drugs.

9.4.2 Dissolution

Dissolution problems, many of which can be traced to changes in moisture content, are frequently encountered during drug development. For example, exposure of delavirdine methanesulfonate tablets to high relative humidity adversely affected their dissolution characteristics [66]. Using solid-state NMR data, an increase in water content in the stressed tablets was shown to cause the salt to dissociate, resulting in the free base and methanesulfonic acid. Not only did the lower solubility of the free base lead to poorer dissolution, but a subsequent acid–base reaction was observed by IR between the methanesulfonic acid and croscarmellose sodium, which compromised the disintegration properties of this excipient in the tablets. Such acid–base interactions can potentially occur when any acid salt is formulated with a basic excipient. Since the water content of the delavirdine methanesulfonate tablets was critical for the initial dissociation of the salt, desiccants or special packaging could be used to protect these tablets.

Norfloxacin is an unusual example, where tablet dissolution behavior was adversely affected by *lower* humidities [67]. Here, Hu et al. [68] have determined that water molecules induce proton transfer (Fig. 9.9) between norfloxacin molecules in the solid state as the anhydrate converts into norfloxacin dihydrate at higher RHs [69]. The change from a neutral crystal to an ionic crystal after exposure to 75% RH at room temperature resulted in a higher dissolution rate. Stabilization of the ionic structure by water was based on earlier conclusions of Remko and Scheiner [70] that water can reduce the energy barrier to proton transfer reactions. While proton transfer may not be considered a major change in the molecular structure, its effect on the solid-state properties was significant.

Bauer et al. [71] investigated the slowdown in dissolution rate for erythromycin tablets using XRD, NMR, and NIR, and found that the dissolution rate was compromised by overdrying the tablet formulation, which effectively dehydrated erythromycin dihydrate. The constant low level of water in the coated tablets appeared to prevent re-hydration of the isomorphic desolvate of erythromycin dihydrate, maintaining the API in an activated state. XRD and NIR evidence suggested that the tablet excipient $Mg(OH)_2$ eventually associated with the erythro-

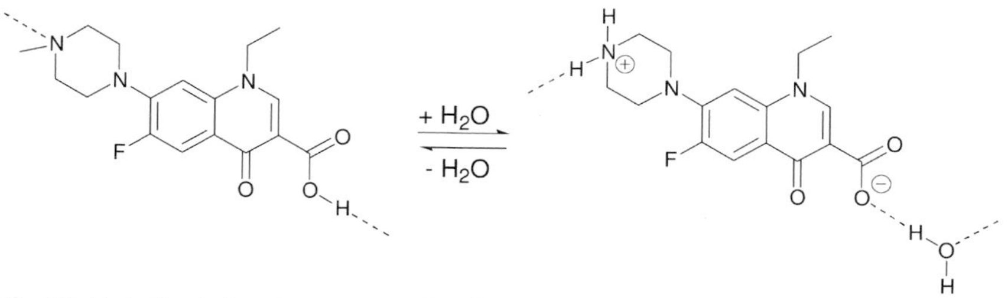

Fig. 9.9 Hydration-induced proton transfer of norfloxacin.

mycin at the same sites in the crystal structure as water in the dihydrate. In this case, the binding process could be reversed and the dissolution rate increased by humidifying the tablets.

9.4.3 Physical-mechanical Characteristics

Tableting and compression properties of formulations, which are commonly determined by excipients, can be affected by water sorption. An excellent review by Hancock and Shamblin [72] discusses the effects water can have on various sugars that ultimately affect drug product performance. The compactability of β-cyclodextrin has been reported to be completely lost upon removal of its waters of crystallization [73]. In some cases, the drug substance can also significantly affect compression properties. Sun and Grant [74] have shown that the incorporation of water into the crystal lattice of *p*-hydroxybenzoic acid (PHBA) facilitates plastic deformation, providing much stronger tablets. Both the anhydrate and monohydrate crystal structures of PHBA contain hydrogen-bonded zigzag layers; however, when anhydrate crystals are compressed, the zigzag layers mechanically interlock, inhibiting slip and reducing plasticity. The water molecules in the monohydrate crystals assume a space-filling role, which increases the interlayer spacing, allowing for easier slip and greater plasticity.

In contrast to crystalline materials, which are generally highly elastic and brittle upon exposure to external stress, amorphous materials exhibit varying degrees of viscoelasticity, which is important in creating tablet bonds following compression. Owing to the plasticizing effects of water, moisture sorption into amorphous excipients has been reported to have different effects on tablet hardness. The function of microcrystalline cellulose (MCC), which is $\sim 30\%$ amorphous, appears to be closely related to how water is held within the powder mass and how mechanical processing influences the mobility of water [75]. It has been proposed that MCC and maltodextrin, like β-CD and PHBA, compact by plastic deformation and that an optimum amount (4–6% in the case of MCC) of water will provide optimal compression properties by forming interparticular hydrogen-bond bridges, which prevent elastic recovery [76]. Tablets formulated with maltodextrin, a common binder that has an adhesive effect when exposed to moisture, initially increase in hardness at high relative humidities [77]. Upon further moisture sorption, however, the binding forces of the maltodextrin formulations are reduced and the tablet hardness decreases.

Relative humidity has also been shown to have a significant effect on particle cohesion [78], which can affect the aerosolization properties of micronized drugs [79]. Young et al. [80, 81] have developed applications of AFM to study the effects of RH on the cohesion properties of salbutamol sulfate (SS), triamcinolone acetonide (TAA), and disodium cromoglycate (DSCG). They showed that the calculated force and energy of cohesion required to separate the SS and DSCG particles increased, but that for TAA decreased as the humidity increased from 15 to 75% RH. The former behavior was attributed to capillary forces, which are induced by multilayer adsorption of water and therefore dominate at high RH, while the latter behavior was attributed to long-range attractive electrostatic interactions. Rowley and Mackin [82] used this type of analysis to evaluate the role of moisture in triboelectrification processes. They showed that while moisture sorption isotherms are useful for predicting the charging of pharmaceutical excipient powders, analysis of the moisture distribution more reliably predicts the effect of RH on electrostatic charge for porous solids. In these cases, moisture absorption into the porous particles pulls water away from the particle surface, where triboelectrification occurs.

9.5 Strategies for Dealing with Hygroscopic Systems

Since one of the best ways to deal with hygroscopic solids is, arguably, to avoid them altogether, a non-hygroscopic crystalline form of an API is highly desirable. Hygroscopicity, one of three primary criteria identified by Morris et al. [83] for selecting salt crystal forms, is assessed during the initial stage of salt evaluations. For solid forms with acceptable moisture sorption behavior, controlled crystallization processes should be developed to avoid not only crystal form mixtures, but also regions of amorphous materials. Generally, problems associated with phase transformations can be minimized by identifying the thermodynamically stable crystal form over the temperature/RH range to which the drug will be subjected, and adjusting processing conditions accordingly. If a metastable form is selected, the risk of moisture-induced transformations should be evaluated. In the event that a hygroscopic form is selected, then special manufacturing and storage conditions may be required [84, 85]. Here, having an understanding of water–solid interactions on a molecular level should facilitate the development of control strategies for processing, handling, and packaging hygroscopic solids.

Careful examination of the impact of drying and milling operations on the crystal form and crystallinity of solid materials should be performed. Drying materials to the lowest possible water content may be counterproductive, since materials may undergo substantial changes when the last layer of water is dried off the surface, handling may be aggravated, and the risk of inaccuracies in dry mass correction is enhanced [86]. Non-stoichiometric hydrates present unique challenges in drug development, not only for the analysts, who should be wary of dynamic changes in the water content, but also for the process chemists and formulators, who design drying processes. During drying processes, control methods may be required to determine either when drying is complete or a switch to a humid nitrogen sweep is needed to preserve the desired form of a compound, or to stop constant agitation to reduce shear-induced dehydration at low water levels. Online methods, such as NIR, which can differentiate between surface and bound water, have been particularly valuable for monitoring drying operations [87] and can be used to develop suitable drying protocols. Milder drying processes, such as the fluidized bed fast drying process developed for low melting compounds [88], may also be applicable for hydrated materials. The fluidized bed fast drying process is based on an overall cooling effect that water evaporation produces in the bed. This process allows the inlet air temperature to increase (beyond the melting point of the active) and expedites evaporative drying, without raising the bed temperature or risking the batch.

Organic solvates may be converted into hydrates by direct displacement of the solvent guest by water, affording hydrates that in some cases cannot otherwise be manufactured [89]. Water sorption studies on these types of solvates using automated systems may not be straightforward since the loss of solvent and uptake of water may not be easily differentiated in the isotherm. Studies using RH chambers and other characterization techniques, including quantitation of solvent and water, will provide valuable information on the conditions for the displacement. This type of displacement can also be used to facilitate drying of organic solvates. Pikal [90] has shown that converting certain methanolates into hydrates at ambient humidity and then dehydrating the hydrated materials under vacuum or dry air was faster than vacuum drying the methanolate directly.

When dealing with molecularly disordered (amorphous) solids, strategies must be based on an understanding of their thermodynamic and kinetic properties to ensure stability [44]. At a minimum, the critical humidity should be defined, i.e., the minimum RH at which a crystalline sample must be exposed to eliminate amorphous content and the maximum RH at which an amorphous sample must be stored to stabilize the amorphous state. Of course, the time-dependent nature of nucleation in amorphous solids needs to be considered [91], as does the plasticizing effect of water. Recognizing the potential for activation of solid surfaces when subjected to stressful processes, e.g., milling or lyophilization, is also essential to understanding physicochemical instabilities at seemingly low moisture contents.

9.6 Conclusions

There exists no universal definition of hygroscopicity that covers all aspects of water interactions with pharmaceutical materials. The amount of water sorbed at different relative humidities is commonly obtained; however, a useful classification system for hygroscopicity in pharmaceutical solids needs to include the amount of water sorbed, the rate of water sorption, and the physical/form changes that occur. An overall understanding of where and how water is bonding in a solid should lead to the development of more robust processes and result in fewer physicochemical stability, dissolution, and physical-mechanical property issues during the various stages of drug development. Until a system is established, it is important to define how the term "hygroscopicity" is being used.

References

1 *Lange's Handbook of Chemistry*, N.A. Lange (Ed.), Handbook Publishers, Sandusky, OH, **1944**, p. 1650.
2 *Webster's New Collegiate Dictionary*, G. & C. Merriam Company, Springfield, MA, **1973**, p. 561.
3 S.R. Byrn, R.R. Pfeiffer, J.G. Stowell, *Solid State Chemistry of Drugs*, 2nd edn., SSCI, West Lafayette, IN, **1999**, p. 510.
4 W.H. Nebergall, H.F. Holtzclaw, Jr., W.R. Robinson, *General Chemistry*, 6th edn., D.C. Heath and Co., Lexington, **1980**, p. 286.
5 *Hawley's Condensed Chemical Dictionary*, John Wiley and Sons, New York, **1997**, p. 601.
6 M.W. Scott, H.A. Lieberman, F.S. Chow, *J. Pharm. Sci.* **1963**, *52*, 994–998.
7 C. Giles, T.H. MacEwan, S.N. Nakha, D. Smith, *J. Chem. Soc.* **1960**, 3973–3993.
8 K. Umprayn, R.W. Mendes, *Drug Dev. Ind. Pharm.* **1987**, *13(4 & 5)*, 653–693.
9 L. Van Campen, G. Zografi, J.T. Carstensen, *Int. J. Pharm.* **1980**, *5*, 1–18.
10 M. M. Markowitz, D.A. Boryta, *J. Chem. Eng. Data*, **1961**, 6, 16–18.
11 I.M. Kuvshinnikov, Z.A. Tikhonovich, V.A. Frolkina, *Khim. Prom.* (Moscow) **1971**, *47*, 599–600.
12 E. Shoton, H. Harb, *J. Pharm. Pharmacol.* **1965**, *17*, 504–508.
13 W.C. Griffin, R.W. Behrens, S.T. Cross, *J. Soc. Cosmet. Chem.* **1952**, *3*, 5–29.
14 J.C. Callahan, G.W. Cleary, M. Elefant, G. Kaplan, T. Kensler, R.A. Nash, *Drug Dev. Ind. Pharm.* **1982**, *8(3)*, 355–369.
15 *Handbook of Pharmaceutical Excipients*, eds. R.C. Rowe, P.J. Sheskey, P.J. Weller, Pharmaceutical Press, Chicago, **2003**.
16 M.J. Kontny, G. Zografi, in *Physical Characterization of Pharmaceutical Solids*, ed. H.G. Brittain, Marcel Dekker, New York, **1995**, Chapter 12, pp. 386–418.
17 P.A. Thiel, T. E. Madey, *Surf. Sci. Rep.* **1987**, *7*, 211–385.
18 G.R. Desiraju, *J. Chem. Soc., Chem. Commun.* **1991**, 426–428.
19 G. Ferraris, M. Franchini-Angela, *Acta Crystallogr., Sect. B* **1972**, *28*, 3572–3583.
20 G.A. Jeffrey, H. Maluszynska, *Acta Crystallogr., Sect. B* **1990**, *46*, 546–549.
21 A.L. Gillon, N. Feeder, R.J. Davey, R. Storey, *Cryst. Growth & Des.* **2003**, *3(5)*, 663–673.
22 G. Zografi, *Drug Dev. Ind. Pharm.* **1988**, *14(14)*, 1905–1926.
23 C. Ahlneck, G. Zografi, *Int. J. Pharm.* **1990**, *62*, 87–95.
24 M.U.A. Ahlqvist, L.S. Taylor, *J. Pharm. Sci.*, **2002**, *91(3)*, 690–698.
25 M. Gordon, J.S. Taylor, *J. Appl. Chem.* **1952**, *2*, 493–500.
26 H. Nyqvist, *Int. J. Pharm. Tech. Prod. Mfr.* **1983**, *4*, 47–48.
27 A. Roberts, *J. Therm. Anal. Calorim.* **1999**, *55*, 389–396.

28 L. Stubberud, H. G. Arwidsson, C. Graffner, *Int. J. Pharm.* **1995**, *114*, 55–64.
29 K. R. Morris, N. Rodriguez-Hornedo, *Encyclopedia of Pharmaceutical Technology*, Volume 7, Marcel Dekker, New York, **1993**, pp. 393–440.
30 M. D. Ticehurst, R. A. Storey, C. Watt, *Int. J. Pharm.* **2002**, *247*, 1–10.
31 M. Sachetti, *Int. J. Pharm.* **2004**, *273*, 195–202.
32 E. Shefter, T. Higuchi, *J. Pharm. Sci.* **1963**, *52*, 781–791.
33 R. K. Khankari, D. J. W. Grant, *Thermochim. Acta* **1995**, *248*, 61–79.
34 S. M. Reutzel-Edens, R. L. Kleemann, P. L. Lewellen, A. L. Borghese, L. J. Antoine, *J. Pharm. Sci.* **2003**, *92(6)*, 1196–1205.
35 L. Yu, N. Milton, E. G. Groleau, D. S. Mishra, R. E. Vansickle, *J. Pharm. Sci.* **2000**, *88(2)*, 196–198.
36 M. Sacchetti, *J. Pharm. Sci.* **1998**, *87(8)*, 982–986.
37 R. J. Willson, A. E. Beezer, *Int. J. Pharm.* **2003**, *258*, 77–83.
38 N. Markova, L. Wadsö, *J. Therm. Anal.* **1999**, *57*, 133–139.
39 D. Lechuga-Ballesteros, A. Bakri, D. P. Miller, *Pharm. Res.* **2003**, *20(2)*, 308–318.
40 I. M. Grimsey, J. C. Feeley, P. York, *J. Pharm. Sci.* **2002**, *91(2)*, 571–583.
41 M. R. Sunkersett, I. M. Grimsey, S. W. Doughty, J. C. Osborn, P. York, R. C. Rowe, *Eur. J. Pharm. Sci.* **2001**, *13*, 219–225.
42 M. Ohta, G. Buckton, *Int. J. Pharm.* **2004**, *269*, 81–88.
43 H. E. Newell, G. Buckton, D. A. Butler, F. Thielmann, D. R. Williams, *Int. J. Pharm.* **2001**, *217*, 45–56.
44 B. C. Hancock, G. Zografi, *J. Pharm. Sci.* **1997**, *86(1)*, 1–12.
45 C. L. Jackson, R. G. Bryant, *Biochemistry* **1989**, *28*, 5024–5028.
46 E. R. Andrew, M. Kempka, *Solid State Nucl. Magn. Reson.* **1995**, *4*, 249–253.
47 R. L. Te, U. J. Griesser, K. R. Morris, S. R. Byrn, J. G. Stowell, *Cryst. Growth Des.* **2003**, *3(6)*, 997–1004.
48 T. Steiner, G. Koellner, *J. Am. Chem. Soc.* **1994**, *116*, 5122–5128.
49 T. Steiner, A. M. Moreira da Silva, J. J. C. Teixeira-Dias, J. Müller, W. Saenger, *Angew. Chem. Int. Ed. Engl.* **1995**, *34(13/14)*, 1452–1453.
50 A. M. Amado, P. J. A. Ribeiro-Claro, *J. Chem. Soc., Faraday Trans.* **1997**, *93(14)*, 2387–2390.
51 M. U. A. Ahlqvist, L. S. Taylor, *Int. J. Pharm.* **2002**, *241*, 253–261.
52 P. R. Perrier, S. R. Byrn, *J. Org. Chem.* **1982**, *47*, 4671–4676.
53 L. R. Chen, V. G. Young, D. Lechuga-Ballesteros, D. J. W. Grant, *J. Pharm. Sci.* **1999**, *88*, 1191–1200.
54 S. M. Reutzel, V. A. Russell, *J. Pharm. Sci.* **1998**, *87(12)*, 1568–1571.
55 M. C. Etter, R. C. Hoye, G. M. Vojta, *Cryst. Rev.* **1988**, *1*, 281–338.
56 G. A. Stephenson, E. G. Groleau, R. L. Kleemann, W. Xu, D. R. Rigsbee, *J. Pharm. Sci.* **1998**, *87(5)*, 536–542.
57 M. J. Kontny, G. P. Gandolfi, G. Zografi, *Pharm. Res.* **1987**, *4*, 104–112.
58 E. Y. Shalaev, G. Zografi, *J. Pharm. Sci.* **1996**, *85(11)*, 1137–1141.
59 S. R. Byrn, W. Xu, A. W. Newman, *Adv. Drug Delivery Rev.*, **2001**, *48(1)*, 115–136.
60 S. R. Byrn, R. R. Pfeiffer, J. G. Stowell, *Solid-State Chemistry of Drugs*, SSCI, West Lafayette, IN, **1999**, pp. 331–344.
61 S. R. Byrn, P. A. Sutton, B. Tobias, J. Frye, P. Main, *J. Am. Chem. Soc.* **1988**, *110*, 1609–1614.
62 J. Roche Johnson, L.-H. Wang, M. S. Gordon, Z. T. Chowhan, *J. Pharm. Sci.* **1991**, *80(5)*, 469–471.
63 S. Li, B. Wei, S. Fleres, A. Comfort, A. Royce, *Pharm. Res.* **2004**, *21(4)*, 617–624.
64 B. D. Herman, B. D. Sinclair, N. Milton, S. L. Nail, *Pharm. Res.* **1994**, *11*, 1467–1473.
65 R. Jain, A. S. Railkar, A. W. Malick, C. T. Rhodes, N. H. Shah, *Eur. J. Pharm. Biopharm.* **1998**, *46*, 177–182.
66 B. R. Rohrs, T. J. Thamann, P. Gao, D. J. Stelzer, M. S. Bergren, R. S. Chao, *Pharm. Res.* **1999**, *16*, 1850–1856.
67 A. V. Katdare, J. F. Bavitz, *Drug Dev. Ind. Pharm.* **1984**, *10(5)*, 789–807.
68 A. J. Florence, A. R. Kennedy, N. Shankland, E. Wright, A. Al-Rubayi, *Acta Crystallogr., Sect. C* **2000**, *56*, 1372–1373.

69 T.-C. Hu, S.-L. Wang, T.-F. Chen, S.-Y. Lin, *J. Pharm. Sci.* **2002**, *91(5)*, 1351–1357.

70 M. Remko, S. Scheiner, *J. Pharm. Sci.* **1991**, *80(4)*, 328–332.

71 J. F. Bauer, W. Dziki, J. E. Quick, *J. Pharm. Sci.* **2000**, *88(11)*, 1222–1227.

72 B. Hancock, S. Shamblin, *PSTT,* **1998**, *1(8)*, 345–351.

73 G. S. Pande, R. F. Shangraw, *Int. J. Pharm.* **1995**, *124*, 231–239.

74 C. Sun, D. J. W. Grant, *Pharm. Res.* **2004**, *21(2)*, 382–386.

75 K. E. Fielden, J. M. Newton, P. O'Brien, R. C. Rowe, *J. Pharm. Pharmacol.* **1988**, *40*, 674–678.

76 K. A. Khan, P. Musikabhumma, J. P. Warr, *Drug Dev. Ind. Pharm.* **1981**, *7*, 525–538.

77 L. C. Li, G. E. Peck, *J. Pharm. Pharmacol.* **1990**, *42*, 272–275.

78 G. A. Turner, M. Balasubramanian, *Powder Tech.* **1974**, *10*, 121–127.

79 P. M. Young, R. Price, M. J. Tobyn, M. Buttrum, F. Dey, *Drug Dev. Ind. Pharm.* **2003**, *29*, 959–966.

80 P. M. Young, R. Price, M. J. Tobyn, M. Buttrum, F. Dey, *J. Pharm. Sci.* **2003**, *92(4)*, 815–822.

81 P. M. Young, R. Price, M. J. Tobyn, M. Buttrum, F. Dey, *J. Pharm. Sci.* **2004**, *93(3)*, 753–761.

82 G. Rowley, L. A. Mackin, *Powder Tech.* **2003**, *135–136*, 50–58.

83 K. R. Morris, M. G. Fakes, A. B. Thakur, A. W. Newman, A. K. Singh, J. J. Venit, C. J. Spagnuolo, A. T. M. Serajuddin, *Int. J. Pharm.* **1994**, *105*, 209–217.

84 R. J. Bastin, M. J. Bowker, B. J. Slater, *Org. Proc. Res. Dev.* **2000**, *4*, 427–435.

85 S. Balbach, C. Korn, *Int. J. Pharm.* **2004**, *275*, 1–12.

86 S. Ruckold, K. H. Grobecker, H. D. Isengard, *Fresenius' J. Anal. Chem.* **2001**, *370*, 189–193.

87 G. X. Zhou, Z. Ge, J. Dorwart, B. Izzo, J. Kukura, G. Bicker, J. Wyvratt, *J. Pharm. Sci.* **2003**, *92(5)*, 1058–1065.

88 P. L. D. Wildfong, A. Samy, J. Corfa, G. E. Peck, K. R. Morris, *J. Pharm. Sci.*, **2002**, *91(3)*, 631–639.

89 S. Nordhoff, J. Ulrich, *J. Therm. Anal.* **1999**, *57*, 181–192.

90 M. J. Pikal, J. E. Lang, S. Shah, *Int. J. Pharm.* **1983**, *17*, 237–262.

91 E. Schmitt, C. W. Davis, S. T. Long, *J. Pharm. Sci.* **1996**, *85(11)*, 1215–1219.

10
The Amorphous State

Samuel Petit and Gérard Coquerel

10.1
Introduction

In the context of polymorphism, the study of amorphous solids, defined as non-crystalline solids, has a special status. Whether an amorphous phase can be considered a polymorph depends on the definition of polymorph [1–3]. However, the utility of considering the amorphous state in the present book can be demonstrated from several points of view. First, there is fundamental interest in investigating disordered states, since pure crystals with perfect order and no defects are exceptional. Second, the amorphous state can be an intermediate state in some crystal into crystal transformations. Several studies have discussed this situation [4–10], which can be seen as an unusual pathway to overcome specific kinds of activation energies. Other authors [11, 12] have reported that polymorphic transformations occurring in the solid state could be accompanied by the formation of substantial lattice defects, which are detected by a change of the overall crystallinity.

Finally, many scientists involved in the development of new pharmaceutical products or fine chemicals sometimes have to deal with so-called “difficult-to-crystallize” substances, whereas others concentrate on enhancing the physical and chemical stability of amorphous products selected because of the inadequate behavior of their crystalline counterparts.

Since several excellent reviews devoted to organic and pharmaceutical amorphous solids have appeared (see for instance [13–16]), our purpose here is not to elaborate a systematic and detailed overview of this (large) subject, but rather to summarize the major issues related to the amorphous state, and to discuss in more detail a few topics directly connected to polymorphism, such as the possible existence of polyamorphism and the use of amorphous solids in polymorphic investigations.

Polymorphism: in the Pharmaceutical Industry. Edited by Rolf Hilfiker

ISBN: 3-527-31146-7

10.2
Definition of the Amorphous State

10.2.1
Order, Disorder and Structural Aspects

A fundamental notion allowing differentiation between crystalline and amorphous solids is that of long-range order (LRO), implying simultaneous translational, rotational and conformational orders in crystalline particles. Since molecular units in a crystal lattice are assumed to be repeated according to a three-dimensional pattern along crystallographic directions, the relative location and orientation as well as the interactions between neighboring components can be accurately described at the molecular level. The crystal packing usually corresponds to a high density arrangement and therefore to a minimal molar volume.

By contrast, the absence of translational and rotational order in amorphous solids could be ascribed, at first sight, to a random distribution in the relative orientation of neighboring molecular units, implying that only the molar volume could give an estimate of the probability of finding a molecule at a given distance from another. In reality, however, a local or short-range order exists in amorphous solids [17], and has been experimentally established for inorganic glasses, for instance, by means of X-ray spectroscopy [18]. There is much less experimental evidence of local ordered structures in glassy states of molecular compounds, but recent studies devoted to indomethacin [19, 20] or sucrose [21] have suggested the existence of "structural elements" in amorphous solids. Therefore, the immediate environment of a molecule may be similar or even identical in crystalline and amorphous phases [14], and considering that non-covalent interactions (mainly hydrogen bonds, but also hydrophobic and polarization interactions) could have the same self-organizing role in both types of solids, a recent suggestion is that the amorphous state may be considered as a precursor to the crystalline state [22].

The molecular pattern of amorphous solids is, however, more often depicted as that of a frozen liquid, with the rheological properties of a solid [13, 23]. For flexible molecular compounds that contain many internal degrees of freedom, this implies the probable existence of a conformational diversity [24]. From this definition, the proportion of the various conformers should then be determined mainly by the preparation route of each sample and, possibly, by its history. Furthermore, this conformational diversity (and disorder) has been assumed to be responsible for the slow crystallization rate of molecular families such as polysaccharides and synthetic polymers [23] and could account in some cases for the empirical distinction between good and poor glass formers [14].

The idea that a continuum exists from 100% crystalline (perfect LRO) to 100% amorphous (complete absence of LRO) has often been postulated [11, 13, 23], the extremes being considered as idealized limit situations. This continuity can be explained from at least three points of view (Fig. 10.1), associated with different physical states of non purely crystalline solids. The first notion lies in

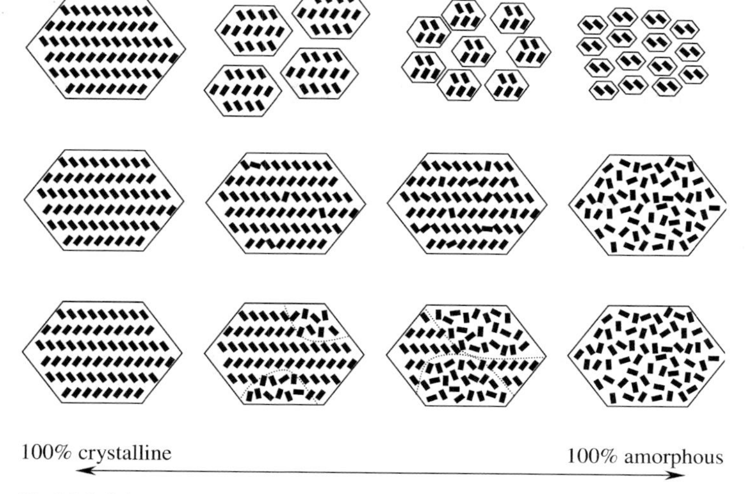

Fig. 10.1 Schematic representation of the three types of continuity between 100% crystalline and 100% amorphous solids.

a continuous decrease of crystalline particle size from macroscopic (0.1 mm to several mm) to atomic dimensions (10 to 100 Å) [25]. Nevertheless, Yu [14] has suggested that one should not confuse a microcrystalline state with a "truly" amorphous material. This is supported by experimental investigations using DSC, which have allowed these two physical states to be distinguished [26]. Assuming that no change in particle size occurs, a second type of continuity from crystalline to amorphous may involve a progressive "decrystallization" [27] through a progressive increase of the proportion of lattice defects and/or crystal distortions [11], with a simultaneous and continuous decrease of the degree of crystallinity. This evolution leads to a one-state model of the amorphous state [13], and may be further understood by a careful characterization of crystal disruptions and defects [28]. A third way of considering a continuity from crystalline to amorphous states corresponds to the two-state model, which is defined as a physical mixture of purely amorphous or purely crystalline regions that would coexist in a same solid particle. This description is related in some respects to the concept of structural heterogeneity highlighted by Zografi and coworkers [13, 17] and to that of microheterogeneity [14]. Experimental investigations have, for instance, revealed different ordering tendencies in supercooled liquids [29]. Furthermore, the surface of the particles is more sensitive to external conditions of storage (for instance RH), so that the superficial layers may differ from the core of the particle.

Hence, no clear-cut definition can be established to differentiate between crystalline and amorphous solids, and this is reinforced by the existence of several families of solid phases corresponding to defined intermediate states with different types of disorders [23]. These partially crystalline solids are called mesophases.

Depending on the type of disorder, mesophases can be classified in three categories of condensed matter. In liquid crystals, the molecular shape induces an orientational order, but the packing lacks three-dimensional translational and conformational order [30]. The second category corresponds to plastic crystals, which are sometimes called glassy crystals [31]. In these crystals, only a translational order exists, and the absence of orientational and/or conformational order is often caused by the rounded shape of the molecules. Finally, conformationally disordered crystals are well-known among organic and pharmaceutical compounds. For all these partially disordered phases, it may be useful to distinguish between static and dynamic disorder [32]. The main properties of mesophases, some of the experimental techniques useful for their identification, and several illustrative examples can be found in recent reviews [17, 33].

10.2.2 Energetic Aspects: Thermodynamics and Kinetics

Thermodynamically, amorphous solids are commonly defined as out of equilibrium states since they necessarily contain an excess of Gibbs energy with reference to the crystalline phases. At least theoretically, the various excess properties (enthalpy, entropy and free energy) can be quantified from heat capacities determined over a same temperature range for both the crystalline and the amorphous phases $[C_p = (\partial H/\partial T)_p = T(\partial S/\partial T)_p]$. The stored internal energy means that an amorphous solid is, by definition, an unstable state, which can release its energy excess either completely through crystallization associated with $\Delta G < 0$ or partially by means of irreversible relaxation processes.

Practically and qualitatively, a convenient way to visualize some energetic features of the amorphous state is to use a schematic representation of enthalpy (or volume) variations as a function of temperature (Fig. 10.2). The slope of each segment in this diagram represents the heat capacity of the corresponding state. By cooling a liquid to the melting point (T_m) of the crystalline phase, a first-order phase transition should be depicted, which is associated with a decrease of free volume because of thermal contraction effects (water is an exception). Since the heat exchange associated with crystallization (or melting) and T_m are of purely thermodynamic origin, their determination can be accurately achieved, and these values can be used for instance for the calibration of thermal analysis equipments.

If crystallization does not occur on cooling a liquid below T_m, a supercooled liquid is obtained, which is similar to the equilibrium state of the liquid phase, and which is often called a rubbery state because of its physical aspect and mechanical properties. Further cooling induces a change from the supercooled liquid to a glassy state at the glass transition temperature (T_g), accompanied by a decrease of the slope and therefore of the heat capacity of the material. Formation of a glass is not associated with a heat transfer, but is accompanied by a dramatic decrease in molecular mobility since translational (or diffusive) and rotational motions are almost frozen. In first approximation, it is assumed that

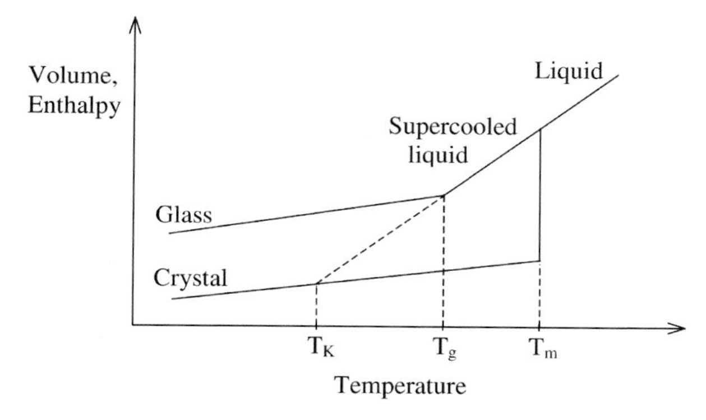

Fig. 10.2 Schematic representation of enthalpy or volume variations as a function of temperature for condensed materials, allowing us to visualize T_K (Kauzmann temperature) T_g (glass transition temperature) and T_m (melting temperature).

the molecular arrangement in a glass is similar to that of the liquid. In contrast to T_m, T_g can fluctuate with operating conditions and as a function of the history of the sample, therefore indicating that the glass transition is a thermal event affected by kinetic factors. At a macroscopic scale, the material becomes brittle at temperatures below T_g, and its viscosity reaches a high value ($>10^{12}$ Pa s).

Although glassy and crystalline solids exhibit different physical and thermodynamic features, their C_p values, depicted by their slopes in the $H=f(T)$ diagram, are, notably, quite similar. From the statements and paradox highlighted by Kauzmann [34] regarding the entropy of glassy solids, it has been suggested that there exists a lower limit to the T_g, known as the Kauzmann temperature (T_K). Its value corresponds to the temperature at which the enthalpy of the supercooled liquid attains that of the crystalline state, so that there is no more configurational entropy in the system. Entropically, T_K should actually be considered as a mathematical extrapolation, rather than a physically significant temperature. However, T_K, when determined by extrapolation, can be a valuable indicator for the long-term storage stability of amorphous solids since almost no risk of spontaneous evolution towards a more stable state should exist below this temperature.

10.3
Preparation of Amorphous Solids

Amorphous or partially amorphous states can be formed in different situations and result from various circumstances. First, some compounds in the pharmaceutical industry are known only or mainly in amorphous form, such as poly(vinylpyrrolidone) (PVP) or high molecular weight PEGs [35]. Second, an amorphous character can be induced inadvertently by using unsuitable crystallization conditions or by a processing operation such as drying, desolvation, milling or compression. Third, an amorphous solid may be prepared deliberately because of the necessity to enhance some of the physical or biological properties of the substance.

In this last situation, several preparative strategies can be envisaged, corresponding to different ways to reach a disordered, high energy and unstable amorphous solid. The experimental procedure is of major importance since amorphous, glassy or microcrystalline samples prepared according to different routes may exhibit different chemical stabilities, physical properties and possibly structural arrangements (polyamorphism, see Section 10.4.6). The various strategies that can induce some amorphous character to a solid are derived from the definitions of amorphous solids. Some consist in preventing crystallization, so that the matter cannot reach its lowest energy state during solidification. As a consequence, an excess of free energy and entropy is incorporated into the solids during amorphization. The principle of other strategies is to provide directly this excess of energy, e.g., by using suitable mechanical treatments.

Several methods commonly employed to produce amorphous solids are summarized below. Some of them may not be applicable in all cases, either because of practical difficulties or due to side effects such as chemical degradation. Nevertheless, it can be of interest to compare the properties and behaviors of solids obtained by different procedures in order to gain greater understanding of amorphous solids and their potential applicability.

10.3.1
Preparation from a Liquid Phase: Quench-cooling

This method, also called vitrification, is the most usual procedure for preparing glassy solids. It consists of cooling the molten phase (the initial temperature is therefore above T_m) of a pure compound below its glass-transition temperature (T_g) without crystallization. A decisive prerequisite is that no chemical decomposition occurs during melting. Decomposition occurs, for example, with sugars such as lactose. In other cases, it can be avoided if the melting step is carried out under nitrogen [36]. Progressive reduction of thermal motions and molecular mobility can be observed at a macroscopic scale by a continuous evolution of the liquid towards a syrup with increasing viscoelasticity, and a rubbery or malleable state characteristic of the supercooled liquid. Below T_g, the viscosity is usually high [13], and the frozen liquid then appears as a brittle glass.

The cooling rate that should be applied to prepare a glassy solid depends on the spontaneous tendency of each compound to crystallize, which is partially determined by the conformational diversity existing in the liquid phase [24] and by the nucleation ability. Quite often, it is advisable to perform a quench-cooling within a few seconds from above T_m to liquid nitrogen temperature [37], as has been reported for well-known pharmaceutical compounds such as indomethacin [20], felodipine [38], nifedipine [39] and trehalose [40]. However, the quench step is not always necessary, and a slower cooling rate to room temperature is for instance sufficient for permethylated β-cyclodextrin and benperidol [41]. For very good glass-forming compounds, a cooling rate as low as 10 K day^{-1} has been reported [42]. Kuhnert-Brandstaetter [43] has given other examples where vitrification was observed by means of hot stage microscopy.

10.3.2
From a Solution: Rapid Precipitation

Provided very high supersaturation can be reached, standard crystallization or precipitation methods can induce the formation of solids exhibiting an extremely low crystallinity, or even being truly amorphous, as established by the existence of a glass transition. This situation is encountered with "difficult-to-crystallize" compounds (Section 10.6.1), but is rarely published! A few examples can, however, be found, and can result from the use of anti-solvents, as with calcium D,L-pantothenate [44], or from a swift change in pH. This last phenomenon has been reported for piretanide [45] and iopanoic acid [46] when an acidic solution is added to a NaOH solution containing the solute.

10.3.3
From a Frozen Solution: Freeze-drying (Lyophilization)

Freeze-drying has been widely used during the last fifty years in the food and pharmaceutical industries to prepare solid forms able to produce almost instantaneously an aqueous solution just before or during human administration. Although experimental conditions have most often been determined empirically, several reviews [15, 47] have detailed some issues, therapeutic applications and risks associated with the use of a lyophilization process. The three successive steps that constitute this process are (a) freezing a solution (usually in water but organic solvents can be added) below the temperature of the eutectic invariant formed between the two constituents, (b) primary drying, obtained by sublimation of the frozen water (or vaporization of the organic solvent) at reduced pressure, and (c) secondary drying, consisting of the desorption of residual water or solvent at low pressure and higher temperature. Amorphization is more likely when the freezing step is rapid and performed at liquid nitrogen temperature, so that nucleation can be avoided. The other critical step is secondary drying, since crystallization can occur in the presence of residual solvent if the temperature is high enough. The strong influence of operating conditions during freeze-drying is illustrated by the case of pentamidine isothionate [10], showing that an amorphous form, a trihydrate or various polymorphic forms can occur during the successive steps. A usual strategy employed to inhibit crystallization consists in adding excipients such as cyclodextrins [48] or polymers, which can also prevent the chemical degradation of drug substances [49]. Guillory has summarized other examples of amorphous solids of pharmaceutical compounds obtained by lyophilization with or without additives [50].

10.3.4
From an Atomized Solution: Spray-drying

Spray-drying technology employs rapid drying from a concentrated solution (sometimes a suspension) by exposure to a heated atmosphere. Since drying should occur within seconds, the liquid phase is first atomized to obtain a maximal surface area in contact with the drying gas [50]. As in freeze-drying procedures, a glassy solid can be produced, but a major difference lies in the possibility of producing uniform spherical particles of desired size (usually a few μm), so that the obtained powder presents good flow properties [51]. Some of the properties and thermal behavior of spray-dried products have been reviewed by Corrigan [52].

Using amorphous samples prepared by spray-drying, the thermal and physical properties of important pharmaceutical excipients and food additives such as lactose and sucrose have been characterized [53]. However, amorphous or glassy solids obtained with this process may be of low physical stability and spontaneously crystallize, which is observed, e.g., for indomethacin [54]. It may be necessary to add suitable additives to inhibit recrystallization (magnesium stearate or microcrystalline cellulose are efficient in the case of lactose [55]).

Other techniques designed to prepare micro- or nano-particles have appeared more recently. For instance, in the RESS process (Rapid Expansion of Supercritical Solution) a supercritical solution is decompressed quickly. It can be used if the solute can be "dissolved" in supercritical CO_2. As with any technique in which kinetics prevail, RESS may yield amorphous materials, and this vitrification process can be adjusted to obtain a desired particle size distribution.

10.3.5
From a Crystalline Phase: Grinding and Milling

Reduction of particle size is a widespread process and is often one of the last steps during the production of pharmaceuticals. Grinding or milling can also be used at other stages of industrial development and manufacturing procedures to deliberately produce non-crystalline solids and to influence their properties and behaviors. A difficulty that can arise during characterization of these solids is the unambiguous distinction between microcrystalline forms, with mean particle size below or equal to 100 Å, and true amorphous solids having the structure of a frozen liquid. Indeed, X-ray powder diffraction might be unable to differentiate between these two states [25], and the detection of a glass transition by means of routine DSC might be difficult. This distinction is actually connected to the hypothetical mechanisms proposed to account for amorphization of organic compounds by grinding or milling, corresponding either to a stage-by-stage process of crystal lattice disorganization or to a contact melting mechanism followed by quenching of the melt [11].

Many authors, however, have reported that grinding or milling was prone to produce glassy solids, as exemplified by the case of trehalose [56] and other pharmaceutical compounds [57, 58]. Although some properties may differ when a glass is prepared by quench-cooling or by milling [59], these differences may be determined by distinct relaxation behaviors [17]. As with other methods of producing amorphous solids, experimental parameters such as the type of milling apparatus (vibratory ball mill, jet mill, etc.), the intensity, duration or "dose" of mechanical stress, the temperature and the use of suitable excipients can affect the physical (and sometimes the chemical) state of the ground product. For instance, it has been demonstrated that the percentage of disorder could be influenced by the milling equipment [60] and, for indomethacin, the temperature could determine the physical form of the resulting material [61, 62]. Other examples illustrating the diversity of experimental procedures have been collected [50].

In the 1970s and 1980s, Hüttenrauch [27] elaborated a self-consistent model devoted to the mechanical activation of organic and pharmaceutical systems. He postulated that the mechanical energy (so-called activation) transferred to the solids had two types of consequences: a minor part (about 10%) could generate a surface enlarging effect, whereas the major portion could contribute to the development of thermodynamically unstable states through the formation of lattice defects (dislocations, distortions) and structural disordering. He also highlighted that a mechanical activation process could be subdivided into two contributions, a dynamic part, which ceases as soon as the mechanical treatment is stopped, and a static activation that actually consists of a residual disorder, usually corresponding to an amorphous solid. In other words, it can be imagined that the material adapts to the stress by a recovery process and reaches a new steady state of "equilibrium" with new variables. The pertinent variable connected to the damage/recovery effect is the "dose", corresponding to the same concept as that used in radiology.

Successive surface effects and crystal lattice disruptions generated by milling and leading to a glass-like structure have been established in the case of sucrose [63]. Experimental investigations carried out with griseofulvin have indicated that the increase of free energy and the disordering of the solid structure could be limited to thin surface layers [64]. Often, surface alterations are the root cause of observed particle aggregation after mechanical treatments [65, 66]. More generally, milled samples are more reactive than native solids, as illustrated by the higher dissolution rates of cephalexin [67] and griseofulvin [64] in water. For pharmaceutical hydrates, numerous studies have revealed that grinding could decrease the dehydration temperature, or induce a combined partial dehydration and amorphization. A few examples of such effects are hydrated forms of ampicillin [68], lactitol [69], cephalexin [67], cefixime [70], citric acid and α-lactose [71]. For this last compound, Otsuka and coworkers have also noticed that grinding is associated with isomerization in the solid state [72]. More recently, however, it has been demonstrated that this chemical effect could be avoided during a ball milling treatment by careful preliminary dehydration of α-lactose monohydrate, so that the glassy state of pure α-lactose could be accurately characterized [73]. Notably, since large amounts of energy can be involved and transferred to solid phases during milling operations, chemical degradation is also likely to occur among organic compounds. This has been both studied and interpreted in the case of *p*-aminosalicylic acid [74]. Therefore, a fundamental requirement is that the chemical integrity of amorphous solids produced by milling and grinding should be controlled before further investigations, characterizations and interpretations.

10.3.6
From a Crystalline Solvate: Desolvation/Dehydration

The previous section outlined how mechanical treatments could be responsible for partial dehydration concomitant to amorphization. In some cases, such treatments are not necessary since dehydration (or desolvation) itself, obtained by heating, possibly at reduced pressure [44], can be employed to induce the formation of an amorphous solid. For carbamazepine, the detection of a glass transition is evidence that the solid obtained from the dihydrate by thermal dehydration under a nitrogen purge was in a glassy state [75].

Morris has suggested that dehydration-induced amorphization is likely to occur when solvent molecules are fully surrounded by drug molecules in the initial hydrate structure (isolated site hydrates) [76], and this situation was exemplified with cephradir dihydrate [77]. Yu has stated that dehydration is a "feasible and gentle route" for preparing amorphous solids [14], which can be illustrated with Tranilast, rendered amorphous by dehydration of its monohydrated form over P_2O_5 [78]. Comparison of raffinose amorphous solids obtained by freeze-drying and dehydration of the pentahydrate revealed that these two solids were identical [79]. A more recent investigation of the dehydration mechanism of this compound revealed the formation of an intermediate tetrahydrated form prior to structural collapse when further water molecules are released from the crystal lattice [80]. This situation differs from that of α,α-trehalose dihydrate, which can be rendered either amorphous [81] or transformed into a crystalline anhydrous form by careful removal of water followed by annealing [82]. Willart and coworkers [83] have further investigated the influence of operating conditions and kinetics of dehydration, showing also that the glassy trehalose obtained by rapid dehydration was similar to that resulting from a quench-cooling procedure. More recently, these authors suggested that the pertinent parameter allowing one to account for the formation of a glassy solid was the instantaneous rate of water release rather than the thermal treatment used [84].

A complementary approach that can be helpful in understanding amorphization resulting from a dehydration (or desolvation) process is in the detailed analysis of the mechanisms involved in dehydrations at a molecular scale. The unified model for the dehydration of molecular crystals published by the present authors suggests that amorphization arising from destruction of the crystal lattice can be a consequence of the departure of solvent molecules. This is consistent with the proposal by Morris [76], but it can also result from the lack of a reorganization step of the so-called "new anhydrous material" if appropriate criteria are not satisfied after a cooperative departure of the solvent [85].

10.3.7
Physical Mixture with Amorphous Excipients

In the pharmaceutical industry, it is quite common to prepare, usually by empirical procedures, physical mixtures of a drug substance with one or several excipients or drug carriers. These mixtures are called solid dispersions and are usually designed to obtain a pharmaceutical product exhibiting desired properties in terms of tableting, stability, ageing, dissolution, bioavailability, etc. Since these properties can be considered as satisfactory only if the physical mixture is non-crystalline, the most widespread excipients (PEGs, PVP, cellulose derivatives, etc.) in these multiple-component systems consist of polymers and amorphous solids. Rodriguez-Spong et al. [22] have recently overviewed the preparation and characterization of amorphous molecular dispersions.

The mechanisms responsible for kinetic stabilization of the amorphous state of drug compounds in solid dispersions, or for the amorphization of crystalline solids during the preparation of physical mixtures, are poorly understood. It is usually assumed that the amorphous material could "contaminate" the crystalline solid by means of intimate contact during grinding with excipients [50]. Nevertheless, the mechanisms resulting in inhibition of crystallization should be understood through investigations designed to identify the specific molecular interactions between drug and excipient molecules. This methodology has been nicely illustrated by Zografi and coworkers, who have established that nucleation and subsequent crystal growth of indomethacin in PVP solid dispersions can be prevented at room temperature by the formation of hydrogen bonds between PVP and indomethacin. As a result, the dimeric associations existing in glassy indomethacin as well as in the γ polymorph can no longer form in solid dispersions, which explains why the γ polymorph (usually obtained at room temperature) cannot crystallize from solid dispersions [19, 86].

10.4
Properties and Reactivity

Amorphous solids are high-energy states containing no long-range order and presenting molecular arrangements similar to that of liquids. Consequently, their properties and behaviors cannot be characterized according to the rules and methods used for crystalline solids. During the last fifteen years, theoretical concepts specific to the amorphous state have been developed to provide a more rational approach to the study of non-crystalline solids.

Generally, the properties of amorphous solids are relevant for the evaluation of their kinetic stability. The study of these properties may also be of major practical interest for the determination of adequate storage conditions and, if required, for the research of stabilization procedures.

10.4.1
The Glass Transition

As outlined in Section 10.2.2, the glass transition corresponds, by definition, to a change in physical state from a glassy (rigid, highly viscous, brittle) material to a more fluid or malleable rubbery state. Upon heating a glass, this transition may have dramatic consequences on the physical and chemical stability of the amorphous solid since the probability (or risk) of both crystallization and chemical degradation phenomena strongly increases. Knowledge of T_g is also a prerequisite for the design and monitoring of processes such as freeze-drying, spray-drying or the production of solid dispersions. The value of T_g relative to ambient temperature has also long ago been recognized to be of major importance for storage stability. Several techniques are useful for the detection of one or several changes of material properties during the glass transition [13]. Standard DSC remains the most popular and widely used method [87], but modulated DSC presents a significant advantage over conventional DSC due to its ability to separate the reversing and non-reversing components of the enthalpy contributions.

Although the glass transition is considered a thermodynamic requirement, ensuring that amorphous material cannot acquire a lower enthalpy than the crystalline state (the Kauzmann paradox [34]), this transition is a kinetic event. On heating, primary (C_p change) and secondary (overshoot) relaxation processes are observed in a narrow temperature range. The latter is assumed to be of molecular origin [88]. T_g may be significantly affected by kinetic parameters such as cooling and heating rates, whereas the relaxation endotherm [commonly observed upon heating and corresponding to a secondary (β) process], should be non-existent in the case of identical cooling and heating rates. The difficulty in identifying a reproducible T_g can be illustrated by the case of α,α-trehalose, for which published data rage from 79 to 115 °C [83]. Similarly, the measured T_g of spray-dried lactose varies by about 35 °C, depending on pan type [89]. With frusemide, different spray drying protocols induced a T_g change of about 10 °C [90]. One of the difficulties in the accurate determination of experimental T_gs is caused by the ubiquitous presence of water, since the presence of moisture in samples usually decreases T_g [91] significantly (Section 10.4.5).

The glass transition, as studied by thermal analysis methods, has been the subject of intense discussions; the main aspects of current knowledge can be found in several reviews [14–16]. Comparison of melting and glass transition temperatures for organic and pharmaceutical compounds has revealed that the ratio T_g/T_m is usually close to 2/3, whereas this ratio is about 1/2 for many polymers [16]. For ideal binary mixtures (no specific interactions) of amorphous components, the glass transition temperature of the mixture ($T_{g,mix}$) can be estimated by using the Gordon–Taylor equation (Eq. 1) [92]; w_1 and w_2 are the weight fractions, and $T_{g,1}$ and $T_{g,2}$ the glass transition temperatures of the components; k is a model dependent constant. Often $k = T_{g,1}/T_{g,2}$ is used.

$$T_{g,mix} = \frac{w_1 T_{g,1} + k w_2 T_{g,2}}{w_1 + k w_2} \quad (1)$$

Illustrative examples of this model have been discussed [93, 94].

10.4.2
Molecular Mobility and Structural Relaxation

The concept of molecular mobility is one of the most important aspects for a comprehensive approach to the amorphous state. It is directly connected to the notion of free volume or activation volume [27], since the mean molecular volume is larger within an amorphous solid than in the corresponding crystalline phase. Molecules are therefore in some ways loosely packed, and the resulting enhanced molecular mobility in disordered solids affects their chemical and physical reactivity [32]. Another major notion included in molecular mobility is that amorphous solids can never be considered as static solids, but consist, by definition, of dynamic systems, which means that their stability (or lifetime) is actually determined by kinetic aspects of their transformation towards a more stable state.

Molecular motions are obviously temperature dependent, and they undergo large changes in the vicinity of the glass transition. Estimated time scales of molecular motions are lower than 100 s above T_g, and exceed this limit value below T_g [13]. Their experimental determination is associated with the study of structural relaxation processes, which consist of a continuous evolution of real amorphous solids towards the more stable equilibrium liquid (or glass) at a given temperature. Structural relaxation is therefore determined by its degree of non-equilibrium, which in turn results from its conditions of formation.

Typically, the molecular relaxation time (τ) is used to characterize structural relaxation. For organic compounds, which usually consist of fragile glass formers (Section 10.4.3), variations of τ as a function of temperature in the region of the glass transition can be described on the basis of the Vogel–Tammann–Fulcher (VTF) equation, and the activation enthalpy of structural relaxation (ΔH^*) can be estimated from the slope of a $\ln(\tau)$ versus $1/T$ plot [13, 14]. When molecular motions are investigated by means of DSC, the measurement of enthalpy relaxation can be used to evaluate, for instance, the influence of isothermal ageing (or annealing) below T_g on the rate of relaxation [95, 96].

Studies of relaxation processes and the non-exponential behavior of relaxation rates have led to the development of the concept of heterogeneity among organic glasses [17]. Indeed, a homogeneous spatial distribution at the molecular scale in amorphous solids should lead to a single relaxation process, so that a distribution of relaxation times provides strong support for spatial heterogeneity, consisting for instance of clusters and density fluctuations [97, 98].

From a practical point of view determinations of relaxation times are of interest to evaluate the upper limit at which a glassy material is expected to be stable for months or even for a few years. This limit is assumed to be the Kauzmann "temperature" (point of negligible molecular mobility), and can either be determined experimentally from heat capacity and enthalpy of fusion data or predicted by using an adapted form of the VTF equation [99]. A rough approximation of T_K can also easily be obtained from $T_m/2$ or the well-known empirical rule T_g–50 K [13].

10.4.3 Strong/Fragile Classification of Angell

While molecular mobility is often related to the stability of an amorphous material below T_g [99, 100], the concept of fragility is an essential indicator for the characterization of physical changes above T_g. These two notions are important tools in the study of structural relaxation kinetics and mechanisms. The strong/fragile classification has been developed by plotting viscosities ($\log\eta$) or relaxation times ($\log\tau$) versus T_g/T for $T > T_g$ (Fig. 10.3) [101, 102]. If a linear relationship (quasi-Arrhenius behavior) is observed, the material presents a strong behavior, characterized by a small change of molecular mobility at T_g, a limited shift in heat capacity and is often associated with a self-reinforcing network in the liquid state. SiO_2 and GeO_2 are well-known strong glass formers, but some proteins also exhibit such behavior [103, 104].

A non-Arrhenius relationship is generally obtained with glasses of small organic molecules, which are classified as fragile [105, 106]. Structurally, these systems are characterized by non-directional intermolecular bonds and a conformational diversity, which may explain why fragile glass-forming compounds undergo marked changes during the glass transition, with larger changes in heat capacity.

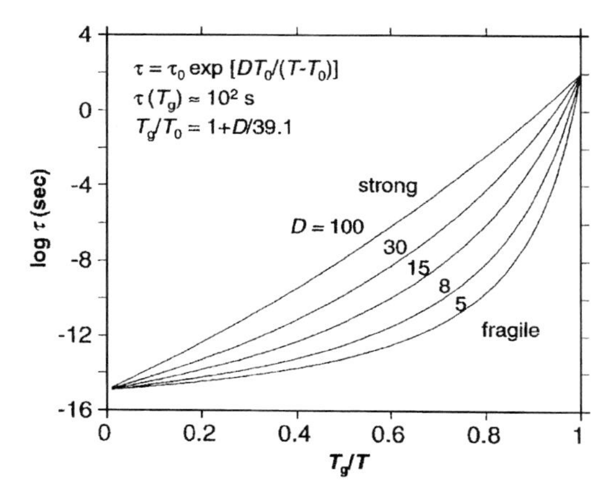

Fig. 10.3 Calculated strong/fragile pattern, showing the temperature dependence of structural relaxation times (τ) for supercooled liquids. The equation (upper-left corner) gives a good reproduction of experimental "Angell plots". (Reproduced from [14] with permission of Elsevier).

Although the strong/fragile character of a glass is supposed to be predictable from the values of T_m and T_g in Kelvin (fragile if the T_m/T_g ratio is less than 1.5; strong otherwise), the experimental approach to this property may require evaluation of viscoelastic changes, performed for instance by dynamic mechanical analysis (DMA). This method allows us to investigate the decrease in shear modulus through the glass transition, as well as the magnitude and phase relationship of the stress and strain values [35]. An example of the applicability of DMA is the study of the glass transition for amorphous indomethacin [107].

10.4.4
Mixing with Solvents/"Dissolution" Behavior

Note: Strictly speaking, the concept of solubility refers to an equilibrium between a saturated solution and a crystallized material M: $\langle M \rangle \Leftrightarrow (M)_{solvated}$. This steady state is temperature dependent and can correspond either to a metastable or to a stable thermodynamic equilibrium. When the two opposite mass transfers compensate each other and when the stable polymorphic form of the solid is considered, the state of the suspension is time independent. In other words, without modification of temperature, the suspension will remain unchanged for a long time. Therefore, the concept of solubility should not be applied to noncrystalline solids. Unfortunately, the term "solubility" is commonly used to define the maximum amount of amorphous solid that can be "dissolved" in a solvent within a given period. The terms solubility and dissolution are not appropriate in this context and will be used hereafter between inverted comas to remind the reader that this is by no means a heterogeneous equilibrium. Actually, the amount of solute in solution evolves with time towards a stable or a metastable equilibrium. Therefore the "solubility" of an amorphous material can be considered as a kinetically driven state that is in constant evolution. It can also be considered as the boundary, on the solvent side, of the miscibility gap between the solvent and the liquid-like solute. This latter interpretation refers to the limits of mixing of liquids and establishes a link with thermodynamics of heterogeneous equilibria.

Despite the associated difficulties, it is sometimes desirable to develop a drug substance as an amorphous form, in particular when the solubility and the dissolution rate of the crystalline form in aqueous medium are so poor that a therapeutic activity cannot be achieved. Novobiocin is a well-known example of such behavior; no detectable amount of the drug could be found in plasma when the crystalline form was administrated. By contrast, the quantity of amorphous form that can be "dissolved" was 70× higher if the particle size is lower than 10 μm, so that novobiocin could be rapidly absorbed [108].

Hancock and Parks [109] have investigated the "solubility" ratios between amorphous and crystalline forms of pharmaceutical compounds, and have compared these values with ratios obtained for different crystalline forms. Solubility ratios are in the range 1.1 to 2.8 for crystalline forms of the same compound. By contrast, the "solubility" ratios can vary, according to predicted values, from

10 to about 1600 when comparing an amorphous material with its crystalline counterpart. Experimental "solubilities" revealed much lower ratios, and this was attributed to difficulties in the determination of the "solubilities" for amorphous substances.

Interestingly, this study also confirmed that significant "solubility" enhancements and increased "dissolution" rates could also be established for only partially disorganized materials. From a study of milled griseofulvin, the effect of mechanical treatments on its "solubility" behavior revealed that an increased apparent "solubility" could be reached even though the disordering of the crystal lattice was limited to a thin external surface layer representing less than 6% of the sample mass [64]. However, the supersaturated solution slowly evolved to the stable solubility, indicating that the enhanced "dissolution" obtained by mechanical activation was a purely kinetic effect. A similar effect was observed with a macrolid antibiotic prepared as an amorphous solid [110].

10.4.5
Influence of Water Content: Plasticization and Chemical Degradation

The widespread hygroscopicity of organic amorphous solids is a result of the disorganized structural arrangement and the free volume of amorphous materials. Whereas only a surface adsorption effect is assumed to exist for crystalline solids, the hygroscopic character of amorphous materials may involve a deeper penetration of water, so that the weight fraction of absorbed water may reach much higher values [111]. Furthermore, the proportion of residual moisture may also increase with time, depending on storage temperature and atmospheric or surrounding humidity. Notably, the chemical affinity of the solute to water may enhance the hygroscopic character of a given compound.

It is well-known among scientists interested in the amorphous state of organic compounds that the water content may dramatically influence several properties that often determine the physical and chemical stability of amorphous solids [91]. A common and major effect of moisture uptake is a decrease in T_g [112, 113]. As stated above, T_g values of binary mixtures can be estimated by using the Gordon–Taylor equation (Eq. 1) [114]. Since this equation assumes no specific interactions between the two components (ideal behavior), comparison of measured and predicted T_gs may provide some insight into the existence of chemical or binding effects of water. Use of the Gordon–Taylor equation was shown, for instance, to be applicable to the trehalose–water binary system [115].

Owing to the low T_g of water (136 K), a dramatic decrease of T_g may be observed even at low levels of residual moisture [116], in particular for glasses with a T_g higher than room temperature. For such compounds, the decrease in T_g is usually associated with a plasticization that corresponds to an increase of the free volume and of molecular mobility. Exposure of these materials to humidity may reduce T_g to ambient temperature (the water content is then denoted w_g), which may lead to a rapid crystallization process. Both devitrification and crystallization have been investigated in detail for sucrose and lactose [53].

A study of the crystallization behavior of freeze-dried lactose at constant water content and humidity showed that the amount of water in the amorphous matrix was increased by the crystallization process, thereby further decreasing T_g and accelerating crystallization [117]. Since water may act as a plasticizer for many pharmaceutical glasses [116], it is important for storage purposes to take into account the combined effects of temperature and water content to limit the risk of a devitrification process, usually followed by crystallization [47, 111].

A less studied aspect related to absorbed moisture concerns the spatial distribution of water in amorphous solids. Shalaev and Zografi have recently highlighted that structural heterogeneities could imply that the concentration of water may strongly vary within an amorphous material [17]. This assumption is supported by a NMR study of the amorphous sucrose–water system, indicating that "pools" or "pockets" of water may exist, depending on temperature and sucrose/water proportions [118].

10.4.6 Polyamorphism

Debate about the existence of "true" or "apparent" polyamorphism is an opportunity for scientists to investigate several fundamentals and practical aspects related to the amorphous state. Indeed, it can be a stimulating challenge to establish the nature and origin of different thermal, physical or kinetic behaviors of two amorphous samples prepared by distinct routes. This understanding may generate new ideas for the optimization of manufacturing and stabilization procedures. For instance, one can imagine preparing, using suitable procedures, two amorphous samples of the same compound that present different molecular arrangements (monomers, dimers, cyclic systems, etc.). These two solids could be considered to show polyamorphism.

The possibility of different amorphous structures or phases can be envisaged from several points of view, including the existence of distinct molecular arrangements, of different thermodynamic and energetic properties, and of specific physical, thermal and kinetic behaviors. Polyamorphism also implies the possible observation and characterization of solid–solid transitions occurring between amorphous materials. In the definition proposed by Angell [106, 119], polyamorphism refers to the existence of two glassy forms of the same compound differentiated by their fragility: one is a strong glass former with a low density and the other is a fragile glass former of higher density. The existence of true polyamorphic compounds, characterized by a first-order transition between the two amorphous states, has been exemplified with inorganic compounds, in particular water [120, 121], silica [122] and carbon [123].

Such solid–solid transitions between glassy solids have not yet been reported for organic compounds or polymers, and are probably rare in the case of pharmaceutical substances. However, Shalaev and Zografi have recently discussed the possibility of first-order-like transitions between liquid phases [17], and have illustrated this phenomenon with polymers [124] and triphenyl phosphite [125].

For this last compound, a temperature range of the transition could be determined between T_g and T_m, different physical properties (viscosity, relaxation times, density) and an induction period could also be established, and phase boundaries were deduced from microscopy observations. This true polyamorphic behavior in the supercooled liquid state could be of general relevance among organic compounds, in particular carbohydrates since it was postulated by Shalaev and Zografi that it could be related to different conformational states of the molecules in amorphous solids [17]. Further experimental investigations are, however, required to improve our understanding of such first-order-like transitions.

By contrast, many authors have reported that amorphous forms produced or stored in different conditions could exhibit distinct properties. Changing, for instance, the cooling rate during the formation of glassy felodipine affected the relaxation endotherm observed by DSC at a constant heating rate [126]. The influence of annealing at a temperature slightly below T_g was also established and can be exemplified by the study of indapamide [127]. Comparison of amorphous permethylated β-cyclodextrin prepared either by quench-cooling or by milling revealed different relaxation enthalpies and crystallization rates despite similar T_g and heat capacity values [59].

The numerous examples of such thermal and kinetic effects can only be interpreted as a consequence of different kinetic states of amorphous materials, associated with continuous changes of various properties. The so-called "relaxation polyamorphism" should be clearly distinguished from true polyamorphism since no transition can be expected in this case. The relaxation process actually depicts the dynamics of the strain release (produced for instance by rapid cooling) that may occur when the molecular mobility of the glass is sufficient to allow an evolution of the system towards the equilibrium state of the liquid or glassy phase.

10.5 Characterization and Quantification

The two usual methods employed to assess the absence of crystallinity in a solid are optical microscopy and X-ray powder diffraction (XRPD). Using cross-polarized light, the absence of birefringence is a strong indicator of an amorphous state, although birefringence may also be non-existent for high symmetry (cubic) crystals, which are rare among organic molecules. Conventional XRPD analysis is also an easy way to establish that partial or complete amorphization has occurred. For further characterization of amorphous solids, many other techniques may be used, including small-angle X-ray diffraction, neutron scattering, pycnometry, viscosity measurements, diffusion cells, solid-state NMR, molecular probes, Raman and IR spectroscopies, electron spin resonance, DSC, microcalorimetry and vapor sorption studies. The advantages and limitations of these methods have been discussed by Hancock and Zografi [13], who also highlighted that only a combination of several techniques may provide a sufficient degree of confidence for satisfying interpretations. Recently, Reid [128] has also pointed out several key issues related to the characterization of amorphous aqueous systems, in terms of size, representativity and uniformity of the studied sample, as well as the possible "distortions of reality" resulting from the inherent characteristics of the coupling between sample and measuring system.

10.5.1 Thermal Analysis and Spectroscopic Methods

Energetic aspects are traditionally investigated by means of thermal analysis techniques [87]. The influence of various factors on the glass transition has been frequently reviewed [14–16, 52]. However, DSC and DTA may also be used to detect and quantify crystal defects and disruptions [28].

Although the interpretation of data provided by techniques such as IR and Raman spectroscopies may be difficult, these methods can constitute complementary tools for the identification of amorphous materials. Dielectric relaxation and dynamical mechanical spectroscopy can also be helpful for investigating both the glass transition and the molecular motions involved in secondary relaxation processes [129, 130]. However, one of the most important methods for the analysis of molecular mobility is solid-state NMR, since the determination of spin–lattice relaxation times (T_1) as a function of temperature can provide a good estimate of changes in rotational and translational motions existing in the studied amorphous system [32, 118].

10.5.2 Detection and Quantification of Small Amorphous Contents

When processing may induce a partial amorphization of a highly crystalline material (often on the surface of particles), the detection and quantification of small amounts of amorphous product may be critical for pharmaceutical applications. Most quantitative techniques require the construction of a calibration curve obtained from physical mixtures of purely amorphous and purely crystalline samples. Using X-ray diffraction methods, the limit of detection is generally in the range 5–10% [131]. Conventional DSC can also be employed to identify proportions of about 5%, either by detecting the crystallization peak of the amorphous part or by comparing melting enthalpies of the crystalline form [87]. Although DSC is assumed to provide more accurate results [132], the limit of detection remains relatively high.

Actually, the most suitable method for the detection of smaller proportions of amorphous content is isothermal microcalorimetry, which permits an accurate quantification of 1–2% disorder in a crystalline sample [133]. This method is based on measurement of the heat released when the amorphous part crystallizes after absorption of water under high humidity. Although time consuming, microcalorimetry has been validated as an efficient analytical tool for several drug substances [134], and has also been used to investigate the influence of small amounts of amorphous lactose on solid–solid transformations occurring between crystalline forms of this excipient [135].

10.6
Crystallization of Amorphous Solids

Thermodynamically, the intrinsic instability of amorphous solids should induce, in principle, their irreversible evolution towards a (more stable) crystalline state. The crystallization process, which can be envisaged as a deactivation phenomenon [27, 53], implies the release of excess energy stored in the amorphous material, associated with an entropy loss. As for any crystallization, successive nucleation and growth steps are involved, and can be solvent mediated [136]. In connection with the physical stability of the material, the crystallization rate may be affected by numerous, possibly interdependent parameters, among which temperature plays a key role. In particular, the nucleation rate in a supercooled liquid increases exponentially when the temperature decreases from T_m towards T_g, whereas molecular motions required for the growth step vary in an opposite way in the same temperature range (Fig. 10.4) [13, 137]. However, many different crystallization behaviors can be encountered among amorphous solids, and general tendencies as well as illustrative examples have been depicted in previous reviews ([13, 14] and references therein). Provided chemical degradation does not occur, amorphous or glassy materials can also constitute a valuable starting point for screening procedures designed to crystallize new polymorphs of a compound.

Depending on the specific objectives and issues within the framework of a particular piece of research or development project, it may be advisable to promote the crystallization of an amorphous substance or, conversely, to prevent

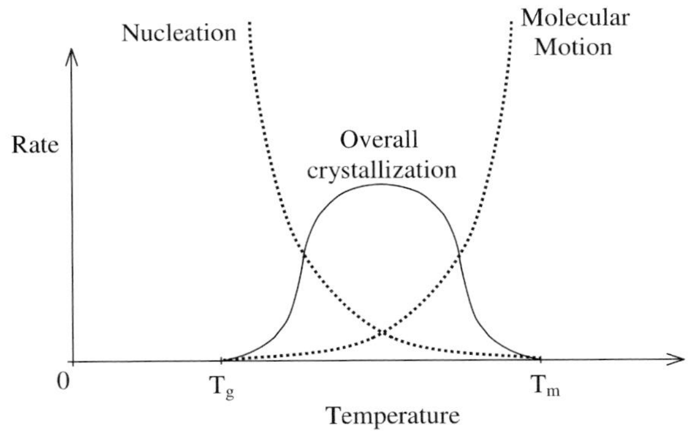

Fig. 10.4 Influence of temperature on the nucleation rate, magnitude of molecular motions and crystallization rate for supercooled liquids. (After [13], adapted from [137]).

any spontaneous crystallization during processing and storage. These two opposite situations are highly relevant in the pharmaceutical and fine chemicals industries since they may be very time consuming. Although no systematic procedure exists, it may be helpful to propose some strategies and guidelines based on knowledge of the amorphous state.

10.6.1 "Difficult-to-crystallize" Compounds

It is not always easy to discern why a particular compound readily crystallizes or not. Comparison between the crystallization behaviors of mannitol and sorbitol is often used to illustrate that closely related compounds may exhibit opposite crystallization tendencies [138]. To a certain extent, the difficulties encountered when trying to crystallize an amorphous solid can also be analyzed in a similar way as so-called "disappearing polymorphs" [139] since a key issue lies in finding suitable conditions for the spontaneous nucleation and growth of a crystalline solid. The present authors have come across an unusual example, namely the case of a pharmaceutical compound that could only be crystallized as an ethanol solvate. In any other medium, and whatever the experimental conditions, only gels could be prepared.

As a general rule, it can be assumed that two main factors may be involved in crystallization difficulties. The first is related to the presence of chemical impurities. The formation of one or several by-products that cannot be easily removed by standard purification procedures is not uncommon during the last stages of chemical synthesis. These impurities may act as nucleation inhibitors, and can therefore be responsible for amorphization upon solidification. In such cases, an improved purification of the crude samples can be sufficient and is sometimes the only way to remove this impurity effect.

The second factor that is often responsible for difficulties in crystallizing an organic compound involves its conformational diversity in solution or in the melt. Polysaccharides as well as synthetic polymers are molecular families known to exhibit numerous conformational possibilities, which is usually associated with an amorphous character of the corresponding solids. When this situation is suspected, the preferred strategy is to search for crystallization conditions (mainly a solvent or a mixture of solvents) that could limit the conformational distribution. Another possibility is to perform seeding experiments with related compounds.

Finally, attempts designed to crystallize an amorphous solid require preliminary investigations concerning the chemical and physical behavior of the studied compound. In particular, the main properties, such as the T_g, and the chemical degradation rate, for instance upon heating, should be carefully determined before further crystallization attempts.

10.6.2
Inadvertent Crystallization

Our limited knowledge of the molecular arrangement of amorphous solids that can undergo spontaneous crystallization makes it difficult to evaluate and predict the conditions that promote this crystallization. Actually, numerous parameters may be involved, including the history of each sample, the manufacturing process, the physical stimuli (ultrasound, cycling temperature programs, electromagnetic radiation, electric field, very high pressure, and so on) as well as the storage temperature. Just as for "difficult-to-crystallize" compounds, the physical characterization of the studied amorphous solid and knowledge of its thermal and hygroscopic behaviors are prerequisites for understanding its crystallization. Although in the temperature range between T_g and T_m the probability for spontaneous crystallization is highest, it must be kept in mind that an amorphous solid may crystallize below T_g, e.g., as reported for indomethacin [140].

In connection with the free volume existing in amorphous compounds, a critical parameter for inadvertent crystallization is the molecular mobility at a given temperature. Despite the lack of experimental data, it can also be envisaged that structural heterogeneities as well as organized molecular clusters could play a decisive role in spontaneous nucleation occurring in the solid state. Recent investigations devoted to the crystallization behavior of indomethacin support this statement [20].

Several authors have noted that inadvertent crystallization is likely to be avoided if the sample is maintained at $T_g - 50$ K or at the "zero-mobility" (T_o) temperature, which is assumed to be close to the Kauzmann temperature (T_K) [13, 14]. However, these rules can only be applied for long-term storage if technical conditions allow the prevention of any moisture absorption. It has been often pointed out that even limited water uptake can dramatically decrease the T_g, and therefore increase the risk of inadvertent crystallization.

When possible, the preferred strategy for stabilizing amorphous compounds susceptible to undergoing crystallization upon storage is to prepare mixtures with excipients of high T_g, such as PVP, trehalose, sorbitol, etc. [14]. However, this type of procedure remains empirical and, at present, cannot replace careful stability studies.

10.6.3
Crystallization as a Tool for Insight into the Amorphous State

During the last twenty years, both theoretical and technological advances have allowed important progress in understanding the crystalline state. In particular, the use of crystal structure determinations by X-ray diffraction techniques as a routine tool has stimulated the development of the "crystal engineering" approach, which is based mainly on a rational description of structural elements among crystal packings of organic compounds [141, 142]. The accuracy of structural descriptions has also enabled improved development strategies, e.g., through an easier, more reliable interpretation of physicochemical and thermal behaviors. This statement is also illustrated by the possibility to predict, to control and to modify the crystal shape of particles, in some cases by using adapted crystallization procedures [143, 144].

For amorphous solids, an equivalent accuracy in structural data cannot be envisaged. However, one may hope that our understanding of the physicochemical features of glassy solids can be improved by using knowledge deduced from detailed crystallization or devitrification studies. The crystallization process is usually considered to be comparable to that occurring from a melt, and the main factors involved have been described by several authors, together with the fundamental formulae that allow expression of the associated energetic quantities [13–15].

Several recent papers illustrate the possibility of obtaining insight into the amorphous state from the study of crystallization behaviors, sometimes in connection with polymorphic behavior and the formation of hydrates. For instance, Legrand and coworkers [42] have investigated the nucleation and growth of *m*-toluidine, finding high nucleation rates more than 30 K below T_g, whereas nuclei growth was mainly detected about 20 K below the melting temperature, i.e., more than 70 K above the nucleation temperature. Another interesting aspect of this study is that nucleation was promoted by the formation of macroscopic cracks in the glass, indicating a heterogeneous nucleation process, followed by a growth step for a few nuclei, occurring at a temperature at which the cracks can no longer be observed. It was therefore suggested that a decrease of the nucleation barrier was induced by the formation of cracks at low temperature. The importance of such "hot spots" has been highlighted by other authors [53]. However, the parameters most often invoked for their influence on the crystallization process are the moisture content, its effect on T_g, the possible interactions between various components in solid dispersions, and the conformational distribution [91]. The study of carbohydrates has also revealed that the existence of several crystal forms (polymorphs or hydrates) could complicate the understanding of their crystallization, but could also provide some important insights about their physicochemical properties [14]. In this regard, research conducted by Zografi and coworkers on indomethacin represents a unique and nice example, illustrating how spectroscopic, structural and crystallization studies can lead to a progressive elucidation of some structural features of the glassy state. In particular, a rational explanation for the crystallization behavior of indomethacin could be proposed, based on the pre-existence or absence of structural elements (dimers) in the amorphous state [19, 20].

Other examples of attempts to gain a deeper knowledge of the amorphous state based on the study of local structures both in the amorphous and crystalline states have been reviewed recently by Shalaev and Zografi [17]. These developments constitute, in our opinion, one of the most promising aspects of research devoted to the amorphous state.

References

1 W.C. McCrone, Polymorphism, in *Physics and Chemistry of the Organic Solid State*, D. Fox, M.M. Labes, A. Weissberger (Eds.), Interscience, New York, Vol. 2, pp. 725–767, **1965**.
2 J.K. Haleblian, W.C. McCrone, *J. Pharm. Sci.*, **1969**, *58*, 911–929.
3 M.J. Buerger, *Trans. Am. Crystallogr. Assoc.*, **1971**, *7*, 1–23.
4 M. Otsuka, N. Kaneniwa, *Int. J. Pharm.*, **1990**, *124*, 307–310.
5 T.P. Shakhtshneider, V.V. Boldyrev, *Drug Dev. Ind. Pharm.*, **1993**, *19*, 2055–2067.
6 T.P. Shakhtshneider, *Solid State Ionics*, **1997**, *101–103*, 851–856.
7 A. Miyamae, S. Kitamura, T. Tada, S. Koda, T. Yasuda, *J. Pharm. Sci.*, **1991**, *80*, 995–1000.
8 K.R. Morris, A.W. Newman, D.E. Bugay, S.A. Ranadive, A.K. Singh, M. Szyper, S.A. Varia, H.G. Brittain, A.T.M. Serajuddin, *Int. J. Pharm.*, **1994**, *108*, 195–206.
9 Y. Matsuda, S. Kawaguchi, H. Kobayashi, J. Nishijo, *J. Pharm. Sci.*, **1984**, *73*, 173–179.
10 S. Chongprasert, U.J. Griesser, A.T. Bottorff, N.A. Williams, S.R. Byrn, S.L. Nail, *J. Pharm. Sci.*, **1998**, *87*, 1155–1160.
11 T.P. Shakhtshneider, V.V. Boldyrev, Mechanochemical synthesis and mechanical activation of drugs, *Reactivity of Molecular Solids*, E. Boldyreva, V. Boldyrev, (Eds.), John Wiley & Sons, New York, Chapter 8, pp. 271–311, **1999**.
12 K.R. Morris, U.J. Griesser, C.J. Eckhardt, J.G. Stowell, *Adv. Drug Deliv. Rev.*, **2001**, *48*, 91–114.
13 B.C. Hancock, G. Zografi, *J. Pharm. Sci.*, **1997**, *86*, 1–12.
14 L. Yu, *Adv. Drug Delivery Rev.*, **2001**, *48*, 27–42.
15 D.Q.M. Craig, P.G. Royall, V.L. Kett, M.L. Hopton, *Int. J. Pharm.*, **1999**, *179*, 179–207.
16 J. Kerč, S. Srčič, *Thermochim. Acta*, **1995**, *248*, 81–95.
17 E. Shalaev, G. Zografi, The concept of structure in amorphous solids from the perspective of the pharmaceutical sciences, in *Amorphous Food and Pharmaceutical Systems*, The Royal Society of Chemistry, Cambridge, pp. 11–30, **2002**.
18 K.J. Rao, B.G. Rao, *Bull. Mater. Sci.*, **1985**, *7*, 353–365.
19 L.S. Taylor, G. Zografi, *Pharm. Res.*, **1997**, *14*, 1691–1698.
20 V. Andronis, G. Zografi, *J. Non-Cryst. Solids*, **2000**, *271*, 236–248.
21 J.J. Seyer, P.E. Luner, M.S. Kemper, *J. Pharm. Sci.*, **2000**, *89*, 1305–1316.
22 B. Rodriguez-Spong, C.P. Price, A. Jayasankar, A.J. Matzger, N. Rodriguez-Hornedo, *Adv. Drug Delivery Rev.*, **2004**, *56*, 241–274.
23 T.L. Threlfall, *Analyst*, **1995**, *120*, 2435–2460.
24 L. Yu, S.M. Reutzel-Edens, C.A. Mitchell, *Org. Proc. Res. Dev.*, **2000**, *4*, 396–402.
25 S.R. Byrn, *Solid-State Chemistry of Drugs*, Academic Press, New York, **1982**.
26 G.P. Johari, S. Ram, G. Astl, E. Mayer, *J. Non-Cryst. Solids*, **1990**, *116*, 282–285.
27 R. Hüttenrauch, S. Fricke, P. Zielke, *Pharm. Res.*, **1985**, 302–306.
28 S.P. Duddu, D.J.W. Grant, *Thermochim. Acta*, **1995**, *248*, 131–145.
29 H. Tanaka, *J. Chem. Phys.*, **1999**, *111*, 3163–3174 and 3175–3182.
30 S. Chandrasekhar, *Liquid Crystals*, 2nd edn., Cambridge University Press, Cambridge, **1992**.
31 H. Szwarc, C. Bessada, *Thermochim. Acta*, **1995**, *266*, 1–8.
32 S.R. Byrn, R.R. Pfeiffer, G. Stephenson, D.J.W. Grant, W.B. Gleason, *Chem. Mater.*, **1994**, *6*, 1148–1158.
33 B. Wunderlich, *Thermochim. Acta*, **1999**, *340–341*, 37–52.
34 W. Kauzmann, *Chem. Rev.*, **1948**, *43*, 219–256.
35 D.Q.M. Craig, *Thermochim. Acta*, **1995**, *248*, 189–203.
36 F. Giordano, A. Rossi, I. Pasquali, R. Bettini, E. Frigo, A. Gazzaniga, M.E. Sangalli, V. Mileo, S. Catinella, *J. Therm. Anal. Cal.*, **2003**, *73*, 509–518.
37 M.D. Ediger, C.A. Angell, S.R. Nagel, *J. Phys. Chem.*, **1996**, *100*, 13200–13212.

38 S. Srčič, J. Kerč, U. Urleb, I. Zupančič, G. Lahajnar, B. Kofler, J. Šmid-Korbar, *Int. J. Pharm.* **1992**, *87*, 1–10.
39 Y. Aso, S. Yoshioka, T. Otsuka, S. Kojima, *Chem. Pharm. Bull.*, **1995**, *43*, 300–303.
40 J. L. Green, C. A. Angell, *J. Phys. Chem.*, **1989**, *93*, 2880–2882.
41 A. E. H. Gassim, P. G. Takla, K. C. James, *Int. J. Pharm.* **1986**, *34*, 23–28.
42 V. Legrand, M. Descamps, C. Alba-Simionesco, *Thermochim. Acta*, **1997**, *307*, 77–83.
43 M. Kuhnert-Brandstaetter, *Thermomicroscopy in the Analysis of Pharmaceuticals*, International Series of Monographs in Analytical Chemistry, Pergamon Press, Oxford, Vol. 45, **1971**.
44 M. Inagaki, *Chem. Pharm. Bull.*, **1977**, *25*, 1001–1009.
45 Y. Chikaraishi, M. Otsuka, Y. Matsuda, *Chem. Pharm. Bull.*, **1996**, *44*, 1614–1617.
46 W. C. Stagner, J. K. Guillory, *J. Pharm. Sci.*, **1979**, *68*, 1005–1009.
47 M. J. Pikal, Impact of polymorphism on the quality of lyophilized products, in *Polymorphism in Pharmaceutical Solids* (Drugs and the Pharmaceutical Sciences, Vol. 95), H. G. Brittain (Ed.), Marcel Dekker, New York, Chapter 10, pp. 395–419, **1999**.
48 T. Oguchi, M. Okada, E. Yonemochi, K. Yamamoto, Y. Nakai, *Int. J. Pharm.*, **1990**, *61*, 27–34.
49 T. J. Anchordoquy, J. F. Carpenter, *Arch. Biochem. Biophys.*, **1996**, *332*, 231–238.
50 J. K. Guillory, Generation of polymorphs, hydrates, solvates and amorphous solids, in *Polymorphism in Pharmaceutical Solids* (Drugs and the Pharmaceutical Sciences, Vol. 95), H. G. Brittain (Ed.), Marcel Dekker, New York, Chapter 5, pp. 183–226, **1999**.
51 H. G. Brittain, S. J. Bogdanowich, D. E. Bugay, J. De Vincentis, G. Lewen, A. W. Newman, *Pharm. Res.*, **1991**, *8*, 963–973.
52 O. I. Corrigan, *Thermochim. Acta*, **1995**, *248*, 245–258.
53 A. A. Elamin, T. Sebhatu, C. Ahlneck, *Int. J. Pharm.*, **1995**, *119*, 25–36.
54 O. I. Corrigan, E. M. Holohan, M. R. Reilly, *Drug Dev. Ind. Pharm.*, **1985**, *11*, 677–695.
55 G. Buckton, P. Darcy, *Int. J. Pharm.*, **1995**, *121*, 81–87.
56 J. F. Willart, A. De Gusseme, S. Hemon, G. Odou, F. Danede, M. Descamps, *Solid State Commun.*, **2001**, *119*, 501–505.
57 J. Font, J. Muntasell, E. Cesari, *Mater. Res. Bull.*, **1997**, *32*, 1691–1696.
58 I. Tsukushi, O. Yamamuro, T. Matsuo, *Solid State Commun.*, **1995**, *94*, 1013–1018.
59 I. Tsukushi, O. Yamamuro, H. Suga, *J. Non-Cryst. Solids*, **1994**, *175*, 187–194.
60 M. L. Ezerskii, A. V. Savitskaya, *Zh. Fizicheskoi Khim.*, **1992**, *66*, 3109–3114.
61 M. Otsuka, T. Matsumoto, N. Kaneniwa, *Chem. Pharm. Bull.*, **1986**, *34*, 1784–1793.
62 M. Otsuka, K. Otsuka, N. Kaneniwa, *Drug Dev. Ind. Pharm.*, **1994**, *20*, 1649–1660.
63 K. Dialer, K. Kuessner, *Kolloid Z. Z. Polym.*, **1973**, *251*, 710–715.
64 A. A. Elamin, C. Ahlneck, G. Alderborn, C. Nystroem, *Int. J. Pharm.*, **1994**, *111*, 159–170.
65 A. V. Gubskaya, Y. V. Lisnyak, Y. P. Blagoy, *Drug Dev. Ind. Pharm.*, **1995**, *21*, 1953–1964.
66 A. V. Gubskaya, K. A. Chishko, Y. V. Lisnyak, Y. P. Blagoy, *Drug Dev. Ind. Pharm.*, **1995**, *21*, 1965–1974.
67 M. Otsuka, N. Kaneniwa, *Chem. Pharm. Bull.*, **1984**, *32*, 1071–1079.
68 Y. Takahashi, K. Nakashima, H. Nakagawa, I. Sugimoto, *Chem. Pharm. Bull.*, **1984**, *32*, 4963–4970.
69 K. Yajima, A. Okahira, M. Hoshino, *Chem. Pharm. Bull.*, **1997**, *45*, 1677–1682.
70 S. Kitamura, A. Miyamae, S. Koda, Y. Morimoto, *Int. J. Pharm.*, **1989**, *56*, 125–134.
71 R. Huettenrauch, S. Fricke, *Int. J. Pharm. Technol. Prod. Manuf.*, **1981**, *2*, 35–37.
72 M. Otsuka, H. Ohtani, N. Kaneniwa, S. Higuchi, *J. Pharm. Pharmacol.*, **1991**, *43*, 148–153.
73 J. F. Willart, V. Caron, R. Lefort, F. Danède, D. Prévost, M. Descamps, *Solid State Commun.*, **2004**, *132*, 693–696.

74 Y. Nakai, S. Nakajima, K. Yamamoto, K. Terada, M. Suenaga, T. Kudoh, *Chem. Pharm. Bull.*, **1982**, *30*, 734–738.

75 Y. Li, J. Han, G. G. Z. Zhang, D. J. W. Grant, R. Suryanarayanan, *Pharm. Dev. Technol.*, **2000**, *5*, 257–266.

76 K. R. Morris, Structural aspects of hydrates and solvates, in *Polymorphism in Pharmaceutical Solids* (Drugs and the Pharmaceutical Sciences, Vol. 95), H. G. Brittain (Ed.), Marcel Dekker, New York, Chapter 4, pp. 125–181, **1999**.

77 K. Florey, Cephradine, in *Analytical Profiles of Drug Substances*, Academic Press, New York, Vol. 5, pp. 21–59, **1976**.

78 Y. Kawashima, T. Niwa, H. Takeuchi, T. Hino, Y. Itoh, S. Furuyama, *J. Pharm. Sci.*, **1991**, *80*, 472–478.

79 A. Saleki-Gerhardt, J. G. Stowell, S. R. Byrn, G. Zografi, *J. Pharm. Sci.*, **1995**, *84*, 318–323.

80 K. Kajiwara, A. Motegi, M. Sugie, F. Franks, S. Munekawa, T. Igarashi, A. Kishi, Studies on raffinose hydrates, in *Amorphous Food and Pharmaceutical Systems*, The Royal Society of Chemistry, Cambridge, pp. 121–130, **2002**.

81 S. Ding, J. Fan, L. Green, Q. Lu, C. A. Angell, *J. Thermal Anal.*, **1996**, *47*, 1391–1405.

82 F. Shafizadeh, R. A. Susott, *J. Org. Chem.*, **1973**, *38*, 3710–3715.

83 J. F. Willart, A. De Gusseme, S. Hemon, M. Descamps, F. Leveiller, A. Rameau, *J. Phys. Chem. B*, **2002**, *106*, 3365–3370.

84 J. F. Willart, F. Danede, A. De Gusseme, M. Descamps, C. Neves, *J. Phys. Chem. B*, **2003**, *107*, 11 158–11 162.

85 S. Petit, G. Coquerel, *Chem. Mater.*, **1996**, *8*, 2247–2258.

86 T. Matsumoto, G. Zografi, *Pharm. Res.*, **1999**, *16*, 1722–1728.

87 D. Giron, *Pharm. Sci. Technol. Today*, **1998**, *1*, 191–199.

88 G. P. Johari, M. Goldstein, *J. Chem. Phys.*, **1970**, *53*, 2372–2388.

89 V. L. Hill, D. Q. M. Craig, L. C. Feely, *Int. J. Pharm.*, **1998**, *161*, 95–107.

90 Y. Matsuda, M. Otsuka, M. Onoe, E. Tatsumi, *J. Pharm. Pharmacol.*, **1992**, *44*, 627–633.

91 D. Lechuga-Ballesteros, D. P. Miller, J. Zhang, Residual water in amorphous solids: measurement and effects on stability, in *Amorphous Food and Pharmaceutical Systems*, The Royal Society of Chemistry, Cambridge, pp. 275–316, **2002**.

92 J. L. Ford, *Drug Dev. Ind. Pharm.*, **1987**, *13*, 1741–1777.

93 E. Fukuoka, M. Makita, S. Yamamura, *Chem. Pharm. Bull.*, **1989**, *37*, 1047–1050.

94 R. J. Timko, N. G. Lordi, *Drug Dev. Ind. Pharm.*, **1984**, *10*, 425–451.

95 E. Fukuoka, M. Makita, S. Yamamura, *Chem. Pharm. Bull.*, **1986**, *34*, 4314–4321.

96 E. Fukuoka, M. Makita, Y. Nakamura, *Chem. Pharm. Bull.*, **1991**, *39*, 2087–2090.

97 M. D. Ediger, J. L. Skinner, *Science*, **2001**, *292*, 233–234.

98 E. W. Fischer, C. Becker, J. U. Hagenah, *Prog. Colloid Polym. Sci.*, **1989**, *80*, 198–208.

99 S. L. Shamblin, X. Tang, L. Chang, B. C. Hancock, M. J. Pikal, *J. Phys. Chem. B*, **1999**, *103*, 4113–4121.

100 B. C. Hancock, S. L. Shamblin, G. Zografi, *Pharm. Res.*, **1995**, *12*, 799–806.

101 C. A. Angell, *J. Non-Cryst. Solids*, **1991**, *131–133*, 13–31.

102 J. Wong, C. A. Angell, *Glass: Structure by Spectroscopy*, Marcel Dekker, New York, **1976**.

103 J. L. Green, J. Fan, C. A. Angell, *J. Phys. Chem.*, **1994**, *98*, 13780–13790.

104 P. F. Green, *J. Non-Cryst. Solids*, **1994**, *172–174*, 815–822.

105 C. A. Angell, *Proc. Natl. Acad. Sci. USA*, **1995**, *92*, 6675–6682.

106 C. A. Angell, *Science*, **1995**, *267*, 1924–1935.

107 V. Andronis, G. Zografi, *Pharm. Res.*, **1997**, *14*, 410–419.

108 J. K. Haleblian, *J. Pharm. Sci.*, **1975**, *64*, 1269–1288.

109 B. C. Hancock, M. Parks, *Pharm. Res.*, **2000**, *17*, 397–404.

110 T. Sato, A. Okada, K. Sekiguchi, Y. Tsuda, *Chem. Pharm. Bull.*, **1981**, *29*, 2675–2682.

111 C. Ahlneck, G. Zografi, *Int.J. Pharm.*, **1990**, *62*, 87–95.

112 L. Slade, H. Levine, *Pure Appl. Chem.*, **1988**, *60*, 1841–1864.

113 C.A. Oksanen, G. Zografi, *Pharm. Res.*, **1990**, *7*, 654–657.

114 M. Gordon, J.S. Taylor, *J. Appl. Chem.*, **1952**, *2*, 493–500.

115 T. Chen, A. Fowler, M. Toner, *Cryobiology*, **2000**, *40*, 277–282.

116 B.C. Hancock, G. Zografi, *Pharm. Res.*, **1994**, *11*, 471–477.

117 Y. Roos, M. Karel, *J. Food Sci.*, **1992**, *57*, 775–776.

118 D. Girlich, H.D. Luedemann, *Z. Naturforsch.*, **1994**, *49*, 250–257.

119 P.H. Poole, T. Grande, C.A. Angell, P.F. McMillan, *Science*, **1997**, *275*, 322–323.

120 O. Mishima, L.D. Calvert, E. Whalley, *Nature*, **1985**, *314*, 76–78.

121 O. Mishima, K. Takemura, K. Aoki, *Science*, **1991**, *254*, 406–408.

122 M. Grimsditch, *Phys. Rev. Lett.*, **1984**, *52*, 2379–2381.

123 M. van Thiel, F.H. Ree, *Phys. Rev. B*, **1993**, *48*, 3591–3599.

124 R.F. Boyer, *Plastics Eng.*, **1992**, *25*, 1–52.

125 I. Cohen, A. Ha, X. Zhao, M. Lee, T. Fischer, D. Kivelson, *J. Phys. Chem.*, **1996**, *100*, 8518–8526.

126 J. Kerč, S. Srčič, M. Mohar, J. Šmid-Korbar, *Int.J. Pharm.*, **1991**, *68*, 25–33.

127 A.T.M. Serajuddin, M. Rosoff, D. Mufson, *J. Pharm. Pharmacol.*, **1986**, *38*, 219–220.

128 D.S. Reid, Use, misuse and abuse of experimental approaches to studies of amorphous aqueous systems, in *Amorphous Food and Pharmaceutical Systems*, The Royal Society of Chemistry, Cambridge, pp. 325–338, **2002**.

129 D.Q.M. Craig, F.A. Johnson, *Thermochim. Acta*, **1995**, *248*, 97–115.

130 J. Alie, J. Menegotto, P. Cardon, H. Duplaa, A. Caron, C. Lacabanne, M. Bauer, *J. Pharm. Sci.*, **2004**, *93*, 218–233.

131 A. Saleki-Gerhardt, C. Ahlneck, G. Zografi, *Int.J. Pharm.*, **1994**, *101*, 237–247.

132 M.J. Pikal, A.L. Lukes, J.E. Lang, K. Gaines, *J. Pharm. Sci.*, **1978**, *67*, 767–769.

133 T. Sebhatu, M. Angberg, C. Ahlneck, *Int.J. Pharm.*, **1994**, *104*, 135–144.

134 D. Giron, P. Remy, S. Thomas, E. Vilette, *J. Thermal Anal.*, **1997**, *48*, 465–472.

135 M. Angberg, *Thermochim. Acta*, **1995**, *248*, 161–176.

136 J.W. Mullin, *Crystallization*, 3rd edn., Butterworth-Heinemann, Oxford, **1997**.

137 J.E. Jolley, *Photogr. Sci. Eng.*, **1970**, *14*, 169–177.

138 M. Siniti, S. Jabrane, J.M. Letoffe, *Thermochim. Acta*, **1999**, *325*, 171–180.

139 J.D. Dunitz, J. Bernstein, *Acc. Chem. Res.*, **1995**, *28*, 193–200.

140 M. Yoshioka, B.C. Hancock, G. Zografi, *J. Pharm. Sci.*, **1994**, *83*, 1700–1705.

141 M.C. Etter, J.C. MacDonald, J. Bernstein, *Acta Crystallogr. B, Struct. Commun.*, **1990**, *46*, 256–262.

142 G.R. Desiraju, *Angew. Chem. Int. Ed. Engl.*, **1995**, *34*, 2311–2327.

143 G. Clydesdale, K.J. Roberts, R. Docherty, *J. Crystal Growth*, **1996**, *166*, 78–83.

144 I. Weissbuch, R. Popovitz-Biro, M. Lahav, L. Leiserowitz, *Acta Crystallogr. B, Struct. Commun.*, **1995**, *51*, 115–148.

11
Approaches to Polymorphism Screening*

Rolf Hilfiker, Susan M. De Paul, and Martin Szelagiewicz

11.1
Introduction

We have seen in previous chapters and will see in the following ones that choosing, manufacturing and developing the optimal solid form is of paramount importance in order to obtain a product that has the desired physicochemical properties, is convenient and cheap to manufacture, is easy to handle, is stable, fulfills regulatory requirements, can be protected by a patent and does not infringe on existing patents.

Obviously, the first step in selecting the optimal form is to identify all relevant solid forms that a new compound (be it an uncharged molecule or a salt) can exist in. The probability that a compound can exist in several solid forms (polymorphs, solvates, hydrates, amorphous form, co-crystals) is probably close to 100%, considering that 56–87% of all small organic molecules can form solvates and polymorphs alone (Chapter 1). In the pharmaceutical industry, polymorphs and hydrates play a special role among possible solid forms since they are the most common ones used in finished products.

Different polymorphs can be formed when the crystallization conditions are changed [1–3]. This is explained by the famous Ostwald rule of stages, which states that "When leaving a given state and in transforming to another state, the state which is sought out is not the thermodynamically stable one, but the state nearest in stability to the original state" [4]. This implies that, during a crystallization process, a series of polymorphs can be obtained, starting from the thermodynamically least stable one and ending up with the stable polymorph. Which polymorph would be isolated in a certain crystallization process would then depend on when the crystals are harvested, i.e., a fast crystallization process would preferentially lead to a metastable form and a slow crystallization process to the stable form. Ostwald's rule of stages is not universal [5–7] and can be violated, depending on the relative widths and positions of the metastable zones. Nevertheless, it is a

* A list of abbreviations used is located at the end of this chapter.

Polymorphism: in the Pharmaceutical Industry. Edited by Rolf Hilfiker

ISBN: 3-527-31146-7

useful principle for planning polymorphism screenings. A more mechanistic view of the formation of different polymorphs takes into account cluster formation, nucleation and crystal growth (Chapter 2) [8–12]. This model also readily explains the influence of the solvent (if crystallization is performed from a solution) and additives [13] in addition to the crystallization speed. Clusters and nuclei are small and, therefore, their surface free energy significantly influences their total free energy. Solvent and/or additive molecules can preferentially adsorb on the surface of the nuclei and, thus, alter the heights of the activation barriers $\Delta G^*_{c,A}$ and $\Delta G^*_{c,B}$ in Fig. 2.6 (Chapter 2) so that in one solvent form A might be obtained preferentially while form B might be obtained in a second solvent using exactly the same crystallization conditions. Impurities [14–16] with a similar structure to the main component can have a particularly large effect since they may adsorb very strongly on the surface of certain nuclei due to their chemical similarity. Such impurities often arise from the final steps of the synthetic process. It is therefore crucial to perform crystallizations with various solvents to increase the chance of successful competition of solvent molecules with impurity molecules for preferential adsorption.

The relative stability of polymorphs (excluding stoichiometric and non-stoichiometric solvates), however, is not affected by the solvent since the solvent can only interact with the surface and the surface contribution to the free energy of a macroscopic entity such as a crystal is generally negligible. What is affected to a very large degree by the solvent is the conversion rate from a less into a more stable polymorph. This so-called solvent-mediated transition is controlled by (a) nucleation of the more stable polymorph, (b) crystal growth of the more stable polymorph and (c) dissolution of the less stable polymorph [17]. Depending on both the properties of the molecule that undergoes the transition and the solvent, (a), (b) or (c) can be the rate-limiting step. Often, the conversion is faster in solvents where the solubility is higher, but the relationship is not monotonic [17]. Other solvent parameters can also play a large role (Section 11.3). Finally, the solvent in which the crystallization is carried out is the deciding factor for the outcome of the crystallization process if a solvate can be formed with that particular solvent. So, in summary, the type of solvent can play a pivotal role for the outcome of a crystallization experiment from solution and should be chosen very carefully. For a reliable polymorphism screening both crystallization conditions and solvent type have to be varied as broadly as possible.

Needless to say, one can never be sure that one has identified all relevant polymorphs unless one carries out an infinite number of experiments, which essentially paraphrases McCrone's famous statement: "It is at least this author's opinion that every compound has different polymorphic forms and that, in general, the number of forms known for a given compound is proportional to the time and money spent in research on that compound" [18]. Nevertheless, there are a few examples of substances that have been crystallized countless times and for which no polymorphs have been found so far, such as naphthalene, sucrose [5] and aspirin [19].

11.2 Crystallization Methods

The choice of crystallization method has a major influence on which form is produced, and it therefore clearly makes sense to perform crystallizations using various methods when looking for polymorphs [19, 20]. Classical crystallization methods have been reviewed by Guillory [21]. They are listed in Table 11.1, together with the degrees of freedom available for each process.

Many of these processes [i.e., (i) to (v), (xi)] are also influenced by the initial solid form (i.e., polymorph, solvate, hydrate, or the amorphous form) that is used as this can affect the solubility and hence the degree of supersaturation.

In (i) to (v) appropriate solvents or solvent mixtures have to be chosen. It must be ensured that the substance is chemically stable in the given solvent or solvent mixture and that the solubility lies in an appropriate range for the given method. At the same time the solvent properties (Section 11.3) should be as diverse as possible.

For (i) the solubility should ideally be in the range 10–100 mg mL^{-1}. If the solubility is too low, yields will be impracticably small, and if it is too high, the

Table 11.1 "Classical" crystallization methods.

	Method	Degrees of freedom
(i)	Crystallization by cooling a solution	Solvent or solvent mixture type, cooling profile, temperature at start, temperature at end, concentration
(ii)	Evaporation	Solvent or solvent mixture type, initial concentration, evaporation rate, temperature, pressure, ambient relative humidity, surface area of evaporation vessel
(iii)	Precipitation	Solvent, anti-solvent, rate of addition, order of mixing, temperature
(iv)	Vapor diffusion	Solvents, rate and extent of diffusion (i.e., vapor pressure of non-solvent), temperature, concentration
(v)	Suspension equilibration (often also called "slurry ripening")	Solvent or solvent mixture type, temperature, ratio of solvent to solid, solubility, temperature programs, stirring/shaking rate, incubation time
(vi)	Crystallization from the melt	Temperature programs (min, max, gradients)
(vii)	Heat induced transformations	Temperature programs
(viii)	Sublimation	Temperature hot side, temperature cold side, gradient, pressure, surface type
(ix)	Desolvation of solvates	(a) "Dry": temperature, pressure (b) In suspension: see (v)
(x)	Salting out	Type of salt, amount and rate of addition, temperature, solvent or solvent mixture, concentration
(xi)	pH change	Temperature, rate of change, acid/base ratio, method: acid/base added as solution or in gaseous form
(xii)	Lyophilization	Solvent, initial concentration, temperature and pressure programs

viscosity might get so large that crystal growth is very slow or the resulting product becomes gel-like and hard to handle. Occasionally, there are very few individual solvents that fulfill these conditions. In such cases, solvent mixtures are highly advisable. Evaporation requires, preferably, solubilities of $>10\ mg\ mL^{-1}$, while for precipitation a solvent with a solubility $>10\ mg\ mL^{-1}$ and a second one with a solubility $<1\ mg\ mL^{-1}$ is, generally, required, and the two solvents should also be miscible. For vapor diffusion, the same conditions apply as for precipitation with the added requirement that the non-solvent has to be more volatile than the solvent. In suspension equilibration, where a solvent-mediated transition is aimed for, solubility should again be reasonably high (>1 but $<100\ mg\ mL^{-1}$), but one has to keep in mind that solubility is not the only deciding factor, as discussed above. Solvent mixtures are also often effective in this context.

Crystallization from the melt is of course only an option if the melt is thermally stable, and heat-induced transformations can also only be carried out within the thermal stability range. Sublimation requires a non-negligible vapor pressure of >0.1 Pa. According to Kuhnert-Brandstätter [22] this applies to about two-thirds of drug substances. For modern drug substances this number is significantly smaller since the average molecular weight of drug substances has increased over the years, with a concomitant decrease in vapor pressure. Obviously, (ix) requires the existence of a solvate, and (x) is only useful for compounds that are charged in solution. In (i) to (vi) and (x) seeds may be used very efficiently. They could consist of the desired solid form, a crystalline form of a similar compound or inorganic, organic or polymeric heteroseeds [23–26]. Of course, owing to regulatory requirements, heteroseeds would never be used for production of a drug substance. However, they can be used for producing seed crystals, which can subsequently be purified by a series of crystallizations.

In view of later processing and formulation steps, crystallization under pressure plays an important role, too. Often, grinding can change the polymorphic form [15]. Pressure-induced changes can be brought about by, for example, subjecting the sample to pressure in an IR-press, or grinding it in a mortar, a ball mill or a small-scale jet mill. Grinding a mixture of two components can also be an efficient way to produce co-crystals even in the absence of solvent [27–30].

In addition to these classical methods, newer methods such as capillary crystallization [31], laser-induced nucleation [32], supercritical fluid crystallization [33], epitaxial matching [34, 35], and ultrasound [36] have been successfully used to create elusive metastable as well as stable polymorphs.

11.3 Solvent Parameters

The importance of the solvent for the outcome of the crystallization process has been discussed in the introduction, and one of the first questions for a polymorphism screening is which solvents should be used. Ideally, solvents as diverse as possible should be included, which leads to the question of relevant solvent parameters. Clearly, in addition to molecular solvent–solute interactions, bulk properties of solvents may play a role. Gu [37] has examined 96 solvents in terms of eight relevant solvent properties: hydrogen bond acceptor propensity, hydrogen bond donor propensity, polarity/dipolarity, dipole moment, dielectric constant, viscosity, surface tension and cohesive energy density (calculated from the heat of vaporization). Based on all eight properties, the 96 solvents were sorted into 15 groups according to their statistical similarity. Some groups only contain one solvent (e.g., water), some several (e.g., *N*-methyl-2-pyrrolidone, dimethylformamide, dimethylacetamide, dimethyl sulfoxide), while others contain up to 16. Molecules with the same functional groups often were grouped together, as one would intuitively assume. Such an approach can be of great help for designing experiments and might suggest that it is advisable to use solvents from each group to maximize diversity. Since certain crystallization methods can only be used if the solvents meet certain solubility or boiling point requirements, groups containing several solvents are more likely to be utilized. If the classification is made by neglecting viscosity and surface tension, a slightly different grouping is obtained [37]. This demonstrates that it is important to choose the "right" parameters for classification. Other solvent descriptors such as molecular volume, shape, density, vapor pressure, melting point, boiling point, miscibility, solubility parameters, etc. may play a role, too. A recently proposed method for classification of solvents based on atomic electronegativity [38] allowed researchers to predict the outcome of certain crystallization experiments in terms of polymorphic form. Other approaches to solvent clustering have been suggested [39, 40]. Blomsma et al. used an approach where they characterized 413 solvents with 46 descriptors each, from which 16 principal components were derived. The matrix was then further reduced to five principal components and 22 solvents [41]. Clearly, some solvent properties have a greater influence than others on the outcome of crystallization experiments, and hydrogen donor and acceptor propensities seem to be of particular importance [17].

In summary, the classification of solvents into groups is a valuable approach for rational polymorphism screening. On top of that, it is also advisable to include solvents that are intended to be used in the formulation process, e.g., poly(ethylene glycol).

11.4 Systematic Polymorphism Screening

While conducting a polymorphism screening is required by regulatory guidelines (Q6A decision tree #4, Chapter 15) and is also highly advisable for other practical reasons, there are no universally accepted recipes for how to proceed. Clearly, a well structured and somewhat standardized way makes sense, while recognizing that the procedure has to be tailored with respect to the substance investigated. Every substance is different [42], and an approach not adapted to the substance is prone to be inefficient and to overlook relevant forms. The procedure used should also be tunable to permit various degrees of thoroughness in the search for all relevant polymorphs since at an early stage in development a smaller screening might make economical sense while at a later stage one wants to be highly confident that all relevant solid forms have been found (Chapter 1). In 1965 McCrone had already described a procedure for polymorph screening that included melt crystallization, heating, sublimation, slurries, and crystallization by rapid cooling [18]. Burger [43, 44] first suggested a flow-diagram for polymorphism screening. His method relied on hot-stage microscopy as a first step and crystallizations from solutions as a secondary approach. Various slow and quick crystallization methods from solution are recommended, such as shock cooling of a hot, saturated solution in ice water, slow cooling of a solution in an insulated device, precipitation and evaporation. All of the obtained forms are thoroughly characterized, and energy–temperature diagrams (Chapter 2) are constructed.

In our opinion, a polymorphism screening should at least contain the following elements:

- As a reference, the starting material should be characterized by several methods, such as XRD (powder or single-crystal), DSC, TG-FTIR or TG-MS, DVS, Raman or IR spectroscopy, magic-angle spinning (MAS) NMR, solubility measurements, microscopy, and HPLC (purity).

- Since hot-stage optical microscopy (often combined with a second method such as Raman spectroscopy) can be an efficient, fast way of generating new polymorphs, and since it provides a wealth of other information (Chapter 7), it is highly recommended [45]. Generally, more true polymorphic forms can be obtained by crystallizing from the super-cooled melt than by any other single method [19]. Sometimes, it is also only possible to obtain certain polymorphs from the melt. An example is form III of paracetamol [45, 46]. Obviously, this method is of very limited use if the molten substance is chemically unstable.

- The most time-consuming step in a classical polymorphism screening (as opposed to a high-throughput screening, see Section 11.5) is crystallization from solution. The "parameter space" both in terms of solvent properties and crystallization methods should be covered as broadly as possible to increase the probability of discovering all relevant forms. Solvents with highly diverse properties should be chosen – the concepts outlined in Section 11.3 are helpful. Often, solvent mixtures are useful in obtaining systems with suitable solubilities, polarities, etc.

  Just as important as solvent choice is the crystallization method, where both fast (e.g., precipitation, fast evaporation) and slow (e.g., slow cooling, suspension equilibration, vapor diffusion) methods should be chosen. Not all forms are equally important! Highly metastable forms with very low kinetic stability, for example, are often of little practical interest. Particularly impor-

tant, however, are the thermodynamically stable form (at room temperature) and hydrates – it would be disastrous to miss them in a screening. Thus, an emphasis on slow crystallization methods such as suspension equilibration is extremely advisable.
- Owing to their practical importance, hydrates deserve a special role in screening as well. Dynamic vapor sorption (DVS) and crystallizations from water and water–solvent mixtures, including suspension equilibration experiments, are suitable methods to create them.
- The new solid forms have to be characterized (Section 11.8).

In the case of suspension equilibration, experiments both where the initial form is crystalline and where it is amorphous (or at least highly metastable) are also recommended. For thermodynamic reasons, the system can only evolve towards more stable forms in suspension equilibration experiments. Therefore, if the starting material is crystalline, it is impossible to obtain a less stable polymorph than the initial one. If the starting material is amorphous, however, a much greater variety of forms may be obtained [47]. The amorphous form, which should be characterized anyway, can often be quite easily obtained by freeze-drying, quench cooling of the melt or very fast precipitation or evaporation. Solvates can also be useful high-energy forms for suspension equilibration experiments. Obviously, for that purpose a solvate with a different solvent than the one used for the suspension must be chosen.

The purity of the material used can strongly influence the outcome of the screening [48]. Impurities may hinder (or facilitate) nucleation and growth of certain forms [13]. A wide choice of solvents in the polymorphism screening can reduce the impact of impurities since a particular solvent may prevent adsorption of impurities. Nevertheless, it is advisable to repeat at least a few crucial steps of the polymorphism screening with a batch of final production quality (Chapter 1, Section 1.3.2). It is also a very good idea to confirm that no chemical degradation occurs during the crystallizations to avoid incorrect interpretation of the data [49].

It is important to be clear about the goals of the screening [50]. As mentioned in Chapter 1, it makes sense to adapt the depth of screening to the actual pharmaceutical development phase. Often, a screening performed early in development simply aims to identify the thermodynamically stable form and hydrates, and thus slow crystallization experiments should be favored over fast ones. Later pharmaceutical development phases require different depths of screening. Identification of metastable polymorphs and of solvates allows the determination of key parameters for designing the final crystallization procedure and ensures that the best possible intellectual property protection can be obtained.

11.5
High-throughput Methods

Because many crystallizations have to be performed, a reliable classical polymorphism screening can be rather time-consuming. In addition, about 5–10 g of substance is typically needed for such a study. This can be a limiting factor for studies in the late research or early development phase. Therefore, automation and miniaturization are needed to perform crystallizations in a high-throughput format. When designing such a system, it is critical to incorporate the best practices developed earlier and not to choose particular crystallization methods just because they can be easily automated. Even if experiments can be performed quickly and in parallel, they should still be designed carefully and adapted to the properties of the substance investigated. In particular, the choice of solvents and solvent mixtures should normally be different for every substance, taking into account solubilities, possible chemical reactions and specific solvent–solute interactions.

High-throughput polymorphism screening systems have been developed by research foundations [51], big pharma companies [48, 52, 53] and companies that specialize in solid-state research and development [23, 54–58]. Some systems also allow salt screening (Chapter 12), combined salt and polymorphism screening, and searches for co-crystals.

Both hardware and software components make up a high-throughput screening system.

Hardware components can include:
- a crystallization array, such as a modified 96-well microtiter plate, set of test tubes or similar;
- liquid and solid handling systems capable of dispensing solids and/or solvents, retracting solutions, etc.;
- tools to manipulate the samples in the array by shaking, heating/cooling, evaporating, filtering, sonicating, etc.;
- a mechanism to isolate or suitably prepare the samples for detection;
- detection system(s).

Software components can include:
- workflow design,
- classification of measured characteristics into groups,
- reporting.

Depending on requirements and objectives, the various procedures may be run by a fully robotic system or performed partially manually.

Important aspects for the design of the crystallization array are that cross-contamination between crystallization vessels and therefore unwanted seeding are avoided and that the containers are adequately and quickly sealed so that no uncontrolled solvent evaporation takes place. Ideally, the manipulation systems should allow the performance of all solvent-based crystallization methods described in Section 11.2. Commonly found in HTS systems are crystallization by cooling, evaporation, precipitation and suspension equilibration. Sometimes heteroseeding, e.g., with polymer templates, is used [25]. Primary detection methods for the solid forms are normally Raman microscopy or powder X-ray diffraction or both. Raman microscopy offers the advantage of spatial resolution within a single crystallization vessel and allows faster recording of spectra of small sample amounts [23]. It also does not require sample preparation, which could potentially change the polymorphic form (e.g., desolvation of labile solvates). Furthermore, is not very sensitive to orientation effects and allows measurements through quartz or high quality glass, permitting convenient measurement in a solvent-saturated atmosphere. When using X-ray diffraction, care must be taken to eliminate strong orientation effects, e.g., by rocking the sample during measurement. In addition to the primary detection methods, which are meant to give a fingerprint of the solid form, other characterization techniques are sometimes incorporated, such as melting point detection and coupling to HPLC equipment to measure the concentration of the solute in the supernatant and/or to detect chemical degradation.

A critical component of an HTS system is the software used for classifying the measured Raman spectra or X-ray powder diffractograms into groups. Each group should, of course, correspond to a distinct solid form (polymorph, solvate, amorphous form) although it may also represent a mixture of polymorphic forms or degradation products in some cases. Algorithms can be developed to compare the similarity of spectra according to peak positions or to perform principal component analysis of the spectra and rate their similarity according to the distances in the principal component space. Most companies that have developed HTS systems have also developed their own proprietary software. A commercial computer program (PolySNAP [59]) is available to sort powder diffractograms based on whole pattern comparison, using a weighted mean of parametric and non-parametric correlation coefficients [60]. All the calculations use the full measured data profile, and the effects of common shortcomings of the measured spectral quality due to poor signal-to-noise ratio, strong backgrounds, broad peaks, etc. are minimized.

Perhaps the key element of an HTS system is the intelligent selection of crystallization techniques and conditions as it is very easy to perform thousands of crystallizations and always end up with the same solid form. Irrespective of automation and miniaturization, it is never a good idea to perform useless experiments. The quality of crystallization experiments is at least as important as the quantity! A moderate number of well chosen experiments gives a better result than a huge number of meaningless ones.

Results of an HTS can form the basis for more complete polymorphism studies if the product proceeds in development. Such a comprehensive polymorphism investigation involves scaling up the HTS crystallizations and confirming the corresponding results, performing additional complementary crystallization experiments, carrying out extensive physicochemical characterization of the various forms (Section 11.8), and determining which form is thermodynamically stable at room temperature.

Storey summarizes the current status of HTS very well, "There is still a role for manual experimentation – high-throughput is not a substitute for thinking!" [52]. The combination of HTS with carefully designed supplementary experiments offers the greatest likelihood that all important polymorphs, solvates, and hydrates will be discovered.

11.6 An Example of a High-throughput Screening Approach

To illustrate the principles developed in the previous sections, an example of a high-throughput polymorphism screening application will be presented [58].

11.6.1 Model Substance

Carbamazepine (CBZ, Fig. 11.1), a drug used in the treatment of epilepsy and trigeminal neuralgia, was selected as a challenging model substance to demonstrate the efficacy of the chosen HTS method. Four polymorphs as well as several solvates of CBZ had previously been reported [44, 61–71] but were often described by conflicting nomenclature. In this chapter, we use the nomenclature of Griesser [44], in which the four true polymorphs are identified as forms I, IIa, IIb, and III (see also Table 11.3). Literature data show that form III is the thermodynamically stable polymorph in the technically relevant temperature range [62, 64, 72–75]. Highly pure form III carbamazepine was used as the starting material for all HTS experiments.

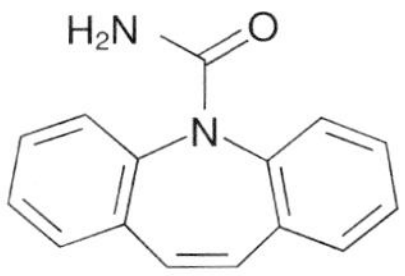

H_2N O N

Fig. 11.1 Carbamazepine (CBZ).

11.6.2 Solubility

To design meaningful crystallization experiments, the approximate solubility of the substance in the solvents and solvent mixtures of interest has to be known. The solubility of carbamazepine in 50 solvents and solvent mixtures was determined by an UV/VIS assay. This method is applicable to molecules showing sufficient UV/VIS absorbance.

11.6.3
Crystallization Experiments

Two types of crystallization methods were implemented in this HTS: (1) crystallization from solution by controlled evaporation of the solvent and (2) crystallization by suspension equilibration. Furthermore, in suspension equilibration experiments, both wet and dried crystals were analyzed, i.e., desolvation of solvates and hydrates was used as a third method for generating new polymorphs. The use of various solvents and solvent mixtures creates different nucleation and crystal growth conditions so that both stable and metastable forms can be produced. In addition, the stability of hydrates and solvates as a function of water or solvent activity can be investigated by equilibration in a homologous series of solvent mixtures.

Both suspension and evaporation experiments are easily miniaturized and require only approximate solubility data to design. Ninety-six-well microtiter plates made from quartz with PTFE seals were chosen to make the system leak-proof, to be compatible with most solvents, and to permit detection by Raman microscopy without an additional sample preparation step. The plates can be heated, cooled and shaken. For evaporation experiments and for drying the products of suspension equilibration experiments, the sealing was modified to allow a flow of dry N_2 to pass through the individual wells [76, 77].

As mentioned in Section 11.4, suspension experiments play a key role, as they are most likely to lead to the thermodynamically stable form [17]. Not finding the most stable form is probably the worst outcome in a polymorphism screening. For the suspension experiments, a microtiter plate containing slurries of CBZ each in 100 μL of solvent was shaken at 400 rpm for 20 h at room temperature. The suspensions were analyzed both in the presence of the solvent and after complete evaporation of the solvent, permitting detection of labile solvates. Evaporations were started with 100 μL of saturated solution per well. They were carried out both under nitrogen flow at two different rates as well as in the ambient atmosphere (and thus in the presence of water vapor). For the complete set of about 200 suspension and evaporation experiments, less than 1 g CBZ was needed. A total of 43 solvents and solvent mixtures was used in this HTS.

11.6.4
Data Acquisition

Raman microscopy was selected as the method to detect the form of the crystals. An advantage of its spatial resolution is that mixtures of several forms within the same well are normally easily identified, permitting the study of concomitant polymorphs and the conditions that produce them [6, 78]. In our experience, Raman spectroscopy is an excellent method to discriminate among various polymorphs, solvates, and hydrates of a single substance. We found that the probability of failure to discriminate between solid forms is similar for Raman and X-ray diffraction, provided that the Raman spectra are analyzed carefully. Further information that can be gained by visual inspection, such as crystal morphology, may also be helpful. The amorphous form of a substance is readily identified by such methods. A Renishaw RM 1000 dispersive Raman microscope system equipped with a 785-nm laser was used in this study.

For suspension equilibration experiments, spectra were collected automatically using a predefined grid. Four different positions per well were sampled in the presence of the solvent and three different positions after evaporation of the solvent. The automatic mapping of the plate, consisting of up to 200 measurements, was completed within less than five hours. Raman spectra of the crystals obtained by evaporation were measured manually by visual selection of the crystals. Three crystals per well were investigated.

11.6.5
Data Analysis

All data acquired during the screening experiment were stored in the “HTS data manager”, a central database, which was developed by Solvias. For each well of the plate, the database contains results of the Raman microscopy measurements, including spectra, images and acquisition parameters, as well as information about the experimental conditions, e.g., solvents, temperature and humidity.

Based on their spectral similarity, the Raman spectra were divided automatically into 12 classes using the computer program “Peak Compare” developed by Solvias. The program automatically subtracts the contributions from residual solvent. Spectra are compared on the basis of peak position, and reference spectra are defined for each class of spectra. Alternatively, spectra can be compared with predefined references. All relevant parameters for spectral comparison can be manually adjusted and optimized for each substance.

Reference spectra for each class are also stored in the “HTS data manager”.

Some results from the automatic evaluation are shown in Table 11.2. Visual inspection of the reference spectra identified class 7 as a mixture of the classes 1 and 6 (Table 11.3). Conditions where two classes are listed also represent cases where two forms crystallized concomitantly in the same well.

Crystallization experiments with CBZ on a larger scale and subsequent analysis by Raman spectroscopy and X-ray powder diffraction allowed the assignment of six of the classes to solid forms described in the literature (Table 11.3). A detailed region of the Raman spectra of forms I, IIa, IIb, and III is presented in Fig. 11.2, where differences among the forms can readily be seen.

Recently, a new single-crystal structure was published as form IV [66]. These single-crystal data were converted into powder XRD data by using the program Powder Cell 2.4 (Federal Institute for Materials Research and Testing, Berlin, Germany) and demonstrated to be identical to the previously identified form IIa [58]. The group that crystallized form IV subsequently acknowledged [67] that the form bears significant similarities to polymorphs previously observed by Krahn et al. [62] and Rustichelli et al. [65] (Table 11.3).

Table 11.2 Part of the matrix of carbamazepine experiments with the corresponding identified forms.

Solvents/mixtures	Crystallization conditions				
	Suspension, in solvent	Suspension, dried	Fast evaporation, N_2-flow: 0.4 L min^{-1}	Slow evaporation, N_2-flow: 0.03 L min^{-1}	Very slow evaporation (ambient atmosphere)
Acetic acid	3	3	8	8	8, 3
Acetone	12	1	1	1	2, 11
Anisole	1	1	8	8	8
N,N-Dimethylformamide	1	1	–	8	8
Dimethyl sulfoxide	4	–	–	–	–
1,4-Dioxane	5	5	9	9	8, 9
Ethanol	1	1	8	8	2
Isopropyl ether	1	1	8	8	8
Methanol	1	1	6	6, 11	2
N-Methylpyrrolidone	1	1	–	8, 10	–
tert-Butyl methyl ether	1	1	8	8	8
Tetrahydrofuran	1	1	2	8	8
Toluene	1	1	8	8	8
Acetone/water 1:1	2	1	2	6	6
Ethanol/water 1:1	2	1, 7	6	6	2
Methanol/water 4:1	1	1	2	6	6
Tetrahydrofuran/water 1:1	2	1	6	6	2

Table 11.3 Correspondence between classes and forms described in the literature.

Classes	Form [a)]	Alternative nomenclature
1	III [61, 62, 64, 65]	Beta-form [63], monoclinic form [64], P-monoclinic form [67]
2	Dihydrate [44, 61–63]	
3	Acetic acid solvate	
4	Dimethyl sulfoxide solvate	
5	Dioxane solvate	
6	II a	II [62, 65]; IV [66], C-monoclinic form [67]
7	Mixture of II a and III	
8	II b	II [61], alpha-Form [63], trigonal form [64]
9	Dioxane solvate?	
10	*N*-Methylpyrrolidone solvate?	
11	I [44, 61, 62, 64, 65]	Gamma-form [64], triclinic form [64]
12	Acetone solvate [44, 63]	

a) Forms of the true polymorphs are designated using the nomenclature of [44].

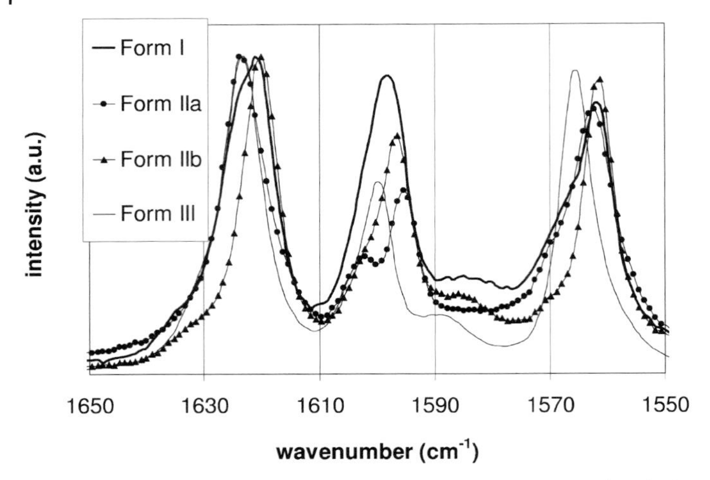

Fig. 11.2 Selected region of the Raman spectra of some of the identified forms of carbamazepine.

Most of the new classes were tentatively assigned to solvates. No new form that is more stable at room temperature than form III was found. Further investigations would be necessary to characterize the new forms.

The results demonstrate the suitability of HTS in both discovering the relevant polymorphic forms and identifying the conditions under which they can be produced. Further investigations, such as elucidation of the thermodynamic relationships between the forms [79–82], extensive physicochemical characterization of all forms, etc., would subsequently be required (Section 11.8).

11.7 Theoretical Methods

"Are crystal structures predictable?" [83, 84] is one of the most important questions. If the answer is "yes", then most of the experimental polymorphism screening effort could be replaced by computational work and the patentability of polymorphic forms would be questionable since "nonobviousness" (Chapter 14) could probably no longer be claimed.

Computational models for crystal structure prediction have been developed [83, 85–89] and commercialized, e.g., in the program Polymorph Predictor by Accelrys [90]. Such programs seek to determine the crystal structure that corresponds to the global minimum in lattice energy. That is to say, these methods should provide the thermodynamically stable crystal structure at 0 K since entropic effects are not considered. If other local minima are found in the lattice energy hypersurface, these should correspond to metastable polymorphs. Many simple crystal structure programs are restricted to the use of rigid molecules

[91]. It is assumed that forces between molecules are much weaker than intramolecular forces and that the structure of the molecule in the crystal lattice is the same as that in the gas phase. For conformationally flexible molecules, these assumptions break down, and one can either perform calculations with various different hypothesized initial gas phase structures or one can use a force-field that includes intramolecular energy terms [90] stemming from bond rotations and distortions. The search algorithm to find the optimal crystal structure is a critical element of all computational methods. A huge multi-dimensional surface has to be examined, where cell parameters and the orientation and position of the molecule in the unit cell have to be optimized. If there is more than one molecule in the asymmetric unit cell or if conformational flexibility has to be considered as well, the number of adjustable parameters multiplies (see [91] for a review).

Two blind tests organized by the Cambridge Crystallographic Data Centre (CCDC) tried to determine whether it is possible to predict the most stable known crystal structure of an organic molecule [92, 93]. More than a dozen participants used several computer programs and methodologies. The results showed that it is not possible to predict the stable polymorph reliably. Often the correct structure was present in the list of calculated polymorphs, but it was one among many and not the one with the lowest calculated energy. This result is not surprising since the energy difference between various polymorphs of the same molecule is generally only a few kJ mol^{-1}, which is smaller than the accuracy of current calculations. Also, the neglect of entropic contributions can make predictions at ambient temperatures ineffective.

Nevertheless, the combined application of computational crystal structure prediction and targeted experiments can be an effective tool [94]. Davey et al. [12, 95] analyzed many calculated structures of 2-amino-4-nitrophenol in terms of hydrogen-bonding motifs. For crystallization experiments they then chose solvents that were expected to either promote or inhibit the formation of the hydrogen bonds required for a particular motif. Four solvents (i.e., nitromethane, methanol, pentylamine, and toluene) were chosen, and three solid forms (one polymorph, two solvates) with crystal structures that contained the desired hydrogen bonding motifs were obtained. A study on diflunisal, a fluorinated aromatic carboxylic acid, produced similar findings [94]. Here, calculated structures were classified automatically by computer programs according to graph sets [5, 96]. Again, four solvents and one solvent mixture were chosen rationally and four new crystal forms (two polymorphs, two solvates) were obtained. The authors conclude, "Overall, this combined approach of structure prediction and directed crystallization has yielded a significant number of new crystal forms from a limited experimental screen" – an interesting antithesis to high-throughput screening. However, they also conclude that "Much fundamental work remains to be done before we can begin to understand and have control over this process."

Another valuable application of structure prediction is structural identification of unknown forms. For form III of paracetamol, a powder X-ray diffraction pattern could be recorded [46], but its quality was insufficient for structure determination. It matched the calculated powder diffraction pattern of a crystal structure obtained from crystal structure prediction, however, so that the crystal structure of paracetamol form III could be assigned with confidence [89].

So while the answer to "Are crystal structures predictable?" is still "no" [14, 81, 82, 89], the usefulness of theoretical predictions for experimental design has been clearly demonstrated.

11.8 Characterization

Identification of the polymorphic forms, solvates, and hydrates of a substance through a high-throughput screening (possibly complemented by additional experiments inspired by computation or intuition) is only the first step in understanding a drug's solid-state behavior. Of equal importance are the physical properties of the various forms. A thorough characterization of these forms generally requires more material than is available in an HTS. High-throughput screenings are therefore typically followed by laboratory-scale experiments designed to produce 50 to 100 mg of each polymorph, solvate, and hydrate. The results obtained in the HTS generally provide a good indication of what types of experiments can produce a given form and even what that form is likely to be (e.g., forms that only appear when water is present in the solvent mixture are likely to be hydrates). In some unfavorable cases it might be almost impossible to reproduce forms encountered during HTS experimentation. However, microscopic experiments represent somewhat special conditions; if it is not possible to obtain macroscopic amounts of a solid form encountered during an HTS with reasonable effort, such a form would probably be irrelevant for practical purposes.

Once a sample of each polymorph has been obtained, the forms are generally characterized using various physicochemical methods. Important parameters include the form's crystallinity (as determined by XRD and DSC), melting point (also determined by DSC), hygroscopicity (determined by DVS), and bioavailability (inferred by solubility and/or dissolution rate measurements along with measurements of $\log D$ and $\mathrm{p}K_a$). X-ray powder diffraction and spectroscopic methods (Raman, IR, and/or NMR) provide a fingerprint of each form, reveal fundamental structural information and are very useful for patent filing. TG-FTIR is useful for determining whether a solid is a solvate or hydrate, and both TG-FTIR and DSC provide information about thermal stability and decomposition. Particle-size distribution measurements and identification of possible crystal habits for a given polymorph can give an indication of whether this polymorph can be easily processed and manufactured. These techniques should be supplemented by HPLC purity analysis.

Single-crystal structures of the various forms may help to elucidate their solid-state behavior. The formation of isomorphous solvates or the moisture sorption behavior can in favorable cases be explained by knowing the three-dimensional arrangements of the constituent molecules. Channel or sheet-like structures, for example, might easily form non-stoichiometric solvates and hydrates. Single-crystal structures also permit simulation of the powder X-ray diffraction patterns of the respective crystals and subsequent comparison of such a pattern with patterns obtained from the polymorph screening.

Thermal measurements, suspension equilibration experiments, and solubility/dissolution rate measurements can all provide information about the relative stabilities of the different forms. Such information can be used to construct ET (energy vs temperature) diagrams, which identify which form is thermodynamically stable at a given temperature. Monotropic and enantiotropic relationships between pairs of forms can be discerned, and such diagrams can be used as guidelines for directed crystallization of a specific form. If a metastable form is to be developed (e.g., for intellectual property reasons or due to undesirable properties of the thermodynamically stable form), such studies must be supplemented by an investigation of the kinetics of polymorph conversion.

Once the solids are characterized, the intellectual property aspect becomes important. The choices of whether and when to apply for patent protection for selected or for all polymorphs encountered should be carefully made and well analyzed, and any decision for patent life cycle management should be taken with all relevant data in hand.

Finally, all characterization data and information on the solid-state behavior of a molecule will have to enter the final decision matrix where, on a side-by-side basis, the solid forms of all entities of interest, i.e., the uncharged molecule and the various salts (or co-crystals), are listed with their respective profiles. Besides physicochemical data gained by characterization of the solid forms, data concerning chemical stability, pharmacokinetics and, for example, the patent situation enter into such a matrix.

11.9 Conclusions

Identifying and characterizing all relevant solid forms is an important step in an integrated approach to the solid-state issues discussed in this book. The search for new solid forms should be both systematic and flexible. Different stages of development and product maturity may require different degrees of thoroughness for such a search. While it is generally the goal to enter clinical development Phase I with the thermodynamically stable form, it is highly advisable to identify all relevant polymorphic forms during further pharmaceutical development. Intellectual property aspects as well as the design of robust larger-scale crystallization processes rely on an exact understanding of the polymorphic behavior of a drug substance.

A good strategy for polymorph screening should encompass the characterization of the starting material, thermoanalytical investigations of the solid, a solvent-based crystallization screening with both fast and slow crystallization techniques and the search for hydrates. Microscopic, multi-parallel approaches allow for rapid, low substance consuming and efficient ways of identifying the tendency of solids to show polymorphism. In addition, such approaches can, in general, be used for combined salt and polymorph screenings. But even though it may be very easy to carry out numerous experiments quickly, such quantity should not replace the quality of careful experimental planning. Very often, additional rational crystallization methods are needed to get a complete picture of the behavior of the substance.

Polymorph screening, while being very important, is only a first step. Almost equally important and, at least in some cases, challenging is polymorph characterization as it is the basis of every selection process. Characteristics of all solid forms of interest of the potential drug substances, i.e., the neutral molecule and the various salts and co-crystals, have to be compared. Physicochemical data, data on chemical stability and pharmacokinetics, assessment of the patent situation, etc., enter into the process of deciding which solid form to promote for further development. The final selection of the optimal solid might be a tedious exercise in some cases, and the pros and cons of each form have to be considered in a well balanced way as they will gravely influence the future of the drug. This decision should be made early in the development process (Chapter 1) to avoid the necessity of making changes in the solid form later on in the process, which would lead to both additional costs and delays in market introduction.

In the end, only the answers to the following three questions matter:

- Which form should be developed?
- How can that form be produced reliably?
- How can it be assured that no undesired changes occur during formulation and the shelf life of the product?

The quality of a polymorphism screening and characterization process can be judged on its ability to answer these questions reliably.

Acknowledgments

We thank Jörg Berghausen for pointing out important references.

Abbreviations

CBZ	Carbamazepine
CCDC	Cambridge Crystallographic Data Centre
DSC	Differential scanning calorimetry
DVS	Dynamic vapor sorption
ET diagram	Energy–temperature diagram
HPLC	High-performance liquid chromatography
HTS	High-throughput screening

IR	Infrared
log*D*	Logarithm of the octanol–water distribution coefficient at a fixed pH
MAS	Magic-angle spinning
NMR	Nuclear magnetic resonance
PTFE	Polytetrafluoroethylene (Teflon)
TG-FTIR	Thermogravimetry coupled to Fourier transform IR spectroscopy
TG-MS	Thermogravimetry coupled to mass spectroscopy
UV/VIS	Ultraviolet/visible
XRD	X-ray diffraction

References

1 Bernstein, J., *J. Phys. D: Appl. Phys.*, 26 (**1993**) B66–B76.

2 Brittain, H.G., *Polymorphism in Pharmaceutical Solids*, Marcel Dekker, Inc., New York (**1999**).

3 Byrn, S.R., Pfeiffer, R., Stowell, J.G., *Solid State Chemistry of Drugs*, 2nd edn., SSCI Inc., West Lafayette (**1999**).

4 Ostwald, W., *Z. Phys. Chem.*, 22 (**1897**) 289–330.

5 Bernstein, J. *Polymorphism in Molecular Crystals*, Oxford Science Publications, Oxford (**2002**).

6 Threlfall, T., *Org. Process Res. Dev.*, 4 (**2000**) 384–390.

7 Threlfall, T., *Org. Process Res. Dev.*, 7 (**2003**) 1017–1027.

8 Etter, M.C., *J. Phys. Chem.*, 95 (**1991**) 4601–4610.

9 Gavezzotti, A., Filippini, G., *Chem. Commun.*, (**1998**) 287–294.

10 Rodriguez-Hornedo, N., Murphy, D., *J. Pharm. Sci.*, 88 (**1999**) 651–660.

11 Datta, S., Grant, D.J.W., *Crystal Growth & Design*, 5 (**2005**) 1351–1357.

12 Davey, R.J., Allen, K., Blagden, N., Cross, W.I., Lieberman, H.F., Quayle, M.J., Righini, S., Seton, L., Tiddy, G.J.T., *Cryst. Eng. Commun.*, 4 (**2002**) 257–264.

13 Gu, C.-H., Chatterjee, K., Young Jr., V., Grant, D.J.W., *J. Crystal Growth*, 235 (**2002**) 471–481.

14 Blagden, N., Davey, R., *Chem. Br.*, 35 (**1999**) 44–47.

15 Beckmann, W., Otto, W., Budde, U., *Org. Process Res. Dev.*, 5 (**2001**) 387–392.

16 Mukuta, T., Lee, A.Y., Kawakami, T., Myerson, A.S., *Crystal Growth & Design*, 5 (**2005**) 1429–1436.

17 Gu, C.-H., Young Jr., V., Grant, D.J.W., *J. Pharm. Sci.*, 90 (**2001**) 1878–1890.

18 McCrone, W.C., *Phys. Chem. Org. Solid State*, 2 (**1965**) 725–767.

19 Griesser, U.J., Stowell, J.G. in *Pharmaceutical Analysis* (eds. Lee, D.C., Webb, M.L.), Blackwell Publishing Ltd., Oxford, UK (**2003**).

20 Newman, A.W., Stahly, G.P., *Drugs Pharmaceutical Sci.*, 117 (**2002**) 1–57.

21 Guillory, J.K. in *Polymorphism in Pharmaceutical Solids* (ed. Brittain, H.G.), Marcel Dekker, New York (**1999**) pp. 125–181.

22 Kuhnert-Brandstätter, M., *Thermomicroscopy in the Analysis of Pharmaceuticals*, Pergamon Press, Oxford (**1971**).

23 Morissette, S.L., Almarsson, Ö., Peterson, M.L., Remenar, J.F., Read, M.J., Lemmo, A.V., Ellis, S., Cima, M.J., Gardner, C.R., *Adv. Drug Delivery Rev.*, 56 (**2004**) 275–300.

24 Price, C.P., Grzesiak, A.L., Matzger, A.J., *J. Am. Chem. Soc.*, 127 (**2005**) 5512–5517.

25 Lang, M., Grzesiak, A.L., Matzger, A.J., *J. Am. Chem. Soc.*, 124 (**2002**) 14834–14835.

26 Cacciuto, A., Auer, S., Frenkel, D., *Nature*, 428 (**2004**) 404–406.

27 Trask, A.V., Motherwell, W.D.S., Jones, W., *Chem. Commun.*, (**2004**) 890–891.

28 Trask, A.V., Motherwell, W.D.S., Jones, W., *Crystal Growth Design*, 5 (**2005**) 1013–1021.

29 Almarsson, Ö., Zaworotko, M.J., *Chem. Commun.*, (**2004**) 1889–1896.

30 Childs, S.L., *WO 2004/064762 A2*, Novel Cocrystallization.

31 Childs, S.L., Chyall, L.J., Dunlap, J.T., Coates, D.A., Stahly, B.C., Stahly, G.P., *Cryst. Growth Design*, 4 (**2004**) 441–449.

32 Zaccaro, J., Matic, J., Myerson, A.S., Garetz, B.A., *Crystal Growth Design*, 1 (**2001**) 5–8.

33 Beach, S., Latham, D., Sidgwick, C., Hanna, M., York, P., *Org. Process Res. Dev.*, 3 (**1999**) 370–376.

34 Bonafede, S.J., Ward, M.D., *J. Am. Chem. Soc.*, 117 (**1995**) 7853–7861.

35 Rodriguez-Spong, B., Price, C.P., Jayasankar, A., Matzger, A.J., Rodriguez-Hornedo, N., *Adv. Drug Delivery Rev.*, 56 (**2004**) 241–274.

36 Dennehy, R.D., *Org. Process Res. Dev.*, 7 (**2003**) 1002–1006.

37 Gu, C.-H., Li, H., Gandhi, R.B., Raghavan, K., *Int. J. Pharmaceutics*, 283 (**2004**) 117–125.

38 Mirmehrabi, M., Rohani, S., *J. Pharm. Sci.*, 94 (**2005**) 1560–1576.

39 Snyder, L.R., *J. Chromatogr. Sci.*, 16 (**1978**) 223–234.

40 Carlson, E.D., *Design and Optimization in Organic Synthesis*, Elsevier, Amsterdam (**1992**).

41 Blomsma, E., van Langevelde, A., Conference on Polymorphism and Crystallization, Barnett International, Brussels, September 29–30 (**2003**).

42 Singhal, D., Curatolo, W., *Adv. Drug Delivery Rev.*, 56 (**2004**) 335–347.

43 Burger, A., *Zur Auffindung und Herstellung neuer Kristallformen von Arzneistoffen*, Vortrag zum 30. Jahrestag der Arbeitsgemeinschaft für Pharmazeutische Verfahrenstechnik e.V. (A.P.V.), Mainz (**1984**).

44 Griesser, U.J., *PhD Thesis*, Universität Innsbruck (**1991**).

45 Szelagiewicz, M., Marcolli, C., Cianferani, S., Hard, A.P., Vit, A., Burkhard, A., von Raumer, M., Hofmeier, U.C., Zilian, A., Francotte, E., Schenker, R., *J. Therm. Anal. Cal.*, 57 (**1999**) 23–43.

46 Peterson, M.L., Morissette, S.L., McNulty, C., Goldsweig, A., Shaw, P., LeQuesne, M., Monagle, J., Encina, N., Marchionna, J., Johnson, A., Gonzalez-Zugasti, J., Lemmo, A.V., Ellis, S.J., Cima, M.J., Almarsson, O., *J. Am. Chem. Soc.*, 124 (**2002**) 10 958–10 959.

47 Blatter, F., Szelagiewicz, M., von Raumer, M., *WO 2005/037424 A1*, Process for the Parallel Detection of Crystalline Forms of Molecular Solids.

48 Balbach, S., Korn, C., *Int. J. Pharmaceutics*, 275 (**2004**) 1–12.

49 Giron, D., *Eng. Life Sci.*, 3 (**2003**) 103–112.

50 Huang, L.-F., Tong, W.-Q., *Adv. Drug Delivery Rev.*, 56 (**2004**) 321–334.

51 Maier, W.-F., Klein, J., Lehmann, C., Schmidt, H.W., Offenlegungsschrift *DE 198 22 077 A1*, Kombinatorisches Verfahren zur Herstellung und Charakterisierung von kristallinen und amorphen Materialbibliotheken im Mikrogramm-Maßstab (**1999**).

52 Storey, R., Docherty, R., Higginson, P., Dallman, C., Gilmore, C., Barr, G., Dong, W., *Crystallogr. Rev.*, 10 (**2004**) 45–56.

53 Carlton, D.L., Dhingra, O.P., Waters, P.W., *WO 00/67872*, High Throughput Crystal Form Screening Workstation and Method of Use.

54 Gardner, C.R., Walsh, C.T., Almarsson, Ö., *Nat. Rev.*, 3 (**2004**) 926–934.

55 Van Langevelde, A., Blomsma, E., *Acta Crystallogr., Sect. A*, 58 (**2002**) C9 (supplement).

56 Desrosiers, P., Carlson, E., Chandler, W., Chau, H., Cong, P., Doolen, R., Freitag, C., Lin, S., Masui, C., Wu, E., Crevier, T., Mullins, D., Song, L., Lou, R., Zhan, J., Tangkilisan, A., Ung, Q., Phan, K., *Acta Crystallogr., Sect. A*, 58 (**2002**) C9 (supplement).

57 Carlson, E.D., Cong, P., Chandler Jr., W.H., Chau, H.K., Crevier, T., Desrosiers, P.J., Doolen, R.D., Freitag, C., Hall, L.A., Kudla, T., Luo, R., Masui, C., Rogers, J., Song, L., Tangkilisan, A., Ung, K.Q., Wu, L., *Pharmachem.* (**2003**) 10–15.

58 Hilfiker, R., Berghausen, J., Blatter, F., Burkhard, A., De Paul, S. M., Freiermuth, B., Geoffroy, A., Hofmeier, U., Marcolli, C., Siebenhaar, B., Szelagiewicz, M., Vit, A., von Raumer, M., *J. Therm. Anal. Cal.*, 73 (**2003**) 429–440.

59 Barr, G., Dong, W., Gilmore, C. J., *J. Appl. Crystallogr.*, 37 (**2004**) 658–664.

60 Gilmore, C. J., Barr, G., Paisley, J., *J. Appl. Crystallogr.*, 37 (**2004**) 231–242.

61 Kala, H., Haack, U., Pollandt, P., Brezesinski, G., *Acta Pharm. Technol.*, 32 (**1986**) 72–77.

62 Krahn, F. U., Mielck, J. B., *Pharm. Acta Helv.*, 62 (**1987**) 247–254.

63 Lowes, M. M. J., Caira, M. R., Lötter, A. P., van der Watt, J. G., *J. Pharm. Sci.*, 76 (**1987**) 744–752.

64 Edwards, A. D., Shekunov, B. Y., Forbes, R. T., Grossmann, J. G., York, P., *J. Pharm. Sci.*, 90 (**2001**) 1106–1114.

65 Rustichelli, C., Gamberini, G., Ferioli, V., Gamberini, M. C., Ficarra, R., Tommasini, S., *J. Pharm. Biomed. Anal.*, 23 (**2000**) 41–54.

66 Lang, M., Kampf, J. W., Matzger, A. J., *J. Pharm. Sci.*, 91 (**2002**) 1186–1190.

67 Grzesiak, A. L., Lang, M., Kim, K., Matzger, A. J., *J. Pharm. Sci.*, 92 (**2003**) 2260–2271.

68 Fleischmann, S. G., Kuduva, S. S., McMahon, J. A., Moulton, B., Bailey Walsh, R. D., Rodriguez-Hornedo, N., Zaworotko, M. J., *Crystal Growth Design*, 3 (**2003**) 909–919.

69 Johnston, A., Florence, A. J., Kennedy, A. R., *Acta Crystallogr., Sect. E*, 61 (**2005**) o1777–o1779.

70 Johnston, A., Florence, A. J., Kennedy, A. R., *Acta Crystallogr., Sect. E*, 61 (**2005**) o1509–o1511.

71 Lohani, S., Zhang, Y., Chyall, L. J., Mougin-Andres, P., Muller, F. X., Grant, D. J. W., *Acta Crystallogr., Sect. E*, 61 (**2005**) o1310–o1312.

72 Behme, R. J., Brooke, D., *J. Pharm. Sci.*, 80 (**1991**) 986–990.

73 Ceolin, R., Toscani, S., Gardette, M.-F., Agafonov, V. N., Dzyabchenko, A. V., Bachet, B., *J. Pharm. Sci.*, 86 (**1997**) 1062–1065.

74 Griesser, U. J., Szelagiewicz, M., Hofmeier, U. C., Pitt, C., Cianferani, S., *J. Therm. Anal. Cal.*, 57 (**1999**) 45–60.

75 Anquetil, P. A., Brenan, C. J. H., Marcolli, C., Hunter, I. W., *J. Pharm. Sci.*, 92 (**2003**) 149–160.

76 Blatter, F., Cron-Eckhardt, B., Hofmeier, U. C., Koller, P., Marcolli, C., Szelagiewicz, M., *WO 03/026797 A2*, Sealing System with Flow Channels.

77 Szelagiewicz, M., Marcolli, C., Berghausen, J., Cron-Eckhardt, B., Hofmeier, U. C., Blatter, F., *WO 2004/045769 A1*, Multiple Sealing System for Screening Studies.

78 Bernstein, J., Davey, R. J., Henck, J.-O., *Angew. Chem. Int. Ed.*, 38 (**1999**) 3441–3461.

79 Grunenberg, A., Henck, J.-O., Siesler, H. W., *Int. J. Pharmaceutics*, 129 (**1996**) 147–158.

80 Marti, E., *J. Therm. Anal. Cal.*, 33 (**1998**) 37–45.

81 Gu, C.-H., Grant, D. J. W., *J. Pharm. Sci.*, 90 (**2001**) 1277–1287.

82 Yu, L., Reutzel, S. M., Stephenson, G. A., *Pharmaceutical Sci. Technol. Today*, 1 (**1998**) 118–127.

83 Gavezzotti, A., *Acc. Chem. Res.*, 27 (**1994**) 309–314.

84 Dunitz, J. D., *Chem. Commun.*, (**2003**) 545–548.

85 Karfunkel, H. R., Gdanitz, R. J., *J. Comput. Chem.*, 13 (**1992**) 1171–1183.

86 Gavezzotti, A., Fillipini, G., *J. Am. Chem. Soc.*, 117 (**1995**) 12299–12305.

87 Price, S. L., Wibley, K. S., *J. Phys. Chem. A*, 101 (**1997**) 2198–2206.

88 Beyer, T., Day, G. M., Price, S. L., *J. Am. Chem. Soc.*, 123 (**2001**) 5086–5094.

89 Verwer, P., Leusen, F. J. J. in *Reviews in Computational Chemistry* (eds. Lipowitz, K. B., Boyd, D. B.), Wiley-VCH, New York (**1998**), Vol. 12, Chapter 7, p. 327.

90 Accelrys Inc., *Cerius2 Modelling Environment*, Accelrys Inc., San Diego (**1999**).

91 Price, S. L., *Adv. Drug Delivery Rev.*, 56 (**2004**) 301–319.

92 Lommerse, J. P. M., Motherwell, W. D. S., Ammon, H. L., Dunitz, J. D., Gavezzotti, A., Hofmann, D. W. M., Leusen, F. J. J., Mooij, W. T. M., Price, S. L., Schweizer, B., Schmidt, M. U., van Eijck, B. P., Verwer, P., Williams, D. E., *Acta Crystallogr., Sect. B*, 56 (**2000**) 697–714.

93 Motherwell, W. D. S., Ammon, H. L., Dunitz, J. D., Dzyabchenko, A., Erk, P., Gavezzotti, A., Hofmann, D. W. M., Leusen, F. J. J., Lommerse, J. P. M., Mooij, W. T. M., Price, S. L., Scheraga, H., Schweizer, B., Schmidt, M.U., van Eijck, B.P., Verwer, P., Williams, D.E., *Acta Crystallogr., Sect. B*, 58 (**2002**) 647–661.

94 Cross, W. I., Blagden, N., Davey, R. J., Pritchard, R. G., Neumann, M. A., Roberts, R. J., Rowe, R. C., *Crystal Growth Design*, 3 (**2003**) 151–158.

95 Blagden, N., Cross, W. I., Davey, R. J., Broderick, M., Pritchard, R. G., Roberts, R. J., Rowe, R. C., *Phys. Chem. Chem. Phys.*, 3 (**2001**) 3819–3825.

96 Bernstein, J., Davis, R. E., Shimoni, L., Chang, N.-L., *Angew. Chem. Int. Ed. Engl.*, 34 (**1995**) 1555–1573.

12 Salt Selection

Peter Heinrich Stahl and Bertrand Sutter

12.1 Introduction

The objective of drug discovery is to identify substances that are highly active in biological test systems. Screening for potential drug candidates is performed in solution at micromolar and nanomolar concentration levels. Hence there is no concern about the solid-state properties of the substances at the discovery phase. However, these come into focus as soon as promising drug candidates are identified and selected for further biological and safety studies *in vivo*. Concomitant with the biological evaluation of a drug candidate during the preclinical phase of drug development, it is necessary to come up with a decision about the physical nature of the selected drug candidate material. It will be the form to be reproducibly provided for GLP-conforming safety studies, for the development and the manufacture of dosage forms intended for clinical studies.

12.2 Salt Formation and Polymorphism

Polymorphism implies the potential of a limited variety of solid-state manifestations inherent in a given substance. Only one of these is characterized by definitive thermodynamical stability within a given range of environmental conditions. Most drug substances are weak electrolytes capable of forming salts. Without altering the chemical integrity of the same active entity, salt formation can appreciably widen the selection basis by creation of new chemical entities. This is because each of the possible salts of a drug substance is characterized by its individual profile of physicochemical properties, generally covering much broader ranges than the limited set of polymorphs of the bare drug entity. Generations of pharmacists and chemists have even changed the state of aggregation of their compounds of interest from liquids to solids by the preparation of salts for easier handling: the viscous liquid scopolamine base forms a hydrobro-

Polymorphism: in the Pharmaceutical Industry. Edited by Rolf Hilfiker

ISBN: 3-527-31146-7

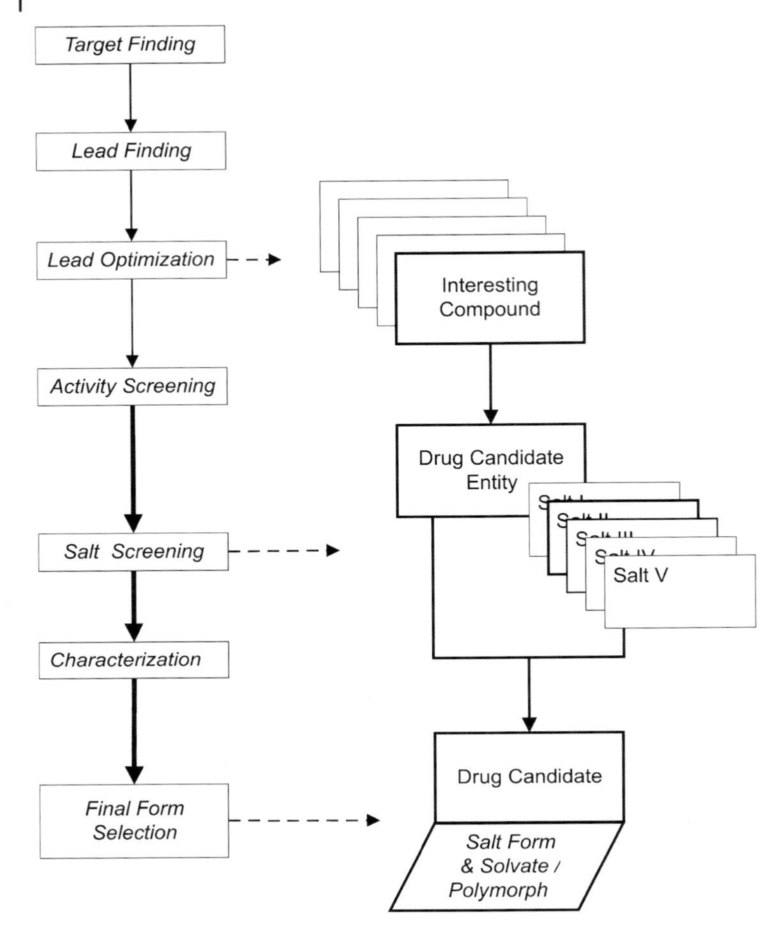

Fig. 12.1 Salt and solid-state form selection in the traditional course of drug development.

mide with mp 195 °C, and the solid sodium and magnesium salts of the liquid valproic acid are much more convenient to produce a solid dosage form for anticonvulsant therapy. Therefore, during drug development, salt formation precedes the thorough study of the morphic states and propensities (Fig. 12.1).

These activities will then converge to a final assessment and the definition of the preferred chemical as well as the particular solid-state form of the drug substance. This final form constitutes the active principle of the pharmaceutical drug product to be developed and produced for administration in humans.

12.3 Target Properties of Active Substances for Drug Products

Drug substances may be administered to the body by various routes of administration: oral – the most frequently chosen route, parenteral by several modes of injection [intramuscular (i.m.), intravenous (i.v.), subcutaneous (s.c.) and others], intranasal, by inhalation, topical to the various areas of the body surface, to name just the most important ones. For these administration routes various pharmaceutical dosage forms can be provided. Each of them requires that a typical set of relevant parameters of the active substance lie within a certain range, beyond which the development of said dosage form becomes increasingly difficult and eventually impossible. Also, in a mutual interdependence, the selection of dosage form and the applicable technology depend on the properties profile of the chosen material.

For a better understanding, the most significant properties are discussed in relation to the technological background of the dosage forms delivered via the most important routes of administration. It is generally accepted that solubility in aqueous media, such as pharmaceutical vehicles and body fluids, is of uppermost interest.

12.3.1 Injectables

For injectable solutions, high solubility under the conditions close to physiological pH 7.4 is essential. For small-volume injectables such as i.m. and s.c. the solubility should be as high as possible to accommodate the dose to be administered in 0.5–2 mL for s.c. or up to 5 mL for i.m. administration. The solubility requirement for i.v. injections is less stringent because volumes up to 20 mL can be administered. With lower solubility, one has to resort to infusions of volumes of up to 1000 mL.

Injectable solutions require particularly high chemical stability. Ideally, a drug substance must withstand heat sterilization in solution and subsequent storage for up to five years. For those drug substances lacking such optimum stability it is possible to circumvent heat stress by sterile filtration. There are drug substances that cannot be stored in solution but only in the dry state. This means that solutions must be prepared, sterile-filtered and subsequently freeze-dried. Thus, they need to be sufficiently stable to survive at least these processing steps, essentially without decomposition.

Naturally, for injectables, solid-state drug properties are of minor importance as long as they do not hamper processing or the dissolution of a lyophilizate.

12.3.2
Solid Dosage Forms

For solid dosage forms the most critical step after swallowing a unit dose is the release of the drug substance. Solubility and hence dissolution can control or limit, respectively, this important process. Therefore, a solubility that is reasonably high in relation to the drug dose is desirable. In the Biopharmaceutics Classification System (BCS) for drug substances, suggested by Amidon and co-workers and adopted by the FDA, the dose number D, which is defined as

$$D = \frac{M_0}{V_0 c_s}$$

indicates this relationship, where M_0 is the dose to be administered, V_0 is 250 mL, by definition the standardized stomach volume, and c_s is the solubility in mg L^{-1}. Substances with $D > 20$ are regarded as "low soluble and dissolution rate limited" drugs.

As far as technologically relevant properties are concerned, the chemical stability during storage of the bulk substance, during processing and during storage of the drug product must be warranted. In addition, the drug substance must be chemically and physically compatible with a set of excipients necessary for producing the dosage form. A further requirement for manufacturing tablets is that the drug substance does not corrode the tools of tablet machinery. Such damage can occur with salts of very weak drug bases with mineral acids.

The physical state of the drug substance must remain unchanged during storage of the bulk material under appropriate conditions. Particle size, particle size distribution, the morphic state (e.g., polymorph, degree of crystallinity and solvation, in particular hydrate or anhydrous state) must be under control during processing and has to arrive at a well defined and reproducible state within the final product.

Sparingly soluble drug substances require comminution by mechanical means to improve their dissolution. To assure content uniformity this applies also for very low dosed drugs regardless of their solubility. Milling is possible only if the melting point is high enough, i.e., above about 100 °C. Deliquescence due to extreme hygroscopicity can lead to significant problems during handling and processing.

12.3.3
Dosage Forms for Other Routes of Application

12.3.3.1 Inhalation

Three pharmaceutical principles are available for administering drugs by inhalation: (a) The classical pressurized Metered Dose Inhalation (MDI) generates an aerosol of fine droplets of a solution, or of a microparticulate suspension of drug substance, by means of a pressurized gas (liquefied propellant phase).

(b) An increasingly appreciated technique is Dry Powder Inhalation (DPI); the drug particles are dispersed directly by a stream of the air being inhaled. Single dose and multiple dose inhaler types are available. (c) Nebulization of aqueous drug solutions is also in use but is rarely considered in modern product development.

Solid-state and particulate properties are of major importance for both MDI, which is used in conjunction with suspensions, and DPI, because the drug substance needs to be micronized. Consequently, the degree of comminution must be maintained for the product's lifetime; particle growth and agglomeration would be deleterious to product quality.

The sensitivity of the bronchial mucosa requires that the drug substance lacks any irritative potential. In addition, unpleasant taste must be absent. Both requirements can be met by drugs with rather low water solubilities.

12.3.3.2 **Topical Products and the Transdermal Route**

In selecting a salt form for a product that is applied onto the skin it is a prerequisite that it must not cause local irritation, whether the target of treatment is the skin surface, dermal tissue, or underlying organs (joints, muscles, tendons) or even if systemic effects are intended. Weakly acidic to neutral pHs (~5–7.5) are tolerated best and deviations from this range should be avoided.

For percutaneous application into the living dermis and adjacent tissues to be treated, a drug substance must cross several barriers. The principal resistance to skin penetration resides in the *stratum corneum*. The ideal prerequisite for good penetration is good aqueous solubility and, in addition, good lipid solubility. Such an ideal compound can utilize both available routes through the *stratum corneum* for penetration, i.e., the transcellular route through the hydrated keratocytes and the tortuous route along the lipid layers separating the keratocytes.

The percutaneous permeability of ionizable drug substances can be from one to more than two orders of magnitude higher in the uncharged form than in the ionized state. The normalized transdermal flux of ephedrine base is 6×10^{-3} cm h^{-1}, whereas that of the ephedrine cation is only 3.3×10^{-4} cm h^{-1}; for chlorpheniramine the respective values are 2.2×10^{-3} and 8×10^{-6} cm h^{-1} [1]. Hence, a salt form, rather than the free base or free acid, appears to reduce percutaneous penetration. Nevertheless, a suitable salt former can transfer desirable solubility properties to the salt ("ion pair") that may lead to enhanced skin penetration. Cases in point are the diethylamine and the *N*-hydroxyethylpyrrolidine (HEP) salts of the antiinflammatory drug diclofenac [2].

12.4
Basics of Salt Formation

12.4.1
Ionization Constant

In physicochemical terms, about two-thirds of all existing drug entities are weak electrolytes, which in aqueous solution are at least partly present as ions. They are formed by releasing protons (in the case of acids) into, or by accepting protons (for bases) from, an aqueous environment. The resulting ions are easily hydrated and hence are generally more water soluble than the non-ionized molecules. The aqueous solubility of weak electrolytes can be controlled by adjusting the pH of the solution via the equilibrium between the non-ionized and ionized species. The strength of acids, i.e., the power to release a proton into the aqueous environment, is quantitatively expressed by the ionization constant.

Dissociation of a monoprotic acid HA can be described by the equilibrium in Eq. (1).

$$[\mathrm{HA}] \underset{}{\overset{K_a}{\rightleftarrows}} [\mathrm{H^+}] + [\mathrm{A^-}] \tag{1}$$

The equilibrium constant K_a of this reaction is the ionization constant, defined in Eq. (2).

$$K_a = \frac{[\mathrm{H^+}][\mathrm{A^-}]}{[\mathrm{HA}]} \tag{2}$$

For more convenient use, ionization constants are converted into their negative decadic logarithms (Eq. 3).

$$\mathrm{p}K_a = -(\log_{10} K_a) \tag{3}$$

As an example, for benzoic acid $K_a = 6.4 \times 10^{-5}$ and the corresponding pK_a is 4.19.

Dissociation of a monobasic compound equilibrium may be expressed as follows:

$$[\mathrm{BH^+}] \overset{K_a}{\rightleftarrows} [\mathrm{H^+}][\mathrm{B}] \tag{4}$$

with the dissociation constant given by Eq. (5).

$$K_a = \frac{[\mathrm{H^+}][\mathrm{B}]}{[\mathrm{BH}]} \tag{5}$$

To arrive at a uniform system, the protonated base is regarded as the corresponding acid of the free base. Thus the strength of both acids and bases can

Table 12.1 Classification of acids and bases according to strength.

Attribute	Acids pK_a	Bases pK_a
Very strong	<0	>14
Strong	0–4.5	9.5–14
Weak	4.5–9.5	4.5–9.5
Very weak	9.5–14	0–4.5
Extremely weak	>14	<0

now be expressed with the same strength-indicating parameter, i.e., the acid ionization constant pK_a (Table 12.1).

12.4.2
Ionization and pH

After dissolving an acidic drug substance HA in water, the total mass concentration of A in solution, whether ionized ($[A^-]$) or not ($[HA]$), remains constant irrespective of the extent of the dissociation reaction and is the sum all of the species containing A (Eq. 6).

$$[A_T] = [HA] + [A^-] \tag{6}$$

To describe the extent of an ionization equilibrium, Eq. (6) is substituted into Eq. (2) for [HA] which yields Eq. (7),

$$K_a([A_T] - [A^-]) = [H^+][A^-] \tag{7}$$

and the solution for $[A^-]$ is given by Eq. (8).

$$[A^-] = \frac{K_a[A_T]}{K_a + [H^+]} \tag{8}$$

Thus, the fraction of the ionized species is given by Eq. (9),

$$f_{A^-} = \frac{[A^-]}{[A_T]} = \frac{K_a}{[H^+] + K_a} \tag{9}$$

and, as the mass balance is maintained, the fraction of the unionized species is

$$f_{HA} = 1 - f_{A^-} = \frac{[HA]}{[A_T]} = \frac{[H^+]}{[H^+]+K_a} \tag{10}$$

Transformed into an expression with convenient logarithmic terms, where $[H^+]$ is replaced by the definition of pH,

$$pH = -(\log_{10}[H^+]) \tag{11}$$

we have

$$f_{A^-} = \frac{1}{1 + 10^{pK_a - pH}} \tag{12}$$

and for the undissociated acid

$$f_{HA} = \frac{1}{1 + 10^{pH - pK_a}} \tag{13}$$

For bases the corresponding equations may be derived from Eqs. (4) and (5) in the same way. For the fractions of protonated base, f_{BH+}, and of the unionized base species, f_B, respectively, we obtain Eqs. (14) and (15).

$$f_{BH^+} = \frac{1}{1 + 10^{pH - pK_a}} \tag{14}$$

$$f_B = \frac{1}{1 + 10^{pK_a - pH}} \tag{15}$$

These equations describe in a most condensed form the interrelationship between pK_a and solution pH. They also illustrate the function of pK_a as the key parameter providing a measure of the strength of acids and bases.

12.4.3
Solubility

Most drug salts are prepared to improve the aqueous solubility of a drug substance. To assess the solubility improvement, the interrelationship of pK_a, solution pH and intrinsic solubility of simple weak electrolytes is reviewed, based on theoretical solubility–pH profiles of a base, an acid and a zwitterionic compound (Figs. 12.2 to 12.4, respectively). The diagrams show pH regions where solubility is independent of pH, and other regions below or above the pK_a where the solubility increases exponentially with pH up to a cut-off point. These are the regions where solubility improvement by salt formation can be expected. However, for each salt the solubility levels off at a particular pH, beyond which no further increase in solubility is observed. This pH is referred to as "pH_{max}". In a drug salt suspension, pH_{max} not only marks this change of pH–solubility relationship but also indicates a change in the nature of excess solid in equilibrium with the saturated solution. This statement becomes quite important for the preparation and isolation of solid salts: For salt B of the basic compound in

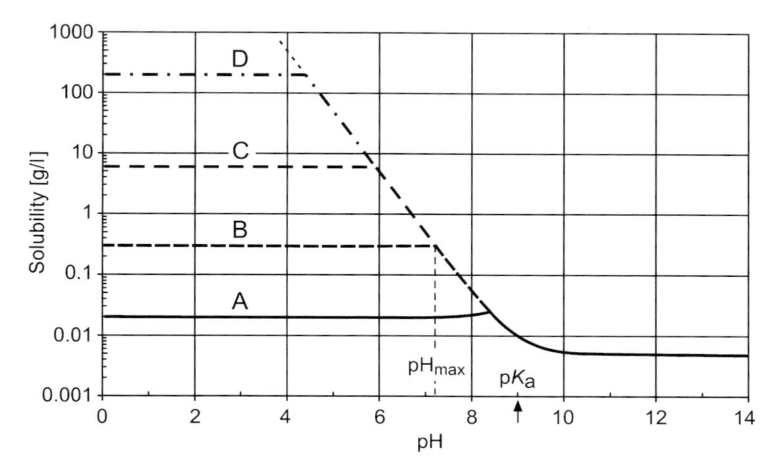

Fig. 12.2 Theoretical solubility diagram of a basic substance, $pK_a = 9.0$, and four of its salts characterized by different solubility products. pH_{max} is indicated for salt B.

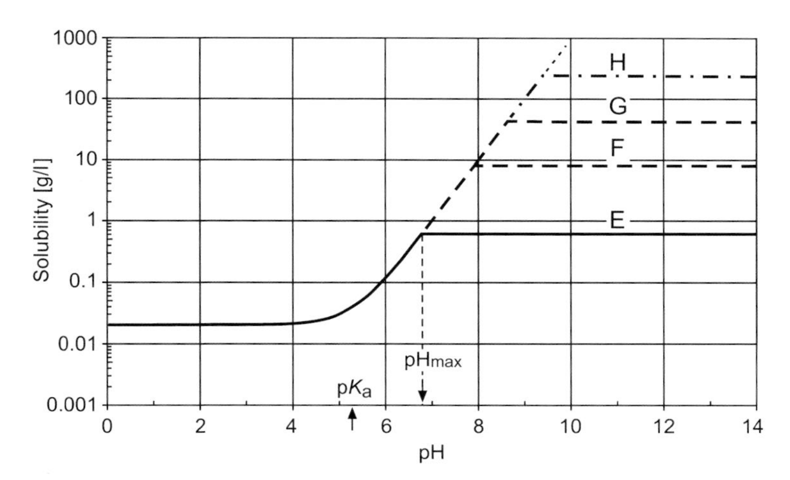

Fig. 12.3 Theoretical solubility diagram of an acidic substance, $pK_a = 5.5$, and four of its salts characterized by different solubility products. pH_{max} is indicated for salt E.

Fig. 12.2, above pH_{max} (7.2) the excess solid is the free base, whereas below pH_{max} the undissolved solid is the salt. pH_{max} is the only condition where both solid salt and solid free base can co-exist in equilibrium with the solution.

Figures 12.2 and 12.3 show the solubility branches for four different salts, A–D for the basic drug and E–H for the acidic drug. Each is characterized by the level of its maximum solubility plateau, associated with a specific pH_{max}. The region of the unionized drug substance remains uninfluenced by salt for-

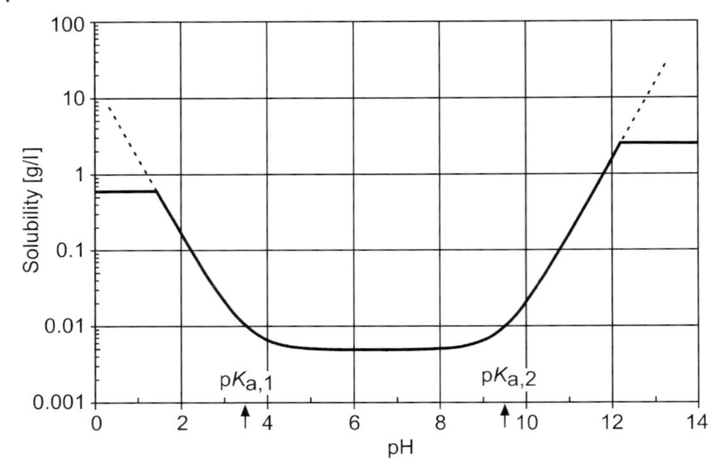

Fig. 12.4 Theoretical solubility diagram of an amphoteric substance, $pK_{a,1} = 3.5$, $pK_{a,2} = 9.5$.

mation. The amphoteric substance (Fig. 12.4) has a deep solubility plateau in the pH range between the pK_a values. Beyond these, it can behave either as a base or as an acid, so that salts can be formed with either acids or bases of appropriate strength, and different salt solubility levels are achieved as described above for the monovalent basic or acidic substance. Theoretical equations for the above pH–solubility relationships can be found in [3].

Supersaturation phenomena can interfere with an experimental search for salts of high solubility. A practical approach is to titrate aqueous suspensions of a sparingly soluble basic drug with one of a selected series of acids, e.g., hydrochloric, tartaric, fumaric, citric, or phosphoric acid. After incremental additions of acid, samples of the supernatant solution are analyzed for drug concentration. Increasing concentrations are found as the pH is progressively lowed. However, frequently, much higher values may be found initially than can be reproduced in later experiments. This is because, initially, there is no trace of the new salt present in the excess solid as the pH falls below pH_{max} by the addition of acid, hence supersaturation is likely to occur and more base would dissolve along the rising solubility line of the base. Depending on the nucleation and kinetics of crystallization of the new salt, the system equilibrates sooner or later, and the concentration eventually arrives at the lower level typical for the solubility of the respective salt. However, although fast equilibration is common, the time scales can in principle vary widely, and at times a desired solid salt does not crystallize within a reasonable time.

12.5 Approaches to Salt Screening

12.5.1 Initial Data

Before starting experimental work in search of salts of a drug entity, two parameters need to be known. As outlined above, the pK_a(s) of the drug substance must be determined experimentally. Alternatively, they can be predicted from any of the available estimation programs. In addition, the aqueous solubility of the non-ionized form – free base or free acid – can be estimated at least to an order of magnitude by Yalkowsky's general solubility equation [4, 5]. These data are useful in constructing a "first-guess" pH–solubility diagram.

12.5.2 Selection of Salt Formers

Based on the pK_a values, a series of salt formers can then be selected from those frequently used pharmaceutically (Tables 12.2 and 12.3) while others must be rejected. From experience, to form a stable salt the pK_a values of an acid–base pair should differ by at least two pK_a units, i.e., the pK_a of the acid should be at least two units lower than that of the base. This corresponds to a situation in which both compounds, brought together in water, are ionized to a degree of at least 90%. Strong mineral acids such as HCl ($pK_a \approx -6$) or H_2SO_4 ($pK_a = -3$) can form solid salts with the weak base atazanavir, having a pK_a as low as 4.25,

Table 12.2 Acids frequently used as salt formers.

Acid	$pK_{a,1}$	$pK_{a,2}$	$pK_{a,3}$
Hydroiodic	–8		
Hydrochloric	–6		
Hydrobromic	–6		
Sulfuric	–3	1.92	
p-Toluenesulfonic	–1.34		
Nitric	–1.32		
Methanesulfonic	–1.2		
Maleic	1.92	6.23	
Phosphoric	1.96	7.12	12.32
Embonic	2.51	3.1	
L(+)-Tartaric	3.02	4.36	
Fumaric	3.03	4.38	
Citric	3.13	4.76	6.40
D,L-Lactic	3.86		
Succinic	4.19	5.48	
Acetic	4.76		

Table 12.3 Bases frequently used as salt formers.

Base, cation	$pK_{a,1}$	$pK_{a,2}$	$pK_{a,3}$
Sodium	14		
Potassium	14		
Zinc	~14		
Choline	13.9		
L-Arginine	13.2	9.09	2.18[a)]
Calcium	12.7		
Magnesium	11.4		
Diethylamine	10.93		
L-Lysine	10.79	9.18	2.16[a)]
Ammonium	9.27		
Meglumine	8.03		
Tromethamine	8.02		

a) pK_a of the carboxylic function of the amino acids.

whereas attempts to isolate a salt with either acetic acid (pK_a=4.76) or benzoic acid (pK_a=4.19) would fail with such a weak base. Instead, reducing the volume of a slurry or solutions of the base in solutions of the acids to dryness would simply return the base because acetic acid evaporates completely or a mixture of base and benzoic acid, respectively. An annotated compilation of salt formers, including less frequently encountered ones and those used to solve special problems, is available in [6].

12.5.3 Automated Salt Screening

The preparation of salts from a basic or acidic drug candidate usually follows classical chemical procedures. However, for feasibility studies these have been scaled down to milligram and microgram formats, and for screening purposes the traditional procedures have been replaced by partly or fully automated medium or high-throughput methods. Several companies active in pharmaceutical research (e.g., Merck, Novartis, Pfizer), chemical parallel synthesis (Symyx), scientific services (Solvias, Transform) or instrument manufacturing (Zinsser, Chemspeed, Autodose) have developed processes and instruments suitable for automated polymorphism screening. Basically, these systems consist of dispensing modules for liquid and solid reagents, shaking, stirring and tempering possibilities built around a central proprietary assembly that allows parallel processing of multiple samples. The core device often uses the footprint of a 96-well plate since this format is a quasi-universal standard for instrument manufacturers. With the success of high-throughput screening there are a huge number of commercially available dispensation and analytical instruments compatible with the 96-well format at affordable prices. Special devices are designed for *in situ* monitoring of one or several solid-state properties – the most commonly investigated are morphology, birefringence and Raman spectrometry. In addition, the assembly must support parallel preparation and isolation of multiple samples for a suitable solid-state characterization (e.g., 96 samples processed as one single manual trial). Here again the most frequently applied characterization is based on Raman microscopy and on X-ray powder diffractometry. Some instrument manufacturers of Raman microscopes and X-ray powder diffractometers offer optional motorized x,y,z positioning of samples. This feature allows scanning predefined fields of a multiple sample holder, e.g., the coordinates of 96 different wells, and storage of the results together with the position data within the holder. In addition to the standard evaluation software supplied with these instruments, algorithms are available for comparing and clustering multiple spectra. Because the samples are small, nondestructive analytical methods are preferred. The amount of solid material used for, or isolated in, a single experiment is in the milligram range.

In essence, the same instrumentation as is used for polymorphism screening can efficiently be applied for salt formation research. A mandatory, critical operation for automation of the techniques of solubility determination and polymorphism screening is the capability to distribute minute amounts of solid samples. However, this capability is not needed for salt screening, although high quantitative accuracy is required. The processing variables investigated in salt screening in addition to those studied in polymorphism screening are the nature and stoichiometry of the salt-forming agent. Quantitative precision is more critical in salt research than in polymorphism investigations because of stoichiometry issues. No commercial solid dispensation system is currently available that would allow dosing into a parallel assembly, weighing over one hundred grams, of one milligram of unspecified solid powder with an accuracy as low as 1%. Therefore, the drug candidate under investigation is delivered to the sample wells as a solution that may either be directly processed further or evaporated to remove the carrier solvent. Hence, users prefer the dispensation of a bulk solution, which is more precise, more robust, easier, and faster to process automatically than any presently available solid dispensation technique.

The drawback of using polymorphism screening systems for performing salt formation screening is the limited analytical information generated automatically. For polymorphism screening, chemical structural information is not critical since the experiments focus on the solid-state organization of a single chemical entity. Here the only changes in chemical composition of the solid would result from the formation of solvates. As already mentioned, salt formation screening requires the chemical nature of the solid isolated and the stoichiometry of its components to be determined. X-ray powder diffraction patterns can give information on the crystallinity of the resulting solid residue. Matching diffraction patterns may give the additional information that a residue is the same solid as the native starting material. However, this would simply mean that the formation of a salt failed. If several different patterns are generated, the user still is left without evidence of whether this is due to either polymorphic variations of a common single chemical entity or the presence of different chemical entities, e.g., salts, or a combination of both. In some cases Raman spectroscopy can generate more information. Only combined traditional methods like NMR, elemental analysis and thermogravimetry can provide final evidence as to whether a useful salt has been generated. Such additional analyses require individual, manual processing of each sample to be characterized. The latter time consuming step has not yet been successfully automated. However, to limit the overall effort and also to obtain sample sizes that are more convenient to handle manually, the results of the automated processing phase can be used to pool samples representing the same solid-state structure before such chemical analysis.

12.6 Selection Procedures and Strategies

12.6.1 Points to be Considered

The final decision on the salt and solid-state form of a drug substance is based on data collected from analytical and physical characterization of a certain number of prepared salts. The data are evaluated primarily against the requirements set by biopharmaceutical and therapeutical considerations for the use of the drug, then pharmaceutical and chemical-technological aspects in view of the development and manufacturing of dosage forms are considered, as shown in Section 12.3, as well as the synthesis and isolation of the drug substance. It is not possible to present here a complete catalogue of all the data to be provided for a full assessment, and there are cases where a minimum of information may lead to a reliable decision, but the most fundamental data can be found in the tables of the case reports presented in Section 12.7. Further categories of properties may have to be studied, depending on the nature of the substance and on the therapeutic application of the future drug product. Some typical and also some less frequently encountered issues, in brief, are:

- Solubility and dissolution rate.
- Chemical stability, including potential interaction between drug entity and counterion; stability in the presence of pharmaceutical excipients (drug/excipient compatibility).
- Stability of the morphic state in bulk form; in solid and suspension dosage forms.
- Hydrate formation and stability during storage and processing; chemical stability of hydrated versus anhydrous forms (influence of released hydrate water during storage); uptake of water of hydration by anhydrates in solid dosage forms with consequences for mechanical properties.
- Molecular weight: large counterions may surmount the tolerable drug amount to be packed into a solid dose unit; in contrast, for extremely low-dose drugs a larger molecular weight can improve handling and content uniformity.

With drug candidates showing absorption difficulties, investigations may even include *in vivo* studies in animals (rat, dog) with experimental formulations to identify the most suitable salt form or solid-state form. Sometimes such a search may address the state of distribution of the active substance in an experimental formulation (e.g., micronized, nano-sized; amorphous, in suspension, emulsion, microemulsion) and so the borderlines between chemical technical operations and pharmaceutical formulation and processing techniques may become quite diffuse.

Choosing suitable salt forms can have the following advantages:
- annoying polymorphism problems can be circumvented;
- high hygroscopicity resulting in deliquescence may be avoided;
- amorphous material may be turned to a crystalline salt;
- taste and smell problems may be minimized;
- the melting point may be raised to improve mechanical properties (e.g., for milling), to the extent that liquid bases or acids are turned into solids;
- a drug substance can be purified (as the last or penultimate synthesis step);
- absorption rate may be controlled: retardation of gastrointestinal absorption by salts with low solubility/low dissolution rate; transdermal absorption may be enhanced;
- irritation may be avoided (e.g., for inhalation);
- two pharmacological principles can be combined (though in stoichiometrically fixed ratios).

12.6.2 Final Decision

Up to the late 1960s there was no question that the inventing chemist who synthesized the substance determined the salt form for safety investigations and for subsequent dosage form development. During the decades thereafter, feedback from Pharmaceutical Development into Medicinal Chemistry has shown that certain substance properties may be unfavorable up to the point that further work with the drug candidate in the form provided could slow down or even halt the process of developing dosage forms for clinical studies. Technical problems or insufficient bioavailability could require the chemist to come up with other salt forms. Later, a preformulation phase was introduced to establish the physicochemical and technological properties of the candidate beforehand while biological and pharmacological work was still going on in Discovery. Based on a first assessment the chemist was asked to prepare small amounts of a few salts. These were studied for critical properties, and a final recommendation for the salt form to proceed with was made. Figure 12.1 reflects this route from discovery to development in a concise scheme. The present paradigm in identifying the final salt and solid-state form has shifted from screening out only one candidate entity to studying several interesting compounds in parallel, including the search for suitable chemical and physical forms, and to selecting

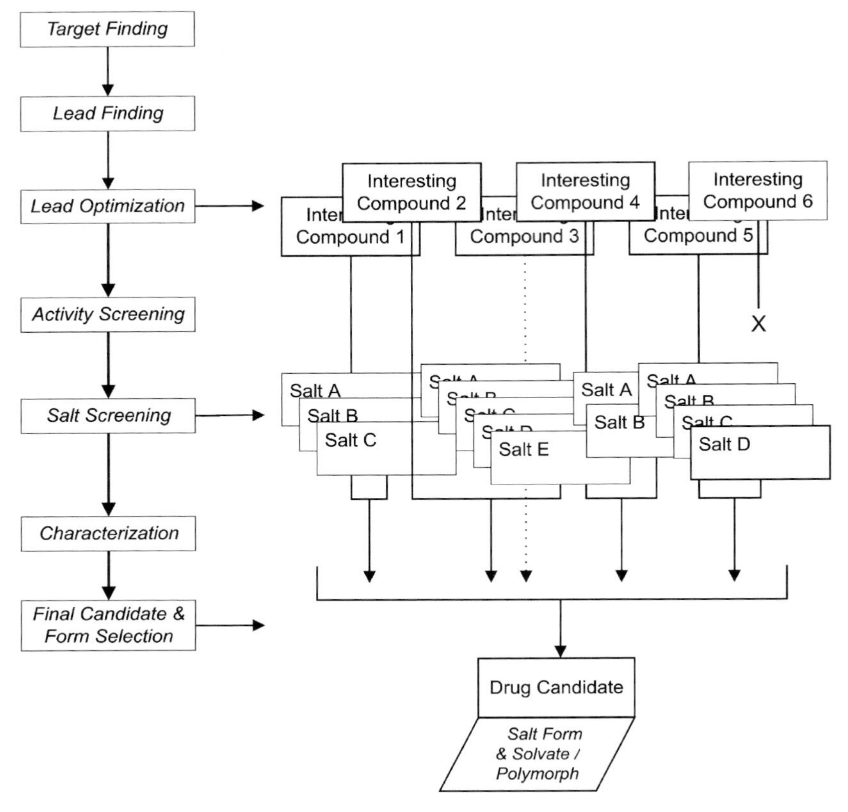

Fig. 12.5 Organizational sequence of activities in industrial Research and Development for identifying drug candidates and their suitable salt and solid-state form.

the final candidate in its final salt and solid-state form from this group of fully characterized pharmacologically similar entities. Thus, the decision point along the time-line of the discovery–development process of a drug product has shifted backwards into the Research Phase (Fig. 12.5).

The final decision calls representatives from Chemistry, Biology, Chemical Process Development, Pharmaceutical Development, Drug Metabolism, Safety, Analytical Development and Regulatory Affairs, and Patent Affairs to assess the data presented and to propose the candidate for development. Depending on the critical issues that will be discussed, some units have more weight in the decision while others play a more advisory role. An example may illustrate such a situation:

An NCE has been characterized with the conclusion that an oral product would be equally feasible with the free base or with the mesylate salt. While the latter could be used also for manufacturing an injectable solution, such a solution can also be prepared by dissolving the base together with the stoichiometric amount of methanesulfonic acid. Thus, in this case, Dosage Form Development had no preference. The final decision here was taken by Chemical Process Development with the argument that they preferred the salification as a further purification step for the final product.

12.6.3
Salt Form and Life Cycle Management of Drug Products

Salt issues reappear regularly during a drug's life cycle. In most cases a new drug is launched as a solid dosage form for oral use. Once established in therapy in its primary indication, other routes of administration are investigated; also, the field of indications may be widened and, consequently, new products with the same drug substance are added to the product line. Such line extensions usually require other salt and solid-state forms that give the drug a better match for the requirements of the additional products and fields of use. Last but not least, extension of proprietary protection of drug substances demands intensive search for additional salts and solid-state forms that may secure the market for the originator for additional years beyond the normal patent life.

12.7
Case Reports

12.7.1
Overview of Salt Forms Selected

Of 43 ionizable new drug substances entering development in Novartis between 2000 and 2003, 19 (44%) were suitable for development in their native form. Fourteen of them were free bases, four were free acids and one was the internal salt of an amphoteric substance. The remaining 24 candidates were more suitable as a salt. Among them, the most frequent salt selected was the hydrochloride (four candidates) followed by the fumarate, maleate and succinate (three candidates each) and the benzenesulfonate (two candidates). Finally, only one candidate entered development in each of the following salt forms: calcium, potassium, arginine salt, aspartate, lactate, tartrate, methanesulfonate, sulfate and pamoate. Six substances, whether native form or a salt, were selected in a hydrated form as the most suitable entity for development.

12.7.2
Salt Selection Process

The selection of these salts was performed in one, two or three steps, depending on the physicochemical characteristics of the native form involved. In the most clear-cut situation, the properties of the unionized form were so favorable (1 out of 43) that there was no need to look for a more suitable solid form.

Whenever the native drug substance properties did not appear to satisfy all physicochemical criteria expected for the development of a suitable drug product, a salt pre-screening was performed. This allowed the identification of which counter ions out of the largest possible pool satisfying pharmaceutical safety requirements were able to form a salt with the new drug substance. In the third step, a selection of the most promising salts was prepared and more extensively characterized to allow the final selection of the form to be developed.

The crystallinity and chemical stability under stress conditions, such as light exposure, high temperature and high humidity, were investigated first.

Water vapor sorption, the potential for polymorphic transformations, for hydrate formation, the chemical compatibility with selected excipients, chemical stability in solution, melting range and aqueous solubility at physiological pH in regard to the anticipated doses are systematically studied.

The following three example case studies indicate the width of the field of salt and solid state selection of drug substances. Case 1 is a brief report on a quite uncomplicated substance, Case 2 illustrates a typical and most frequently encountered pattern, while in Case 3 polymorphism and chirality play rather subtle roles.

12.7.3
Case 1: NVP-BS001

For NVP-BS001 an oral solid dosage form with immediate release was foreseen, with an upper single unit strength of around 100 mg. The free base, mp 220 °C, crystallizes as plates, 90% of them less than 30 μm in diameter. The aqueous solubility in buffered systems below pH 7 is very high, ranging from 20 to more than 100 mg mL^{-1}, in contrast to the relatively low solubility in pure water (0.2 mg mL^{-1} with a resulting pH of 8.2). The stability in aqueous solution is excellent, with less than 1% degradation after 3 days at 100 °C in 0.1 M HCl and in 0.1 M NaOH. Significant degradation in solution is only achieved by treatment with hydrogen peroxide. No detectable decomposition is found in the bulk powder after three days at 100 °C. The crystalline form remains unchanged after equilibration experiments in a series of relevant solvents (a suspension of the solid in the respective solvent is subjected to vibration at a given temperature, usually room temperature) as well as after mechanical stress. The powder is not hygroscopic, its water uptake at 94% r.h. being only 0.24% by weight.

Summarizing the essential characteristics, the free base of NVP-BS001 displays high aqueous solubility and excellent solid-state stability. With the demonstration of rapid dissolution of the powder, the project team decided that the manufacture of a salt would offer no improvement of physical properties but rather would add to the complexity of chemical synthesis and product formulation.

Clearly, such an ideal situation is an exception, representing the rare case where the unionized form of an ionizable drug displays excellent aqueous solubility over the whole physiological pH range. In addition, the solid-state properties of free base crystals were also outstanding. Usually, compromises on physicochemical properties need to be drawn since one single solid form rarely displays exclusively favorable properties, and the decision requires careful comparison of the biopharmaceutical, processing and storage relevant parameters of the alternative forms.

12.7.4
Case 2: NVP-BS002

NVP-BS002 is a weak diprotic base ($pK_{a,1}=11.5$; $pK_{a,2}=4$). The intended administration route was oral with an immediate release dosage form. The strength of a single dose unit was aimed at up to 150 mg. An investigation for crystallization with possible counter ions without toxicity potential showed that the only suitable salt candidates were the hydrochloride and the maleate. Both salts were further characterized and their physicochemical properties compared with those of the free base. The data are compiled in Table 12.4.

- *Crystallinity*: According to X-ray powder diffraction the crystallinity was medium for all three forms.
- *Melting point*: All melting points were above 185 °C, and no problems for milling were expected from this aspect.
- *Morphology*: The hydrochloride and the maleate salts were isolated as needle shaped crystals whereas the free base crystallized as column shaped primary particles.
- *Hygroscopicity*: Exposure to 92% relative humidity caused weight gain, which was highest for the hydrochloride with 0.6% after one day and increased to 5.4% after three days. Although this water uptake is significant, it is slow enough to allow powder processing without problems.
- *Solubility*: The aqueous solubility in water is best for the hydrochloride, but this salt has the lowest solubility in 0.1 M HCl, as expected by a common-ion effect. Whereas the intrinsic dissolution rate of the three candidates is identical at pH 4, the maleate and hydrochloride salts dissolve faster than the free base in 0.1 M HCl.
- *Crystal modification after equilibration in solvents*: After 24 h equilibration by vibrating a suspension at room temperature in ethanol, isopropanol, ethyl acetate and acetone no changes were observed in the X-ray powder diffraction patterns, DSC and TG. The same result was observed in water with the maleate. When the hydrochloride was suspended in water a small amount of free base was formed. When the same procedure was applied to the free base a small additional peak appeared on the X-ray powder diffraction pattern, indicating a potential polymorphism issue.
- *Chemical stability*: In the solid state, after one week stress at 80 °C in a sealed container and at 75% relative humidity (either alone, or in mixture with two

different excipients mixtures for standard oral formulations) the stability was good for all three candidates, with a slight advantage for the hydrochloride. The stability with regard to light exposure delivered a slightly different ranking, but still with the best results for the hydrochloride.

Table 12.4 Physicochemical properties of NVP-BS002 solid forms.

Item	Free base	Hydrochloride	Maleate
Salt : base mass ratio	1.000	1.094	1.299
Crystallinity (XRPD)	Medium	Medium	Medium
Melting point (DSC) (°C)	203	243	185
Morphology	Column	Needles	Needles
Loss on drying (%) by TG			
Initial	0.11	0.10	0.42
After 1 day 80% r.h.	0.12	0.11	0.35
After 1 day 92% r.h.	0.15	0.68	0.56
After 3 days 92% r.h.	0.12	5.51	0.62
Solubility at 25 °C (mg mL^{-1})			
0.1 M HCl	0.22	0.17	0.29
Buffer pH 4.0	<0.01	<0.01	0.03
Buffer pH 6.8	<0.01	<0.01	<0.01
Water	<0.01	0.77	0.36
Methanol	34	3.7	18
Ethanol	20	0.41	8.9
Microemulsion concentrate	50	4.3	12
Intrinsic dissolution rate (mg min^{-1} cm^2)			
0.1 M HCl	0.0008	0.0066	0.0061
Buffer pH 4.0	0.0040	0.0035	0.0033
Degradation (%)			
• After 1 week 80 °C in a sealed container			
Bulk	0.60	0.52	0.75
In excipient mix 1 (1%)	0.68	0.52	0.72
In excipient mix 2 (1%)	0.68	0.54	0.64
• After 1 week 80 °C at 75% r.h.			
Bulk	0.62	0.54	0.69
In excipient mix 1 (1%)	0.66	0.56	0.68
In excipient mix 2 (1%)	0.66	0.53	0.66
• Xenon light (1200 kLuxh)			
Bulk	2.09	0.91	1.23

12.7.4.1 **Discussion and Decision**

As the behavior of the three compounds was quite similar, the solubility in aqueous systems was considered to be the most important issue with regard to bioavailability. All three candidates are slightly soluble in 0.1 M HCl, whereas in water only the salts are soluble, with an advantage for the hydrochloride, which has twice the solubility of the maleate. The intrinsic dissolution rate is also unfavorable for the base. Thus, the hydrochloride and maleate were proposed for further development. The pharmacists opted for the hydrochloride after a study in rats compared the relative bioavailability of different oral formulation principles. The conventional powder mixture using the hydrochloride at therapeutic dose displayed performance close to a solution and its absolute bioavailability of 30% was considered acceptable. It was, however, proposed that close attention be paid to the system, to find out whether the absorption was solubility or dissolution rate limited, but this aspect was not clarified before the project was discontinued for toxicity reasons.

12.7.5
Case 3: NVP-BS003

In the late 1980s the racemic nootropic NVP-BS000 entered development. Because the relatively strong base ($pK_a=9$) was oily at room temperature, Chemistry Research delivered the hydrochloride salt. The substance is the methyl ester prodrug of the pharmacologically active acid, is fairly stable in the solid state, and its chemical stability in solution is just sufficient to allow passage of the gastrointestinal membrane unhydrolyzed.

While the aqueous solubility of the free base is 0.7 mg mL^{-1}, that of the hydrochloride salt is 40 mg mL^{-1}. The salt melts at 195 °C and is not hygroscopic. An endothermic peak at 120 °C was observed in DSC, but raised no concerns as the signal was still present in a second DSC scan after heating above 120 °C and cooling down before starting the second scan, indicating that this signal is associated with a reversible event. Two years later it was decided to replace the racemic substance by the same salt of the corresponding active enantiomer, NVP-BS003, which could be synthesized via enzymatic reaction. The intention was to limit potential toxicity problems by disposing of the burden of the pharmacologically inactive portion of the racemic mixture. No polymorphic risks were anticipated with this change although it had not been investigated whether NVP-BS000 hydrochloride crystallizes as a *racemic conglomerate* (equimolecular mixture of homochiral crystals of each antipode), *racemic compound* (crystals in which two enantiomeric molecules of opposite chirality are paired up in the unit cell of the crystal lattice) or as a *pseudoracemate* (the two antipodes are arranged more or less randomly in the same crystal lattice).

Thus, two batches of the hydrochloride salt of the pure active enantiomer were synthesized. Their processing differed by the start temperature of the crystallization and the final drying temperature. Solid state characterization of both batches showed that one batch consisted of a single crystal form whilst the other one was contaminated with about 50% of a different polymorph. Both polymorphs of NVP-BS003 hydrochloride displayed different X-ray powder diffraction patterns, which again differed from the pattern of the racemic drug substance NVP-BS000 hydrochloride. This indicates that polymorphs A and B were not present in the racemic mixture; hence it may be concluded that the racemate crystallizes either as a racemic compound or as a racemic conglomerate of a third polymorphic variant. Notably, the aqueous solubility of the racemate NVP-BS000 hydrochloride was 40 mg mL^{-1} whereas it exceeded 100 mg mL^{-1} for the active enantiomer NVP-BS003 hydrochloride.

DSC analysis of NVP-BS003 samples from the mixed polymorphs batch (A + B) displayed three endothermic peaks (34, 64, 102 °C) below the melting peak (mp 187 °C) whereas the pure polymorph B displayed only two endothermic peaks, at 56 and 102 °C, before melting at 187 °C (scan trace 2 in Fig. 12.6). The DSC events were assigned to the following solid-state changes: the endotherm at 34 °C resulted from the solid-state conversion of the crystal form A into C followed by the conversion of C into D at 102 °C. The endotherm at 56 °C was associated with the transition of B into C (see DSC traces and conversion pathways in Fig. 12.6). Cooling caused form D to revert to A. Thus, the mixture of polymorphs A and B could be converted into pure polymorph A by thermal treatment at 90 °C followed by cooling to room temperature.

An additional complexity occurred in the presence of water. When both polymorphs A and B were dissolved in water, their initial solubility rose transiently above 115 g L^{-1} and stabilized for both forms at 104 g L^{-1} after 3 h. This suggests that either another new polymorph or a hydrate was formed in water *in*

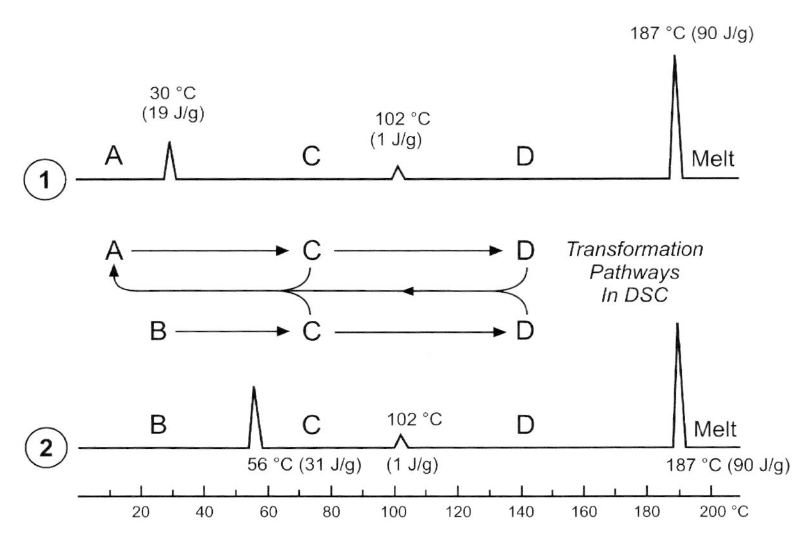

Fig. 12.6 DSC scans of NVP-BS003 samples. Trace 1: Polymorph A; trace 2: polymorph B. The intermediate diagram shows the thermal interconversion pathways of the polymorphs A, B, C and D.

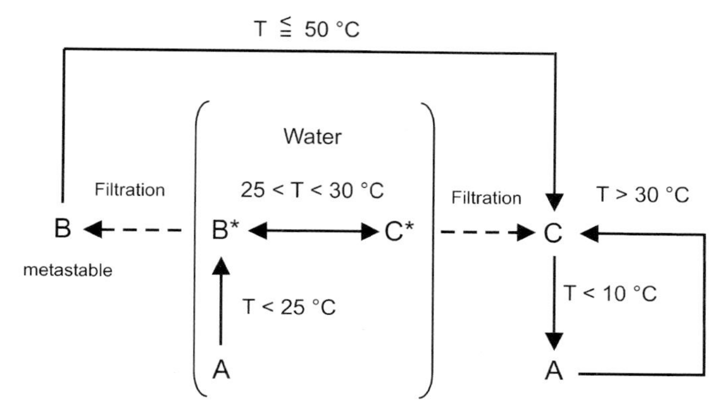

Fig. 12.7 Transformation pathways of NVP-BS003 (see text).

situ. Depending on the equilibration temperature in water the crystal forms obtained after drying differed. Starting from either A or B or a mixture of both at 25 °C and below, the resulting dry crystal form was polymorph B. At 30 and 37 °C, B or a mixture of A and B were transformed into A. The transformation scheme is shown in Fig. 12.7. This indicates that there was at least one aqueous precursor transforming into B after drying. The precursor was formed by equilibration in water of either A or B or a mixture of both polymorphs at 25 °C or below. Above 30 °C in water, this precursor was transformed into either C or a precursor of C that was converted into C after drying. In the dry state, form C was reversibly converted into A in a DSC pan by cooling to below 10 °C whilst C reappeared after heating above 30 °C. These transformation pathways were confirmed and refined with experiments in which bulk powder samples of forms A and B were stored under isothermal conditions at several temperatures. After one month the change in polymorph content was assessed. Thus A and C presented an enantiotropic relationship that could not be demonstrated using the traditional comparison of melting temperature and melting enthalpy since both polymorphs transformed into another common form (form D) before melting.

Technical concerns arose because the conversion temperature of both enantiomorphs A and C occurred within a temperature range around 30 °C, and the situation was worsened by the potential formation of a metastable polymorph B formed by drying of a precursor obtained by treatment of A or C in water below 30 °C. This meant that it was practically impossible to avoid at least partial transformation of the crystal modification of the active ingredient during processing, handling or storage of the final dosage form.

In this particular case polymorphic transitions may hardly have had any biopharmaceutical consequences. The solubility of the polymorphs is relatively high, and it is reasonable to assume that dissolution from the solid would not be negatively affected. However, these transitions have several potential effects on the mechanics of the drug product, depending on the dosage form involved. Because the oral doses of NVP-BS003 were anticipated to be rather high, potential modification of the mechanical properties associated with polymorphic transitions may have had a severe influence on cohesion, hardness, friability and disintegration of tablets. Therefore, the development strategy primarily focused on the exploration of an oral dosage form, which is not so much affected by physical changes of the drug's solid-state properties, e.g., hard gelatine capsules or sachets. As a backup, the exploration of other salts was also suggested. However, the project was discontinued for other reasons before completion of dosage form development.

References

1 S.K. Chandrasekaran, in *Transdermal Delivery of Drugs*. Vol. III, A.F. Kydonieus, B. Berner (Eds.), CRC Press, Boca Raton, FL, **1987**, pp. 35–38.

2 A. Fini, G. Fazio, I. Rapaport, presented at *DHEP Plasters: A new delivery system for a topical NSAID*, Geneva, **1993**. Abstract published in *Drugs Exp. Clin. Res.* **1993**, *19*, 81–88.

3 P.H. Stahl, Preparation of water-soluble compounds through salt formation, in *The Practice of Medicinal Chemistry*, C.G. Wermuth (Ed.), Academic Press/Elsevier Science, Amsterdam, Boston, **2003**, p. 605.

4 S.H. Yalkowsky, Solubility and solubilization of nonelectrolytes, in *Techniques of Solubilization of Drugs*, S.H. Yalkowsky (Ed.), Marcel Dekker, New York, **1985**, pp. 1–14.

5 N. Jain, S.H. Yalkowsky, Estimation of the aqueous solubility. I: Application to organic nonelectrolytes, *J. Pharm. Sci.* **2001**, *90*, 234–252.

6 Handbook of Pharmaceutical Salts. Properties, Selection, and Use, P.H. Stahl, C.G. Wermuth (Eds.), Verlag Helvetica Chimica Acta, Zürich/Wiley-VCH, Weinheim, **2002**.

13
Processing-induced Phase Transformations and Their Implications on Pharmaceutical Product Quality

Ramprakash Govindarajan and Raj Suryanarayanan

13.1
Introduction

The importance of physical characterization of pharmaceutical solids has long been recognized [1, 2]. The physical form (polymorphic form, degree of crystallinity and state of solvation) of the active pharmaceutical ingredient (API) can significantly influence the stability and performance of the dosage form. Even though an appropriate physical form of the API may be selected, it may not be retained in the final pharmaceutical product. Stresses experienced during processing and interactions with formulation components can result in phase transformations [3]. The behavior of the resulting pharmaceutical composition will be influenced by the extent of conversion and the properties of the transformed phase. It is therefore important to detect, quantify if necessary, and understand the implications of such phase changes on product quality and performance.

Physical characterization of the final product would reveal the overall effect of processing. However, multiple phase transformations can occur during the sequence of pharmaceutical processing steps. The effect of each processing step on physical form, and the behavior of the transformed phase, can be understood only by monitoring the phases *during* processing. This information can then be used to design processes and optimize formulation and processing variables, with the objective of controlling the physical form, not only during manufacture but also during the shelf-life of the final product.

Scheme 13.1 outlines the possible phase transitions during pharmaceutical processing. Although not an exhaustive summary, it lists the common transitions of interest to a drug formulator. Changes, either in the arrangement of molecules in the crystal lattice or in lattice order, can lead to physical transformations such as amorphization, crystallization or polymorphic transitions. Complex formation, incorporation or removal of solvent molecules from the crystal lattice and interconversions between salt and free acid/base forms constitute changes in the chemical composition of the phase. Hence, the latter do not fit

Polymorphism: in the Pharmaceutical Industry. Edited by Rolf Hilfiker

ISBN: 3-527-31146-7

under the conventional definition of "physical solid-state transformations". However, in light of their importance, we will include these under the broad definition of processing-induced phase transitions (Scheme 13.1), and discuss some pharmaceutically relevant systems. The product phase formed as a result of a change in the chemical composition may in turn undergo further transformations. For example, stoichiometric incorporation of water into the crystal lattice of an anhydrous phase results in the formation of a hydrate. Subsequent dehydration of the hydrate may result in an amorphous anhydrate. This hydration–dehydration cycling can also result in a stable or metastable crystalline anhydrate, as has been demonstrated for theophylline [4].

Scheme 13.2, based on the stress–relaxation concept of Morris et al. [3], shows how thermal or mechanical stresses or interaction with other formulation components (e.g., water) can induce phase transformations. Thus, a transition may be thermodynamically favored, resulting in a product phase that is stable under the stressed conditions. The transition of an anhydrate into a hydrate during wet granulation and the formation of a higher-temperature-stable polymorph on heating are instances of such conversions. The stress might also result in a kinetically stabilized metastable phase (e.g., partial or complete amorphization due to mechanical processing, vitrification of solutes in frozen solutions). These constitute "trapping" of the system under the applied stress (Scheme 13.2). When the stress is partially or completely removed (e.g., when the wet-massed granules are dried, or when the compressive force is removed during tableting or when a frozen solution is heated to an annealing or drying temperature), the system may "relax" back to the original state. If the phase generated under stress is metastable under the de-stressed conditions, it will tend to convert into the stable form. The kinetics of this transformation will determine the metastable phase concentration in the final product, over its shelf-life. During manufacture of a dosage form involving several unit operations, the formulation components can undergo multiple stress–relaxation cycles.

Phase transformations during pharmaceutical processing have been the subject of several reviews [3, 5–7]. A recent one deals with phase changes during the manufacture of solid oral dosage forms and their implications on product performance [7]. Our objective is to focus on pharmaceutical processes, leading from drug substance to end product, and to provide a broad overview of (a) phase transformations *during* various pharmaceutical unit operations, (b) their detection, and (c) potential repercussions on the final product quality. We emphasize the potential advantages and disadvantages of such transitions. Very often, a transition can be beneficial in some respects and detrimental in others, as will be evident from the discussed examples.

While phase transformations of the API can have a direct effect on the stability and performance of the final product, any such changes in excipients can also affect product quality. In this chapter, we consider such processing-induced changes in drugs as well as excipients, and discuss (a) solid-state manipulation intentionally brought about by pharmaceutical processing and (b) unintended transformations that may have a major impact on product properties.

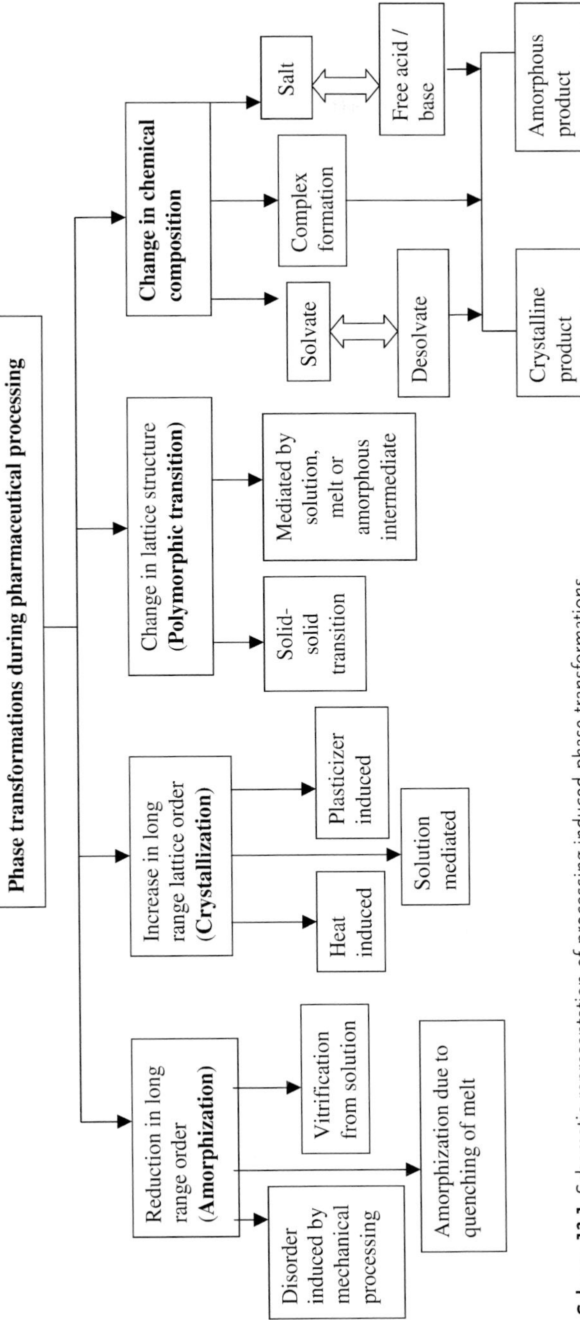

Scheme 13.1 Schematic representation of processing-induced phase transformations.

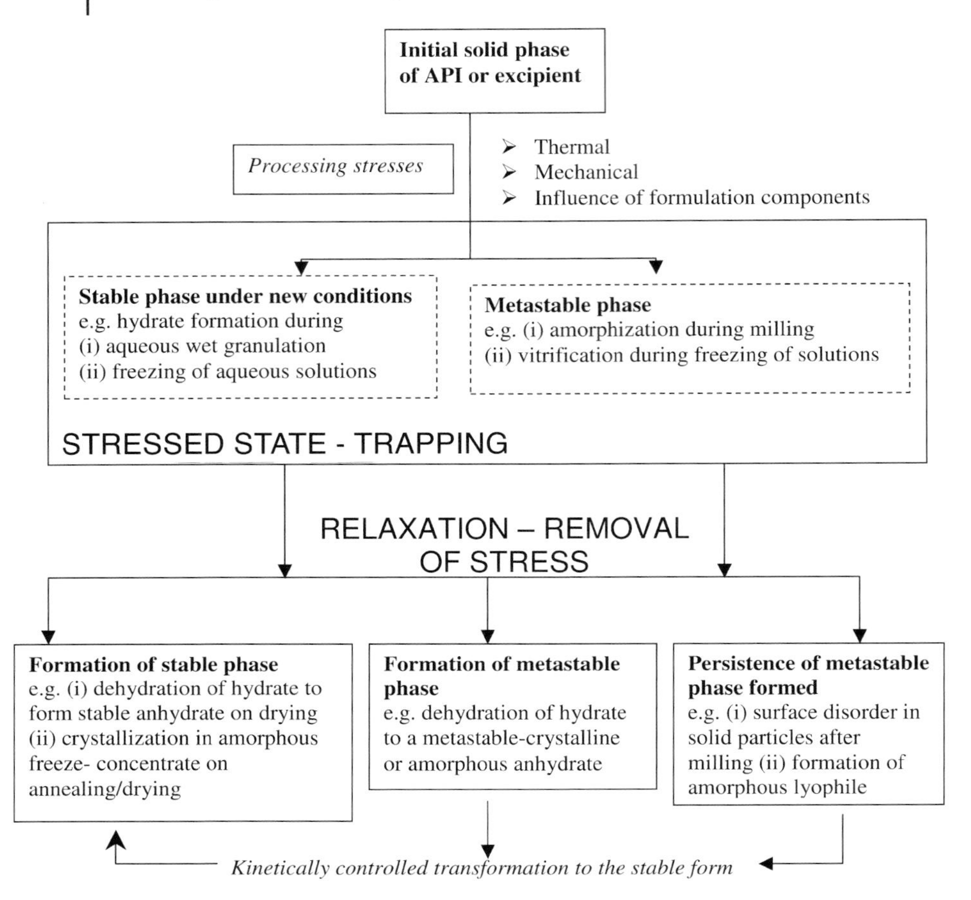

Scheme 13.2 Schematic representation of the mechanism of phase transformations as a result of processing stresses. (Adapted from [3]).

13.2 Processing-related Stress

During pharmaceutical processing, various formulation-specific processes are carried out using a wide variety of techniques and equipment. However, the major processing-related stresses include mechanical, thermal and those due to interaction with other components [3]. The effect of each of these will be discussed while recognizing that pharmaceutical unit operations can exert multiple stresses. For example, while thermal and mechanical stress, during the freezing of aqueous solutions and during wet granulation respectively, can lead to phase transformations, the presence of water in both cases can cause hydrate crystallization. Multiple operations can be carried out using specialized equipment, as in a fluid-bed granulator, wherein granulation and drying are simultaneously accomplished. The material may therefore be exposed to several stress–relaxation processes in a single step. The final physical form of the material is expected to be influenced by the type, intensity and duration of each stress.

13.2.1 Mechanical Stress

In many pharmaceutical processes, mechanical energy is used for the physical manipulation of formulation components, and can possibly lead to phase transitions [8]. More common among these are milling for reduction of particle size, mechanical dispersive techniques, high shear mixing during granulation, and compression of powders to obtain tablets.

13.2.1.1 Milling

Milling, by imparting mechanical energy, can lead to particle fracture and size reduction. The heat and vibrational energy generated during grinding, coupled with the mechanical energy, may also lead to amorphization. The extent of lattice disruption may vary from surface amorphization [9–12] to complete lattice disorder [13, 14].

Indomethacin (IMC) polymorphs and solvates were subjected to cryogrinding [14] to evaluate the effect of mechanical stress on the solid-state properties, at temperatures well below the melting point of the crystalline phase, thereby excluding the possibility of amorphization due to a "melt-quench mechanism". The α-, δ- and γ-polymorphic forms of IMC were rendered X-ray amorphous [glass transition temperature (T_g) ~43 °C] by cryomilling for 60 min. The amorphous phase obtained by milling the methanol solvate was plasticized by the released solvent, leading to crystallization as a mixture of the anhydrous γ- and α-forms. The t-butanol solvate, however, did not undergo desolvation and retained its lattice structure under the mechanical stress.

Otsuka et al. have demonstrated reduction in crystallinity on milling of cephalothin [15], lactose [16], and nitrofurantoin [17]. An amorphous product can be obtained if the processing temperature is substantially below the T_g of the product phase. However, crystallization of the product phase is favored under conditions that enhance its molecular mobility, namely increase in temperature and the presence of plasticizers such as water. Nitrofurantoin monohydrate Form I when ground at 5% relative humidity (RH), dehydrated to an amorphous product. When the process was carried out at 50% or at 75% RH, however, a polymorphic transformation into Form II monohydrate occurred [17].

13.2.1.2 Compression

Dry granulation (roller compaction or slugging) and tableting are two unit operations wherein powder blends are subjected to compression. Compression can induce dissolution of crystals in adsorbed water or crystal melting at points of particle contact. Recrystallization on decompression is one of the mechanisms of tablet consolidation [18]. Thus, phase changes can occur via solution or by melt-recrystallization mechanisms.

Chan and Doelker studied the compression-induced conversion of unstable crystal forms of caffeine, maprotiline hydrochloride and sulfabenzamide into their corresponding stable forms, as a function of powder particle size and compression pressure. Crystal dislocations can act as nucleation sites for the stable form [19]. Repeated compression in a tableting machine caused interconversion of the polymorphic forms, of chlorpropamide. After 30 compression cycles, the polymorphic composition was constant, and the amorphous content was 30%, suggesting that the interconversion between the forms could have occurred via an amorphous intermediate [20]. As the compression pressure increased, a larger fraction of TAT-59, an anticancer agent, was amorphized [21].

The influence of excipients on compression-induced phase transformations of the API has been investigated [22–24]. Dehydration of theophylline monohydrate and crystallization of amorphous indomethacin were minimized when carrageenan was used as a tableting excipient. This was attributed to the pronounced elastic recovery exhibited by carrageenan during decompression.

13.2.2 Thermal and Pressure Stresses

Drying, mixing, dispersion, dissolution, and melting are instances of pharmaceutical processes often requiring input of thermal energy. Conversely, in the lyophilization cycle, where solutions are frozen before drying, thermal energy is removed from the system. Processes such as vacuum drying and lyophilization are carried out under sub-ambient pressures. These thermal and pressure stresses can induce phase changes in the formulation components.

13.2.2.1 Freezing

Freezing of aqueous solutions, the first step in the manufacture of freeze-dried products, invariably results in crystallization of ice, leading to solute freeze-concentration. Non-crystallizing solutes, including sucrose and trehalose, are retained amorphous during freeze-drying whereas mannitol and glycine are examples of solutes that tend to crystallize. The processing conditions and formulation composition influence the crystallization behavior.

The influence of the cooling rate of the pre-lyophilization solution on the polymorphic composition of mannitol in the final lyophile has been investigated. While a mixture of anhydrous polymorphs was found in all the samples, the δ-form predominated when the solution was cooled slowly, and the α-polymorph was the major solid phase under conditions of uncontrolled rapid cooling [25]. The crystallization of bulking agents such as glycine and mannitol in the frozen state is desired and can be facilitated by annealing the frozen solutions [26, 27]. Annealing, by increasing the molecular mobility in the freeze-concentrated amorphous phase, facilitates solute crystallization. However, the extent of crystallization will be affected by the processing history and formulation composition [28]. The presence of non-crystallizing as well as crystallizing solutes can inhibit the crystallization of mannitol during freezing [28–33], which can lead to amorphous mannitol during the drying stage. Cooling of an aqueous solution can also lead to crystallization of solute as a hydrate, e.g., with cefazolin sodium, nafcillin sodium, dibasic sodium phosphate and mannitol [34–37]. Thus, the physical form of the solute in the frozen solution can be different from that used to prepare the pre-lyophilization solution.

13.2.2.2 **Drying**

Drying is often carried out at elevated temperatures, as in spray-drying or the drying of granules following wet-massing. Based on a model proposed by Davis et al. [38], if the rate of drying of wet granules is higher than the rate of solution-mediated metastable → stable transformation (dissolution of the metastable form followed by crystallization of the stable form), then the metastable polymorph may be formed during drying. When wet granulated γ-glycine was subjected to fluid-bed drying, a mixture of the stable (γ-) and metastable (α-) forms were obtained. Faster drying at 80 °C resulted in a higher α-content in the granules compared with drying at 60 °C. Similarly, granulations subjected to slower drying in a tray-dryer, at 21 °C, had much lower α-glycine content than the fluid-bed-dried granules [38]. Rapid solvent removal (e.g., in spray-drying), therefore, can yield crystalline metastable or amorphous forms [39–41]. While spray-drying of prednisolone resulted in crystallization of the metastable Form III, hydroflumethazide was retained amorphous [42].

The polymorphic or salt form of glycine that crystallized during spray-drying was influenced by the solution pH. In the absence of pH control, an aqueous glycine solution (pH 6.2) yielded α-glycine. When the solution pH was adjusted to 4.0 or 8.0, γ-glycine was the predominant crystalline phase [43].

A drug or an excipient existing as a hydrate in a wet mass may undergo dehydration during drying, with the physical form of the product phase being influenced by the drying conditions [44–47]. Theophylline monohydrate granules, when dried under ambient pressure at 50 °C, yielded the stable anhydrous phase that was 75% crystalline. Drying at 90 °C under reduced pressure resulted in a highly crystalline product with no detectable amorphous character [46]. Similarly, isothermal dehydration of carbamazepine hydrate at 44 °C resulted in a crystalline anhydrate, when the water vapor pressure was $\geq$12 torr. The product was amorphous at water vapor pressures $\leq$5.1 torr [47].

The water content of the dryer effluent formed the basis for the endpoint of effective drying of an organic monohydrate while still preventing dehydration [48]. Airaksinen and coworkers investigated the combined effects of temperature and water vapor pressure during the drying of theophylline monohydrate wet mass in a multichamber microscale fluidized bed dryer [49]. At an inlet air water content of 7.6 g m^{-3}, while no dehydration occurred at 30 °C the anhy-

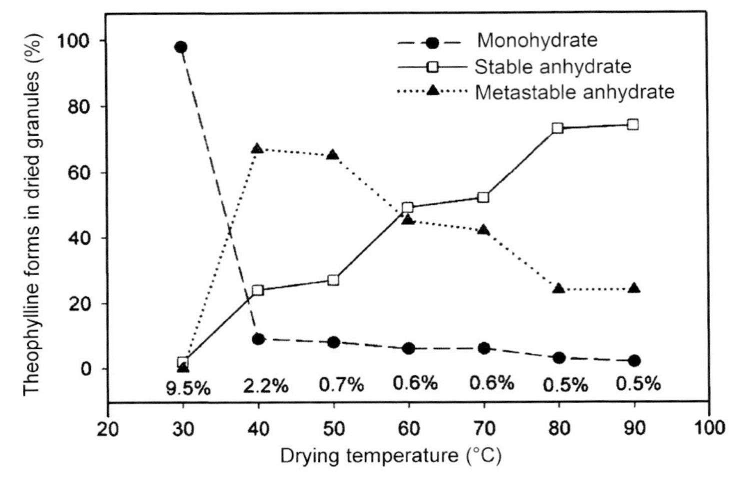

Fig. 13.1 Phase composition of theophylline granules as a function of drying temperature. The wet massed granules were dried in a multichamber microscale fluid-bed dryer using an inlet air with a water content of 7.6 g m^{-3}. The % values at the bottom represent the water content of the dried granules [49].

drous phase was formed at ≥40 °C (Fig. 13.1). Interestingly, the anhydrous phase composition was influenced by the temperature. At ≤50 °C, the metastable polymorph predominated. With increasing temperature, this ratio progressively changed in favor of the stable anhydrate. Even when dried at 90 °C, 25% of the product phase was composed of the metastable anhydrate. When the water content in the inlet air was decreased (0.5 g m^{-3}), dehydration occurred even at 30 °C, but the metastable to stable anhydrate ratio had a similar dependence on drying temperature.

In lyophilization, a frozen solution is heated to a sub-ambient primary drying temperature and dried under reduced pressures. Solutes that are retained amorphous during freezing can crystallize during heating and/or subsequent drying. When rapidly frozen aqueous buffer salt solutions were heated, while there was no crystallization from the citrate and malate freeze-concentrates, the succinate and tartrate buffer systems exhibited pH-dependent crystallization [50]. Water removal under reduced pressure, in primary-drying, can result in partial or complete dehydration of hydrates formed during freezing. While a poorly crystalline hemihydrate of sodium nafcillin was obtained, cefazolin pentahydrate dehydrated to a poorly crystalline anhydrate [34, 35].

13.2.2.3 **Melting**

During the melt-extrusion process, one or more formulation components are in the molten state prior to extrusion. While the intention is to retain the drug in the amorphous state, the physical state of the drug depends on the formulation and process variables. In solid dispersions of 17β-estradiol [melting point (mp) 175 °C] made by melt extrusion with poly(vinyl pyrrolidone) (PVP) and saccharose monopalmitate as carriers, the drug retained its crystallinity when extruded at 60 °C, while it was completely amorphous when extruded at 180 °C. At intermediate temperatures of 100 and 160 °C, the drug was partially crystalline [51]. Similar results were obtained on extrusion of itraconazole (mp 158 °C) with a cationic acrylic copolymer, Eudragit E100®. The effect of drug loading was investigated at an extrusion temperature of 180 °C, which retained the drug amorphous. While itraconazole was molecularly dispersed in the carrier matrix at concentrations <13% w/w, higher drug concentrations resulted in phase separation and the existence of an amorphous drug phase [52].

13.2.3 Interaction with Other Components

The δ-form of mannitol, metastable under ambient conditions, is commonly encountered in freeze-dried formulations [32, 53]. While grinding and compression did not induce any phase transformations, it converted into the stable β-form on exposure to 97% RH at 25 °C [54]. The δ-polymorph has a rigid H-bond network that stabilizes it against mechanically induced $\delta \rightarrow \beta$ conversion. This conversion into the β-form, which has a relatively loose H-bond network, would require disruption of the H-bonds in the crystal lattice. The presence of solvents with hydroxyl groups, such as water or ethanol, facilitated disruption of the hydrogen bond network of the δ-form, followed by lattice restructuring to yield the β-form. Interaction with the solvent molecule, which acts as a "molecular loosener", facilitated conversion into the stable polymorph [54].

Interaction with other formulation components can also result in changes in the chemical composition of the phase (Scheme 13.1); relevant examples are discussed below.

13.2.3.1 Hydrate Formation

As pointed out earlier, the physical form of the solute crystallizing in frozen aqueous solutions can be different from that used to prepare the pre-lyophilization solution. In several instances, while the solution was prepared using the anhydrous form, the corresponding hydrate crystallized from the freeze-concentrate [36, 37]. Similarly, during aqueous wet granulation, anhydrous forms of theophylline, carbamazepine and chlorpromazine HCl transformed into their respective hydrates [44, 46, 55]. Physical mixtures of anhydrous theophylline with either lactose monohydrate or silicified microcrystalline cellulose (SMCC) were wet-massed and equilibrated at 5 °C [56]. The extent of theophylline monohydrate formation was much higher in the presence of lactose. Hydrophilic polymers such as SMCC, by strongly binding to water, limit water availability, thereby retarding hydrate formation [56, 57]. A similar mechanism was invoked for the inhibition of theophylline hydrate formation by PVP in anhydrous theophylline–PVP physical mixtures exposed to high RH [58].

13.2.3.2 Complexation

Cyclodextrin (CD) complexation is one of the most widely investigated approaches for enhancing aqueous drug solubility. The complexation can be brought about by different processing techniques. β-CD, a crystalline solid, forms crystalline complexes; S-ibuprofen forms a 1:2 crystalline inclusion compound with β-CD [59]. When eflucimibe, a potential drug for the treatment of dyslipidemia, was subjected to an aqueous kneading process with γ-CD there was a progressive decrease in drug crystallinity, owing to the formation of a non-crystallizing complex [60]. Amorphization of drugs in CD-based formulations has often been interpreted as formation of an inclusion complex. However, the loss in crystallinity could also be due to the existence of the drug as a discrete amorphous phase, or to a molecular dispersion of the drug in the CD matrix and not necessarily within the CD cavity [61].

13.2.3.3 Salt–Free-acid/Base Conversions

Acid–base reactions and salt hydrolysis during processing can result in salt formation, interconversion between salts, and formation of free acid/base.

A recent example is the conversion of the crystalline hydrochloride salt of an investigational drug into the amorphous free base during a wet granulation process [62]. The extent of conversion was affected by the water content in the granulation fluid, which was an ethanol–water mixture. Granulation with 96 or 90% v/v ethanol, followed by storage for 4 h, resulted in ~4 and ~7% conversion, respectively. The conversion of the maleate salt of enalapril into the stable sodium salt, during pharmaceutical processing, is the subject of patents [63, 64].

13.2.3.4 Metastable Phase Formation

Co-grinding with excipients not only facilitated drug amorphization but also the stabilization of the amorphous product [65–67]. This may be brought about by specific drug–excipient interactions and a possible antiplasticizing effect. Co-grinding with PVP facilitated amorphization of glisentide [67]. Mechanical activation of piroxicam and sulphathiazole in the presence of PVP stabilized the resulting amorphous or metastable polymorphic form [66].

Stable amorphous forms of various drugs were obtained by co-milling with amorphous magnesium aluminosilicate (Neusilin®) [65]. Interestingly, amorphization if incomplete at the end of milling was found to progress to completion on storage. The amorphization was attributed to an acid–base reaction between the acidic drugs (indomethacin, naproxen and ketoprofen) and the silanol groups on Neusilin®. Milling of progesterone with Neusilin® resulted in complete amorphization due to H-bond interactions between the drug and the carrier [65]. Similarly, stable amorphous indomethacin was prepared by co-grinding with silica [68].

13.2.3.5 Multiple Interactions

The extent of phase transformation may also be influenced by multiple interactions. Although extrusion-granulation of anhydrous carbamazepine, form I, with water did not produce any changes in the solid state of the drug, use of 50% aqueous ethanol as the granulating liquid resulted in formation of the dihydrate. Since the phase transformation is likely to be solution-mediated, the higher carbamazepine solubility in the ethanol–water mixture is expected to facilitate the transition [69]. When a physical mixture of flurbiprofen and hydroxypropyl β-cyclodextrin was kneaded with water, the addition of ammonia enhanced the aqueous drug solubility. As a result, the interaction of the drug with hydroxypropyl β-cyclodextrin was facilitated, leading to complete loss in drug crystallinity in the final dried solid [61].

13.3 Detection and Quantification of Phase Transformations

Numerous techniques are available to probe the solid-phase composition of pharmaceuticals and can be used to monitor the effects of processing. Powder X-ray diffractometry, thermoanalytical techniques, including differential scanning calorimetry, isothermal microcalorimetry and thermogravimetric analysis, and spectroscopic techniques such as IR, Raman, NMR, find widespread use. Solid–vapor interaction (using vapor sorption microbalance), thermoelectrical and microscopic techniques (scanning electron microscopy, hot stage microscopy, atomic force microscopy) also form the basis for monitoring phase transformations.

Solid-state characterization techniques have been the subject of numerous recent publications [70–73] (see also Chapters 2 to 7). This section will therefore focus on characterization of processing-induced phase transformations and is subdivided according to the nature of the product phase formed. In addition to the commonly used methods, some emerging sensitive characterization techniques are also presented.

13.3.1 Generation (Creation) of Lattice Disorder

From both chemical and physical stability considerations, the stable crystalline form of a drug is desired in formulations. Even if a highly crystalline form is used as the starting material, processing is known to introduce lattice disorder. If there is pronounced lattice disruption, this becomes evident from the broadening of XRD peaks, and can be quantified based on peak width measurements (width at half maximum). Peak broadening is often accompanied by the appearance of an amorphous halo, as was evident in ball-milled triamcinolone acetonide [74] and jet-milled fluticasone propionate [9]. The amorphous regions generated by processing will also be characterized by a glass transition, detectable by DSC. When heated above the glass transition temperature, crystallization of the amorphous regions is evident in the DSC as an exothermic event, as was observed in ground glisentide [67]. However, most of these "bulk techniques" may not be suitable for detection of low levels of disorder (<5% w/w) [75–77]. In the rest of this section, we will therefore emphasize techniques that can quantify low levels of lattice disorder and, thus, can reveal the first evidence of processing-induced lattice disruption in highly crystalline materials.

Amorphous phases have a much stronger tendency than their crystalline counterparts to sorb solvents from the vapor phase. The amount sorbed can be used for quantification of the amorphous content. Solvent sorption can also cause sufficient plasticization and consequently crystallization of disordered regions. This is accompanied by loss of the sorbed solvent, and the amount desorbed is a measure of the extent of crystallization. Acetone sorption formed the basis for the quantification of the amorphous content in micronized samples of an experimental drug [78]. Physical mixtures of the crystalline and amorphous drug were prepared and subjected to acetone sorption studies. The amount of acetone sorbed, at an acetone partial pressure (p/p_0) of 0.3, was measured before (Fig. 13.2; stage I) and after crystallization (Fig. 13.2; stage III). The crystallization (region II), carried out at an acetone partial pressure of 0.85, was accompa-

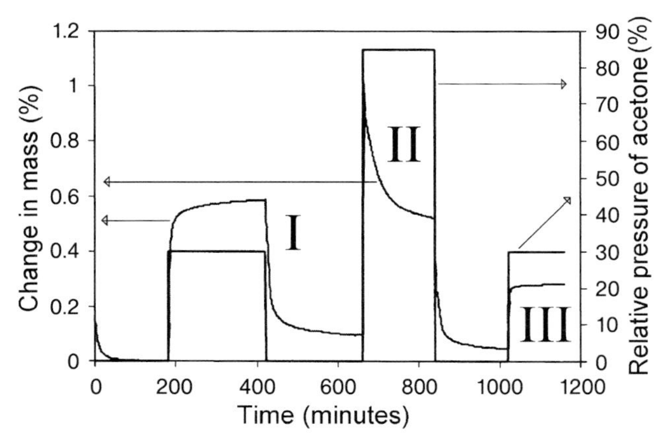

Fig. 13.2 Mass change profile for a sample of an experimental drug with a 4.8% w/w amorphous content, when exposed to acetone vapor, before (I) and after (III) crystallization of the amorphous phase (II). The acetone partial pressure in stages I and III was 0.3 and crystallization was carried out at a partial pressure of 0.85. The difference in mass change between stages I and III is a measure of the amorphous content [78].

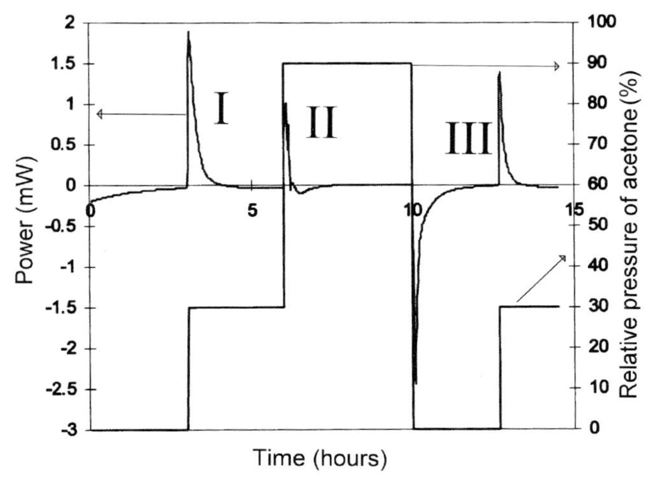

Fig. 13.3 Heat response by isothermal microcalorimetry for a sample of an experimental drug with an amorphous content of 4.4% w/w. The difference between heat of interaction of the sample with acetone vapor, before (I) and after (III) crystallization of the amorphous phase, was a measure of the amorphous content. The acetone partial pressure in stages I and III was 0.3 and crystallization (stage II) was carried out at a partial pressure of 0.9 [78].

nied by a weight loss due to desorption. The difference in acetone sorbed between stages III and I was directly proportional to amorphous content over the range 0.2–9.8% w/w. This enabled quantification of low levels of amorphous character in micronized drug powder [78].

Isothermal microcalorimetry can be used to quantify amorphous content based on heat of crystallization measurements under controlled temperature and humidity conditions [10, 12, 79, 80]. Levels of disorder as low as 2% w/w, generated by micronization of revatropate hydrobromide, could be quantified as a function of micronization grind pressure [10]. A major limitation of this technique, however, is its non-specificity. Extraneous physical processes such as wetting of the sample, collapse of the powder bed and chemical changes can contribute significantly to the heat flow signal. A careful choice of experimental conditions and appropriate data analysis is required to obtain reliable measures of amorphous content [80].

Heat of sorption measurements can also form the basis for amorphous phase quantification. The amorphous regions in a micronized drug sample could be crystallized by exposure to acetone vapor at a partial pressure (p/p_0) of >0.6, in a vapor perfusion microcalorimeter. The heat of acetone sorption (at $p/p_0=0.3$) was determined before and after crystallization of the disordered regions. The difference between the two was used as a measure of the lattice disorder (Fig. 13.3). Amorphous content as low as 0.5% w/w could be detected [78].

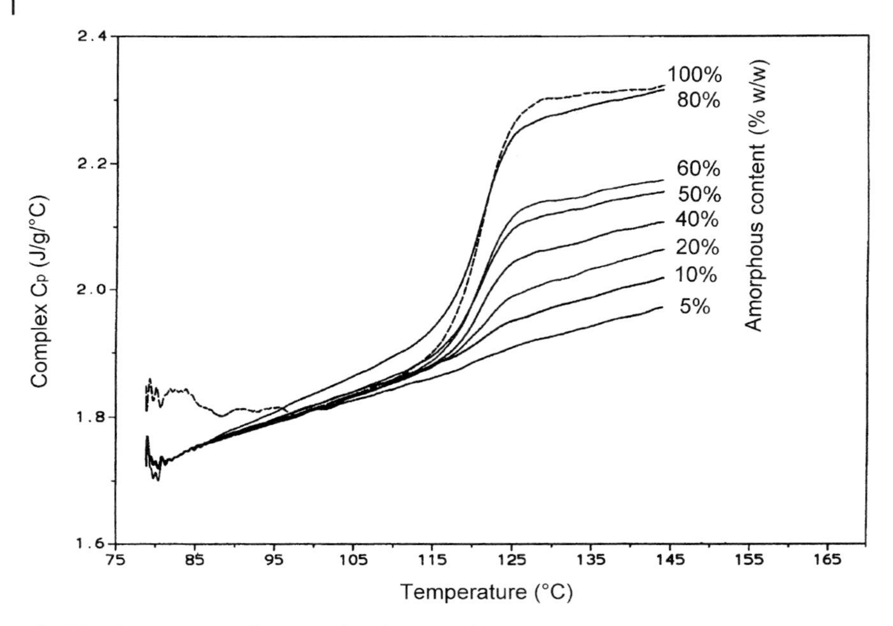

Fig. 13.4 Heat capacity change at the glass transition temperature, as seen in the reversible heat flow curves, of an experimental antibiotic. Physical mixtures of the crystalline and amorphous antibiotic (amorphous content in % w/w) were subjected to modulated DSC. The heat capacity change at T_g (ΔC_p) varied linearly with amorphous content according to the relationship ΔC_p (J g^{-1} °C^{-1}) = 0.0057 + 0.0038 (weight % of amorphous phase) [82].

Modulated temperature DSC, by separating the glass transition event from the other overlapping thermal events such as enthalpic recovery and crystallization, enables a precise determination of the heat capacity change (ΔC_p) at T_g. The magnitude of this step change was used as a measure of the amorphous content of an experimental drug [81] and of an investigational antibiotic [82]. Since the rate of crystallization of the amorphous antibiotic was unacceptably slow and it also degraded during melting, other calorimetric techniques were rendered unsuitable. Figure 13.4 shows the heat capacity change in the reversible heat flow curves, at amorphous contents ranging from 5 to 100%. The limit of detection and quantification of amorphous content were 0.9% and 3.0% w/w, respectively. Disorder introduced by micronization of the antibiotic was quantified by this technique [82].

Several other techniques provide unique information about disordered lattice regions but are not necessarily used quantitatively. Milling of salbutamol sulfate resulted in surface non-uniformity, as evidenced by "pits or craters" in atomic force microscope (AFM) amplitude images. Simultaneous AFM phase-shift imaging revealed that these regions had different physico-mechanical properties, suggesting amorphization [83].

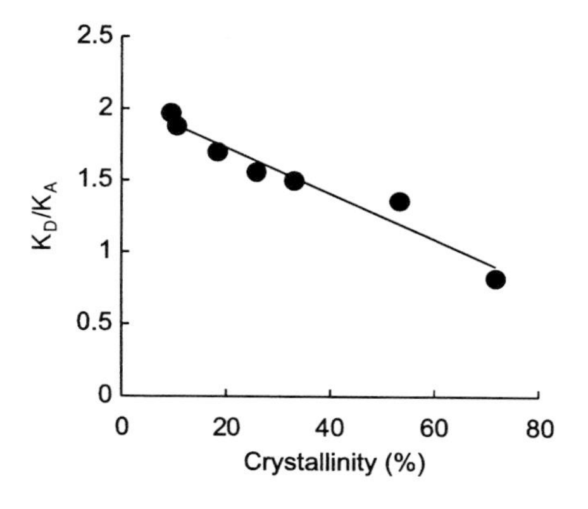

Fig. 13.5 Linear relationship between the basic/acidic parameter ratio (K_D/K_A) and crystallinity of milled cefditoren pivoxil, indicating an increase in basic nature of the sample surface as amorphization proceeds [13].

The dispersive surface energies and the acidic/basic properties of milled cefditoren pivoxil were investigated by inverse gas chromatography (IGC) [13]. The acidic or electron-accepting parameter (K_A) and the basic or the electron-donating parameter (K_D) of the powder surface were determined as a function of milling time. The intact crystalline surface was relatively acidic ($K_D/K_A < 1$, $= 0.82$). Milling for 30 min resulted in complete amorphization. This exposed the electron-donating (basic) carbonyl groups, rendering the surface basic ($K_D/K_A = 1.97$). The K_D/K_A ratio increased linearly as a function of the amorphous content, and was therefore a measure of crystallinity (Fig. 13.5).

While the XRD patterns of both melt-quenched ketoprofen and that of the drug co-milled with Neusilin® revealed the absence of long-range order, their FTIR spectra revealed pronounced differences at the molecular level [65]. Melt-quenched ketoprofen exhibited the carbonyl stretching vibrations characteristic of the dimeric form, which were also found in the crystalline state. These vibrations were absent in the ketoprofen ball-milled with Neusilin®. However, the presence of a peak characteristic of the carboxylate group suggested the formation of a salt with the carrier. An indomethacin : Neusilin® (1 : 5 w/w) mixture, when ball-milled for 24 h, resulted in conversion of the drug into an X-ray amorphous form. However, the dimer peak in the FTIR spectrum of this sample suggested a low level of residual order [65].

13.3.2
Crystallization – Anhydrous Phase

Processing can not only induce polymorphic transformations, but can also cause an amorphous to crystalline transition. When the existence of an API or excipient in the amorphous state is necessary for product quality and performance, characterization of low levels of lattice order is of great practical importance. Using conventional X-ray powder diffractometry, the limit of quantification of crystalline content, in physical mixtures of amorphous and crystalline sucrose, was 1.8% w/w [84]. The use of synchrotron radiation and a two-dimensional detector allowed diffraction patterns to be acquired with a high time resolution during an *in situ* crystallization event. The *in situ* crystallization approach circumvented the problem of non-homogeneity in mixing – a potentially serious issue at extreme mixture compositions. The estimated limit of detection of crystalline sucrose in an amorphous matrix was 0.2% w/w [85]. However, in other systems, accurate quantification of low levels of crystallinity, especially in the presence of other formulation components, may be challenging.

The random noise in the halo pattern of the amorphous phase also poses a major challenge. An approach was developed to extract the noise from the amorphous halo [86]. The relative standard deviation (RSD) in the resulting crystallinity measurement was plotted as a function of the degree of crystallinity. The limit of detection (LOD) of crystalline cefditoren pivoxil, expressed as the crystalline content at 30% RSD, was 4.8% w/w.

In situ XRD has been used to monitor phase transformations during various stages of pharmaceutical processing [35, 87]. During wet granulation, the polymorphic transformation of the metastable form I of flufenamic acid into the stable form III was monitored. By simulating the freeze-drying process in the sample chamber of the XRD, phase transformations of mannitol [53], glycine [88], and the crystallization behavior in mannitol–glycine systems [32] were investigated. Figure 13.6 demonstrates the ability of the technique to probe the phase composition of the formulation during various stages of the freeze-drying

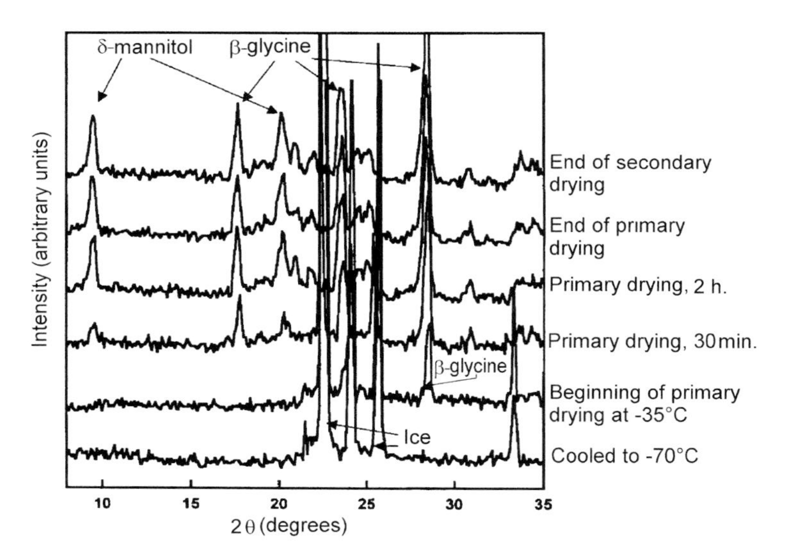

Fig. 13.6 *In situ* XRD during freeze-drying of aqueous solution containing 5% w/w mannitol and 4% w/w glycine. The solution was cooled to –70 °C at 20 °C min^{-1} and then heated to a primary drying temperature of –35 °C at 5 °C min^{-1}. Primary drying was conducted at a chamber pressure of 100 mtorr for 160 min. Secondary drying was carried out at –10 °C [32].

process. Peaks of δ-mannitol and β-glycine indicate crystallization of these polymorphic modifications during drying.

Electrical resistance measurements suggested crystallization of monosodium succinate salt in frozen succinate buffer solution [89]. An irreversible transition was observed at ~−30 °C when a succinate buffer solution was subjected to electrical thermal analysis. The absence of this transition in frozen solutions of succinic acid or its disodium salt indicated that the crystallizing species was the monosodium salt.

Spectroscopic techniques have great potential for both physical and chemical characterization of pharmaceutical compositions during processing. The sensitivity of FT-Raman spectroscopy has been demonstrated in physical mixtures. The technique has been reported to detect down to <1% w/w of either crystalline or amorphous content in binary powder mixtures of crystalline and amorphous indomethacin [90]. Its utility in detection of different polymorphic forms of drugs, in intact tablets, at levels of 1% w/w has also been demonstrated [91]. Compression-induced crystallization of amorphous indomethacin was monitored by Raman spectroscopy [23]. The benzoic stretching vibrations of indomethacin, appearing at 1698 cm^{-1} for crystalline drug and at 1680 cm^{-1} for the amorphous form, were used to determine the extent of crystallization after compression. The polymorphic behavior of glycine, during the drying of granules, was also studied using near-infrared (NIR) spectroscopy [38].

13.3.3
Hydrates – Formation and Dehydration

Our discussion here will be restricted to hydrates though it is recognized that solvents other than water can be incorporated into the crystal lattice. In contrast to hydrates, wherein water is stoichiometrically incorporated in the crystal lattice, in pharmaceutical compositions water can also be physically adsorbed on crystal surfaces, absorbed into amorphous regions or remain as bulk water (e.g., during wet granulation). Conventional techniques such as Karl Fischer titrimetry and thermogravimetric analysis (TGA) are suited to determine the total water content and may not distinguish between the different states of water in a solid.

NIR can differentiate water bound in the crystal lattice from free water in the formulation [73, 92]. The absorption maxima of water, in hydrates of lactose and theophylline, were observed at 1933 and 1970 nm, respectively, whereas that of water sorbed in the amorphous silicified microcrystalline cellulose (SMMC) appeared as a wide band at 1920 nm [56]. The NIR signals at 1970 nm agreed well with the theophylline hydrate content quantified in wet masses, by XRD. Figure 13.7 contains the second-derivative NIR spectra of a wet granulated 1:1(w/w) mixture of theophylline and SMCC [57]. Crystallization of theophylline monohydrate was evident from the increase in the signal at 1970 nm as a function of time. The second derivative of the absorbance at 1970 nm divided by the signal

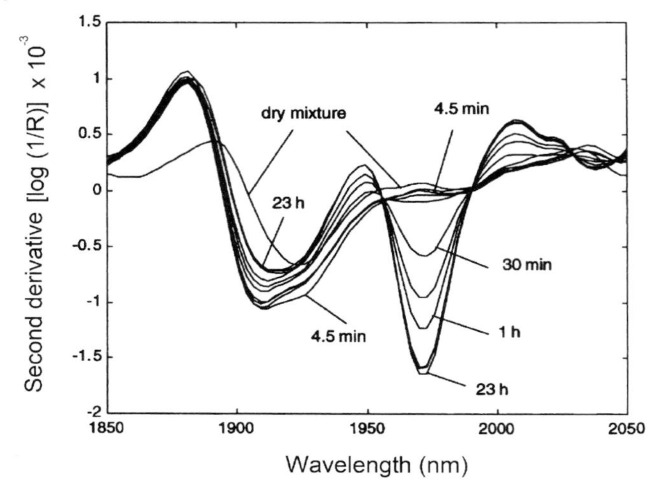

Fig. 13.7 Second derivative NIR spectra of wet mass containing theophylline and silicified microcrystalline cellulose (1:1). Time points refer to time from the beginning of water addition (10% w/w). The "dry mixture" refers to the powder mixture before water addition [57].

at a correction wavelength (1810 nm) was used as a measure of the hydrate content.

The hydrate formed during wet granulation has the potential to dehydrate during subsequent drying. An in-line NIR probe was used to monitor the drying kinetics of theophylline monohydrate in a fluid-bed dryer. The NIR signals were converted into apparent water absorption units (AWA) based on NIR signals at 1998 (water indicator), 1813 (for baseline correction) and 2214 nm (for normalization). The AWA values were a measure of the total water in the system. The AWA curves clearly exhibited a two-stage water loss – (a) the first, rapid, heat-transfer-limited loss of free water by evaporation and (b) a slower, mass transfer limited, dehydration of the monohydrate (Fig. 13.8). As seen from X-ray data (Fig. 13.1), drying at 30 °C with an inlet air water content of 7.6 g m^{-3} resulted in the retention of the monohydrate. As expected, the rates of both the water loss steps increased with temperature (Fig. 13.8) [49].

Although water is a weak Raman scatterer, changes produced in the theophylline crystal lattice as a consequence of hydrate formation could be detected by Raman spectroscopy. There was a shift in the –O=C–N bending peak from 550 cm^{-1} for the anhydrate to 570 cm^{-1} for the monohydrate [56]. The technique also allowed in-line monitoring of the extremely rapid theophylline hydrate formation during aqueous wet granulation [93]. The loss of channel water from risedronate sodium hemipentahydrate granules during fluid bed drying has been monitored by in-line Raman spectroscopy [94]. The compression-induced dehydration of theophylline monohydrate was monitored during tablet storage. The

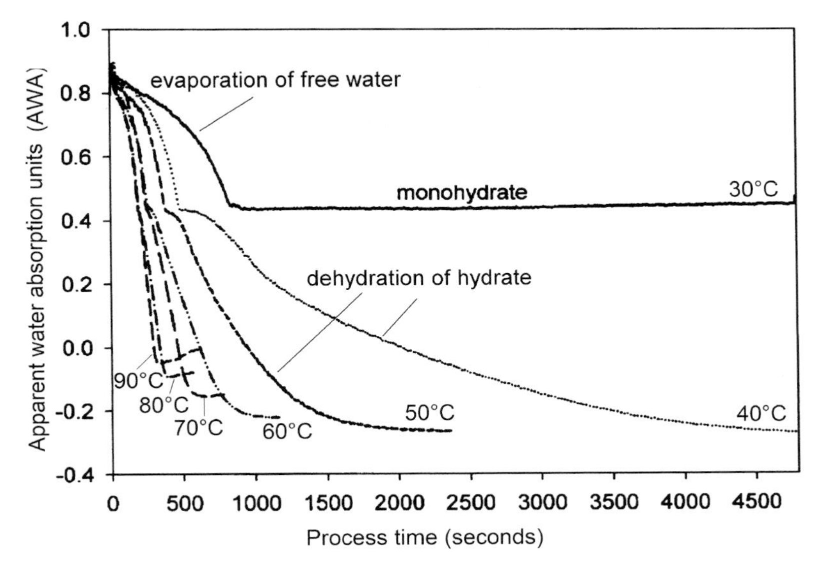

Fig. 13.8 Drying process of wet-massed theophylline containing granules at different drying temperatures. The wet-massed granules were dried in a multi-chamber microscale fluid-bed dryer using inlet air with a water content of 7.6 g m^{-3}. The apparent water absorption (AWA) was obtained from NIR spectral data using an in-line probe [49].

=CH stretching vibration at 3121 and 3108 cm^{-1} for the monohydrate and the anhydrate, respectively, formed the basis for the quantification [23].

TGA, often in conjunction with other techniques, is a powerful tool for the characterization of hydrates. An endothermic event at ~50 °C, observed in the differential thermal analysis curve of freeze-dried mannitol, corresponded to a step-like weight loss in the TGA curve. XRD revealed the presence of a new phase, and the thermoanalytical techniques confirmed the existence of mannitol hydrate [37].

In situ freeze-drying in an X-ray diffractometer enabled detection of hydrates of mannitol, disodium hydrogen phosphate, cefazolin sodium and nafcillin sodium in frozen solutions [35, 53]. Hydrates formed in frozen aqueous solutions can dehydrate during the subsequent drying steps. The technique also revealed the nature of the dehydrated phase formed during subsequent drying [34–36].

13.3.4
Salt–Free-acid/Base Transformations

As discussed earlier, the interaction of a crystalline hydrochloride salt with water resulted in conversion into the amorphous free base [62]. The extent of conversion was insufficient to be detected by either XRD or IR spectroscopy. However, from the characteristic Raman peak of the salt at 777 cm^{-1} and the base at 852 cm^{-1} and spectral subtraction of the excipient contributions, the free base could be quantified even when it constituted only 2% of the tablet weight.

The conversion of delavirdine mesylate into the free base, following exposure of tablets to 40 °C/75% RH conditions, has been quantified by solid-state NMR [95]. Infrared spectroscopy revealed that the disintegrant, croscarmellose sodium, acted as the proton acceptor in the acid–base reaction. The salt to free base conversion was not seen in the absence of the disintegrant.

13.4 Implications of Phase Changes

The influence of the physical form of formulation components on their pharmaceutically relevant properties has been documented [96–98]. The functionality of excipients can also depend on their physical form. A lyoprotectant in a freeze-dried formulation, to be effective, must exist in the amorphous state. Phase transformation induced by processing can therefore influence behavior during subsequent manufacturing processes as well as the final product quality [44, 99].

Metastable phases formed during processing may convert into their respective stable forms during storage. Transformations in the final dosage form, involving crystal growth or polymorphic transitions, can affect the particulate properties and the microstructure in solid formulations. This can not only influence the physical and mechanical attributes of the resulting product, but can also compromise the function of other formulation components. In the following sections we provide overview of the diverse effects of phase transformations.

13.4.1 Amorphization

Pharmaceutical processing, aimed at introducing amorphous character in the API, is a popular approach for developing formulations of poorly soluble drugs. Amorphization increases the apparent solubility, accelerates dissolution and hence has the potential to enhance bioavailability, if the absorption is dissolution rate-limited. The bioavailability of a 10% w/w amorphous dispersion of ritonavir, in poly(ethylene glycol) (PEG) 8000, was 22 times that of the crystalline drug [100]. Surface amorphous character, estimated to be about 5.8% of the total mass, generated by milling griseofulvin resulted in increased "metastable solubility" [101]. If this amorphous character is retained in the final formulation it can potentially influence *in vivo* performance.

However, a decrease in crystallinity will not necessarily translate into an increase in dissolution rate. The enhancement in dissolution, brought about by amorphization, can lead to supersaturation with respect to the stable form and hence to its crystallization. This can negate the solubility advantage of the amorphous form [102]. In many instances, the thermodynamically stable physical form in contact with the aqueous medium is a hydrate. For example, with theophylline, the monohydrate is the stable phase in contact with water at $<66\,^{\circ}C$ [103]. Amorphous regions in anhydrous theophylline particles, because of their higher free energy, underwent faster conversion to the monohydrate than their crystalline counterparts [46]. Similarly, in carbamazepine, milling caused surface amorphization, thereby accelerating hydrate formation in the presence of water [104]. Under physiologically relevant conditions, hydrated forms of numerous drugs possess lower aqueous solubility than their corresponding anhydrates [105]. The enhanced solubility advantage of the metastable form can only be realized when the rate of crystallization of the stable form is less than the metastable dissolution rate.

Lattice disorder introduced during processing may not be uniformly distributed in solid particles. When amorphization is induced by milling, lattice disruption is expected to occur preferentially in the surface regions. Though only a small weight fraction of the solid may have undergone amorphization, this could represent a significant fraction of the particle surface. While conventional bulk characterization may not even reveal the lattice disruption, the surface properties may be profoundly affected. Thus, small variations in processing, while having relatively little influence on the overall degree of crystallinity, may have much larger effects on the surface. Micronization, by causing surface disorder, increased the surface energy (determined by IGC) of salbutamol sulfate, leading to undesirable bulk properties, namely an increase in powder cohesiveness and poor flow [106].

During processing or storage, if the molecular mobility of the disordered region is increased, either due to an increase in product temperature or by plasticization, surface recrystallization will be facilitated. While this amorphous → crystalline transition is a reversal of the lattice disordering process, there could still be implications for the particle size distribution and microstructure [79]. The fraction of coarse particles of salbutamol sulfate increased with ball-milling time. This was attributed to simultaneous recrystallization of the disordered regions, leading to particle growth and agglomeration [12].

Recrystallization of the surface may not reverse the effect of amorphization. The surface energy of such recrystallized xemilofiban powder was higher than that of the original unmilled powder, as evidenced by higher water adsorption enthalpies [107]. The content uniformity was poor in drug-excipient blends made with both milled (amorphous surface) and milled–recrystallized (crystalline surface) xemilofiban (Table 13.1). This was attributed to the surface alteration of xemilofiban particles (first by milling and then by recrystallization), which did not favor homogeneous mixing with the other blend components [107].

We will next consider the effect of a substantial reduction in crystallinity induced by processing. Milling of microcrystalline cellulose (MCC) in a vibrational rod mill reduced its crystallinity from 65 to 23%. This affected granule growth rate and granule structure during wet granulation [108]. Based on scanning electron microscopy (SEM) studies, the authors proposed that MCC particles

Table 13.1 Effect of surface modification of xemilofiban powder on the content uniformity of drug–excipient blends. (Data taken from [107]).

Xemilofiban sample	Drug content in blend (%)			
	Mean	% RSD	Range	
			Max.	Min.
Control (unmilled)	100.0	1.3	102.8	98.1
Apex milled (amorphous surface)	101.7	8.9	138.8	97.9
Apex milled and recrystallized	97.3	6.3	106.2	72.8

with lower crystallinity swell due to water uptake during granulation, and can be easily broken down by the impeller blade of the granulator. This allowed better binding of MCC with corn starch (CS), the other excipient, leading to homogeneous granule formation and acceptable drug content uniformity. Conversely, unmilled MCC, being harder, could not be broken down during granulation, affecting its ability to effectively bridge with CS. SEM revealed heterogeneity in granule structure, with a core consisting of CS and large MCC particles. This core was layered with particles of CS. The lower size fraction of the resulting material consisted of a considerable amount of ungranulated particles. The non-uniform drug distribution led to poor drug content uniformity in the lower size fraction [108]. In a later publication, the authors also demonstrated that the degree of crystallinity of milled MCC influenced the water uptake, swelling and drug release behavior of the resulting tablets [109].

When processed at elevated temperatures, there is scope for amorphization by melt-quench mechanism. Hot-melt granulation of ibuprofen (IBU, mp 75 °C) at ~80 °C with ammonio methacrylate copolymer (Eudragit RS PO®; T_g~55 °C) resulted in an amorphous mixture. IBU (T_g=–43.6 °C [110]) plasticized the polymer and caused a decrease in the glass transition temperature. As a result, the effectiveness of Eudragit® as a thermal binder was improved [111].

Sugars are used as lyoprotectants during the freeze-drying of protein formulations. Various mechanisms have been proposed for their stabilizing effect. The sugar may H-bond with the polar groups on the protein, thus conserving the structure of the protein during drying [112]. The amorphous sugars, by diluting the protein may minimize intermolecular contacts and hence protein aggregation [113]. Finally, the protein mobility in the glassy matrix may be restricted [114]. These stabilizing effects require the lyoprotectant to exist in an amorphous state [115]. The dependence of protein stability on the concentration of amorphous stabilizers has been extensively demonstrated [30–32, 115].

Crystallization of the sugar, either during the freeze-drying process or during storage, would exclude it from the protein-containing amorphous phase, thereby rendering it ineffective. Phosphate buffer salts inhibited mannitol crystallization in freeze-dried systems. Retention of mannitol in the amorphous state in the

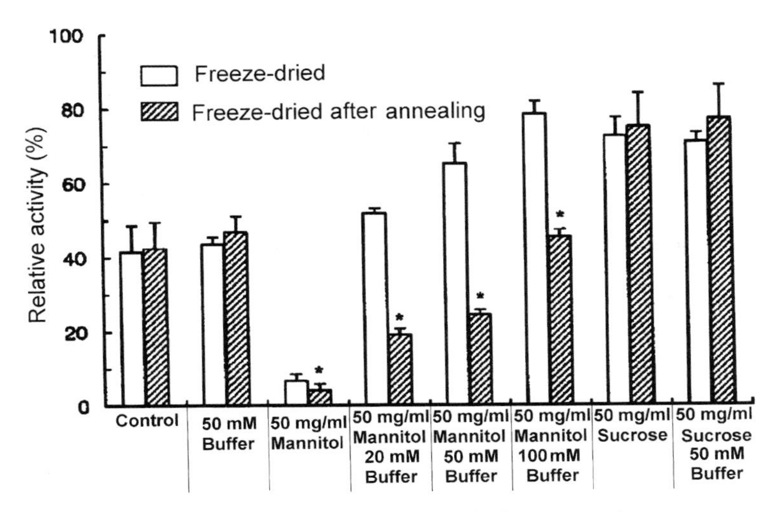

Fig. 13.9 Percent activity of freeze-dried lactate dehydrogenase (relative to original solution without any cosolute) as a function of prelyophilization solution composition [29].

freeze-dried product was accompanied by increased preservation of protein secondary structure [29]. There was also increased retention of lactate dehydrogenase (LDH) activity in freeze-dried formulations where mannitol crystallization was inhibited. Figure 13.9 compares the activity of LDH freeze-dried with different cosolutes. In all cases, the effect of annealing the frozen solutions (at $-10\,^{\circ}C$ for 1 h) before primary drying (shaded bars in Fig. 13.9) was also evaluated. When the protein was freeze-dried with mannitol, the activity recovered was substantially lower than the control (protein alone; no cosolutes). This indicates the additional stress induced by mannitol crystallization on the protein structure. Phosphate buffer exhibited a concentration-dependent inhibition of mannitol crystallization, which correlated with increased retention of protein activity. Annealing facilitated mannitol crystallization and decreased the recovery of activity [29]. In freeze-dried formulations containing sucrose and PEG, sucrose (when present at a sufficiently high concentration) inhibited the crystallization of PEG and increased the recovery of LDH activity in the lyophile [116].

However, retention of solutes in the amorphous state can have other implications. If, for example, mannitol or glycine is retained amorphous on freezing, the lower glass transition temperature of these solutes might necessitate primary drying at lower temperatures and hence for a longer period [117]. When retained amorphous, these solutes may crystallize during storage and release sorbed water. The water would be available for redistribution to the other formulation components, including the API, and cause drug instability [118].

Generation of lattice disorder in the API can also lower the chemical stability [119]. Milling and compression resulted in a decrease in the crystallinity of an anticancer drug, TAT-59. Crystallinity decreased with an increase in the grinding time or compressive force and this was accompanied by an increase in the concentration of a hydrolytic degradation product [21]. The photodegradation rate of the amorphous phase produced by grinding the β-polymorph of nicardipine HCl was ~3.5 times that of the crystalline phase [120].

13.4.2 Crystallization

As seen above, crystallization of lyoprotectants can adversely affect protein stability. Mannitol, a bulking agent in freeze-dried formulations, protected proteins against structural damage only if retained amorphous [29, 30]. Alkaline phosphatase, dispersed in a trehalose glass matrix, was compressed into tablets. Compaction of the glass under humid conditions was believed to have induced trehalose crystallization and a consequent loss of enzyme activity [121]. However, solute crystallization during freeze-drying may not necessarily be detrimental to protein stability. Mannitol crystallization during freeze-drying decreased hemoglobin structural damage [122]. The damage was induced by the separation of poly(ethylene glycol) and dextran-rich phases in the amorphous freeze-concentrate. Crystallization of mannitol resulted in microscopic freeze-concentrated regions separated by mannitol crystals. Consequently, nucleation and proliferation of the phase separation was inhibited and the native secondary structure of the protein was stabilized [122].

Aerosol formulations of a monoclonal antibody, rhuMAbE25, were prepared by spray-drying antibody solutions containing mannitol. Mannitol crystallized during the drying process when the rhuMAbE25 : mannitol ratio was ≤ 1.5. Crystallization led to an increase in the mean particle size of the spray-dried powder and a consequent decrease in the fine particle fraction (FPF), potentially decreasing aerosol performance (Table 13.2) [123]. Addition of sodium phosphate to the solution inhibited mannitol crystallization, resulting in lower mean particle size and higher % FPF (superior aerosol performance). The API stability during storage was also enhanced, as evident from the lower rates of protein aggregate formation (Table 13.2).

Crystallization of solutes during freeze-drying, in addition to affecting API stability, can also influence product behavior during processing. Crystallization of glycine and mannitol during freezing allows primary drying at elevated temperatures, and thereby reduces drying time and also yields an elegant freeze-dried cake [117]. If mannitol is retained amorphous during freezing it might crystallize during primary drying and cause vial breakage [124]. Recently, ampoule breakage during freeze-drying of recombinant human interleukin-11 was attributed to crystallization of dibasic sodium phosphate during primary drying [125]. The ampoule breakage could be prevented by annealing the frozen solutions, which resulted in crystallization of the sodium phosphate before drying. Besides, the amorphous salt in the freeze-concentrate was believed to cause a protein–phosphate interaction, yielding insoluble aggregates during reconstitution. All these problems were avoided by crystallizing the salt during annealing.

Table 13.2 Properties of spray-dried rhuMAbE25 powder as a function of solution composition; the protein concentration was 10 mg mL^{-1} in each case. (Data taken from [123]).

rhuMAbE25/ mannitol ratio (w/w)	Sodium phosphate concentration (mM)	Fine particle fraction (FPF) (%)	Volume mean particle diameter (μm)	k (10^{-3} day^{-1})[a]
No excipient	0	27	3.2	0.50±0.05
9	0	35	3.3	0.12±0.03
4	0	28	3.7	0.04±0.01
2.3	0	27	4.0	0.08±0.01
1.5	**0**	**8.5**	**8.7**	**0.12±0.01**
1.5	10	22	6.3	0.11±0.02
1.5	25	20	4.4	0.08±0.01
1.5	50	17	3.7	0.07±0.02

a) k = Pseudo-first order rate constant of soluble aggregate formation in the spray-dried powder at 5 °C and 38% RH.

Buffer salt crystallization during freezing, although beneficial in the above example, might be disastrous if the stability of the API is pH-dependent. Pronounced pH shifts can occur due to selective crystallization of a buffer component [50, 89, 117]. Interferon-γ was lyophilized from solutions buffered to pH 5 using succinate or glycolate buffers, and stored at 25 °C. The decrease in protein bioactivity was more rapid in lyophiles buffered with succinate, possibly due to crystallization of monosodium succinate during freezing. The resulting increase in acidity of the freeze-concentrate could cause protein unfolding, leading ultimately to lowered solid-state protein stability [89].

13.4.3 Polymorphic Transitions

Polymorphic transformations during processing can result in marked changes in the pharmaceutically relevant properties and can have major implications on subsequent processing and final product quality.

The δ-polymorphic form of mannitol, when subjected to aqueous wet granulation, was converted into the β-polymorph (stable under ambient conditions) [126]. Figure 13.10 compares the effect of compression pressure on the properties of tablets obtained after wet granulation of the β- and δ-polymorphs. Although, in both cases, the granules were composed of the β-form, the polymorphic δ- → β-transition *during* wet granulation of the δ-form appears to have conferred improved compaction properties to the resulting granules. Granulation and vacuum drying of the δ-crystals resulted in a pronounced decrease in particle size and an increase in the specific surface area from 0.4 to 3.4 m^2 g^{-1}. As a result, there was increased particle contact during compression. The in-

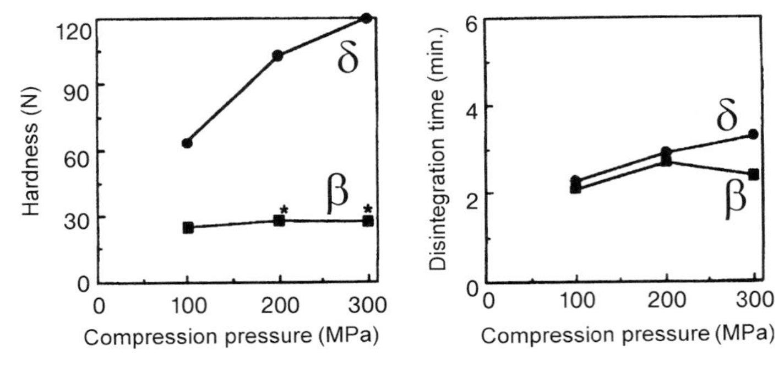

Fig. 13.10 Comparison of hardness and disintegration time of tablets made by wet granulation of the δ- (●) and the β- (■) crystalline forms of mannitol; * indicates compression pressures at which capping of the tablets was observed [126].

crease in plastic deformation and decrease in elastic recovery during compression yielded harder tablets at lower compression pressures with no appreciable increase in disintegration times.

13.4.4
Hydration/Dehydration

We discussed earlier the processing induced anhydrate → hydrate transition. This would influence the properties of the resulting material and subsequent processing parameters. The extent of conversion of anhydrous carbamazepine (CBZ) into carbamazepine dihydrate during wet granulation depended on the polymorphic form of the anhydrate (Table 13.3). While form I converted to a very small extent (2.5%), the degree of conversion was much higher for forms II and III [44]. Incorporation of water into the lattice, following granulation of

Table 13.3 Phase transformation during wet granulation of different polymorphs of anhydrous carbamazepine (CBZ) and the properties of the resulting granules and tablets. (Data taken from [44]).

Original crystal form	Extent of dihydrate formation after wet granulation (%)	Specific surface area of granules ($m^2\ g^{-1}$) [a]	Tablet hardness (kg)
Form I	2.5	0.95 ± 0.02	7.7 ± 0.2
Form II [b]	82.5	2.87 ± 0.02	14.6 ± 0.3
Form III [b]	38.1	2.08 ± 0.03	12.1 ± 0.2

a) Granules were obtained after drying the wet mass at 60 °C for 24 h. The dihydrate formed during wet granulation dehydrated to anhydrous form III.

b) Additional water was needed to obtain granules because of uptake of water into the crystal lattice.

forms II and III, necessitated addition of extra water for the formation of satisfactory granules. The presence of hydrate in the final formulation can have an impact on the product stability and dissolution performance [105].

The hydrate formed during wet massing may be dehydrated during further processing. In the above example of granulation of CBZ forms II and III, dehydration during drying yielded form III. In case of form I, there was no appreciable phase transformation either during wet massing or during subsequent drying. The hydration–dehydration cycle resulted in an increased specific surface area of the granules, which translated into harder tablets (Table 13.3). The highest granule surface area and the tablets with maximum hardness were obtained by wet granulation of form II, the form that showed the highest extent of anhydrate → hydrate → anhydrate transition [44].

The drying conditions influence the physical state of the dehydrated phase, and hence the product attributes [46, 47]. As mentioned earlier, theophylline underwent a hydration–dehydration cycle, during a wet granulation-drying process, yielding an anhydrate with reduced crystallinity. The amorphous regions underwent plastic deformation during granule compression and resulted in hard tablets. These tablets exhibited prolonged disintegration times and slower drug dissolution than tablets prepared with granules containing highly crystalline theophylline [46].

The dehydrated phase produced during drying may show a propensity to convert back into the hydrate on storage. Risedronate sodium hemipentahydrate is a mixed hydrate containing both lattice water as well as channel water. Fluid-bed drying of the granules, to a water content below a critical value, resulted in loss of the channel water [94]. When tablets made from granules containing the dehydrated crystals were stored at 60 or 70% RH, uptake of water into the channels resulted in an increase in the tablet weight and thickness and, in extreme cases, led to a loss in tablet integrity. The channel hydrate is reported to exhibit "flexible lattice characteristics" on dehydration [127]. Expansion of the crystal lattice during rehydration of the channels was believed to cause the adverse effects on tablet physical stability [94].

Thiamine HCl monohydrate underwent a series of dehydration–hydration reactions during processing. The anhydrate formed during the drying of granules transformed back into the monohydrate when the granules were compressed under humid conditions. In tablets or granules containing the monohydrate, storage at room temperature resulted in partial dehydration to yield a hemihydrate, causing caking of the granules or tablet hardening [45].

The formation of mannitol hydrate during freeze-drying was discussed earlier. The hydrate was formed during the freezing of mannitol solutions and survived dehydration during drying [37]. Formation of the crystalline hydrate during freezing will significantly reduce the rate of water removal during the drying stage, thereby prolonging the drying time or necessitating higher drying temperatures. Incomplete dehydration during the primary and secondary drying stages will increase the final lyophile water content. More importantly, dehydration of the hydrate can occur during storage and the water released from the crystal lattice can be available for interaction with the API or the other formulation components, with undesirable consequences.

Low-temperature X-ray diffractometry of frozen aqueous phosphate buffer solution revealed crystallization of the disodium salt as $Na_2HPO_4 \cdot 12H_2O$ [36]. Selective crystallization of the basic salt will shift the pH of the freeze-concentrate, which can destabilize the API [36, 128]. Dehydration of the dodecahydrate during primary drying yielded the amorphous anhydrate in the final product. Characterization of the final product therefore would not reveal the phase transformations during different stages of processing.

Thus, although the disodium salt was amorphous in the final lyophile, it crystallized during freezing, and therefore phase-separated from the freeze-concentrate. Although this crystalline phase is eventually rendered amorphous by dehydration, the phase separation brought about as a result of crystallization will not be reversed. Therefore, the amorphous salt is not expected to be homogeneously distributed in the lyophile. The inhomogeneity in the lyophile could be detrimental to API stability.

13.4.5 Salt–Free-acid/Base Conversion

As discussed earlier, the acid–base reaction between crystalline anhydrous delavirdine mesylate and a basic tablet excipient yielded the amorphous free delavirdine base. This occurred when the tablets were stored under high RH conditions, and caused a drastic decrease in dissolution rate [95]. There is a potential for such phase transformations in the presence of water, e.g., during wet granulation. If the free acid/base formed is less stable, it can adversely affect final product quality [129, 130].

13.5 Summary

The components of pharmaceutical formulations experience various stresses during manufacture that could lead to physical transformations. The type and intensity of these stresses will dictate the nature of the product phase. The major transformations can be classified as: amorphous $\leftrightarrow$ crystalline, anhydrate $\leftrightarrow$ hydrate and polymorphic transitions. Multiple phase transformations may be observed, e.g., transformation from the crystalline anhydrate into hydrate followed by dehydration to an amorphous anhydrate. The final product characteristics could be influenced by the phase behavior of the API *as well as* the excipients during the various processing steps.

Although only a small fraction of a formulation component may undergo a phase change (e.g., surface amorphization of API), the implications on product quality can be significant. The characterization of such transitions is complicated by the presence of multiple formulation components and the limited sensitivity of commonly used analytical techniques.

Conventionally, a thorough physical characterization of the API is conducted, before it is formulated into a dosage form. In selected instances, the physical form of the API may also be characterized in the final formulation. Evidently, based on the examples presented, the physical form of certain excipients can influence not only the processability, but also the final product quality. The characterization of the final product may be one of the critical components in the validation of the formulation strategy and manufacturing process. However, it may not be sufficient to fully understand the effect of processing on product quality. A detailed study of possible changes *during* the entire process is crucial for detection and control of transformations during processing. Such characterization studies will enable rational selection of formulation components and the design of processing parameters to control the physical form of critical formulation components both during processing and storage.

Acknowledgments

We thank Paroma Chakravarty, Sisir Bhattacharya, and Drs Jaidev Tantry, Dushyant Varshney and Xiangmin Liao for their helpful comments.

References

1 J.K. Haleblian, W. McCrone, *J. Pharm. Sci.* **1969**, *58*(*8*), 911–929.
2 J.K. Haleblian, *J. Pharm. Sci.* **1975**, *64*(*8*), 1269–1288.
3 K.R. Morris, U.J. Griesser, C.J. Eckhardt, J.G. Stowell, *Adv. Drug Delivery Rev.* **2001**, *48*(*1*), 91–114.
4 N.V. Phadnis, R. Suryanarayanan, *J. Pharm. Sci.* **1997**, *86*(*11*), 1256–1263.
5 E. Doelker, *Ann. Pharmaceut. Fr.* **2002**, *60*(*3*), 161–176.
6 H.G. Brittain, E.F. Fiese, Effects of pharmaceutical processing on drug polymorphs and solvates, in *Polymorphism in Pharmaceutical Solids*, ed. H.G. Brittain, **1999**, Marcel Dekker, New York, pp. 331–361.
7 G.G.Z. Zhang, D. Law, E.A. Schmitt, Y. Qiu, *Adv. Drug Delivery Rev.* **2004**, *56*(*3*), 371–390.
8 H.G. Brittain, *J. Pharm. Sci.* **2002** *91*(*7*), 1573–1580.
9 H. Steckel, N. Rasenack, P. Villax, B.W. Muller, *Int. J. Pharm.* **2003**, *258*(*1/2*), 65–75.
10 M.D. Ticehurst, P.A. Basford, C.I. Dallman, T.M. Lukas, P.V. Marshall, G. Nichols, D. Smith, *Int. J. Pharm.* **2000**, *193*(*2*), 247–259.
11 G.H. Ward, R.K. Schultz, *Pharm. Res.* **1995**, *12*(*5*), 773–779.
12 K. Brodka-Pfeiffer, P. Langguth, P. Grass, H. Hausler, *Eur. J. Pharm. Biopharm.* **2003**, *56*(*3*), 393–400.
13 M. Ohta, G. Buckton, *Int. J. Pharm.* **2004**, *269*(*1*), 81–88.
14 K.J. Crowley, G. Zografi, *J. Pharm. Sci.* **2002**, *91*(*2*), 492–507.
15 M. Otsuka, N. Kaneniwa, *Int. J. Pharm.* **1990**, *62*(*July 15*), 65–73.
16 M. Otsuka, H. Ohtani, K. Otsuka, N. Kaneniwa, *J. Pharm. Pharmacol.* **1993**, *45*(*1*), 2–5.

17 M. Otsuka, Y. Matsuda, *J. Pharm. Pharmacol.* **1993**, *45(May)*, 406–413.
18 H.A. Lieberman, L. Lachman, J.B. Schwartz, eds. *Pharmaceutical Dosage Forms: Tablets*, Vol. 2. **1989**, Marcel Dekker, New York.
19 H.K. Chan, E. Doelker, *Drug Dev. Ind. Pharm.* **1985**, *11(2/3)*, 315–332.
20 M. Otsuka, T. Matsumoto, N. Kaneniwa, *J. Pharm. Pharmacol.* **1989**, *41(10)*, 665–669.
21 Y. Matsunaga, N. Bando, H. Yuasa, Y. Kanaya, *Chem. Pharm. Bull.* **1996**, *44(10)*, 1931–1934.
22 K.M. Picker, *Pharm. Dev. Technol.* **2004**, *9(1)*, 107–121.
23 A.G. Schmidt, S. Wartewig, K.M. Picker, *J. Raman Spectrosc.* **2004**, *35(5)*, 360–367.
24 A.G. Schmidt, S. Wartewig, K.M. Picker, *Eur. J. Pharm. Biopharm.* **2003**, *56(1)*, 101–110.
25 A.J. Cannon, E.H. Trappler, *PDA J. Pharm. Sci. Tech.* **2000**, *54(1)*, 13–22.
26 J.F. Carpenter, M.J. Pikal, B.S. Chang, T. W. Randolph, *Pharm. Res.* **1997**, *14(8)*, 969–975.
27 N. Milton, S.L. Nail, *Cryo-Lett.* **1997**, *18(6)*, 335–342.
28 A.I. Kim, M.J. Akers, S.L. Nail, *J. Pharm. Sci.* **1998**, *87(8)*, 931–935.
29 K.-I. Izutsu, S. Kojima, *J. Pharm. Pharmacol.* **2002**, *54(8)*, 1033–1039.
30 B. Lueckel, B. Helk, D. Bodmer, H. Leuenberger, *Pharm. Dev. Technol.* **1998**, *3(3)*, 337–346.
31 M.J. Pikal, M.L. Roy, Pharmaceutical formulation comprising human growth hormone in admixture with aggregation stabilising amounts of glycine and mannitol. *AU Patent 617427*, **1991**.
32 A. Pyne, K. Chatterjee, R. Suryanarayanan, *J. Pharm. Sci.* **2003**, *92(11)*, 2272–2283.
33 C. Telang, L. Yu, R. Suryanarayanan, *Pharm. Res.* **2003**, *20(4)*, 660–667.
34 A. Pyne, R. Suryanarayanan, *Pharm. Res.* **2003**, *20(2)*, 283–291.
35 R.K. Cavatur, R. Suryanarayanan, *Pharm. Dev. Technol.* **1998**, *3(4)*, 579–586.
36 A. Pyne, K. Chatterjee, R. Suryanarayanan, *Pharm. Res.* **2003**, *20(5)*, 802–803.
37 L. Yu, N. Milton, E.G. Groleau, D.S. Mishra, R.E. Vansickle, *J. Pharm. Sci.* **1999**, *88(2)*, 196–198.
38 T.D. Davis, G.E. Peck, J.G. Stowell, K.R. Morris, S.R. Byrn, *Pharm. Res.* **2004**, *21(5)*, 860–866.
39 P. Di Martino, M. Scoppa, E. Joiris, G.F. Palmieri, C. Andres, Y. Pourcelot, S. Martelli, *Int. J. Pharm.* **2001**, *213(1/2)*, 209–221.
40 M. Ohta, G. Buckton, *Int. J. Pharm.* **2005**, *289(1/2)*, 31–38.
41 F. Hirayama, M. Usami, K. Kimura, K. Uekama, *Eur. J. Pharm. Sci.* **1997**, *5(1)*, 23–30.
42 O.I. Corrigan, K. Sabra, E.M. Holohan, *Drug Dev. Ind. Pharm.* **1983**, *9(1/2)*, 1–20.
43 L. Yu, K. Ng, *J. Pharm. Sci.* **2002**, *91(11)*, 2367–2375.
44 M. Otsuka, H. Hasegawa, Y. Matsuda, *Chem. Pharm. Bull.* **1999**, *47(6)*, 852–856.
45 K. Wostheinrich, P.C. Schmidt, *Drug Dev. Ind. Pharm.* **2001**, *27(6)*, 481–489.
46 S. Debnath, R. Suryanarayanan, *AAPS PharmSciTech.* **2004**, *5(1)*, Article 8.
47 J. Han, R. Suryanarayanan, *Pharm. Dev. Technol.* **1998**, *3(4)*, 587–596.
48 S.H. Cypes, R.M. Wenslow, S.M. Thomas, A.M. Chen, J.G. Dorwart, J.R. Corte, M. Kaba, *Org. Process Res. Develop.* **2004**, *8(4)*, 576–582.
49 S. Airaksinen, M. Karjalainen, E. Raesaenen, J. Rantanen, J. Yliruusi, *Int. J. Pharm.* **2004**, *276(1/2)*, 129–141.
50 E.Y. Shalaev, T.D. Johnson-Elton, L. Chang, M.J. Pikal, *Pharm. Res.* **2002**, *19(2)*, 195–201.
51 S. Hulsmann, T. Backensfeld, R. Bodmeier, *Pharm. Dev. Technol.* **2001**, *6(2)*, 223–229.
52 K. Six, C. Leuner, J. Dressman, G. Verreck, J. Peeters, N. Blaton, P. Augustijns, R. Kinget, G. van den Mooter, *J. Therm. Anal. Calorim.* **2002**, *68(2)*, 591–601.
53 R.K. Cavatur, N.M. Vemuri, A. Pyne, Z. Chrzan, D. Toledo-Velasquez, R. Suryanarayanan, *Pharm. Res.* **2002**, *19(6)*, 894–900.
54 T. Yoshinari, R.T. Forbes, P. York, Y. Kawashima, *Int. J. Pharm.* **2002**, *247(1/2)*, 69–77.
55 M.W.Y. Wong, A.G. Mitchell, *Int. J. Pharm.* **1992**, *88(1–3)*, 261–273.
56 S. Airaksinen, P. Luukkonen, A. Jorgensen, M. Karjalainen, J. Rantanen, J. Yliruusi, *J. Pharm. Sci.* **2003**, *92(3)*, 516–528.

57 A.C. Jorgensen, S. Airaksinen, M. Karjalainen, P. Luukkonen, J. Rantanen, J. Yliruusi, *Eur. J. Pharm. Sci.* **2004**, *23(1)*, 99–104.
58 J.G. Kesavan, G.E. Peck, *Drug Dev. Ind. Pharm.* **1996**, *22(3)*, 189–199.
59 S.S. Braga, I.S. Goncalves, E. Herdtweck, J.J.C. Teixeira-Dias, *New J. Chem.* **2003**, *27(3)*, 597–601.
60 A. Gil, A. Chamayou, E. Leverd, J. Bougaret, M. Baron, G. Couarraze, *Eur. J. Pharm. Sci.* **2004**, *23(2)*, 123–129.
61 R. Govindarajan, M.S. Nagarsenker, *J. Pharm. Pharmacol.* **2004**, *56(6)*, 725–733.
62 A.C. Williams, V.B. Cooper, L. Thomas, L.J. Griffith, C.R. Petts, S.W. Booth, *Int. J. Pharm.* **2004**, *275(1/2)*, 29–39.
63 B.C. Sherman, Stable solid formulation of enalapril salt and process for preparation thereof. *US Pat. 5573780*, **1996**.
64 M. Merslavic, J. Razen, A. Rotar, Stable formulation of enalapril salt. *EU Pat. 545194*, **1993**.
65 M.K. Gupta, A. Vanwert, R.H. Bogner, *J. Pharm. Sci.* **2003**, *92(3)*, 536–551.
66 T.P. Shakhtshneider, *Solid State Ionics* **1997**, *101–103(Pt. 2)*, 851–856.
67 P. Mura, M. Cirri, M.T. Faucci, J.M. Gines-Dorado, G.P. Bettinetti, *J. Pharm. Biomed. Anal.* **2002**, *30(2)*, 227–237.
68 T. Watanabe, N. Wakiyama, F. Usui, M. Ikeda, T. Isobe, M. Senna, *Int. J. Pharm.* **2001**, *226(1/2)*, 81–91.
69 M. Otsuka, H. Hasegawa, Y. Matsuda, *Chem. Pharm. Bull.* **1997**, *45(May)*, 894–898.
70 D. Giron, C. Goldbronn, M. Mutz, S. Pfeffer, P. Piechon, P. Schwab, *J. Therm. Anal. Calorim.* **2002**, *68(2)*, 453–465.
71 D.E. Bugay, *Adv. Drug Delivery Rev.* **2001**, *48(1)*, 43–65.
72 H.G. Brittain, ed., *Physical Characterization of Pharmaceutical Solids.* Drugs and the Pharmaceutical Sciences, Vol. 70, **1995**, Marcel Dekker, New York, 424 pp.
73 G.A. Stephenson, R.A. Forbes, S.M. Reutzel-Edens, *Adv. Drug Del. Rev.* **2001**, *48(1)*, 67–90.
74 R.O. Williams III, J. Brown, J. Liu, *Pharm. Dev. Technol.* **1999**, *4(2)*, 167–179.
75 G. Buckton, P. Darcy, *Int. J. Pharm.* **1999**, *179(2)*, 141–158.
76 H. Ahmed, G. Buckton, D.A. Rawlins, *Int. J. Pharm.* **1996**, *130(2)*, 195–201.
77 G.M. Venkatesh, M.E. Barnett, C. Owusu-Fordjour, M. Galop, *Pharm. Res.* **2001**, *18(1)*, 98–103.
78 L. Mackin, R. Zanon, J.M. Park, K. Foster, H. Opalenik, M. Demonte, *Int. J. Pharm.* **2002**, *231(2)*, 227–236.
79 K. Brodka-Pfeiffer, H. Haeusler, P. Grass, P. Langguth, *Drug Dev. Ind. Pharm.* **2003**, *29(10)*, 1077–1084.
80 S.E. Dilworth, G. Buckton, S. Gaisford, R. Ramos, *Int. J. Pharm.* **2004**, *284(1–2)*, 83–94.
81 R. Saklatvala, P.G. Royall, D.Q.M. Craig, *Int. J. Pharm.* **1999**, *192(1)*, 55–62.
82 S. Guinot, F. Leveiller, *Int. J. Pharm.* **1999**, *192(1)*, 63–75.
83 P. Begat, P.M. Young, S. Edge, J.S. Kaerger, R. Price, *J. Pharm. Sci.* **2003**, *92(3)*, 611–620.
84 R. Surana, R. Suryanarayanan, *Powder Diffraction* **2000**, *15(1)*, 2–6.
85 C. Nunes, A. Mahendrasingam, R. Suryanarayanan, Crystallinity quantification in substantially amorphous pharmaceuticals using synchrotron X-ray powder diffractometry, AAPS Annual Meeting Abstracts. *The AAPS Journal* **2004**, 6(4), Abstract R6062, available from http://www.aapsj.org.
86 S. Kitahara, T. Ishizuka, T. Kikkoji, R. Matsuda, Y. Hayashi, *Int. J. Pharm.* **2004**, *283(1/2)*, 63–69.
87 T.D. Davis, K.R. Morris, H. Huang, G.E. Peck, J.G. Stowell, B.J. Eisenhauer, J.L. Hilden, D. Gibson, S.R. Byrn, *Pharm. Res.* **2003**, *20(11)*, 1851–1857.
88 A. Pyne, R. Suryanarayanan, *Pharm. Res.* **2001**, *18(10)*, 1448–1454.
89 X.M. Lam, H.R. Costantino, D.E. Overcashier, T.H. Nguyen, C.C. Hsu, *Int. J. Pharm.* **1996**, *142(1)*, 85–95.
90 L.S. Taylor, G. Zografi, *Pharm. Res.* **1998**, *15(5)*, 755–761.
91 L.S. Taylor, F.W. Langkilde, *J. Pharm. Sci.* **2000**, *89(10)*, 1342–1353.
92 E. Rasanen, J. Rantanen, A. Jorgensen, M. Karjalainen, T. Paakkari, J. Yliruusi, *J. Pharm. Sci.* **2001**, *90(3)*, 389–396.
93 H. Wikstroem, P.J. Marsac, L.S. Taylor, *J. Pharm. Sci.* **2005**, *94(1)*, 209–219.
94 D.S. Hausman, R.T. Cambron, A. Sakr, *Int. J. Pharm.* **2005**, *299(1/2)*, 19–33.

95 B. R. Rohrs, T. J. Thamann, P. Gao, D. J. Stelzer, M. S. Bergren, R. S. Chao, *Pharm. Res.* **1999**, *16(12)*, 1850–1856.

96 E. Suihko, V. P. Lehto, J. Ketolainen, E. Laine, P. Paronen, *Int. J. Pharm.* **2001**, *217(1/2)*, 225–236.

97 A. Burger, J. O. Henck, S. Hetz, J. M. Rollinger, A. A. Weissnicht, H. Stottner, *J. Pharm. Sci.* **2000**, *89(4)*, 457–468.

98 V. Busignies, P. Tchoreloff, B. Leclerc, C. Hersen, G. Keller, G. Couarraze, *Eur. J. Pharm. Biopharm.* **2004**, *58(3)*, 577–586.

99 M. Song, M. M. de Villiers, *Pharm. Dev. Technol.* **2004**, *9(4)*, 387–398.

100 D. Law, S. L. Krill, E. A. Schmitt, J. J. Fort, Y. Qiu, W. Wang, W. R. Porter, *J. Pharm. Sci.* **2001**, *90(8)*, 1015–1025.

101 A. A. Elamin, C. Ahlneck, G. Alderborn, C. Nystroem, *Int. J. Pharm.* **1994**, *111(2)*, 159–170.

102 B. C. Hancock, M. Parks, *Pharm. Res.* **2000**, *17(4)*, 397–404.

103 E. Suzuki, K. Shimomura, K. Sekiguchi, *Chem. Pharm. Bull.* **1989**, *37(2)*, 493–497.

104 D. Murphy, F. Rodriguez-Cintron, B. Langevin, R. C. Kelly, N. Rodriguez-Hornedo, *Int. J. Pharm.* **2002**, *246(1/2)*, 121–134.

105 E. Shefter, T. Higuchi, *J. Pharm. Sci.* **1963**, *52(8)*, 781–791.

106 J. C. Feeley, P. York, B. S. Sumby, H. Dicks, *Int. J. Pharm.* **1998**, *172(1/2)*, 89–96.

107 L. Mackin, S. Sartnurak, I. Thomas, S. Moore, *Int. J. Pharm.* **2002**, *231(2)*, 213–226.

108 T. Suzuki, K. Watanabe, S. Kikkawa, H. Nakagami, *Chem. Pharm. Bull.* **1994**, *42(11)*, 2315–2319.

109 T. Suzuki, H. Nakagami, *Eur. J. Pharm. Biopharm.* **1999**, *47(3)*, 225–230.

110 C. De Brabander, G. Van Den Mooter, C. Vervaet, J. P. Remon, *J. Pharm. Sci.* **2002**, *91(7)*, 1678–1685.

111 M. Kidokoro, N. H. Shah, A. W. Malick, M. H. Infeld, J. W. McGinity, *Pharm. Dev. Technol.* **2001**, *6(2)*, 263–275.

112 J. F. Carpenter, J. H. Crowe, *Biochemistry,* **1989**, *28(9)*, 3916–3922.

113 H. R. Costantino, K. Griebenow, P. Mishra, R. Langer, A. M. Klibanov, *Biochim. Biophys. Acta* **1995**, *1253(1)*, 69–74.

114 F. Franks, R. H. M. Hatley, S. F. Mathias, *BioPharm.* **1991**, *4(9)*, 38, 40–42, 55.

115 H. R. Costantino, K. G. Carrasquillo, R. A. Cordero, M. Mumenthaler, C. C. Hsu, K. Griebenow, *J. Pharm. Sci.* **1998**, *87(11)*, 1412–1420.

116 K.-I. Izutsu, S. Yoshioka, S. Kojima, *Pharm. Res.* **1995**, *12(6)*, 838–843.

117 M. J. Pikal, Impact of polymorphism on the quality of lyophilized products, in *Polymorphism in Pharmaceutical Solids*, ed. H. G. Brittain, **1999**, Marcel Dekker, New York, pp. 395–419.

118 B. D. Herman, B. D. Sinclair, N. Milton, S. L. Nail, *Pharm. Res.* **1994**, *11(10)*, 1467–1473.

119 M. J. Pikal, A. L. Lukes, J. E. Lang, K. Gaines, *J. Pharm. Sci.* **1978**, *67(6)*, 767–773.

120 R. Teraoka, M. Otsuka, Y. Matsuda, *Int. J. Pharm.* **2004**, *286(1/2)*, 1–8.

121 H. J. C. Eriksson, W. L. J. Hinrichs, B. van Veen, G. W. Somsen, G. J. de Jong, H. W. Frijlink, *Int. J. Pharm.* **2002**, *249(1/2)*, 59–70.

122 M. C. Heller, J. F. Carpenter, T. W. Randolph, *Biotechnol. Bioeng.* **1999**, *63(2)*, 166–174.

123 H. R. Costantino, J. D. Andya, P.-A. Nguyen, N. Dasovich, T. D. Sweeney, S. J. Shire, C. C. Hsu, Y.-F. Maa, *J. Pharm. Sci.* **1998**, *87(11)*, 1406–1411.

124 N. A. Williams, J. Guglielmo, *J. Parenteral Sci. Technol.* **1993**, *47(3)*, 119–123.

125 Y. Hirakura, S. Kojima, A. Okada, S. Yokohama, S. Yokota, *Int. J. Pharm.* **2004**, *286(1/2)*, 53–67.

126 T. Yoshinari, R. T. Forbes, P. York, Y. Kawashima, *Int. J. Pharm.* **2003**, *258(1/2)*, 121–131.

127 N. Redman-Furey, M. Dicks, A. Bigalow-Kern, R. T. Cambron, G. Lubey, C. Lester, D. Vaughn, *J. Pharm. Sci.* **2005**, *94(4)*, 893–911.

128 G. Gomez, M. J. Pikal, N. Rodriguez-Hornedo, *Pharm. Res.* **2001**, *18(1)*, 90–97.

129 M. L. Cotton, P. Lamarche, S. Motola, E. B. Vadas, *Int. J. Pharm.* **1994**, *109(3)*, 237–249.

130 W. D. Walkling, B. E. Reynolds, B. J. Fegely, C. A. Janicki, *Drug Dev. Ind. Pharm.* **1983**, *9(5)*, 809–819.

14
Polymorphism and Patents from a Chemist's Point of View [1]

Joel Bernstein

"Forget horse racing", says one patent attorney,
"patent litigation is the true sport of kings"
Economist, March 8, 2003

14.1
Introduction

According to the Oxford English Dictionary definition, a patent is

> "a license to manufacture, sell, or deal in an article or commodity, to the exclusion of other persons; in modern times, a grant from the government to a person or persons conferring for a certain definite time the exclusive privilege of making, using, or selling some new invention."

The first known English patent was granted in 1449 to John of Utyman for a method for making stained glass [2]. Today, patents essentially grant the right to exclude; namely they give the inventor exclusive rights for his invention for a limited time in return for the cooperation of the inventor to teach the rest of society how to use his findings or invention [3].

As noted by Maynard and Peters [3] (see also [4]):

> "patent systems reward the competitive, creative drive with a temporary, limited, exclusive right, in return for the cooperation of an inventor in teaching the rest of society how to use his or her findings for all time thereafter."

Polymorphism presents challenging issues to patent systems. Since crystal modifications of a substance represent different crystal structures with potentially different properties, the discovery or preparation of a new crystal modification represents an opportunity to claim an invention that potentially can be recognized in the awarding of a patent. Our frame of reference here will be the patent laws in the USA. The rules and regulations of patents differ from country to country. Although some degree of standardization has resulted from the World Trade Agreement and subsequent legislation there are still important differences in the nuances of the granting and enforcement of patents in various venues. For instance, in the United Kingdom, a new crystal modification is not

Polymorphism: in the Pharmaceutical Industry. Edited by Rolf Hilfiker

ISBN: 3-527-31146-7

prima facia patentable; the inventor must demonstrate that it is an unobvious variant of the previously known material. As seen in other chapters in this book, a particular crystalline modification can possess considerable chemical, physical or biological advantages over its congeners, and the granting and maintenance of patent exclusivity over the rights to particular polymorphic form(s) may have considerable economic consequences.

As a result, unsurprisingly, the past thirty years or so have witnessed a number of patent litigations essentially involving different crystal modifications, many concerning the definitions, descriptions and analytical techniques described in earlier chapters. Partly because of the size and economic impact of the pharmaceutical industry, some of the most visible cases have involved some widely used drugs. In addition, in the United States there are special patent provisions for pharmaceuticals [5], which also may contribute to the frequency and nature of these legal battles [6].

This chapter reviews some of the cases that have served as the meeting ground between the worlds of science and the law. The intention is not to provide legal opinions or precedents, but rather to demonstrate the scientific issues, more specifically the chemical issues, which were raised during these litigations. Many of these cases also involved principles and nuances of patent law, which also will not be covered here, except to quote them in the context of a particular case in point and to demonstrate how the courts view science when dealing with a specific patent or patent issue. Litigations involving patent issues can generate hundreds of thousands of pages of documents and testimony, with the discovery of many facts and the expression of many (often opposing) scientific opinions on both sides of the issue. We limit the descriptions here to essentially what is given in the official records of the cases considered, mainly from the patents, judicial decisions and the reports of them. Of course, even these are subject to controversy, since court rulings can be and are reversed, thus perhaps altering the way a scientific issue is viewed by a court of law and the society that is guided by that law. Moreover, both science and the law are dynamic, and the interpretations and ramifications of any particular case can and do change with time.

14.2
Some Fundamentals of Patents Related to Polymorphism and Some Historical Notes

In terms of the subject of this book, a crystalline polymorph, or for that matter a crystal form, which constitutes a composition of matter, is a patentable invention, since it generally meets three requirements required to obtain a patent in the US: utility, novelty and non-obviousness. Nevertheless, the challenges to patents, both at the prosecution stage and in later litigations, essentially involve the meaning and interpretation of one or more of these three issues with regard to the specific crystal form in question.

In general, new crystal forms can be patented without showing unexpected properties, since one of ordinary skill cannot predict the structure, properties, or method of preparation of that crystal form prior to its discovery. As the US Court of Customs and Patent Appeals pointed out in In re Irani [7] [a case involving the patent prosecution of crystalline anhydrous amino tri(methylenephosphonic acid) (ATMP)]:

> "...[E]ven assuming that one skilled in the art could have predicted with reasonable certainty that crystalline anhydrous ATMP could be produced, we are not convinced by this record that it would also have been obvious how this could be achieved. We note that neither the examiner nor the board has contended that a suitable process would have been obvious. The closest that either has come to such a contention is the examiner's statement, based on the disclosure in the Irani patent, that, as it turns out, 'little modification of the Process will produce a crystalline material'. Obviousness, however, must not be judged by hindsight, and a "little modification" can be a most unobvious one".

Since the social contract aspect of a patent involves teaching how to use the patented invention, the question arises as to whom this teaching is addressed. Normally this person is described as above, as one of ordinary skill in the same art as described in the patent. Such a definition is also open to interpretation and often leads to controversy in the course of litigation.

One of the tests for novelty is whether the proposed invention has been described in the "prior art". The prior art generally defines the earlier work of others, or the inventor himself, that has been used by others, has been patented by others, or appears in a printed publication. The awarding of a patent to a particular crystal form is evaluated in each individual case, but there is a long history of controversy over this subject. The usual argument against such an award is that the same or similar chemical is involved in the patent application as is found in the prior art, and hence the claimed invention is *ipso facto* purported to be present in the prior art without considering the differences in structure, method of preparation or properties.

The 1910 case of Union Carbide Co. v. American Carbide Co. [8] illustrates many of the points raised in the three previous paragraphs that still arise in patent prosecutions and litigations involving crystal forms [9]. The patent dispute involved calcium carbide, CaC_2, whose preparation was first described by Friederich Woehler in a chemistry yearbook of 1862. He noted that upon reaction with water the CaC_2 had

> "the remarkable property to decompose with water into calcium hydrate and acetylene gas...first discovered by Davy and...recently...produced by Berthelot."

As the court pointed out regarding the Woehler publication,

> "There is nothing to indicate that which Woehler did was anything more than to make and describe a laboratory experiment, and, although his work was

generally recognized in treatises upon chemistry, it does not appear that any appreciable amount of calcium carbide was made by any person before the present patentee came into the field."

Union Carbide, the patentee, made a single claim in the patent, namely,

"As a new product, crystalline calcium carbide existing as masses of aggregated crystals, substantially as described."

The court wrestled with the question of prior art and patentability:

"Concededly the Woehler compound was the highest development of the prior art in calcium carbide, and so we recur to the question whether with that compound in the art – assumed to be amorphous for the purposes of the present discussion – there was patentable novelty in the crystalline form.

In determining the question of patentable novelty there can be no hard and fast rule. Each case must be decided upon its own facts. Mere change of form in and of itself does not disclose novelty. A new article of commerce is not necessarily a new article patentable as such. But patentable novelty in a case like the present may be founded upon superior efficiency; upon superior durability, including the ability to retain a permanent form when exposed to the atmosphere; upon a lesser tendency to breakage and loss; upon purity, and, in connection with other things, upon comparative cheapness. So, as supplementing other considerations, commercial success may properly be compared with mere laboratory experiments..."

The court recognized the differences between amorphous and crystalline material:

"...[T]he amorphous substance is less dense, more soluble, has a lower melting point and less hardness...bulk for bulk, the yield of gas in the case of amorphous compound would be smaller, the tendency for breakage would be greater, both because the substance is more porous and less hard, that for such matters as transportation and dangerous dust the amorphous would be inferior material even if equally pure."

By the time this case came before the court, long after Woehler's description, CaC_2 had been put to extensive use for the generation of acetylene for various industrial uses, including widespread street illumination. Recognizing this fact the court went on,

"It is also quite clear that Woehler published a mere result of a laboratory experiment which was put to no practical use. Crystalline carbide, on the other hand, has been a great commercial success, and has furnished the foundation for important industries...

...To hold an important discovery which has given the world a commercially new product – a product the high utility of which must be conceded – not entitled to protection for want of novelty, would, as it seems, be applying the patent statute to defeat its fundamental purposes."

Moreover, the defendant's claim that Woehler's work anticipated the product claimed by the patent applicant was rejected by the court:

> "If the Woehler compound were amorphous, it manifestly did not anticipate the crystalline product in view of the differences between forms already pointed out...[I]t is not shown that any of the Woehler compound which was made before the application for the patent was crystalline, and indeed, we think the testimony tends to show that it was amorphous...Woehler was making note of a laboratory experiment evidently employing minute amounts of material, and seems to have been more interested in the formation of acetylene from the carbide than in the formation of the carbide."

Many of the questions addressed by that court nearly 100 years ago are still argued today in both patent prosecutions and subsequent litigations regarding solid forms of pharmaceutical compounds. Three of these litigations are described below [10].

14.3 Ranitidine Hydrochloride (RHCl)

Ranitidine was developed in the 1970s by Allen & Hanburys Ltd. of the Glaxo Group (later Glaxo Wellcome and now GlaxoSmithKline) in the flurry of activity following the identification of the histamine H_2 receptor [12] and other H_2 antagonists [13] for the treatment of peptic ulcers. In June 1977, David Collin, a Glaxo chemist, first prepared RHCl, and within a month Glaxo filed a US patent application, which resulted in the issue of US Patent No. 4,128,658 [13] (the '658 patent). Example 32 of this patent gives the procedure for the preparation of the hydrochloride (Fig. 14.1).

During subsequent scale up, Glaxo developed a pilot plant process called 3A, and then one called 3B. On April 15, 1980, for unknown reasons [15], the thirteenth batch of RHCl prepared using the latter process produced crystals that gave different IR spectra and X-ray powder diffraction patterns from previous batches. Glaxo concluded that a new polymorph, designated Form 2, had been produced, and the earlier form, described in the '658 patent, was designated Form 1 [16, 17]. Glaxo subsequently developed a process for production of Form 2, referred to as 3C, to manufacture all the RHCl it has sold commercially as the active ingredient in Zantac.

In October 1981, Glaxo filed a patent application on Form 2, from which two patents were eventually granted in June of 1985 as US Patent No. 4,521,431 (the '431 patent) and in June of 1987 as US Patent No. 4,672,133. Although the context was somewhat different, the file wrapper for the prosecution of the '431 patent reveals that many issues similar to those raised above in the calcium carbide case 75 years earlier were considered again. The abstract of the '431 patent states simply,

United States Patent [19] — [11] **4,128,658**

Price et al. — [45] **Dec. 5, 1978**

[54] **AMINOALKYL FURAN DERIVATIVES**

[75] Inventors: **Barry J. Price**, Hertford; **John W. Clitherow**, Sawbridgeworth; **John Bradshaw**, Ware, all of England

[73] Assignee: **Allen & Hanburys Limited**, London, England

[21] Appl. No.: **818,762**

[22] Filed: **Jul. 25, 1977**

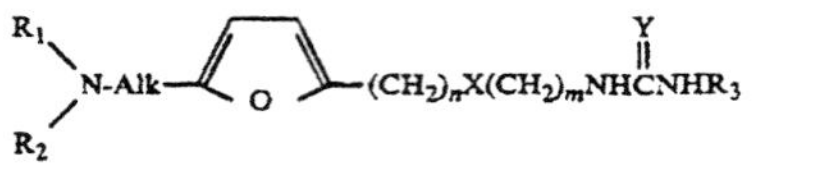

R_1R_2N-Alk–(furan)–$(CH_2)_nX(CH_2)_mNHC(=Y)NHR_3$ (I)

and physiologically acceptable salts thereof and N-oxides and hydrates, in which R_1 and R_2 which may be the same or different represent hydrogen, lower alkyl, cycloalkyl, lower alkenyl, aralkyl or lower alkyl interrupted by an oxygen atom or a group

EXAMPLE 32

N-[2-[[[5-(Dimethylamino)methyl-2-furanyl]methyl]thio]ethyl]-N'-methyl-2-nitro-1,1-ethenediamine hydrochloride

N-[2-[[[5-(Dimethylamino)methyl-2-furanyl]methyl]-thio]ethyl]-N'-methyl-2-nitro-1,1-ethenediamine (50 g, 0.16 mole) was dissolved in industrial methylated spirit 74° o.p. (200 ml) containing 0.16 of an equivalent of hydrogen chloride. Ethyl acetate (200 ml) was added slowly to the solution. The hydrochloride crystallised and was filtered off, washed with a mixture of industrial methylated spirit 74° o.p. (50 ml) and ethyl acetate (50 ml) and was dried at 50°. The product (50 g) was obtained as an off-white solid m.p. 133°–134°.

Fig. 14.1 Example 32 from the '658 ranitidine hydrochloride patent [14].

> "A novel form of ranitidine...hydrochloride, designated Form 2, and having favourable filtration and drying characteristics, is characterized by its infra-red spectrum and/or by its x-ray powder diffraction pattern."

These advantageous filtering and drying characteristics are due, in part, at least, to the fact that Form 2 tends to crystallize as more needle-like crystals than the plate-like Form 1.

By 1991, Zantac sales had reached almost $ 3.5 billion, nearly twice the sales of the next best selling drug. Several generic drug firms undertook efforts (under the provisions of the Waxman-Hatch Law [5]) to prepare to go on the market with Form 1 in 1995, upon anticipated expiration of the '658 patent [18]. One of these generic companies was Novopharm, Ltd., the defendant in the first RHCl litigation that went to trial. Novopharm scientists unsuccessfully attempted to prepare Form 1 faithfully following the procedure of example 32 of the '658 patent and therefore sought approval to market Form 2, claiming that the product is, and always has been, Form 2 RHCl. In November, 1991, Novopharm filed an Abbreviated New Drug Application (ANDA) at the FDA to market Form 2 beginning in 1995. As required by the Waxman-Hatch Act, Novopharm notified

Glaxo of its contention that the '431 patent was invalid. Glaxo sued Novopharm for infringement of the '431 patent. Novopharm admitted infringement of the '431 patent, but contended that it was invalid, claiming that the Form 2 was inherently anticipated in the '658 patent. Novopharm claimed that Glaxo *never* performed Example 32 precisely as written either before or after including it in the '658 patent (emphasis in the Court's original). Novopharm theorized that if Glaxo had performed Example 32, the result would have been Form 2, not Form 1. Therefore the Form 2 patent was invalid.

Glaxo argued, *inter alia,* that Novopharm's experiments were contaminated with "seed" crystals of Form 2 and, therefore, were not faithful replications of Example 32 (quotation marks in Court's original). To make its point on the inherency argument, Glaxo proved that Example 32 does not invariably lead to Form 2, and in fact had led to Form 1 as the basis for the '658 patent. In support of its position Glaxo compared David Collin's original experiments as described in his notebooks in minutiae with Example 32, and also presented evidence that Example 32 as performed in 1993 at Oxford University had led to Form 1. This was sufficient evidence to convince the court that Example 32 does lead to the Form 1 product and that the '431 patent was valid.

Novopharm then examined the possibility of marketing Form 1, and developed a stable, reproducible process for the manufacture of Form 1. In April of 1994 [19], Novopharm filed a new ANDA, this time seeking approval to market Form 1 RHCl upon the expiration of the '658 patent. Shortly thereafter, Glaxo sued Novopharm again, alleging that Novopharm had sought permission to manufacture and market a product that would contain not pure Form 1, but rather a mixture of Form 1 and Form 2, thereby infringing upon Glaxo's Form 2 patents. Novopharm's ANDA, as initially filed, specified that the marketed product be approximately 99% pure Form 1 RHCl (with impurities that may include Form 2 RHCl) as determined by IR. Amended ANDAs filed by Novopharm would have permitted the marketed product to have a Form 1 RHCl of purity as low as 90%. Novopharm, however, submitted X-ray evidence at trial that demonstrated that its actual samples of RHCl did not contain detectable Form 2.

The court found that Novopharm had established that its product would not contain Form 2, and that Glaxo failed to prove infringement. The court thus allowed Novopharm to market mixtures of Form 1 and Form 2.

The Court of Appeals for the Federal Circuit court upheld the district court's decision holding that, based on all the available evidence, Glaxo had not proven the Novopharm was likely to market a product that contained Form 2. Notably, the appeals court pointed to Glaxo's failure to test Novopharm's samples. In reviewing a later case, the same appeals court noted it had explicitly declined to address the question of whether small amounts of Form 2 RHCl in a mixture containing primarily Form 1 would infringe the '431 patent [20].

The two Glaxo v. Novopharm cases involved many aspects of the study and analysis of polymorphic materials described in other chapters of this book: the (often serendipitous) discovery and recognition of polymorphic forms, the role of solvent, heating, stirring and other experimental techniques in attempts to control the polymorph obtained, the use and development of analytical methods for the characterization of polymorphic forms, the relative stability of polymorphic forms, the phenomenon of disappearing polymorphs, the distinction between polymorphic identity and polymorphic purity and the role of seeding, both intentional and unintentional (see also Section 14.6).

14.4 Cefadroxil

Cefadroxil (also sometimes spelled cephadroxil), an antibiotic originally developed by Bristol-Myers (BM) and marketed under the trade names Ultracef and Duricef, had combined sales in the US of $ 100 million in 1988. A crystalline monohydrate (also apparently discovered serendipitously), claimed in US Patent No. 4,504,657 (the '657 patent) issued in March 1985, is described in the patent in terms of its X-ray diffraction pattern (Fig. 14.2). The improvement of this substance over the earlier known material included greater stability and high bulk density, properties that enabled the production of smaller pills.

Prosecution of this patent was in fact quite lengthy, the original application dating from August 1979, and involved questions of similarities and differences in crystalline modifications (see [21] for details of the history of the prosecution of the patent application). The prior art included a 1973 patent (US Patent No. 3,781,282) (the '282 patent) in which a modification of cefadroxil had also been prepared. This was termed the "Micetich form" after one of the inventors. One of the questions surrounding the '657 patent was whether the crystalline material described therein (known as the "Bouzard form" or "Bouzard monohydrate") was the same as or different from the "Micetich form". In the end, the '657 patent was granted, in essence recognizing the difference between the two.

In anticipation of the expiration of '282 patent, there were attempts to make the Micetich form according to Example 19 in that patent. Those attempts invariably led to the Bouzard form, leading the companies involved in carrying out the experiments to claim inherency of the Bouzard form in the '282 patent. Bristol-Myers' explanation of these results included the role of unintentional seeding that tended to favor the formation of the Bouzard form rather than the Micetich form. Several litigations ensued, which involved these issues, e.g., [21, 22]. In the former case, there was also a question of the identity of the various crystal modifications, determined mainly by powder X-ray diffraction, with conflicting opinions by various experts. The court found that the

> "...plaintiff had established a prima facie case of invalidity [of the '657 patent],"

and that

> "...its scientists have replicated the claimed cefadroxil monohydrate according to...Gambrecht Example 7..."

United States Patent [19]
Bouzard et al.

[11] **Patent Number:** **4,504,657**
[45] **Date of Patent:** **Mar. 12, 1985**

[54] CEPHADROXIL MONOHYDRATE

[75] Inventors: **Daniel Bouzard, Franconville; Abraham Weber; Jacques Stemer, both of Paris, all of France**

[73] Assignee: **Bristol-Myers Company, New York, N.Y.**

[21] Appl. No.: **358,567**

[22] Filed: Mar. 16, 1982

Related U.S. Application Data

[60] Continuation of Ser. No. 931,800, Aug. 7, 1978, abandoned, which is a continuation of Ser. No. 874,457, Feb. 2, 1978, Pat. No. 4,160,863, which is a division of Ser. No. 785,392, Apr. 7, 1977, abandoned.

[30] **Foreign Application Priority Data**

Apr. 7, 1977 [GB] United Kingdom 17028/76

[51] **Int. Cl.**[3] **C07D 501/22; C07D 501/12**
[52] **U.S. Cl.** **544/30**
[58] **Field of Search** **544/30**

[56] **References Cited**

U.S. PATENT DOCUMENTS

3,489,752 1/1970 Crast, Jr. 260/243
3,655,656 4/1972 Van Heyningen 260/243
3,781,282 12/1973 Garbrecht 260/243
3,957,773 5/1976 Burton 260/243
3,985,741 10/1976 Crast, Jr. 260/243
4,091,215 5/1978 Bouzard 544/30
4,160,863 7/1979 Bouzard et al. 544/30
4,162,314 7/1979 Gottschlich 544/30

FOREIGN PATENT DOCUMENTS

829758 12/1977 Belgium .
1240687 7/1971 United Kingdom .

OTHER PUBLICATIONS

Dunn et al., *The Journal of Antibiotics*, 29: 65–80, (Jan. 1976).

Primary Examiner—Mark L. Berch
Attorney, Agent, or Firm—Robert E. Carnahan; David M. Morse

[57] ABSTRACT

A novel crystalline monohydrate of 7-[D-α-amino-α-(p-hydroxyphenyl)acetamido]-3-methyl-3-cephem-4-carboxylic acid is prepared and found to be a stable useful form of the cephalosporin antibiotic especially advantageous for pharmaceutical formulations.

1 Claim, 1 Drawing Figure

We claim:

65 1. Crystalline 7-[D-α-amino-α-(p-hydroxyphenyl) acetamido]-3-methyl-3-cephem-4-carboxylic acid monohydrate exhibiting essentially the following x-ray diffraction properties:

19 | 20

Line	Spacing d(A)	Relative Intensity
1	8.84	100
2	7.88	40
3	7.27	42
4	6.89	15
5	6.08	70
6	5.56	5
7	5.35	63
8	4.98	38
9	4.73	26
10	4.43	18
11	4:10	61
12	3.95	5
13	3.79	70
14	3.66	5
15	3.55	12
16	3.45	74
17	3.30	11
18	3.18	14
19	3.09	16
20	3.03	29
21	2.93	8
22	2.85	26
23	2.76	19
24	2.67	9
25	2.59	28
26	2.51	12
27	2.46	13
28	2.41	2
29	2.35	12
30	2.30	2
31	2.20	15
32	2.17	11
33	2.12	7
34	2.05	4
35	1.99	4
36	1.95	14
37	1.90	10

5 10 15 20

* * * * *

Fig. 14.2 Listing of peaks for the diffraction pattern in the '657 patent on cefadroxil [21].

The court also did not accept the defendant's

> "...only challenge, that seeds from its cefadroxil monohydrate acted as a template such that any attempt to repeat Gambrecht Example 7 anywhere in the world would yield Bristol-Myers cefadroxil monohydrate..."

There have been many subsequent litigations and appeals. Many experiments have been done and many chemical/technical issues arose in those cases. Some of them are noted here:

- The X-ray powder pattern of the Bouzard form was indexed to determine that every line in the pattern could be accounted for by that crystal modification.
- Attempts were made to prepare the Micetich form in an unseeded environment, with the measure of success being a matter of contention among the testifying experts.
- Experiments were run in a clean room equipped with filters that would eliminate particles the size of bacteria (which are several microns in size), but could not eliminate seeds that can be hundreds of times smaller.
- A series of side-by-side experiments were run, with one sample sealed from seed crystals and the other not sealed; the seed-free produced the Micetich form and the open flasks produced the Bouzard form [23].

The series of litigations continued over other issues, including the nature of crystal modifications [24]. Zenith prepared and formulated a hemihydrate of cefadroxil, for which it submitted an ANDA to the FDA. BM (now Bristol-Myers Squibb) contended that the FDA should require a much more extensive New Drug Application (NDA) for the hemihydrate, rather than approve it within the framework of the FDA monograph for the monohydrate. BM also alleged that the Zenith product converted into the monohydrate, thereby infringing the '657 patent. BM's theory of infringement was that the Zenith product converts into the Bouzard monohydrate after ingestion in the gullet and the stomach, as it is mixed with liquid. A federal court found in favor of BM. Upon appeal, Zenith contended that the '657 patent does not cover Bouzard form crystals that might form momentarily in a patient's stomach, and there were both legal and scientific grounds for the basis of that contention. One of the main points of contention was whether the X-ray pattern of the material found by BM in a patient's stomach matched the list of X-ray powder diffraction lines given in the claim of the Bouzard patent (Fig. 14.2). BM had identified 15 of the 37 diffraction lines cited in the claim. The court found that,

> "Although the term 'essentially' recited in the claim permits some leeway in the exactness of the comparison with the specified 37 lines of the claim, it does not permit ignoring a substantial number of lines altogether,"

and that

> "...there was a failure of proof as to whether any crystals, assumed to form in the stomach from ingested cefadroxil..., literally infringe the '657 claim."

On this basis, the appeals court reversed the lower court's decision. One result of this decision is that many subsequent patent applications including claims based on substances characterized by, for instance, X-ray diffraction patterns and/or IR spectra have been framed in language that differs from that used in claim 1 of the '657 patent.

14.5 Paroxetine Hydrochloride

Paroxetine hydrochloride is an antidepressant that is a selective inhibitor of the reuptake of serotonin. It is marketed as Paxil® or Seroxat®. Paroxetine in the form of a maleate salt was first prepared in the 1970s by a Danish company (Ferrosan). Ferrosan entered into an agreement with Beecham (now Glaxo SmithKline – GSK) in 1980 concerning paroxetine. Late in 1984, during process development and scale up of paroxetine hydrochloride, a new crystal form appeared serendipitously – the hydrochloride form. The hemihydrate form has been marketed since 1993 by GSK.

The GSK hemihydrate '723 patent is due to expire in 2006. During the period 1995–1998 a Canadian generic company Apotex together with its subsidiary BCI developed a process for making the anhydrate crystalline form of paroxetine hydrochloride; their research started with samples of Paxil samples in which the active ingredient is the hemihydrate.

In May of 1998 Apotex filed an ANDA for approval to market paroxetine hydrochloride anhydrate. On the basis of a 1988 paper [25] and information present in the '723 patent, GSK argued that the factors leading to conversion of the anhydrate into the hemihydrate (water, seeds of hemihydrate, heat and pressure) were present in the BCI facility and would result in the conversion of the BCI/Apotex product. At the February 2003 trial in the US Federal District Court in Chicago GSK presented its own and Apotex documents as well as the results of GSK's analytical testing of the active pharmaceutical ingredient and Apotex's tablets showing conversion. In one such document, Apotex's CEO stated that

> "[P]ersons skilled in the art would have also known and been aware ... [t]hat paroxetine hydrochloride anhydrate converts to paroxetine hydrochloride hemihydrate when exposed to moisture and compacted into tablets..."

GSK also presented evidence that Apotex's crystal form purity specification could not detect less than 5–8% of hemihydrate.

The importance of seeds of the hemihydrate in the conversion process was a subject of considerable debate, with Apotex arguing against their role, calling unintentional seeding "junk science", a phenomenon which is not widely accepted and whose mechanism is not understood. The Court rejected these attacks, noting that Apotex's contention

> "...that there is no scientific basis for believing that seeding occurs... is obviously wrong."

The District Court found that the anhydrate converts into the hemihydrate, citing the factors that had been established in the literature and the '723 patent:

> "[The greater the heat... and the humidity, the likelier is the conversion... [T]he presence of hemihydrate seeds in a batch of anhydrate is likely, provided [normal humidity and temperature]. To produce conversion within a short time."

The court also noted that

> "[G]reater humidity, temperature, or pressure can convert anhydrate to hemihydrate in amounts greater than a few per cent"

and noted that

> "[G]iven enough humidity, heat, etc., conversion [to the hemihydrate form] may continue until it reaches 100 per cent."

The District Court found that Apotex

> "probably will be making at least some hemihydrate crystals and therefore infringing, at least prima facie, patent '723..."

and that

> "*Some* conversion from anhydrate to hemihydrate is likely to occur in a seeded facility in which the anhydrate is exposed to air; the BCI plant is seeded; and the anhydrate manufactured there is exposed to nondehumidified air before it leaves the plant."

The Court concluded that

> "This evidence is sufficient to support an inference that BCI will be making at least tiny amounts of the hemihydrate if it is permitted to manufacture anhydrate."

Nevertheless, the Court ruled against GSK, reasoning as follows: The '723 patent notes that the hemihydrate is not hygroscopic (as opposed to the anhydrate), and thus has certain manufacturing benefits. The Court then limited claim 1 of the '723 patent, which reads "Crystalline paroxetine hydrochloride hemihydrate", to only "commercially significant" amounts of hemihydrate, noting that a product would have to contain

> "high double digits [of hemihydrate] to contribute any commercial value",

and that GSK had not established that Apotex will make "high double digits" amounts of hemihydrate. In short, even though the Court found that Apotex does make the hemihydrate, the Court refused to enjoin Apotex because Apotex does not benefit from the hemihydrate. The District Court also held that GSK was responsible for the seeds of the hemihydrate in Apotex's manufacturing facility, and therefore held that Apotex was not responsible for the hemihydrate in its product.

The decision of the district court was appealed to the Federal Circuit, which hears appeals on patent cases. On appeal GSK argued that claim 1 of the '723 patent contains only the four words "Crystalline paroxetine hydrochloride hemihydrate", and that the claim is not limited to "high double digit" amounts of hemihydrate. In this case (as with other cases related to Waxman-Hatch legislation) the infringement inquiry here is very much hypothetical since the case was argued before Apotex actually went on the market. The district court found that Apotex will likely market "some" hemihydrate, and since the patent law essentially says that making, using or selling the product constitutes infringement, it might appear that GSK should have won.

On appeal Apotex argued that the district court was correct in its claim construction, but also argued that the same court's findings regarding conversion were in error. Apotex also returned to the question of seeding, remarking that,

> "In sum, the district court's apparent fascination with *the seeding theory* led it to a finding that smacks of alchemy, not chemistry."

The three judge panel of the Federal Circuit agreed with GSK's claim construction and disagreed with that of the district court [26]:

> "[N]othing in the '723 patent limits that structural compound to its commercial embodiments,"

concluding,

> "Thus, reading claim 1 in the context of the intrinsic evidence, the conclusion is inescapable that the claim encompasses, without limitation, paroxetine hydrochloride hemihydrate – a crystalline form of paroxetine that contains one molecule of bound water for every two molecules of paroxetine hydrochloride in the crystal structure."

And further, that

> "[T]he specification discusses the superior handling properties of the hemihydrate form that improve the manufacture of [paroxetine hydrochloride]. Those references, however, do not redefine the compound in terms of commercial properties, but emphasize that the new compound exhibits favorable characteristics."

The Federal Circuit agreed with GSK that Apotex would infringe, noting that,

> "SmithKline's experts applied the disappearing polymorph theory to show that Apotex's paroxetine hydrochloride anhydrate tablet inevitably convert to hemihydrate when combined with moisture, pressure, and practically ubiquitous paroxetine hydrochloride seeds." [27, 28]

Thus, the Federal Circuit found in favor of GSK on the issue of claim construction of claim 1 and that Apotex's product would infringe claim 1. However,

GSK had carried out clinical trials more than a year before the application for the patent, and the Federal Circuit found that this constituted a public use, stating,

> "This court has already defined the invention of claim 1 as the paroxetine hydrochloride hemihydrate compound without further limitation regarding efficacy, commercial use, or pharmaceutical viability...[w]ith that definition of the invention in mind, however, clinical trials designed to establish efficacy and safety of the compound as antidepressant for FDA approval are not experimental uses of that claimed invention. In other words, the claim covers the compound regardless of its use as an antidepressant. The antidepressant properties of the compound are simply not claimed features. Consequently, the clinical tests, which measured the efficacy and safety of the compound as an antidepressant, did not involve the claimed features of the invention." [29]

On this basis claim 1 of the patent (the claim at issue) was held invalid by a majority of the court.

A minority of the court disagreed and held that the clinical trials constituted an experimental, not a public, use. This judge also found the patent to be invalid for another reason, namely that paroxetine hydrochloride hemihydrate is acknowledged as

> "...a synthetic compound, created by humans in a laboratory, never before existing in nature, that is nevertheless capable of 'reproducing' itself through a natural process."

This capability of self reproduction and

> "...[t]he implicit concept of 'inevitable infringement'..."

and the

> "...failure of the patent to provide public notice..."

render the claimed subject matter unpatentable:

> "An item reproduced by such a natural process whether an inorganic structure or a life form, must ipso facto be ineligible for patent protection under *Section 101* (of the U.S. Code).
>
> In short, patent claims drawn broadly enough to encompass products that spread, appear and 'reproduce' through natural processes cover subject matter untenable under *Section 101* – and are therefore invalid." [30].

Some of the chemical issues involved in this case included the formal and practical definitions of anhydrate, hemihydrate; the physical and chemical characterization of anhydrate/hemihydrate; the relative stability of anhydrate/hemihydrate; the factors leading to conversion of anhydrate into hemihydrate: moisture, seeds, heat, pressure; the characteristics and role of seeds (intentional and unintentional) in crystallization; the analytical methods – qualitative and quantitative – for determining the composition of mixtures of crystal forms.

14.6 The Importance of Seeding

In all three of the cases described, seeding – more specifically, unintentional seeding – played an important role, both in the crystal chemistry and in the legal battles. Intentional seeding as an integral step of a crystallization procedure or process has been employed as a standard technique by chemists for probably close to 200 years. The phenomenon of unintentional seeding is also familiar to chemists. Most chemists recognize that the crystallization of a newly prepared compound – one that hasn't previously been made in that laboratory – is often a challenging and difficult task, but that once achieved, subsequent crystallizations are considerably more facile. As several authors have noted, that change in difficulty or ease of crystallization is due to the presence of seeds in the laboratory once the first successful crystallization has been carried out [31, 32].

For compounds that exhibit multiple crystal forms, control of the process of crystallization is crucial. We seek a level of control that allows us to obtain the desired crystal form and avoid the crystallization of undesired forms. As seen in the three cases described above, it is the loss of that control, especially when a new crystal form appears, that often leads to misunderstanding and confusion, even among relatively experienced practitioners of this art of crystallization.

Despite chemists' familiarity with these phenomena, parties in the litigations discussed above have tried to discredit the phenomenon of unintentional seeding; it is often presented and/or greeted with skepticism. The words *seeds* or *seeding* are often surrounded by qualifying inverted commas (as above). Parties have argued that since seeds cannot be directly observed they must not exist, suggesting doubt or disbelief about the whole phenomenon.

> "Seed crystals, they're in the air. You can't see them. You can't smell them. You can't taste them. You also can't detect them, they're there, and these seed crystals fall out of the sky, and they're very intelligent because they know when you are running one of these Example 32 experiments. They fall out of the sky and fall in your reaction beaker... Well I submit that if one believes in Santa Claus we might believe in these seed crystals, but if we're beyond that, we're not going to believe in these seed crystals..."

This was part of the opening statement by the counsel for Novopharm in the first ranitidine hydrochloride case against Glaxo in August 1993 [16].

Ten years later, as noted above in the GSK v. Apotex case involving paroxetine hydrochloride, counsel for the defendant referred to seeding as "junk science", a position defined by the court as "obviously wrong".

Yet, the presence of seeds and their influence in crystallizations is no more mysterious or profound than that of bacteria, whose presence and action are accepted despite the fact that most people have never seen them. As Judge Posner noted in his decision in the paroxetine hydrochloride case, above,

> "Many scientific phenomena are identified before their causal mechanism is understood."

Indeed, one of the most commonly recognized of these scientific phenomena is the existence of atoms, first postulated in the modern chemical sense just 200 years ago by Dalton. Nobody has yet ever *seen* an atom, although these days we have quite sophisticated techniques for imaging them; yet, long before these techniques were developed, the existence, indeed the structure, of atoms was quite well understood, and chemists have learned how to make and break bonds between atoms in very sophisticated ways.

However, one need not have waited until 1993 or 2003 for recognition by the courts of the phenomenon of ambient seeding and its role in determining the outcome of a crystallization. In a 1936 litigation involving three patents for the production of crystalline dextrose (grape sugar) as a hydrate, by the acidic hydrolysis of cornstarch, the District Court for the District of Delaware dealt with many of the crystallization issues that arose in the three above cases [33].

Until the 1920s dextrose was produced as the anhydrate, and as the court noted,

> "It was a settled conviction of the industry that the only way of making pure cane sugar was by adopting anhydrous crystallization because anhydrous crystals were of a better form to purge. "

The inventor on all three patents was William B. Newkirk, who carried out hundreds of experiments, varying the conditions to develop a robust method for making dextrose hydrate. Initially, these experiments were designed to determine first whether the use of large amounts of (intentional) seeds influenced control of the process, and having established that they did so, what was the optimum quantity of seeds to use. These were experiments with intentional seeding. One advantage of the hydrate over the anhydrate is that the former tends to crystallize at lower temperatures, and at the higher temperatures the yield of the anhydrous was decreased. A second disadvantage of making the anhydrous, as the court reported was that

> "...it cannot be manufactured in a factory where you are manufacturing hydrate by the process of crystallization in motion. All of the sugar factories are making the pressed sugar or slab sugar [dextrose hydrate], and you have the air filled with hydrate sugar dust. You are beating the hydrate sugar into your crystallizer and seeding the crystallizer with hydrate seed. All efforts which have been made to produce anhydrous sugar by crystallization in motion in an atmosphere laden with hydrate seed have proved failures. Newkirk in carrying out his process ran counter to every settled conviction in the art. He produced hydrate where the settled conviction in the art was that the process lay in the other direction..."

In his experiments he observed on September 13, 1920, probably accidentally and for the first time, hydrate crystals entirely different from the hydrate crystals in clusters that he had been accustomed to see [earlier]...

He experienced a great deal of difficulty in making anhydrous sugar due to the fact that the air was filled with hydrate seed... Finally he obtained permission to give up the process of making anhydrous and to make the hydrate alone. Newkirk says it took a lot of courage to make that recommendation, because everybody in the industry told him that in making pure sugar he would have to make anhydrous.

Interestingly, the same court also dealt with the issue of a natural product raised in the concurring opinion of the Federal Circuit in the paroxetine hydrochloride case:

> "The principal defense against the product [dextrose hydrate] patent is that it is a product of nature. The product of Newkirk is not a product of nature. It is an artificial product. It is made by the conversion of starch through a chemical process by artificially adding water to the starch molecule. You do not find anywhere in nature a separate hydrate crystal because you do not find the crystallization process at work in nature."

14.7 Concluding Remarks

We have presented here but a sampling of several patent issues that involved controversies over the nature, identity, quantity and uniqueness of crystalline modifications. As in other aspects of the law each case has its own character and idiosyncrasies. As evidenced in numerous examples throughout this volume, the same variability exists in the nature of the polymorphic behavior of a substance. That behavior differs because the substances are different, and to be properly understood each substance has to be investigated and characterized on its own, leading to the special features unique to each patent and to each patent litigation. The drafting of patents and the subsequent prosecution and litigation requires a detailed and intimate understanding of the different crystal forms, including their preparation, identification and characteristics. We noted at the outset that the issues discussed here are on the meeting ground of science and the law. Since their background, training, detailed knowledge and even professional culture differ, so the effective drafting and defense of patents for pharmaceutical crystal forms will require a great deal of cooperation between the scientific and legal communities.

References

1 This chapter seeks to describe some aspects and raise some issues for academic discussion in an area that is the meeting ground between science and the law. In the context of the general subject matter of this book it is an area that has generated a remarkable increase in interest and activity in the last thirty years – due, to a great extent, to some of the litigations described herein, and the underlying scientific, legal and economic issues involved. No attempt has been made to be comprehensive, nor does the chapter purport to provide legal advice or make any definitive statements on the current status of the law on the issues raised. The intent is to demonstrate how the chemical and crystallographic issues are argued and interpreted in the specific framework of patent law. Where legal opinions are presented, for the most part they are in the context of direct quotations from publicly available decisions of the courts or transcripts of trials, and they do not necessarily represent the opinion of the author.

2 A. E. Shamoo, D. B. Resnik, *Responsible Conduct of Research*, Oxford University Press, Oxford, **2003**.

3 J. Y. Maynard, H. M. Peters, *Understanding Chemical Patents*, 2nd edn., American Chemical Society, Washington, D.C., **1991**.

4 P. W. Grubb, *Patents in Chemistry and Biotechnology*, Clarendon Press, Oxford, **1986**.

5 A. B. Engelberg, *IDEA: J. Law Technol.*, **1999**, *39*, 389–395.

6 J. H. Barton, *Science*, **2000**, *287*, 1933–1934.

7 Re Riyad R. Irani and Kurt Moedritzer No. 8298 United States Court of Customs and Patent Appeals, *57 C.C.P.A. 1109; 427 F.2d 806*, **1970**.

8 Union Carbide Co. v. American Carbide Co. No. 226, *181 F. 104; 1910 US App.*, **1910**.

9 It is noteworthy that this case and its cited appeal preceded the first experiment of the diffraction of X-rays by crystals in 1911, which led to a tremendous amount of knowledge and understanding of the nature and internal structure of solids, both crystalline and amorphous. More recently, four polymorphic crystalline forms have been identified by synchrotron powder diffraction experiments: M. Knapp, U. Ruschewitz, *Chem. Eur. J.* **2001**, 874–880.

10 The discussions in Sections 14.3 and 14.4 are taken to a large extent from Chapter 10 of Ref. [11].

11 J. Bernstein, *Polymorphism in Molecular Crystals*, Oxford University Press, Oxford, **2002**, Chapter 10.

12 J. W. Black, W. A. M. Duncan, C. J. Durant, C. R. Ganellin, E. M. Parsons, *Nature*, **1972**, *236*, 385–390.

13 J. Bradshaw, in *Annals of Drug Discovery* Vol. 3, D. Lednicer (ed.), American Chemical Society, Washington, D.C., **1993**, pp. 45–81.

14 B. J. Price, J. W. Clitherow, J. W. Bradshaw, *US Pat. 4,128,658*, **1978**.

15 In fact, many of the discoveries of new crystal modifications have been made serendipitously (e.g., H. H. Silvestri, David A. Johnson v. Norman H. Grant and Harvey Alburn, US Court of Customs and Patent Appeals, Patent Appeal No. 8978; 496 F.2d 593.), as have many other important scientific discoveries: R. M. Roberts, *Serendipity. Accidental Discoveries in Science*, Wiley, New York, **1989**. The history of the word "serendipity", and its meaning in a variety of disciplines has been documented recently by R. K. Merton and E. Barber, *The Travels and Adventures of Serendipity*, Princeton University Press, Princeton, **2004**.

16 The following description is taken from the court's opinion in the case of GLAXO, INC. and Glaxo Group Limited v. NOVOPHARM, LTD. No. 91-759-CIV-5-BO.

17 It is understandable that, when a pharmaceutical firm is faced with the unexpected appearance of a new crystal form at some advanced stage of the development of a new drug, management is

reluctant to adopt that new form as the one of choice, since a considerable investment in time, money and resources must be devoted to the changeover. In many cases, however, the changeover becomes simply unavoidable, since the new form often "takes over" from the previous one, a phenomenon that is consistent with Ostwald's Rule of Stages, which states that the latter forming crystal forms tend to be more stable than those that preceded them. In the case of ranitidine hydrochloride the amount of Form 2 increased in the four pilot batches succeeding batch 3B13 to the point where the process yielded essentially a Form 2 product. See also S. R. Chemburkar, J. Bauer, K. Deming, H. Spiwek, K. Patel, J. Morris, R. Henry, S. Spanton, W. Dziki, W. Porter, J. Quick, P. Bauer, J. Donnaubauer, B. A. Narayanan, M. Soldani, D. Rilery, K. McFarland, *Org. Process Res. Dev.*, **2000**, *4*, 413–417.

18 At the time a US patent was valid for 17 years from the date of issue. Subsequent international trade agreements have led to changes in the period of enforcement, and during the changeover period there have been some extensions to the terms of some existing patents.

19 Much of the following description is taken from the original district court opinion in the case of GLAXO, INC. and Glaxo Group Limited v. NOVOPHARM, LTD. No. 5:94-CV-527-BO(1), 931 F.Supp. 1280 and the appeals court's opinion in the case of GLAXO, INC. and Glaxo Group Limited v. NOVOPHARM, LTD. 96-1466, DCT. 94-CV-527.

20 Glaxo, Inc. et al. v. Torpharm Inc. et al. (1998), 153 F.3d 1366.

21 Kalipharma, Inc. v. Bristol-Meyers Company, No. 88 CIV. 4640; 707 F. Supp. 741.

22 Bristol-Meyers Company v. United States International Trade Commission, Gema, S.A., Kalipharma, Inc., Purepac Pharmaceutical Co., Istituto Biochimico Italiano Industria Giovanni Lorenzini, Institut Biochimique, S.A., and Biocraft Laboratories, Inc. (Dec. 8, 1989) No. 89-1530 [See also United States International Trade Commission, Investigation No. 337-TA-293].

23 S. E. Tarling (**2001**), personal communication to author; see also http://www.vino.demon.co.uk/ppxrd/cefad.html

24 Zenith Laboratories, Inc. v. Bristol-Meyers Squibb Co. (1992) Civ. A. No. 91-3423; 1991 WL 267892 (D.N.J.).

25 P. C. Buxton, I. R. Lynch, J. M. Roe, *Int. J. Pharmaceut.*, **1988**, *42*, 135–143.

26 SmithKline Beecham Corporation and Beecham Group P.L.C. v. Apotex Corp., Apotex, Inc. and Torpharm, Inc. 03-1285, 03-1313, *365 F.3d 1306; 2004 US App.*

27 J. D. Dunitz, J. Bernstein, *Acc. Chem. Res.*, **1995**, *28*, 193–200.

28 It is worth noting here that the trial transcript indicates that none of GSK's experts used the term "inevitably" or "ubiquitous" in any of their testimony.

29 Claim 5 of the '723 patent reads, "An antidepressant pharmaceutical composition comprising an effective antidepressant amount of crystalline paroxetine hydrochloride hemihydrate and a pharmaceutically acceptable carrier."

30 This issue of "reproduction" through a natural process apparently did not arise in the *in vivo* conversion issue in the cephadroxil case, Section 14.4.

31 K. B. Wiberg, *Laboratory Techniques in Organic Chemistry*, McGraw Hill Book Company, New York, **1960**.

32 A normal urban environment contains approximately one million airborne particles of 0.5 micron diameter or larger per cubic foot, the number being reduced by an order of magnitude in an uninhabited rural environment. A normal sitting individual generates roughly one million dust particles ($\geq$0.3 micron diameter) per minute (for reference a visible particle is usually 10 microns or greater in diameter). Clean rooms for various purposes (e.g., surgical theaters, biological or pharmaceutical preparations, semiconductor fabrication) employ sophisticated technology to remove these particles and to prevent subsequent contamination. Therefore the possible presence of seeds of even a newly formed crystal form in a laboratory, a manufacturing facility, or any location having been exposed to that form cannot be casually dismissed; indeed its presence would be hard to avoid. http://www.thaihvac.com/knowledge/cleanroom/cleanroo1.htm, **2001**.

33 International Patents Development Co. et al. V. Penick & Ford, Limited, Inc., No. 654, *15 F.Supp. 1038; 1936 U.S. Dist.*

15
Scientific Considerations of Pharmaceutical Solid Polymorphism in Regulatory Applications

Stephen P. F. Miller, Andre S. Raw, and Lawrence X. Yu

15.1
Introduction

Original new drug applications (NDAs) and abbreviated new drug applications (ANDAs) are required to contain information on the chemistry, manufacturing and controls (CMC) used in the manufacture of drug products. This information is evaluated by FDA scientists to assure that the drug product possesses acceptable standards of quality, and hence ensure that the marketed drug product has the critical performance attributes necessary to reflect the product's safety and efficacy that was demonstrated in clinical trials. This assessment of product quality requires the scrutiny of several facets of an applicants' CMC, including solid-state form. This chapter is intended to communicate scientific considerations of pharmaceutical solid polymorphism in the CMC for both NDAs and ANDAs. It opens with a discussion on how pharmaceutical solid polymorphism can have an effect upon drug product safety and efficacy, and hence why it is important from the standpoint of drug product quality. Having shown the significance of polymorphism in this context, we proceed with some perspectives on when and what types of controls on polymorphs may be warranted in NDA or ANDA applications. Finally, we conclude with a discussion of how FDA's process analytical technology (PAT) initiative may provide a new paradigm for pharmaceutical manufacturers wishing to control polymorphic forms.

The opinions expressed in this chapter are those of the authors and do not necessarily reflect the views or policies of the FDA.

15.2
General Principles of Pharmaceutical Solid Polymorphs

Commonly, a drug substance can exist in different solid-state forms. These differing forms can be conceptually divided into three distinct classes. The first in-

Polymorphism: in the Pharmaceutical Industry. Edited by Rolf Hilfiker

ISBN: 3-527-31146-7

cludes crystalline phases that have different arrangements and/or conformations of the molecules in the crystal lattice. The second includes crystalline solvates that contain either stoichiometric or non-stoichiometric amounts of a solvent, and sometimes are referred to as pseudopolymorphs in the literature. In most pharmaceutical solids, the bound solvent is water and these are referred to as hydrates. The last class includes amorphous solids, which consist of disordered arrangements of molecules that do not possess a distinguishable crystal lattice, and can often be described as either a glass or a super-cooled liquid. In the context of this chapter, the term polymorphic forms encompasses all three classes to include crystalline forms, solvates, and amorphous forms as defined in the International Conference on Harmonization (ICH) Guideline Q6A [1]. This broad definition of polymorphism is very useful for pharmaceuticals, where decisions about suitability for use of an amorphous form, a hydrate, or crystalline form rest on the same scientific considerations.

The fundamental importance of pharmaceutical solid polymorphism is based upon the fact that the differing solid-state forms of the same chemical compound may have different chemical and physical properties, including chemical reactivity, apparent solubility, dissolution rate, and mechanical properties [2]. As will be illustrated in subsequent sections, differences in these properties have the potential to affect the ability to process and manufacture the drug product, as well as have an impact upon drug product bioavailability and stability. Thus polymorphism can affect critical drug product performance quality attributes, and hence affect its safety and efficacy. As a result, pharmaceutical solid polymorphism has received much scrutiny throughout various stages of drug development, manufacturing, and regulation [3].

15.3 Influence of Polymorphism on Product Quality and Performance

15.3.1 Effect on Bioavailability (BA) / Bioequivalence (BE)

One of the principle regulatory concerns with regard to pharmaceutical solid polymorphism is based on the effect that it may have upon drug product bioavailability/bioequivalence (BA/BE), particularly for solid oral dosage forms. This stems from the fact that the differing lattice energies of the various polymorphic forms will give rise to differences in their apparent solubilities and dissolution rates. Typically, solubility differences between crystalline polymorphs will be less than several-fold, while hydrates generally exhibit lower aqueous solubilities than the anhydrous crystalline form [4]. Amorphous forms can have solubilities several hundred times that of the crystalline counterparts [5]. When differences in the solubilities of the various polymorphs are sufficiently large, this may alter the drug product *in vivo* dissolution and hence impact drug product bioavailability. An historical example, where polymorphism has been associated with differences in drug product bioavailability, comes from the classic work of Aguiar on chloramphenicol palmitate, in which Form B was shown to exhibit greater oral absorption than Form A due to its enhanced solubility [6].

From an FDA perspective, because drug product BA/BE depends upon several factors that influence the rate and extent of drug absorption, such as gastrointestinal motility, drug dissolution, intestinal permeability, and metabolism, a multi-faceted approach is essential in making a rational assessment of the relative risk of the effect that an inadvertent change in polymorphic form would have on drug product bioavailability. In this context, concepts from the biopharmaceutical classification system (BCS) provide a scientific framework for regulatory decisions regarding drug polymorphism [7, 8]. For drugs exhibiting poor aqueous solubility and high intestinal permeability (BCS Class II), it would be anticipated that dissolution would be the rate-limiting step to drug absorption and it may even be possible to establish an *in vivo–in vitro* correlation. Hence, for such BCS Class II drugs such as carbamazepine, one would anticipate that differences in the solubilities of the various polymorphic forms have the potential to affect drug product BA/BE. In such a situation, because of the relatively high risk that a change in polymorphic form will impact bioavailability, it would be important to incorporate suitable controls on polymorphic forms. Conversely, for drugs that exhibit high aqueous solubility and high intestinal permeability (BCS Class I), or high aqueous solubility and low intestinal permeability (BCS Class III), dissolution would likely proceed rapidly and not be the rate-limiting step to drug absorption. Hence, rapidly dissolving immediate release solid oral dosage forms of such high solubility drugs, such as metoprolol tartrate (BCS I) or ranitidine hydrochloride (BCS Class III), would, in effect, resemble oral solutions *in vivo*. For such drug products, clearly the relative risk of an inadvertent change in polymorphic form to affect bioavailability would be quite low. In such a situation it may not be necessary to incorporate controls on polymorphic forms.

15.3.2 Effect on Stability

An additional concern with regard to pharmaceutical solid polymorphism is the effect that it may have upon drug product stability. This is based upon differences in chemical reactivity between the various polymorphic forms. This stems from various reasons, including differences in thermodynamic stability of the crystal lattices, differing micro-environments of reactive functional groups within the crystal lattice, and greater molecular mobility seen for amorphous forms. These differences in chemical reactivity present concerns as a change to a chemically reactive polymorphic form may result not only in an undesired loss of drug product potency but might also generate elevated levels of degradation impurities. Nonetheless, because drug product stability depends upon not only the intrinsic chemical reactivity of the drug substance polymorphic form but also on other factors, including formulation, manufacturing process, and packaging, many of these facets should be incorporated into the scheme of making a rational determination as to what the relative risk a change in polymorphic form would have upon drug product stability.

Such an approach of assessing the effect of polymorphic forms on drug product stability is nicely illustrated in the case of enalapril maleate. Enalapril maleate is known to exist in two polymorphic modifications (I and II) with Form II being the more thermodynamically stable [9]. Both forms exhibit similar properties, as exemplified by their similar solubilities, dissolution characteristics, heats of solution, IR and Raman spectra, and DSC thermograms. Interestingly, although tablets manufactured via wet granulation using Form I of enalapril maleate and one molar equivalent of sodium bicarbonate are quite stable, tablets manufactured from Form II give rise to unacceptably high levels of the diketopiperazine degradation impurity [10]. Hence, with the given drug product formulation, an inadvertent change in polymorphic form would negatively impact drug product stability, and it would be important to incorporate controls on the drug substance polymorphic forms. Conversely, tablets manufactured via wet granulation from either Form I or II and two molar equivalents of sodium bicarbonate are equally stable [11]. For such a drug product formulation, as there is low risk that a change in solid-state form will impact upon drug product stability, it would not be necessary to incorporate controls on the polymorphic form for incoming batches of the bulk drug substance.

15.3.3
Effect on Manufacturability

A final concern with regard to pharmaceutical solid polymorphism is the effect that it may have upon drug product manufacturability. This is based upon differences in mechanical properties and crystal morphologies between various forms, which may impact powder flow and tablet compressibility. Because product manufacturability depends not only upon the intrinsic mechanical properties and morphologies of the drug substance polymorphic form, but also upon the formulation and manufacturing process, ultimately it is manufacturability of drug product that is the most relevant measure of quality, and it is from this perspective that one can make a rational assessment of the risk a change in polymorphic form will have upon product manufacturability.

This approach of assessing the effect of polymorphic forms on drug product manufacturability is nicely illustrated by paracetamol (acetaminophen). Paracetamol exists in two stable polymorphic modifications (I and II) [12]. By virtue of their differing mechanical properties, Form II is compressible and suitable for the manufacture of tablets via direct compression, whereas Form I does not possess such compression properties [13]. Hence, with paracetamol tablets manufactured by direct compression, there is a risk that an inadvertent change to a polymorphic form will impact drug product manufacturability. To maintain a robust manufacturing process, it would be important to incorporate suitable controls on the polymorphic form of incoming batches of the bulk drug substance. Conversely, by using a wet granulation process in tablet manufacture, the original processing deficits of Form I are "masked", and hence tablets can readily be manufactured from either form. Clearly, for such a process it may not be essential to incorporate controls on the drug substance polymorphic form.

In conclusion, the differing chemical and physical properties of polymorphic forms can certainly impact drug product manufacturability, bioavailability and stability, and hence impact drug product safety, efficacy, and quality. Under these circumstances it is important to have suitable controls on polymorphic forms. However, one must be cognizant of the instances in which polymorphism has little or no effect upon the critical drug product performance quality attributes. In this situation, there is clearly no rational reason to have controls on polymorphs forms. Hence, controls on polymorphs should be incorporated only when appropriate. The following section discusses how Q6A provides a general conceptual framework for determining when such controls are or are not appropriate.

15.4 Pharmaceutical Solid Polymorphism in Drug Substance

As a new drug proceeds through development, studies are performed to identify and characterize the relevant polymorphic forms of the drug substance. These studies aim to develop a mechanistic understanding of how the physical and chemical properties of these polymorphic forms influence the important performance aspects of the drug substance and drug product. Formulation scientists then select appropriate polymorphic form(s) for development and determine what manufacturing controls are necessary to insure the quality of drug substance and drug product so that bioavailability, safety and efficacy are assured.

The amount of investigation into polymorphic forms and the appropriate level of control during manufacturing will depend on the properties of the specific drug substance and the nature of the dosage form. For example, there is a much lower level of concern for a highly soluble drug substance formulated as an immediate release tablet versus a poorly soluble drug substance formulated as an oral suspension. Therefore, largely, the appropriate level of control of polymorphic forms must be individualized for each specific drug substance and dosage form.

While the appropriate level of control will be dependent on the physical and chemical properties of the drug substance as well as the dosage form characteristics, it is vitally important that scientists share a common view of the fundamental approaches to polymorphic form issues. To advance this goal, the International Conference on Harmonization (ICH) included polymorphic form topics in the discussions that led to the guidance "Q6A Specifications: Test Procedures and Acceptance Criteria for New Drug Substances and New Drug Products: Chemical Substances" [1]. This section provides our perspective on how these recommendations may apply to individual drugs or specific situations.

15.4.1
Polymorph Screening

The first fundamental question with regard to polymorphism during drug development is whether the given drug substance even exists as multiple polymorphic forms. Hence, the first section (Part 1) of Q6A's Decision Tree #4 covers the screening for different polymorphic forms of the drug substance and their characterization. Figure 15.1 shows this decision tree, with a leading question added.

Q6A recommends that a polymorph screen be conducted for all drug substances within the scope of the guidance. However, Q6A does not elaborate more specifically on the extent of this screening study, and questions have arisen regarding what constitutes a reasonable screen. It must be acknowledged that a wide diversity of approaches are being taken, ranging from screens that focus only on solvents used in the drug substance and drug product manufacturing process, to screens that also include solvents with a wide range of polarities and hydrogen-bonding potentials. The screening studies would generally include attempts to prepare hydrates, given the nearly universal potential for drug substances and drug products to be exposed to water.

A possible point for scientific discussion is whether the extent of the screen for additional polymorphic forms should be tailored according to properties of the specific drug substance or dosage form. One approach is to perform a more comprehensive screen for situations where polymorphic form changes are more likely to adversely affect the drug product. For example, a simpler screen might be conducted for situations such as:

- Solution drug products with drug load "well below" saturation.
- Solid oral drug products containing a drug substance with high solubility in water.

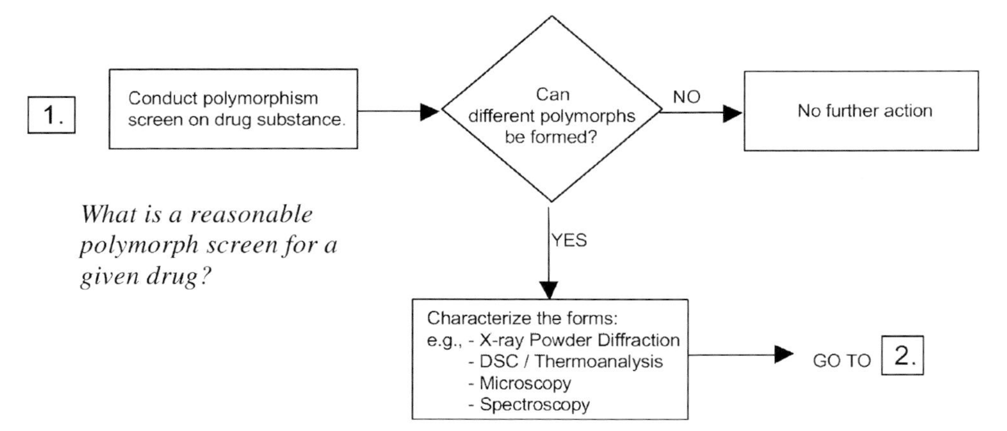

Fig. 15.1 Part 1 of Decision Tree #4 "Investigating the Need to Set Acceptance Criteria for Polymorphism in Drug Substances and Drug Products." (Reproduced from Q6A [1] with a note added; text in italics).

In this approach, a more comprehensive screen might be conducted for situations such as:

- Solution drug products with drug load near saturation.
- Drug products where the formulation can change composition during storage (e.g., liquid-filled soft gelatin capsules).
- Solid oral drug products containing a drug substance with low solubility in water.

These last three cases are examples where a polymorphic form with lower solubility that is discovered late in development might have an adverse effect on product performance (e.g., crystallization within the dosage form, or lowered bioavailability). It may be valuable to identify general situations where the "risk" of a more comprehensive polymorph screen (e.g., delays in selecting the polymorphic form for development) is outweighed by the potential benefits (e.g., early discovery of a polymorph with unexpected physical or chemical properties).

For many solution drug products, it is appropriate to include the drug product vehicle among the solvents used for the polymorph screening studies. This will be especially valuable when the concentration of the drug substance in the product formulation is a significant fraction of its solubility in the vehicle (i.e., the solubility of the least-soluble relevant polymorphic form). The purpose of forced crystallization from the drug product vehicle is to insure that the polymorphic form that results is known and characterized. If this form has a lower solubility in the vehicle, the drug product solution would be more saturated than previously thought, which could have implications for stability during storage conditions.

For generic drugs or new drugs for new dosage forms or indications, polymorph screening is typically not essential since, by the time of development, there is a generally good understanding of polymorph issues of a drug substance, including physical, chemical, and physiological characteristics. Thus their potential effect on product quality and performance can be reasonably predicted. Nevertheless, limited polymorph screening may be necessary under various circumstances. For example, if a softgel capsule dosage form is developed for a drug that was originally developed as an immediate tablet, the potential existence of polymorphs in the new softgel capsule vehicles should be evaluated.

15.4.2
Control of Polymorphism in Drug Substance

After screening studies to identify the polymorphic forms of an active ingredient, knowledge of their physical and chemical properties can be used to determine appropriate strategies for control of polymorphism in the drug substance. The second part of Q6A's decision tree on polymorphism outlines control at the drug substance stage (Fig. 15.2).

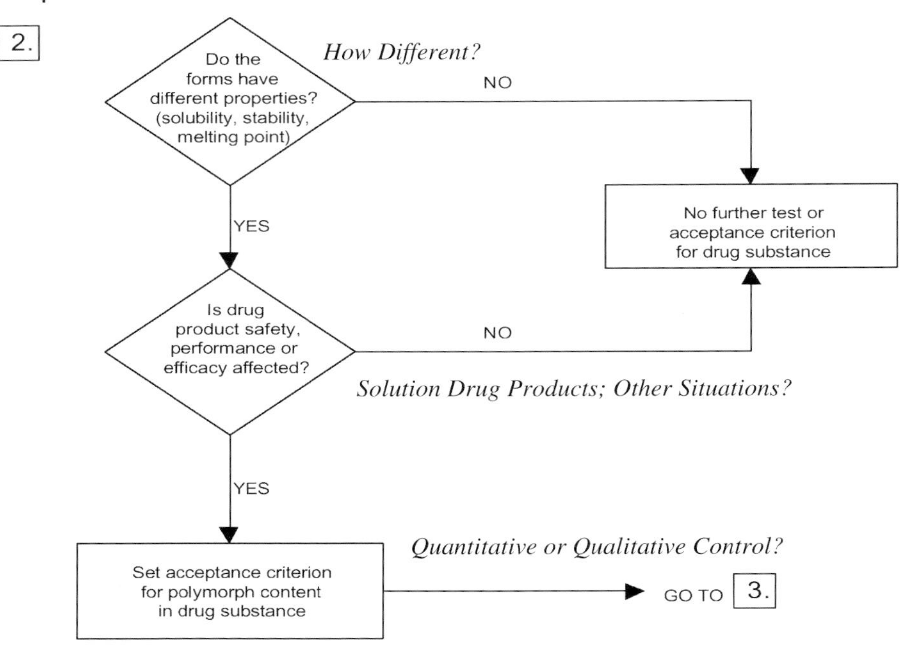

Fig. 15.2 Part 2 of Decision Tree #4 "Investigating the Need to Set Acceptance Criteria for Polymorphism in Drug Substances and Drug Products." (Reproduced from Q6A [1] with notes added; text in italics).

The first decision is whether the identified polymorphic forms have different properties. The parenthetical insert in this decision point serves to focus attention on properties that are directly relevant to drug product quality or performance (e.g., solubility and stability). Because there will always be some difference in properties, it seems appropriate to focus on the properties that would significantly affect the use of the polymorphic forms in the intended dosage form. It may be difficult to develop a universally applicable approach to identifying significantly different properties. For example, a given difference in solubilities might be of little concern for a solid oral product containing a very soluble drug substance, but of significant concern when the drug substance is very insoluble. Therefore, these other factors should be considered when we define different properties.

The second decision point is whether the performance and quality of the dosage form is affected by the polymorphic form of the input drug substance. Should this be interpreted as that performance or quality data would generally be available from drug product manufactured from different polymorphic forms of the drug substance? Such a conservative interpretation does not appear to be appropriate for many new drugs. While that type of data may be needed for situations where there is a high risk of adversely affecting efficacy, such extensive studies would not be justified from risk/benefit considerations for many applications. We believe that it is more reasonable to interpret this decision point as, "would an adverse effect on drug product performance and quality be probable if the polymorphic form of the drug substance is not controlled?" An answer of "No", leading to no routine testing for solid-state form in drug substance batches, might be appropriate for:

- Solution dosage forms where acceptable drug product can be manufactured from all relevant polymorphs.
- Immediate release solid oral dosage forms where all relevant polymorphic forms are highly soluble and the drug substance manufacturing process consistently produces a single polymorph.

Here the aqueous solubility of a drug substance is clearly an important factor in judging the risk/benefit of polymorph controls for a given drug substance. However, the observed aqueous solubility will often vary with the pH, and may also depend on the time-scale of measurement (i.e., apparent solubility versus equilibrium solubility). It is therefore necessary to identify the measurement conditions when discussing the solubility of a drug substance. One approach would be to adopt the solubility criteria from the biopharmaceutical classification system [7]. In this approach the solubility is measured across the pH range that is relevant to oral drug delivery (pH 1–7.5). The minimum solubility within that pH range is then used to calculate the volume needed to dissolve the highest strength dosage form. If this volume is less than 250 mL, a typical volume of fluid taken during dosing, then the drug is considered to be highly soluble. When the primary concern is to insure that polymorphic forms are sufficiently soluble to give rapid dissolution, knowledge that all relevant forms are highly soluble by the BCS definition becomes important for assessing risk/benefit of polymorph controls. In our recent commentaries on polymorphism in ANDAs, no control on polymorphs is proposed when all relevant polymorphs are highly soluble for solid dosage forms or liquids contained undissolved drugs [14, 15].

Finally, it is appropriate to include only these polymorphic forms, which could be generated during drug product manufacture or storage, when assessing the potential impact of relevant forms. Relevant forms usually include those produced by solvents used in the final crystallization of the drug substance or during drug product manufacture, as well as hydrates, and some non-solvated forms originally produced from common solvents not used in manufacture. An example of the latter situation is a polymorph produced by crystallization from acetonitrile, for a drug substance where ethanol is used for the final crystallization. If that polymorph is not solvated, it seems appropriate to determine whether this form will grow if seeded in ethanol solutions before dismissing it as non-relevant. However, an acetonitrile solvate would not be considered a relevant polymorphic form, since it could not be produced by the current drug substance or drug product manufacturing processes.

15.4.3 Acceptance Criterion for Polymorph Content in Drug Substance

The final portion of Part 2 in Q6A's polymorphism decision tree covers situations where more formal control of polymorphic form is appropriate (Fig. 15.2). Building upon the considerations presented above, this outcome could be reached when there are multiple relevant polymorphic forms, and an adverse effect on the dosage form would be reasonably likely unless controls on polymorphism are in place. We believe that "Acceptance Criterion in Drug Substance" should be interpreted in the broad sense, so that it includes process controls as well as the tests performed on drug substance for routine batch release (i.e., the drug substance specification). For example, seeding with the desired polymorph during the final crystallization (with a polymorphic control on the seed crystals) may provide appropriate control while minimizing the need for reprocessing.

Having reached the outcome that an acceptance criterion should be set for polymorph content in the drug substance, it remains to be determined what type of test and acceptance criterion are appropriate. For some situations, a qualitative test (e.g., an identity test) may be sufficient. In other situations a quantitative measurement of polymorphic composition or a limit test may be appropriate. Byrn and coauthors [16] suggest a qualitative test when the drug substance manufacturing process routinely gives a single polymorphic form. An example of such a test would be a spectroscopic procedure that has been shown to differentiate the relevant polymorphic forms, with an acceptance criterion such as "Conforms to Polymorph 1". Conversely, Byrn et al. [16] recommend a quantitative test for polymorph composition when the manufacturing process produces more than one polymorphic form. In addition to methods that provide the quantitative polymorphic composition, another appropriate control option may be a limit test, which can confirm that a particular polymorphic form is below some acceptable amount. When the drug substance is a hydrate, water content can provide useful information with respect to the quality of drug substance.

A final point for discussion in this section is the value of performance data on drug product made from multiple solid-state forms of the drug substance. When a quantitative test for polymorphic composition of the drug substance is appropriate, this type of data can be very useful for setting a meaningful acceptance criterion, by demonstrating what compositions are or are not suitable for manufacture of drug product. Such studies would be especially valuable when the drug substance contains multiple polymorphic forms with sufficiently different properties whose effects on drug product performance could reasonably be anticipated. However, as stated above, we believe that, in most cases, performance studies on experimental batches of drug product made from different polymorphic forms of the drug substance would not be justified from risk/benefit considerations. The stability of the drug substance relative to changes in polymorphic form would need to be considered in this decision, however. For example, when the drug substance is amorphous, manufacture and performance testing of drug product made from drug substance containing some amount of a second relevant polymorphic form might be appropriate. Generally, when the drug substance manufacturing process routinely gives a single polymorphic form, preparation and testing of drug product from other forms of the drug substance would not be necessary.

15.5 Pharmaceutical Solid Polymorphism in Drug Product

The previous section deals with polymorphism in the drug substance, and discusses approaches to finding the appropriate level of control for a specific drug substance. This section will explore the question of when data or controls on polymorphic form within drug product might be appropriate.

15.5.1 Polymorphism Issues in Drug Product Manufacturing

When a drug product is manufactured from the drug substance, it is possible for the polymorphic form to change either during the process or on storage of the dosage form. For example, a change could occur during a wet granulation step or by recrystallization as a suspension ages. Although it is scientifically appealing to understand whether these potential changes are in fact occurring, in practice it is only in a few cases where concerns related to polymorphic form reach a level where such studies are necessary. This is recognized in Q6A by the statement,

> "It is generally technically very difficult to measure polymorphic changes in drug products. A surrogate test (e.g., dissolution) (see Decision Tree #4(3)) can generally be used to monitor product performance, and polymorph content should only be used as a test and acceptance criterion of last resort."

Fortunately, the tightly-controlled nature of both pharmaceutical manufacturing and a drug product's storage conditions adds considerable assurance that changes in polymorphic form will be consistent from batch to batch. Additionally, bioequivalence studies will typically be performed when an intentional change in dosage form composition or manufacturing process is judged to have a significant potential to alter the drug product performance. Finally, tests that measure drug release *in vitro* (e.g., dissolution testing for solid oral products) provide additional evidence of batch-to-batch consistency. There are only a few of cases where the controls in drug substance do not provide an appropriate level of quality assurance and the knowledge of polymorphic composition within the dosage form is necessary. The goal therefore is to identify the subset of drugs where there is a sufficiently high risk that variability in polymorphic composition within the drug product will cause an adverse effect on dosage form performance, such that additional studies or controls are warranted.

15.5.2 Control of Polymorphism in Drug Product

The third and final part of Q6A's Decision Tree #4 outlines studies and controls on polymorphic form at the drug product stage (Fig. 15.3). The text of the guidance states that,

> "These decision trees should be followed sequentially. ...Tree #4(3) should only be applied when polymorphism has been demonstrated for the drug substance, and shown to affect these properties"

(i.e., to affect performance of the drug product). Therefore, when the outcome in Part 2 of Decision Tree #4 is that no further test or acceptance criterion for the drug substance is necessary, it should also be appropriate to conclude that studies on the polymorphic form within the drug product will not be needed. In contrast, where the outcome in Part 2 of Decision Tree #4 is that "an acceptance criterion for polymorphic content in drug substance" should be set, Q6A directs that the issues in Part 3 be considered.

Regarding the first question as to whether "drug product performance testing" provides adequate control of polymorphic ratio changes (e.g., dissolution), we believe that this should not be interpreted as a requirement for performance data on drug product containing different solid-state forms of the drug substance. While studies that allow dosage form performance testing to be correlated with polymorphic composition within the drug product may be needed for situations where there is a high risk of adversely affecting efficacy, such extensive studies would not be justified from risk/benefit considerations for many drugs. What situations would then warrant these studies? Some examples might include:

- A solid oral product where the drug substance contains multiple polymorphic forms with significantly different properties so that the quality and performance of the product will likely be affected.
- A solid oral product where the drug substance is deliberately formulated in a metastable crystalline or amorphous form to enhance bioavailability (see, for example, mefloquine tablets [17, 18] and troglitazone tablets [19, 20]).
- A low-solubility solid oral product where a polymorphic form with properties significantly different from the drug substance would likely be formed during manufacture or storage of the drug product (e.g., during wet granulation, or in the presence of moisture).
- A suspension where a second polymorphic form of the active ingredient with significantly different properties could plausibly form during storage.

In the above cases, it is valuable to know if the polymorphic composition within the drug product is directly related to the composition in the batch of drug sub-

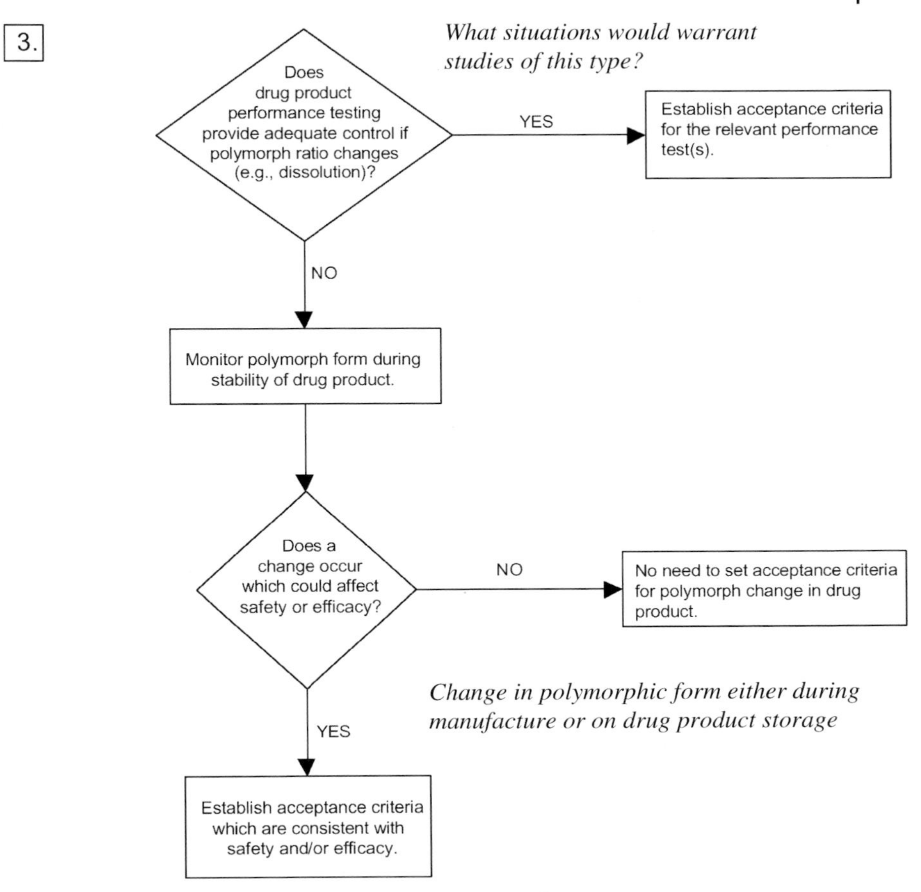

Fig. 15.3 Part 3 of Decision Tree #4 "Investigating the Need to Set Acceptance Criteria for Polymorphism in Drug Substances and Drug Products." (Reproduced from Q6A [1] with notes added; text in italics).

stance from which it was made. In some situations it may be sufficient to know that drug product of acceptable quality can be manufactured from relevant forms of the drug substance. In other situations it may be important to know what polymorphic form(s) are present in the drug product to insure that adequate controls are in place.

When drug product samples made from different solid-state forms of the drug substance are studied to set appropriate controls for the drug substance, these data may show that dissolution testing is sensitive to the solid-state form of the input drug substance. However, when no change in dissolution is seen, it may be that both forms were converted into a similar solid state composition during drug product manufacturing, or that the dissolution procedure is not adequately sensitive. As an example, consider a hypothetical solid oral drug where the synthetic procedure produces the drug substance as a mixture of two polymorphs, which have sufficiently different solubilities that an effect on dosage form performance is plausible. During studies to determine an appropriate acceptance criterion for the level of Polymorph 2 in the drug substance, it might be found that drug product manufactured from drug substance containing a moderate level of Form 2 had adequate performance characteristics, while a high level of Form 2 led to product that did not perform adequately. If the dissolution method can differentiate these two experimental batches of drug product, then a dissolution test seems to "provide adequate control if polymorphic ratio changes". The batch-to-batch dissolution testing of the drug product, combined with an acceptance criterion for Form 2 in the drug substance, therefore, provides a reasonable level of control on polymorphic composition in the drug product. In contrast, if the dissolution method could not differentiate the batches of drug product with adequate and inadequate performance characteristics, it may be appropriate to develop a method for characterizing the polymorphic composition within the drug product, especially in the case of a low-solubility drug. Whether the polymorphic composition within the dosage form should then be measured as part of a one-time study, on a number of commercial batches or as part of the drug product specification, would need to be determined by a risk/benefit consideration, taking into account the data collected during earlier screening and characterization studies.

The outcome of monitoring polymorph during stability of drug product will be reached in Part 3 of Q6A's Decision Tree #4 for situations where (a) there are multiple relevant polymorphic forms with properties that are sufficiently different that effects on drug product performance could reasonably be anticipated, and, in addition, (b) it has not been possible to show that an *in vitro* test such as dissolution is sensitive to the polymorphic composition within the drug product. The four bulleted situations above may be relevant examples, if coupled with the situation that a test such as dissolution is not sufficiently discriminatory. Similarly, this outcome may be reached when during development drug product lots with different bioavailability, but similar dissolution, could plausibly have been related to polymorphic composition.

We consider the decision point as to whether "a change occurs which could affect safety or efficacy" to be dependent upon the magnitude of the change. In instances where such a change is large enough and the polymorphs are poorly soluble and have sufficiently different solubilities, there will certainly be a real potential to cause an adverse effect on the safety or efficacy of the dosage form. Although, in these cases, it may be appropriate to establish an acceptance criterion for polymorphic forms in the drug product, it must be emphasized that this point in Q6A's decision tree will be reached only when tests such as dissolution have been found to be insensitive to changes in polymorphic composition within the dosage form that are likely to adversely affect its performance. In these types of complicated situations, *in vivo* performance data (e.g., bioavailability, etc.) may be needed to establish appropriate controls on polymorphic composition within the dosage form. Fortunately, such complicated situations are quite unusual and, in practice, a specification for polymorphic form within the drug product is generally not necessary, given the multiple levels of control that are generally available (i.e., control of the drug substance manufacturing process, drug substance testing, control of the formulation process, and other appropriate tests of drug product performance).

15.6 Process Analytical Technology

Pharmaceutical production has historically involved the manufacture of the finished product via batch processing, followed by laboratory testing on collected samples to ensure its quality. This approach is implicit from the preceding discussion for determining when testing on polymorphic forms would be recommended to verify quality. While this approach has been successful in providing drug products to the marketplace, there still exists ample opportunity within the pharmaceutical sector to shorten manufacturing cycle times, reduce product variability, and lower the likelihood of product failures and recalls. Within this context, the FDA's process analytical technology (PAT) initiative is a collaborative effort with industry to facilitate the introduction of new and efficient technologies to enhance product quality and manufacturing efficiency in the pharmaceutical industry [21]. PAT is a system for designing, analyzing, and control of manufacturing processes based upon timely measurement during processing of critical quality and performance attributes of raw and in-process materials and products, with the goal of ensuring high quality at the completion of manufacturing. Hence, unlike conventional pharmaceutical manufacturing that relies upon final testing to ensure quality, the central paradigm of PAT is that "quality cannot be tested into products: it should be built-in by design".

Within this PAT framework, and in the particular context of polymorphism, real-time control provides continuous quality assurance that the final product possesses the polymorphic forms that may be critical to product performance. More importantly, real-time measurements with the aid of chemometric tools and appropriate experimental design enable one to quickly identify the critical sources of polymorphic variability that may impact product performance. Identification of critical variables can help design and develop a process, which not only controls polymorphic form variability but also reliably predicts such variability over the design space established for raw materials, process parameters, and other conditions.

15.6.1 Process Analytical Technology and the Crystallization of Polymorphic Forms

When a pharmaceutical manufacturer selects a drug substance polymorphic form that has the desirable characteristics that aide in the manufacture of the drug product formulation, it becomes vital to have a robust crystallization process that consistently produces the desired polymorphic form of the drug substance. For systems that can crystallize in several polymorphic modifications, thermodynamic information on the various polymorphs is generally useful to determine which polymorphs are expected to arise during the crystallization conditions, but crystallization kinetics are likely to be important as well. Hence, in the spirit of PAT, real time controlling of polymorphic forms would be advantageous for several reasons. The *in situ* measurements would minimize possible data artifacts associated with sample isolation and preparation (e.g., evaporation of solvent) that may alter the crystalline form. During development this measurement would provide useful kinetic data and enable a manufacturer to develop a robust process that consistently produces the desired crystalline form by controlling critical process variables. *In situ* monitoring and control during routine production would also be useful in the determination of crystallization or slurry turnover endpoint, and hence optimize operation time to an efficient minimum, and at the same time minimize the probability of allowing an unconverted or "out of specification" batch.

Such continuous monitoring and control requires sophisticated instrumentation and sensor technologies that enable the rapid and quantitative at-line, on-line, or in-line measurement of polymorphs in the crystallization vessels. Over the years, numerous powerful analytical techniques, including Raman and near-IR, have been developed and refined to the point of making this type of *in situ* monitoring a reality. These technologies have been used during the development of crystallization and slurry turnover processes of bulk pharmaceuticals, including progesterone by Pharmacia [22], MK-A by Merck [23], and trovfloxaxacin mesylate by Pfizer [24]. Focused beam reflectance measurements (FBRM) and particle vision measurements (PVM) have also been utilized to indirectly monitor polymorphs during the development of crystallization processes for both model compounds [25–27] and some pharmaceuticals [28], by virtue of the differing crystalline habits that are often (but not always) seen for different polymorphs. Quite impressively, advances in X-ray diffraction have even enabled the "direct" and real-time monitoring of polymorphs in crystallization slurries for several non-pharmaceutical model compounds, including TNT [29], glutamic acid [30], citric acid [30], and glycine [31]. We anticipate that, in the near future, manufacturers of bulk drug substances will increasingly utilize these process analyzers, not only during development but also in routine production that follows the PAT paradigm.

15.6.2
Process Analytical Technology and Polymorphs in Drug Products

Polymorphs may also undergo phase conversions when exposed to the range of unit operations typically used in drug product manufacturing, including drying, milling, micronization, wet granulation, spray-drying, and compaction. Exposure to environmental conditions such as relative humidity and temperature may also induce phase conversion. The rigor used in monitoring such conversions should generally depend upon the relative impact it may have on drug product performance. When the active ingredient is highly soluble and chemically stable, the risk that such process-induced transformations will affect drug product performance would be quite low; in such a situation monitoring of polymorphs may not be warranted. Conversely, this type of monitoring may be important in cases where the drug substance is poorly soluble and the stability of the product is quite sensitive to changes in polymorphic form. In such a scenario and in the spirit of effective PAT implementation, one can envision either a well-defined drug product manufacturing process space where one can reliably be assured that there are no transformations or, alternatively, a manufacturing process where such transformations occur consistently and where critical process variables are well understood and controlled.

While there are numerous examples where the various PAT tools have been used to develop crystallization processes for bulk pharmaceuticals, there are far fewer examples of their application to monitoring polymorphs in drug product manufacture. This is likely due to present limitations in analytical capability, including signal interference from excipients and limits of detection. However, the recent article by Davis et al. [32] demonstrates the feasibility of near-IR for monitoring and modeling the conversion of glycine polymorphs during the drying phase of wet granulation – illustrating a "baby step" in this field. We expect that, with technological advances in polymorph detection and specificity, application of the PAT paradigm to the control of polymorph transformations in drug products manufacturing will become more widespread and, in some cases, routine.

15.7 Summary

Given the rapid pace of drug development, it is vital that compatible approaches to regulatory issues such as control of solid-state form be used by scientists in pharmaceutical companies and regulatory organizations. This chapter briefly reviews the principles of pharmaceutical solid polymorphism and its effect on quality and performance of the drug product, discusses the recommendations on polymorphism in the guidance document "ICH Q6A" and analyzes how these recommendations might be applied to different types of drug substances and drug products. The chapter concludes with a discussion of PAT and its utility in crystallization processes and drug product manufacturing.

Acknowledgments

S.M. would like to acknowledge the discussions that took place during the 2002 AAPS workshop on Specifications, in particular with Tim Wozniak, Ivan Santos, and Jon Clark, as well as many discussions with Chuck Hoiberg, Tony DeCamp, and Chi-wan Chen. The authors also thank David Lin for thoroughly reviewing this chapter.

References

1 International Conference on Harmonization Q6A Guideline: Specifications for New Drug Substances and Products: Chemical Substances, October **1999**. www.fda.gov/cder/guidance/index.htm
2 D.J.W. Grant, Theory and origin of polymorphism, in *Polymorphism in Pharmaceutical Solids*, H.G. Brittain (Ed.), Marcel Dekker, New York, **1999**, pp. 1–34.
3 A.S. Raw, L.X. Yu (Eds.), *Adv. Drug Delivery Rev.* **2004**, *56*, 235–414.
4 L.F. Huang, W. Tony, *Adv. Drug Delivery Rev.* **2004**, *56*, 321–334.
5 B.C. Bruno, M. Parks, *Pharm. Res.* **2000**, *17*, 397–404.
6 A.J. Aguiar, J. Krc, A. W. Kinkel, J.C. Samyn, *J. Pharm. Sci.* **1967**, *56*, 847–853.
7 G.L. Amidon, H. Lennernas, V. P. Shah, J. R. Crison, *Pharm. Res.* **1995**, *12*, 413–420.
8 L.X. Yu, G.L. Amidon, J.E. Polli, H. Zhao, M. Mehta, D.P. Conner, V.P. Shah, L.J. Lesko, M.-L. Chen, V.H.L. Lee, A.S. Hussain, *Pharm. Res.* **2002**, *19*, 921–925.
9 D.P. Ip, G.S. Brenner, J.M. Stevenson, S. Lindenbaum, A.W. Douglas, S.D. Klein, J.A. McCauley, *Int. J. Pharm.* **1986**, *28*, 183–191.
10 R. Eyjolfsson, *Pharmazie* **2002**, *57*, 347–348.
11 R. Eyjolfsson, *Pharmazie* **2003**, *58*, 357.
12 Paracetamol has recently been shown to exist in an additional anhydrous (Form III) and trihydrate crystalline forms. As these polymorphs are highly metastable these are not considered relevant to this discussion on pharmaceutical processing. (See M.L. Peterson, D. McIlroy, P. Shaw, J.P. Mustonen, M. Oliveira, Ö. Almarsson, *Cryst. Growth Des.* **2003**, *3*, 761–765.)
13 E. Joiris, P.D. Martino, C. Berneron, A.M. Guyot-Hermann, J.C. Guyot, *Pharm. Res.* **1998**, *15*, 1122–1130.
14 L.X. Yu, M.S. Furness, A. Raw, K.P. Woodland Outlaw, N. E. Nashed, E. Ramos, S.P.F. Miller, R.C. Adams, F. Fang, R.M. Patel, F.O. Holcombe, Y. Chiu, A.S. Hussain, *Pharm. Res.* **2003**, *20*, 531–536.
15 A.S. Raw, M.S. Furness, D.S. Gill, R.C. Adams, F.O. Holcombe, L.X. Yu, *Adv. Drug Delivery Rev.* **2004**, *56*, 397–414.
16 S. Byrn, R. Pfeiffer, M. Ganey, C. Hoiberg, G. Poochikian, *Pharm. Res.* **1995**, *12*, 945–954.
17 H. Bomches, B. Hardegger, *US Pat. No. 4,579,855*, **1986**.
18 S. Kitamura, L. Chang, J.K. Guillory, *Int. J. Pharm.* **1994**, *101*, 127–144.
19 N. Suzuki, K. Kasahara, Y. Watanabe, S. Kinoshita, H. Hasegawa, T. Kawasaki, *Drug Dev. Ind. Pharm.* **2003**, *29*, 805–812.
20 J.B. Dressnan, C. Reppas, *Eur. J. Pharm. Sci.* **2000**, *11*, S73–S80.
21 Center for Drug Evaluation and Research, Guidance for Industry: PAT – A Framework for Innovative Pharmaceutical Manufacturing and Quality Assurance, September **2003**. www.fda.gov/cder/guidance/index.htm
22 F. Wang, J.A. Wachter, F.J. Antosz, K.A. Berglund, *Org. Process. Res. Dev.* **2000**, *4*, 391–395.

23 C. Starbuck, A. Spartalis, L. Wai, J. Wang, P. Fernandez, C.M. Lindemann, G.X. Zhou, Z. Ge, *Cryst. Growth Des.* **2002**, *2*, 515–522.

24 T. Norris, P.K. Aldridge, S.S. Sekulic, *Analyst* **1997**, *122*, 549–552.

25 B. O'Sillivan, P. Barrett, G. Hsiao, A. Carr, B. Glennon, *Org. Process Res. Dev.* **2003**, *7*, 882–977.

26 D.B. Patience, J.B. Rawlings, *AIChE J.* **2001**, *47*, 212–213.

27 D.B. Patience, Crystal engineering through particle size and shape monitoring, modeling, and control, Ph.D. Thesis, University of Wisconsin, **2002**.

28 Lasentec, Lasentec home page. URL www.lasentec.com

29 R.M. Vrcelj, H.G. Gallagher, J.N. Sherwood, *J. Am. Chem. Soc.* **2001**, *123*, 2291–2295.

30 N. Blagden, R. Davey, M. Song, M. Quayle, S. Clark, D. Taylor, A. Nield, *Cryst. Growth Des.* **2003**, *3*, 197–201.

31 Bede, Bede home page. URL http://www.bede.co.uk

32 T.D. Davis, G.E. Peck, J.G. Stowell, K.R. Morris, S. Byrn, *Pharm. Res.* **2004**, *21*, 860–866.

Subject Index

Polymorphism: in the Pharmaceutical Industry. Edited by Rolf Hilfiker

ISBN: 3-527-31146-7